T0135229

Advances in Intelligent Systems and Computing

Volume 707

Series editor

Janusz Kacprzyk, Systems Research Institute, Polish Academy of Sciences, Warsaw, Poland
e-mail: kacprzyk@ibspan.waw.pl

The series "Advances in Intelligent Systems and Computing" contains publications on theory, applications, and design methods of Intelligent Systems and Intelligent Computing. Virtually all disciplines such as engineering, natural sciences, computer and information science, ICT, economics, business, e-commerce, environment, healthcare, life science are covered. The list of topics spans all the areas of modern intelligent systems and computing such as: computational intelligence, soft computing including neural networks, fuzzy systems, evolutionary computing and the fusion of these paradigms, social intelligence, ambient intelligence, computational neuroscience, artificial life, virtual worlds and society, cognitive science and systems, Perception and Vision, DNA and immune based systems, self-organizing and adaptive systems, e-Learning and teaching, human-centered and human-centric computing, recommender systems, intelligent control, robotics and mechatronics including human-machine teaming, knowledge-based paradigms, learning paradigms, machine ethics, intelligent data analysis, knowledge management, intelligent agents, intelligent decision making and support, intelligent network security, trust management, interactive entertainment, Web intelligence and multimedia.

The publications within "Advances in Intelligent Systems and Computing" are primarily proceedings of important conferences, symposia and congresses. They cover significant recent developments in the field, both of a foundational and applicable character. An important characteristic feature of the series is the short publication time and world-wide distribution. This permits a rapid and broad dissemination of research results.

More information about this series at http://www.springer.com/series/11156

Pankaj Kumar Sa · Sambit Bakshi
Ioannis K. Hatzilygeroudis
Manmath Narayan Sahoo
Editors

Recent Findings in Intelligent Computing Techniques

Proceedings of the 5th ICACNI 2017,
Volume 1

 Springer

Editors
Pankaj Kumar Sa
Department of Computer Science
  and Engineering
National Institute of Technology, Rourkela
Rourkela, Odisha
India

Sambit Bakshi
Department of Computer Science
  and Engineering
National Institute of Technology, Rourkela
Rourkela, Odisha
India

Ioannis K. Hatzilygeroudis
Department of Computer Engineering
  and Informatics
University of Patras
Patras, Greece

Manmath Narayan Sahoo
Department of Computer Science
  and Engineering
National Institute of Technology, Rourkela
Rourkela, Odisha
India

ISSN 2194-5357   ISSN 2194-5365   (electronic)
Advances in Intelligent Systems and Computing
ISBN 978-981-10-8638-0   ISBN 978-981-10-8639-7   (eBook)
https://doi.org/10.1007/978-981-10-8639-7

Library of Congress Control Number: 2018934925

This Springer imprint is published by the registered company Springer Nature Singapore Pte Ltd.
The registered company address is: 152 Beach Road, #21-01/04 Gateway East, Singapore 189721,
Singapore

Foreword

Message from the General Chairs Dr. Modi Chirag Navinchandra and Dr. Pankaj Kumar Sa

Welcome to the 5th International Conference on Advanced Computing, Networking, and Informatics. The conference is hosted by the Department of Computer Science and Engineering at National Institute of Technology Goa, India, and co-organized with Centre for Computer Vision & Pattern Recognition, National Institute of Technology Rourkela, India. For this fifth event, held on June 1–3, 2017, the theme is security and privacy, which is a highly focused research area in different domains.

Having selected 185 articles from more than 500 submissions, we are glad to have the proceedings of the conference published in the *Advances in Intelligent Systems and Computing* series of Springer. We would like to acknowledge the special contribution of Prof. Udaykumar R. Yaragatti, Former Director of NIT Goa, as the chief patron for this conference.

We would like to acknowledge the support from our esteemed keynote speakers, delivering keynotes titled "*On Secret Sharing*" by Prof. Bimal Kumar Roy, Indian Statistical Institute, Kolkata, India; "*Security Issues of Software Defined Networks*" by Prof. Manoj Singh Gaur, Malaviya National Institute of Technology, Jaipur; "*Trust aware Cloud (Computing) Services*" by Prof. K. Chandrasekaran, National Institute of Technology Karnataka, Surathkal, India; and "*Self Driving Cars*" by Prof. Dhiren R. Patel, Director, VJTI, Mumbai, India. They are all highly accomplished researchers and practitioners, and we are very grateful for their time and participation.

We are grateful to advisory board members Prof. Audun Josang from Oslo University, Norway; Prof. Greg Gogolin from Ferris State University, USA; Prof. Ljiljana Brankovic from The University of Newcastle, Australia; Prof. Maode Ma, FIET, SMIEEE from Nanyang Technological University, Singapore; Prof. Rajarajan Muttukrishnan from City, University of London, UK; and Prof. Sanjeevikumar Padmanaban, SMIEEE from University of Johannesburg, South

Africa. We are thankful to technical program committee members from various countries, who have helped us to make a smooth decision of selecting best quality papers. The diversity of countries involved indicates the broad support that ICACNI 2017 has received. A number of important awards will be distributed at this year's event, including Best Paper Awards, Best Student Paper Award, Student Travel Award, and a Distinguished Women Researcher Award.

We would like to thank all of the authors and contributors for their hard work. We would especially like to thank the faculty and staff of National Institute of Technology Goa and National Institute of Technology Rourkela for giving us their constant support. We extend our heartiest thanks to Dr. Sambit Bakshi (Organizing Co-Chair) and Dr. Manmath N. Sahoo (Program Co-Chair) for the smooth conduction of this conference. We would like to specially thank Dr. Pravati Swain (Organizing Co-Chair) from NIT Goa who has supported us to smoothly conduct this conference at NIT Goa.

But the success of this event is truly down to the local organizers, volunteers, local supporters, and various chairs who have done so much work to make this a great event.

We hope you will gain much from ICACNI 2017 and will plan to submit to and participate in the 6th ICACNI 2018.

Best wishes,

Goa, India   Dr. Modi Chirag Navinchandra
Rourkela, India   Dr. Pankaj Kumar Sa
                                                            General Chairs, 5th ICACNI 2017

Preface

It is indeed a pleasure to receive an overwhelming response from academicians and researchers of premier institutes and organizations of the country and abroad for participating in the 5th International Conference on Advanced Computing, Networking, and Informatics (ICACNI 2017), which makes us feel that our endeavor is successful. The conference organized by the Department of Computer Science and Engineering, National Institute of Technology Goa, and Centre for Computer Vision & Pattern Recognition, National Institute of Technology Rourkela, during June 1–3, 2017, certainly marks a success toward bringing researchers, academicians, and practitioners in the same platform. We have received more than 600 articles and very stringently have selected through peer review 185 best articles for presentation and publication. We could not accommodate many promising works as we tried to ensure the highest quality. We are thankful to have the advice of dedicated academicians and experts from industry and the eminent academicians involved in providing technical comments and quality evaluation for organizing the conference in good shape. We thank all people participating and submitting their works and having continued interest in our conference for the fifth year. The articles presented in the three volumes of the proceedings discuss the cutting-edge technologies and recent advances in the domain of the conference.

We conclude with our heartiest thanks to everyone associated with the conference and seeking their support to organize the 6th ICACNI 2018 at National Institute of Technology Silchar, India, during June 4–6, 2018.

Rourkela, India
Rourkela, India
Patras, Greece
Rourkela, India

Pankaj Kumar Sa
Sambit Bakshi
Ioannis K. Hatzilygeroudis
Manmath Narayan Sahoo

In Memoriam: Prof. S. K. Jena (1954–2017)

A man is defined by the deeds he has done and the lives he has touched; he is defined by the people who have been inspired by his actions and the hurdles he has crossed. With his deeds and service, Late Prof. Sanjay Kumar Jena, Department of Computer Science and Engineering, has always remained an epitome of inspiration for many. Born in 1954, he breathed his last on May 17, 2017, due to cardiac arrest. He left for his heavenly abode with peace while on duty. He is survived by his loving wife, beloved son, and cherished daughter.

He is known for his ardent ways of problem-solving right from his early years. Even at 62 years of age, his enthusiasm and dedication took NIT Rourkela community by surprise. From being the Superintendent of S. S. Bhatnagar Hall of Residence to Dean of SRICCE to Head of the Computer Science Department to a second term as the Head of Training and Placement Cell, he not only has contributed to the growth of the institute, but has been a wonderful teacher and researcher guiding a generation of students and scholars. Despite this stature, he was an audience when it came to hearing out problems of students, colleagues, and subordinates, which took them by surprise being unbiased in judgments. His kind and compassionate behavior added splendidly to the beloved teacher who could be approached by all. His ideas and research standards shall continue to inspire generations of students to come. He will also be remembered by the teaching community for the approach and dedication he has gifted to the NIT community.

Committee: ICACNI 2017

Advisory Board Members

Audun Josang, Oslo University, Norway
Greg Gogolin, Ferris State University, USA
Ljiljana Brankovic, The University of Newcastle, Australia
Maode Ma, FIET, SMIEEE, Nanyang Technological University, Singapore
Rajarajan Muttukrishnan, City, University of London, UK
Sanjeevikumar Padmanaban, SMIEEE, University of Johannesburg, South Africa

Chief Patron

Udaykumar Yaragatti, Director, National Institute of Technology Goa, India

Patron

C. Vyjayanthi, National Institute of Technology Goa, India

General Chairs

Chirag N. Modi, National Institute of Technology Goa, India
Pankaj K. Sa, National Institute of Technology Rourkela, India

Organizing Co-chairs

Pravati Swain, National Institute of Technology Goa, India
Sambit Bakshi, National Institute of Technology Rourkela, India

Program Co-chairs

Manmath N. Sahoo, National Institute of Technology Rourkela, India
Shashi Shekhar Jha, SMU Lab, Singapore
Lamia Atma Djoudi, Synchrone Technologies, France
B. N. Keshavamurthy, National Institute of Technology Goa, India
Badri Narayan Subudhi, National Institute of Technology Goa, India

Technical Program Committee

Adam Schmidt, Poznan University of Technology, Poland
Akbar Sheikh Akbari, Leeds Beckett University, UK
Al-Sakib Khan Pathan, SMIEEE, UAP and SEU, Bangladesh/Islamic University in Madinah, KSA
Andrey V. Savchenko, National Research University Higher School of Economics, Russia
B. Annappa, SMIEEE, National Institute of Technology Karnataka, Surathkal, India
Biju Issac, SMIEEE, FHEA, Teesside University, UK
Ediz Saykol, Beykent University, Turkey
Haoxiang Wang, GoPerception Laboratory, USA
Igor Grebennik, Kharkiv National University of Radio Electronics, Ukraine
Jagadeesh Kakarla, Central University of Rajasthan, India
Jerzy Pejas, Technical University of Szczecin, Poland
Laszlo T. Koczy, Szechenyi Istvan University, Hungary
Mithileysh Sathiyanarayanan, City, University of London, UK
Palaniappan Ramaswamy, SMIEEE, University of Kent, UK
Patrick Siarry, SMIEEE, Université de Paris, France
Prasanta K. Jana, SMIEEE, Indian Institute of Technology (ISM), Dhanbad, India
Saman K. Halgamuge, SMIEEE, The University of Melbourne, Australia
Sohail S. Chaudhry, Villanova University, USA
Sotiris Kotsiantis, University of Patras, Greece
Tienfuan Kerh, National Pingtung University of Science and Technology, Taiwan
Valentina E. Balas, SMIEEE, Aurel Vlaicu University of Arad, Romania
Xiaolong Wu, California State University, USA

Organizing Committee

Chirag N. Modi , National Institute of Technology Goa, India
Pravati Swain, National Institute of Technology Goa, India
B. N. Keshavamurthy, National Institute of Technology Goa, India
Damodar Reddy Edla, National Institute of Technology Goa, India

B. R. Purushothama, National Institute of Technology Goa, India
T. Veena, National Institute of Technology Goa, India
S. Mini, National Institute of Technology Goa, India
Venkatanareshbabu Kuppili, National Institute of Technology Goa, India

About the Book

This three-volume book contains the Proceedings of 5th International Conference on Advanced Computing, Networking, and Informatics (ICACNI 2017). The book focuses on the recent advancement of the broad areas of advanced computing, networking, and informatics. It also includes novel approaches devised by researchers from across the globe. This book brings together academic scientists, professors, research scholars, and students to share and disseminate information on knowledge and scientific research works related to computing, networking, and informatics to discuss the practical challenges encountered and the solutions adopted. The book also promotes translation of basic research into applied investigation and converts applied investigation into practice.

Contents

About the Editors

Pankaj Kumar Sa received his Ph.D. degree in Computer Science in 2010. He is currently serving as an assistant professor in the Department of Computer Science and Engineering, National Institute of Technology Rourkela, India. His research interests include computer vision, biometrics, visual surveillance, and robotic perception. He has co-authored a number of research articles in various journals, conferences, and chapters. He has co-investigated some research and development projects that are funded by SERB, DRDOPXE, DeitY, and ISRO. He has received several prestigious awards and honors for his excellence in academics and research. Apart from research and teaching, he conceptualizes and engineers the process of institutional automation.

Sambit Bakshi is currently with Centre for Computer Vision & Pattern Recognition of National Institute of Technology Rourkela, India. He also serves as an assistant professor in the Department of Computer Science and Engineering of the institute. He earned his Ph.D. degree in Computer Science and Engineering. He serves as an associate editor of *International Journal of Biometrics* (2013–), *IEEE Access* (2016–), *Innovations in Systems and Software Engineering* (2016–), *Plos One* (2017–), and *Expert Systems* (2018–). He is a technical committee member of IEEE Computer Society Technical Committee on Pattern Analysis and Machine Intelligence. He received the prestigious Innovative Student Projects Award 2011 from the Indian National Academy of Engineering (INAE) for his master's thesis. He has more than 50 publications in journals, reports, and conferences.

Ioannis K. Hatzilygeroudis is an associate professor in the Department of Computer Engineering and Informatics, University of Patras, Greece. His research interests include knowledge representation (KR) with an emphasis on integrated KR languages/systems; knowledge-based systems, expert systems; theorem proving with an emphasis on classical methods; intelligent tutoring systems; intelligent e-learning; natural language generation; and Semantic Web. He has several papers published in journals, contributed books, and conference proceedings. He has over

25 years of teaching experience. He is an associate editor of *International Journal on AI Tools* (IJAIT), published by World Scientific Publishing Company, and also serving as an editorial board member to *International Journal of Hybrid Intelligent Systems* (IJHIS), IOS Press, and *International Journal of Web-Based Communities* (IJWBC), Inderscience Enterprises Ltd.

Manmath Narayan Sahoo is an assistant professor in the Department of Computer Science and Engineering, National Institute of Technology Rourkela, Rourkela, India. His research interest areas are fault tolerant systems, operating systems, distributed computing, and networking. He is the member of IEEE, Computer Society of India, and The Institution of Engineers, India. He has published several papers in national and international journals.

Part I
Authentication Methods, Cryptography and Security Analysis

Windows Physical Memory Analysis to Detect the Presence of Malicious Code

Dinesh N. Patil and Bandu B. Meshram

Abstract The Windows Physical memory maintains information about the various activities on the system such as processes and their running threads, opened registry key, user authentication details with forensic importance. The cyber attacker modifies the code of the legitimate process to achieve malicious tasks and such malicious codes are not detected by the antivirus program. In order to detect the presence of malicious codes in the legitimate process, this paper suggests a framework. This framework is based on the memory mapped information of a process and its creation time. The techniques discussed in this paper have been verified on the Windows 7 and 8 volatile memory dump.

Keywords Volatile memory · Object · EPROCESS · VAD · Page · Malware

1 Introduction

The Windows Computer system's volatile memory maintains the processes to be executed and their metadata. The memory space is divided into kernel and user space. The system related processes run in kernel memory space whereas the user related process used to switch between user and kernel space in order to gain access to the resources [1]. Therefore malware, in order to gain access to the resources, compromises the running process in the user space. The digital forensic investigation of volatile memory helps in knowing about any malicious process running on the system.

It is essential for the digital forensic investigator to locate the memory mapped file. This has certain advantages. First, it helps in identifying if any unknown information

D. N. Patil (✉) · B. B. Meshram
Department of Computer Engineering, Veermata Jijabai Technological Institute,
Mumbai, India
e-mail: dinesh9371@gmail.com

B. B. Meshram
e-mail: bbmeshram@vjti.org.in

© Springer Nature Singapore Pte Ltd. 2019   3
P. K. Sa et al. (eds.), *Recent Findings in Intelligent Computing Techniques*,
Advances in Intelligent Systems and Computing 707,
https://doi.org/10.1007/978-981-10-8639-7_1

is in the memory. Second, it helps in identifying the presence of malicious code. As cyber attacker alters the code of the memory mapped file for performing their malicious tasks using code injection techniques, by analyzing the code of the memory mapped file it is possible to detect the presence of malicious code. The information extracted about the creation time of a process is helpful to the forensic investigator in understanding the timing at which a process execution has been started on the system. A framework to detect the presence of malicious code in the legitimate process is proposed. This paper is structured as follows: Sect. 2 covers the related work in the volatile memory forensics. Section 3 covers the technique for locating the memory mapped information of a process. A framework to detect the presence of malware is covered in Sect. 4. The experimentation details are specified in Sect. 5. The conclusion and the future work to be carried out are covered in Sect. 6.

2 Related Research

This section details out the existing research on the Windows Physical memory forensics and the malware detection.

2.1 Existing Research

The crucial information regarding the activities going on in a running system can be identified by analyzing the physical memory dump collected from the suspect's system. The extraction of running process list from the dumped volatile memory is performed in [2]. After acquiring the process memory, it was not analyzed for identifying the headers and their fields. As these fields provides the information about the various sections of the memory mapped file.

The importance of forensics of live machines and artifacts which can be found as well as methods and tools which are used for extracting and analyzing data from RAM is discussed in [3]. The paper also has identified the deficiency in the hardware and software method for acquiring the volatile memory without providing any remedial measures.

A method of address translation mechanism is proposed in [4], based on this address running processes, login information, and registry details are obtained. After locating the EPROCESS structure, the procedure for locating the executable image file was not worked out.

The address translation mechanism for converting virtual address to physical address is discussed in [5]. The Windows XP, Windows 2000, Windows Vista memory analysis was performed based on kernel processor control region.

The algorithm for extracting the registry hive files from the dumped memory image is covered in [6]. The paper missed out on the information such as extracting the registry key timestamp from the memory.

A survey of user space virtual address allocations in the Windows XP and Windows 7 operating systems, comprehensively identifying the kernel and user space metadata required to identify such allocations is performed in [7]. A plugin is developed for the volatility tool to provide information about the user allocation represented by the VAD but it has the limitation as the plugin do not explain every allocation.

A set of patches to the Rekall Memory Forensic platform that enables the analysis of page files on all operating systems is presented in [8]. The paper also addressed the issue of page table smear.

A method for recovering mapped files in memory and to link mapped file information process data on Windows XP has been devised in [9]. The locating and parsing of the virtual address descriptor tree structure in case of Window 2000 and XP have been discussed and a plugin developed to construct the VAD tree of the processes which were running during the time of memory dumping in [10].

The existing malware detection techniques have been discussed in [11]. In the static signature-based detection techniques discussed, the program under inspection is examined for sequences of code that would reveal the malicious intent of the program.

2.2   Existing Tool

In order to detect the malicious codes in the legitimate process, the most prominent tools are the Volatility and the Rekall.

Volatility. The Volatility memory forensic framework provides the facility for extracting the forensic information from the dumped volatile memory. The extracted forensic information is about the process running, registry and the related network information. It supports malfind plugin which searches memory sections of the process to look for code injection. The plugin searches for the VirtualProtecEx flag associated with the process so if the flag is not fixed by the attacker then only the presence of malware is identified.

Rekall. The Rekall memory forensic framework provides the facility for extracting and analyzing the forensic information about the process running, registry and the network related artifacts from the dumped volatile memory. It also supports malfind plugin to search memory section of a process to look for code injection. The plugin also searches for the VirtualProtectEx flag (Figs. 1, 2, 3 and 4).

Fig. 1 Translating virtual page number to physical address

32-bit	20-bit	12-bit
Last Page starting Address =	EndingVpn+	000

32-bit	20-bit	12-bit
First Page starting Address =	StartingVpn+	000

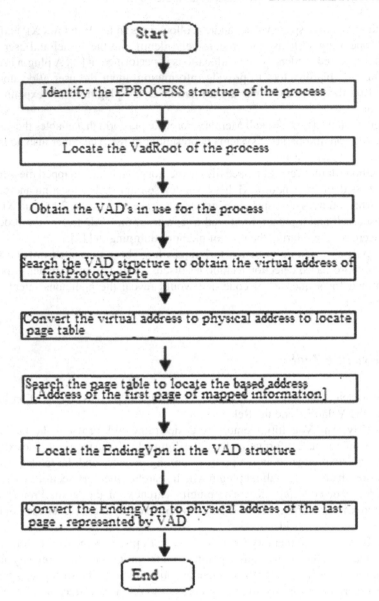

Fig. 2 Locating mapped information of a process

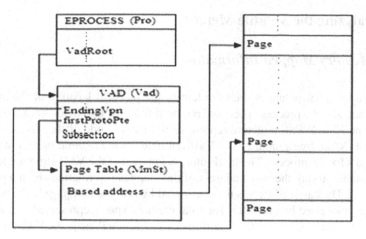

Fig. 3 Link between various memory structures related to mapping of process information in physical pages

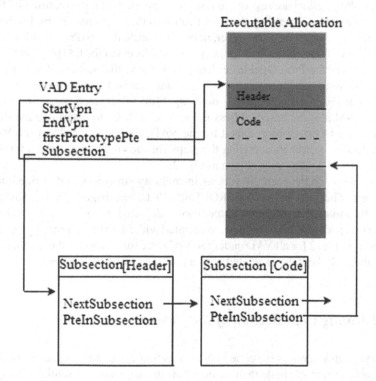

Fig. 4 Process for locating code section of the process

3 Analyzing the Volatile Memory

3.1 Memory Mapped Information

The EPROCESS structure is used to identify the physical location of the mapped information about a process. The VadRoot field found at 632 bytes from the beginning of the EPROCESS structure represents the parent node of the virtual address descriptor (VAD) tree structure. The VadRoot field is used to build the tree structure of the vad's for the process. The firstPrototypePte field of the VAD structure leads to the page table storing the based address of the mapped information about a process as in Fig. 3. The based address is the starting address of the first page in the range of addresses, addressed by the VAD. The total memory space represented by the VAD can be identified by using the field startingVpn and EndingVpn in the VAD structure.

The starting virtual page number (Vpn) and ending virtual page number are the higher 20-bit physical address of the page in case of 32-bit Operating System. The starting 32-bit physical address of the last page and the first page represented by the VAD is obtained by appending the 12-bit (000) to the respective 20-bit EndingVpn and StartingVpn as in Fig. 1. However, in our approach, the firstPrototypePte field of the VAD is used to determine the starting physical address of the first page represented by the VAD. The firstPrototypePte field maintains the virtual address of the page table entry for the first page. On converting this virtual address to the physical address the beginning of the page table having pool tag MmSt is located. At 24 bytes from the pool tag MmSt, the based address is located. In order to determine the starting address of the last page represented by the VAD, EndingVpn field of the VAD is used. A detailed flowchart depicting the steps for locating the mapped information of a process in the memory is shown in Fig. 2.

The relationship between the various in-memory structures as described in [12] is analyzed. The links between EPROCESS, VAD, and page table to identify the mapped information in the physical memory is depicted in Fig. 3. Each VAD represents the mapped information of a file associated with a particular process. Repeating the steps as in Fig. 2 for all VAD under the VAD tree for a process, the mapped information about all the files associated with a particular process is located.

3.2 Locating Code Section of the Process

As the cyber attacker uses the code injection techniques to modify the existing code of a portable executable to do their malicious tasks, it becomes essential to locate the code section of the portable executable and identify the presence of malicious code in it.

Each of the portable executable files is made up of various sections as discussed in [13]. A VAD mapping Portable Executable file can be identified by locating the control area for the VAD which is having pool tag "MmCa". The Control Area structure maintains the field "flag" at 28 bytes from the beginning of the structure as shown in Fig. 5. The Portable Executable has hex value "40A0" in the "flag" field of the Control Area of a VAD. The based address obtained using the firstPrototypePte is also the beginning of the header section. The header section generally occupies one page which is followed by code section. The VAD structure maintains the field Subsections at 36 bytes from the beginning of the structure. The Subsections field a 32-bit pointer maintains the virtual address of the first subsection structure of the Portable Executable file. As each subsection points to the next subsection using the NextSubsection field, therefore, the rest of the subsections of the Portable Executable file are obtained by following the NextSubsection field. The NextSubsection field a 32-bit pointer maintains the virtual address of the next subsection structure. The size of each portable executable file section mapped in the memory is determined based on the value in the field PtesInSubsection of the Subsection structure as shown in Fig. 4. If the PtesInSubsection has the value 1 it means that the number of pages occupied by this section is 1 as only one page entry have been made in the page table related to this particular section. If the PtesInSubsection has the value 2 it indicates that the number of pages occupied by this section is 2 and likewise for the other values in the PteInSubsection. As the size of pages in the 32-bit operating system is 4 KB, therefore, the first 16-bit of the 32-bit address specifies the individual locations in the page. The first page mapped in the VAD has the first 16-bit of the 32-bit address as 0000. Since the size of each page is 4 KB, the next page has the first 16-bit of the 32-bit address as 1000. Adding each time 1000H that is 4096 bytes in decimal to the 32-bit address, the beginning of the next page is obtained.

4   The Framework for Detecting Malicious Process

The traces of the malicious code are detected by performing the forensic investigation of volatile memory. In order to detect the presence of malicious code in the legitimate process, a framework is proposed as shown in Fig. 6. The framework involves three stages: Collecting Evidence, Evidence Analysis, Evidence Generation.

The stage of Collecting Evidence involves identifying all of the running processes on the system at the time of taking the snapshot of the volatile memory along with the locating the code section of the process. The running processes in the system are identified by searching for "Pro" text signature in the dumped memory image which specifies the beginning of an EPROCESS structure. The name of the process in consideration can be located at 364 bytes from the beginning of the EPROCESS structure. The creation time of a process is used to determine the timing of a process execution. The CreateTime field is found at 160 bytes from the beginning of the EPROCESS structure. This field maintains the timestamp of the beginning of a process execution. The code section of a process is located by the technique discussed as in Sect. 3. The code of a process extracted from the dumped volatile memory is then maintained in a database along with the timestamp for its execution.

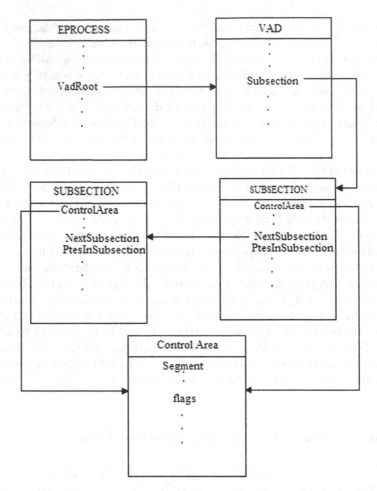

Fig. 5 Locating subsection structure

The stage of Evidence Analysis involves determining the suspicious process based on the comparison of the code of the running process found in the code section. A database of all the code of the legitimate portable executable is maintained. This database is to be created from the original code section content of the various portable executable files, obtained from the dumped memory images. The code of the running process obtained from the dumped volatile memory is then compared byte by byte with the original code of the same process in the database. If the code is found dissimilar then the process is considered as a suspicious process; potentially affected with malicious code. The execution timing of such process is then determined.

The Evidence Generation stage involves generating the report about the running suspicious processes in the system at the time of the volatile memory dumping. The

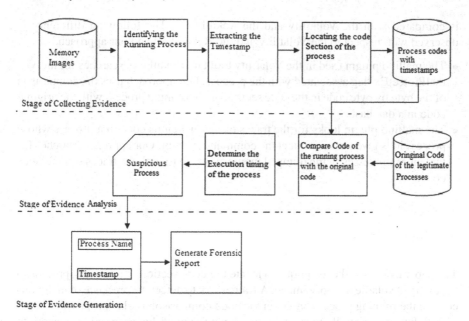

Fig. 6 Framework for detecting malicious process

Table 1 Test machine information	Operating system	32-bit Windows 7, 8 Professional edition
	Memory	2 GB
	Processor	Intel ® Pentium 4 @3.00 GHz

report comprises of the dissimilarity found in the code comparison, process name, and the execution timing of the process in the system.

5 The Experimentation

Our experiments were performed on the memory dump files from 32-bit Windows 7, 8 Professional machine widely used operating systems based on the survey conducted [14]. The dumps were obtained using dumpIt tool which dumps memory completely and accurately as discussed in [15]. Nearly 10 memory dumps files were analyzed. Table 1 provides detailed information about our test machine.

Using the techniques described in Sect. 3, the memory mapped information about a process is obtained. During the experimentation on the memory images, it was observed that the pages represented by a VAD have contiguous virtual memory allocation. The method described in Sect. 3 to detect the presence of malicious code is compared with the Volatility and the Rekall tool. The following difference is observed with respect to the Volatility and the Rekall tool and our approach.

- The malfind plugin used in the Volatility tool and Rekall tool searches for the VirtualProtectEx flag associated with the process. Our approach uses the comparison of the byte by byte code in the code section of the running process with the original code in a database.
- The malfind plugin looks for the traces of the malicious code in a process whose process id is given as an input at the command prompt. Our approach searches for the entire processes in the dumped memory image to look for traces of malicious code.

6   Conclusion

The paper discussed the technique to locate the code section of a running process in the dumped volatile memory image. A framework to detect the presence of malicious code in the running process based on the code comparison is also proposed.

The future work will focus on the implementation of the proposed framework to detect the presence of the suspicious running process having malicious code.

References

1. Soulami, T.: Inside Windows Debugging. Microsoft (2000)
2. Thomas, S., Sherly, K.K., Dija, S.: Extraction of memory forensic artifacts from Windows 7 RAM image. In: Proceedings of the IEEE Conference on Information and Communication Technologies (2013)
3. Hausknecht, K., Foit, D., Buric, J.: RAM Data Significance in Digital Forensics. MIPRO, Opatija, Croatia (2015)
4. Zhang, S., Wang, L., Zhang, R.: Exploratory study on memory analysis of Windows 7 operating system. In: 3rd International Conference on Advanced Computer Theory and Engineering (2010)
5. Zhang, R., Wang, L., Zhang, S.: Windows memory analysis based on KPCR. In: 5th International Conference on Information Assurance and Security (2009)
6. Zhang, S., Wang, L., Zhang, L.: Extracting Windows Registry Information from Physical Memory (2011)
7. White, A., Schatz, B., Foo, E.: Surveying the user space through user allocations. Digit. Investig. S3–S12 (2012)
8. Cohen, M.: Forensic analysis of windows user space applications through heap allocation. In: 3rd IEEE International Workshop on Security and Forensics in Communication Systems (2015)
9. Van Baar, R., Alink, W., VanBallegoojj, A.R.: Forensic memory analysis: file mapped in memory. Digit. Investig. S52–S57 (2008)
10. Dolan-Gavitt, B.: The VAD tree: the process-eye view of physical memory. Digit. Investig. S62–S64 (2007)

11. Idika, N., Mathu, A.P.: A Survey of Malware Detection Techniques (2007)
12. Russinovich, M., Solomon, D., Lonescu, A.: Windows Internals, Part 2, p. 195, 6th edn. Microsoft Press (2009)
13. Microsoft: Microsoft Portable Executable and Common Object File Format Specification. http://courses.cs.washington.edu/courses/cse378/03wi/lectures/LinkerFiles/coff.pdf (2016)
14. Pot, J.: Windows 10 leaps ahead of 7 amongst steam gamers. http://www.digitaltrends.com/computing/steam-users-windows-10-market-share (2016)
15. Ahmed, W., Aslam, B.: A Comparison of Windows Physical Memory Acquisition Tools, Milcom Cyber Security and Trusted Computing (2015)

Signcrypting the Group Signature with Non-transitive Proxy Re-encryption in VANET

Sneha Kanchan and Narendra S. Chaudhari

Abstract SignCryption is the public-key cryptographic tool to guarantee various security paradigms by integrating digital signature and encryption. The traditional digital signature with encryption is very expensive as it requires massive computational overheads. Instead signcryption combines these two into one step, and overhead reduced is significantly observable. Proxy re-encryption is a technique in which a proxy enables a third person Chris to be able to decrypt a message which was sent by Alice for Bob. Instead of first signing and then encrypting, we are sending after just signcrypting the message into the network where proxy will re-encrypt it at the receiver end. We have used Bilinear Diffie-Hellman assumption, i.e., irreversible property of Diffie–Hellman assumption in bilinear groups with digital signatures to provide confidentiality, integrity, and non-repudiation. Later the result is evaluated using AVISPA, Automated Validation of Internet Security Protocols and Applications.

Keywords VANET · Signcryption · Proxy re-encryption · Group signature
Non-transitivity · AVISPA

1 Introduction

From the very beginning, Vehicular Ad hoc Networks (VANETs) have captivated the researchers' attention because of its promising technologies. Tremendous increase in number of vehicles drastically strengthened the VANET momentum that made

S. Kanchan (✉) · N. S. Chaudhari
Department of Computer Science and Engineering, Visvesvaraya National Institute of
Technology, Nagpur, Maharashtra, India
e-mail: sneha.kanchan159@gmail.com

N. S. Chaudhari
e-mail: nsc0183@gmail.com

N. S. Chaudhari
Indian Institute of Technology Indore, Indore, Madhya Pradesh, India

© Springer Nature Singapore Pte Ltd. 2019
P. K. Sa et al. (eds.), *Recent Findings in Intelligent Computing Techniques*,
Advances in Intelligent Systems and Computing 707,
https://doi.org/10.1007/978-981-10-8639-7_2

researchers and companies to provide ITS (intelligent transportation system) in order to improve comfort and safety. The network assures to control the traffic conditions on roads like accidents, jams, etc., by tackling the vehicle-to-vehicle communication. In addition, it can also pose some interesting feature as providing value added services (VAS). Since driving decisions are influenced by the sent traffic message, vehicles in the network are bound to communicate with each other. These messages must be kept secure at any cost to avoid accidents, heavy traffic, and other safety issues. Vehicles are equipped with onboard units (OBUs) to facilitate communication between them. However, with this emerging technology, several challenges are being faced by the network including security issues which should be given priority before implementing VAS into the network. The discussion given below gives the brief idea about recent trends and issues in this area.

1.1 Recent Trends

Recently in VANET, the main focus of researchers is towards providing the data confidentiality and access control [1]. These two security parameters ensures that the data transferred in the network is not accessible to any unauthorized person. The long-established symmetric and public-key cryptography are not much applicable to fulfill cost and time constraints of the vehicular network. In former approach, we need to establish session keys and distribute it securely introducing huge overhead, whereas later one often fails to deliver messages encrypted with separate public keys within given time constraints. Even a single delay in transmission can cause a big accident in real-time scenario. Hence, the alternative methods which are quick but secure enough should be implemented. To accomplish these requirements, group concept came into existence. Vehicles forming a group can exchange keys and messages securely as the vehicles not registered within same group will not be able to access illegitimately. Group signature is followed by encryption to ensure group authentication. Signcryption (signature + encryption) is used to provide same method in less cost and time because both the methods are getting performed in single step and encryption does not have to wait for signature procedure to be completed. During the encryption procedure, the privacy of the message is also preserved. While nodes trust method is used to assign the trust level to each node, with each misbehave; the trust level of that particular node will be decreased by some amount resulting in its elimination from network [2]. In our research, untrustworthy nodes are eliminated from the group to make the group communication more reliable. Signcryption and node trust are the parameters proposed for measuring the security level of the network.

1.2 Our Core Contribution

Presently, VANET cannot perform with its full potential because it is not secure enough. Also, existing technologies are unable to pinpoint a particular vehicle without some serious drawbacks. The proposed research is an extension to our earlier work [3] which integrated the group signature in VANET with Non-Transitive Proxy Re-encryption Scheme (NTPR) [4, 5]. Our core contribution through this paper can be summarized in below points

- Transforming group signature with NTPR into Signcryption with NTPR instead of signature-then-encryption: saves time and space of the network.
- Formal proof using AVISPA (Automated Validation of Internet Security Protocols and Applications).

1.3 Paper Organization

The paper has been organized in six sections. The first section gives the basic introduction of our research. Next section comprises of the literature surveys done by various authors in this area. Next in Sect. 3, we propose our research and formal proof of the same with the outcome is given in Sect. 4. Finally, Sect. 5 concludes the paper following the future scope.

2 Literature Review

Re-encryption authorizes a person to be able to decrypt a message which was originally not encrypted for him. By using this technique, the delegator can provide an authority to a third party to read messages intended for the former one. For example, Alice wants Bob to read her messages which are basically encrypted with her public key. Till 1997, there used to be a trusted third party who was having private keys of the involved parties and was responsible for first decrypting the message using Alice's private key followed by encrypting with Bob's public key. We can clearly sense the severe insecurity level because of sharing the private keys with the third party. To diminish that, in 1998, Blaze, Bleumer and Strauss, introduced the term "proxy" and "re-encryption key" to do the task [6]. Now, there was no need to share the private keys. They proposed the algorithm based on ElGamal cryptosystem in which a re-encryption key could be generated by using private keys of both members but they did not need to share their private keys exclusively with the proxy. The proxy was able to convert the encrypted text by just using the re-encryption key, but was not able to decrypt it for himself. But the main problem was key generation issue for which, one party had to share their secret key either with the other party or with any trusted third party. Similarly in 2003, Ivan and Dodis revisited the proxy cryptography [7] and gave a solution using standard public keys but again this solution involved pre-sharing of secret keys between involved parties which was again a big flaw in the proposal because the involved parties may not be that close to share their secret keys with each other.

Later in 2011, a noble method was put forward by Vijayan, R., and Sumitkumar Singh to secure the network grid using signcryption [8]. The proposed scheme consists of computational server CS (the central part, used for producing keys, controlling data transmission, etc.), vehicular node (the most basic units) and roadside unit RSU (similar to mediator between the two). Each vehicle has a unique Vehicle identity (VID) and it identifies its destination vehicle by destination vehicle identity (DID). Each node in the network transmits signals to CS and nearby nodes containing information like its location, current velocity (CV), etc. It also sends the DID to CS. After the required verifications, required keys are produced and are sent to the VID and DID. This is done with the help of RSUs and network grid model to pinpoint a vehicle exactly. After key reception, Signcryption is performed at VID and sent to CS, which decrypts it and transmits to DID.

In the same year Singh, Sumit kumar, and R. Vijayan focused mainly on trust level for removing disrupting/malicious nodes within the network [9]. As per this scheme, each vehicle is identified as soon as it gets into the VANET. The server provides the vehicle a Unique ID for each transmission, for message security purposes. Only if the ID is found cogent and its trust level is above a given threshold, required keys for VID and DID are created. After each data transmission between a source and the receiver, the receiver sends its feedback only to the server in form of trust level on a scale of 1 to 5 (maximum). Depending upon the trust level of a particular node based on the feedback received from three different nodes, a vehicle may be removed from the network to provide additional degree of security to the network.

3 Our Proposed Scheme

In our previous work [3], we have proposed an algorithm on how to integrate of Group Signature with Non-transitive Proxy Re-encryption in VANET. Group signature denotes the authenticity of being a valid group member. Any message signed with group signature is authentic ensuring that the message has been sent by a legitimate member. Proxy Re-encryption is used to alter the already encrypted message to a new ciphertext which can also be decrypted by the authorized third party. Signcryption combines the two security steps, i.e., encryption and signing into one, and overhead reduced is significantly observable. This paper uses the symmetric properties of the bilinear group which is defined by the computable cost-effective bilinear maps [10]. Since, the base paper [3] is having too many steps and complex calculations, this paper having lesser steps/calculations will firmly improve the efficiency and complexity of the transmissions. Still, our protocol is secured against all four security paradigms, i.e., authentication, integrity, confidentiality and non-repudiation. The flowchart of the basic algorithm can be referenced from that paper.

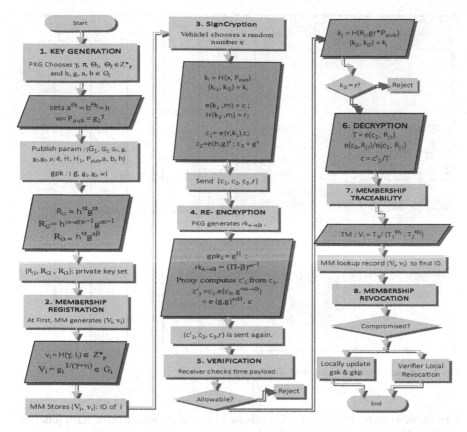

Fig. 1 Signcryption for group signature with proxy re-encryption

Figure 1 summarizes the steps of our algorithm most of which are already explained in the base paper. Avoiding repetitive contents, here we will only include the modified steps. First of all, we need to break our keys in two equal parts k_1 and k_2.

$$k = (k_1, k_2).$$

$$e(k_1, m) = c \; ; \text{and} \, H(k_2, m) = r;$$

The newly defined c and r can be taken in place of ciphertext and signature respectively. Binomial function of k_1 and r makes it more secure because of previously used hash function. So, final ciphertext set will be

$$c_1 = e(r, \, k_1) \cdot c; \; c_2 = e(h, g)^s; \; c_3 = g^s$$

Now the cipher text set $\{c_1, c_2, c_3\}$ will be sent over the network. After receiving the set, proxy calculates the $rk_{A \rightarrow B}$ and then re-encrypts the ciphertext into new

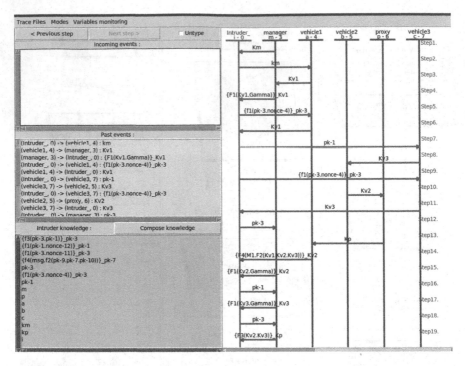

Fig. 2 Intruder simulation window in AVISPA tool

form, which can be opened using C's private keys. Since, we are using hash function instead of just adding the message with the key; the outcome would be more secure.

4   Result and Discussion

The goal of the discussion is to signcrypt a message without violating the security aspects of the network and this goal is verified via AVISPA which is a tool to validate the security of the protocols and applications [11]. If goals are violated, generating event-sequence is displayed. In Fig. 2, an Intruder Simulation window shows the exact message transmission between different nodes. We have coded different scenarios, e.g., managers are generating the keys/signatures, vehicles are encrypting message using those, proxies are re-encrypting using re-encryption keys, the actual receiver is able to decrypt it with his own private key. In the diagram we have presented only starting transactions. As shown in the intruder knowledge section below, no secret information could be gained by the intruder.

4.1 Security Check

- **Confidentiality**: Decryption is impossible for the intruder as he does not have enough knowledge in his knowledge-base. Also, use of group signature hides the sender's individual identity. So, the confidentiality of message as well as node is conserved.
- **Authentication**: Encryption of the message with vehicle A's public key ensures that it can only be open with its private key. But in the middle, proxy converts the message into a re-encrypted cipher which can also be opened by B who is already authenticated by A.
- **Integrity**: Because of end-to-end encryption, no one can open it in middle, and so message cannot be altered hence integrity is achieved.
- **Non-repudiation**: Message is encrypted with group signature which tells that the sender is a valid group member. Nevertheless, MM can always find out the real identity of the member from that signature.

4.2 Result Analysis

The status "SAFE" in Fig. 3 denotes that our protocol has passed the security test. This means our security goal, of keeping the keys/message safe from the intruder's approach, was not violated. Hence, the message transmission is authenticated and no intruder in between is able to hear/disturb the communication.

5 Conclusion and Future Work

The existing re-encryption technique in VANET is modified to signcryption and the results are verified using AVISPA which showed that the proposed protocol is safe. Signcryption poses less complexity with lesser CPU cycles involved. An effective cost management analysis of the system can be a future research direction. The performances of data transmission in VANETs can be evaluated by applying proxy re-signature scheme to provide more flexibility to improve response time with increased throughput and reduced delay.

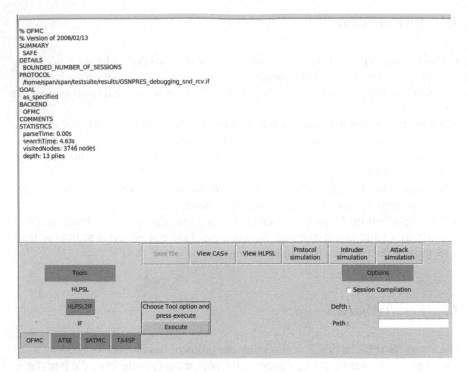

Fig. 3 Simulation result window showing status of the protocol as SAFE

References

1. Liu, X., et al.: SEMD: secure and efficient message dissemination with policy enforcement in VANET. J. Comput. Syst. Sci. (2016)
2. Patel, N.J., Jhaveri, R.H.: Trust based approaches for secure routing in VANET: a survey. Procedia Comput. Sci. **45**, 592–601 (2015)
3. Kanchan, S., Chaudhari, N.S.: Integrating Group Signature scheme with Non-transitive Proxy Re-encryption in VANET (forthcoming)
4. Lin, X., Sun, X., Ho, P.-H., Shen, X.: GSIS: a secure and privacy-preserving protocol for vehicular communications. IEEE Trans. Veh. Technol. **56**, 3442–3456 (2007)
5. Ma, C., Ao, J.: Group-based proxy re-encryption scheme. In: International Conference on Intelligent Computing, pp. 1025–1034 (2009)
6. Blaze, M., Bleumer, G., Strauss, M.: Divertible protocols and atomic proxy cryptography. In: International Conference on the Theory and Applications of Cryptographic Techniques. Springer, Berlin, Heidelberg (1998)
7. Ivan, A.-A., Dodis, Y.: Proxy Cryptography Revisited. In: NDSS (2003)
8. Vijayan, R., Singh, S.: A novel approach for implementing security over vehicular ad hoc network using signcryption through network grid. Int. J. Adv. Comput. Sci. Appl. **2**(4) (2011)

9. Singh, S., Vijayan, R.: Enhanced security for information flow in vanet using signcryption and trust level. Int. J. Comput. Appl. **16**(5), 13–18 (2011)
10. Fu, X.: Unidirectional proxy re-encryption for access structure transformation in attribute-based encryption schemes. Int. J. Netw. Secur. 142 (2005)
11. Von Oheimb, D.: The high-level protocol specification language HLPSL developed in the EU project AVISPA. In: Proceedings of APPSEM 2005 Workshop (2005)

Digital Forensic Approach for Investigation of Cybercrimes in Private Cloud Environment

Ezz El-Din Hemdan and D. H. Manjaiah

Abstract Cloud computing is a revolutionary technology that provide computing and storage resources as a service accessible via network connections. Cloud computing has two foremost models which are services and deployment. One of the significant deployment models is a private cloud which is used by large organizations for building a corporate cloud inside their internal departments to provide computing services such as Software as a Service (SaaS), Platform as a Service (PaaS), and Infrastructure as a Service (IaaS). They adopted the internal private cloud to reduce cost and save time. In last decades, number of severe attacks and threats against the private cloud is increasing which make it facing complex challenges in the security province so that a forensic approach for cybercrime investigation should be systematized. Therefore, this paper presents a digital forensic approach for investigation of cybercrimes in a private cloud environment. Here, focus on using virtualization technology solutions from VMware for building experimental environment. From the experiment analysis, the proposed approach can help digital investigators and practitioners in acquisition and collection of digital evidence from the private cloud infrastructures especially virtual machine which is considered the core element of virtualized cloud systems.

Keywords Cloud computing · Private cloud · Digital forensics and cybercrimes

1 Introduction

Recently, Cloud Computing becomes one of the supreme popular processing and storage paradigms. There are numerous challenges which make the cloud computing cannot be used to store data for various sectors such as business, banks, and

E. E.-D. Hemdan (✉) · D. H. Manjaiah
Department of Computer Science, Mangalore University, Mangalore, India
e-mail: ezzvip@yahoo.com

D. H. Manjaiah
e-mail: manju@mangaloreuniversity.ac.in

© Springer Nature Singapore Pte Ltd. 2019
P. K. Sa et al. (eds.), *Recent Findings in Intelligent Computing Techniques*,
Advances in Intelligent Systems and Computing 707,
https://doi.org/10.1007/978-981-10-8639-7_3

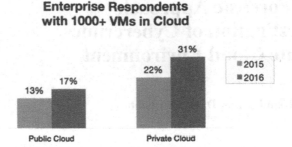

Fig. 1 Enterprises respondents with 1000 + VMS in in private and public cloud [5]

healthcare which require an audit and regulatory compliance. With the growing of cloud computing and related services, the cloud security and privacy have become very critical issues in cloud security area where criminals can use cloud infrastructures which have the exceptional bandwidth, storage, and computing power to launch unlawful activities and severe attacks.

One of the imperative sciences for fighting of cybercrimes is digital forensics, which concern with tracing attackers and discovering a weakness in digital systems such as computers, cell phones, and networks [1]. Digital Forensic process involves four essential steps for performing the digital investigation. These steps are identification, preservation, analysis, and presentation [2]. The process of performing digital forensic in the cloud computing environment is called cloud forensics. Cloud forensics is considered as a discipline that combines cloud computing and digital forensics [3]. In the cloud, there are four deployment models which are private, public, community, and hybrid [4].

Private cloud is one of the cloud deployment models which is dedicated to single organization to reduce costs and save time. This type of cloud is to provide computing services in a dynamic manner with direct control over the infrastructures. More enterprise workloads moved to both public and private cloud over the last year, with private cloud growing faster. The number of enterprises running more than 1,000 virtual machines in public cloud increased from 13 to 17%, while those running more than 1,000 VMs in a private cloud grew from 22 to 31% as shown in Fig. 1. The private cloud growth in workloads also may include long-standing virtualized environments that have been enhanced and relabeled as a private cloud [5].

This paper introduces a digital forensic approach for investigation of cybercrimes in private cloud environment. The proposed approach can accomplish in the private cloud that gives digital investigators some assumptions such as; the investigator already knows the attackers due to in the private cloud, the services are provided to restricted users who have access authority. In addition to this, the investigator has control and authority with assistance from the organization to investigate the crime that has occurred [6].

The rest of this paper is organized as follows: Sect. 2 provides related work while the proposed forensic approach presented in Sect. 3. Section 4 describes the results analysis and discussions while the paper conclusion in this innovative area is presented in Sect. 5.

2 Related Work

Data collection and extraction of digital evidence from cloud environment is a complex challenge step in cloud forensics. Some researchers proposed methods and suggestions for supporting cloud forensics but few provide solution for private cloud. Jeong et al. [6], they proposed digital investigation approach for Virtual Desktop Infrastructure (VDI) solutions. In their paper, they focused on a virtual desktop infrastructure and introduced various desktop virtualization solutions that are widely used like VMware, Citrix, and Microsoft. In the results of the paper, they verified the integrity of the data acquired to ensure that the results of their proposed approach are acceptable as evidence in a court of law. They also observed an error, the reason for this error was from one of the digital forensic tools which failed to mount properly a dynamically allocated virtual disk.

Dykstra and Sherman [7] recommended a cloud management plane for using in Infrastructure as a Service (IaaS) model where Cloud Service Providers (CSPs) can play an important role in data collection by providing a web-based management console. Using this console panel, users, and customers as well as digital investigators can collect and extract related digital evidence from Virtual Machine (VM) image, network, and process in an effective manner. There is only one problem with this solution is that it requires an extra level of trust in the management plane. In traditional evidence collection methods, where we have physical access to the system, this level of trust is not required.

Rani and Geethakumari [8] proposed an efficient digital forensic investigation approach in cloud environment based on a Virtual Machine (VM) snapshots. The basic idea of this approach is that the Cloud Service Provider (CSP) stores snapshots of the VM whose activities are identified as malicious using an Intrusion Detection System (IDS). Simultaneously the CSP should be requested for log files of the suspected VM and digital investigator collects and processes the log files to obtain the evidence.

3   Proposed Approach

The flowchart of proposed approach is shown in Fig. 2.

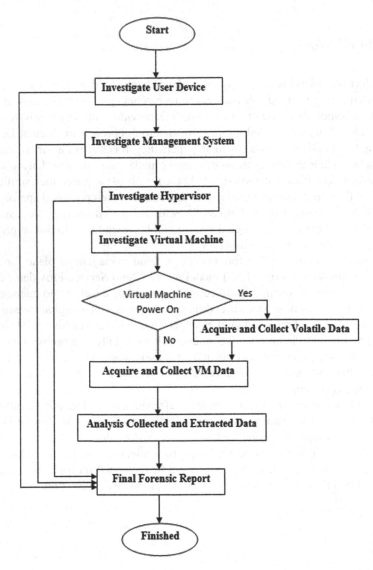

Fig. 2 Proposed forensic approach

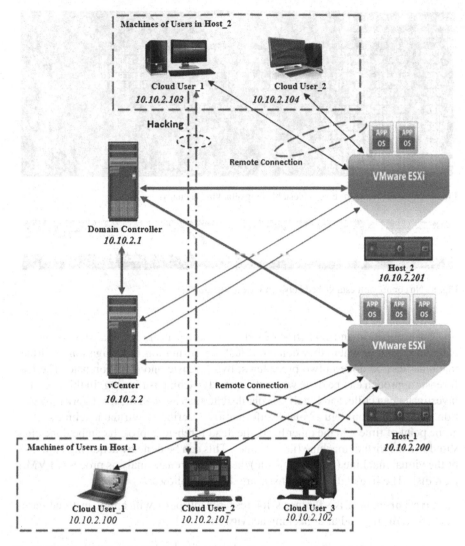

Fig. 3 Experimental environment

4 Results Analysis and Discussions

In this paper, a practical test environment which represents the private cloud environment is setup as shown in Fig. 3 where a hacking scenario is launched between two virtual machines running inside the virtual datacenter to exemplify a malicious activity inside the private cloud. This section discusses the acquisition and extraction of evidential data from virtual machine to find any traces and footprints of the committed crime in the private cloud.

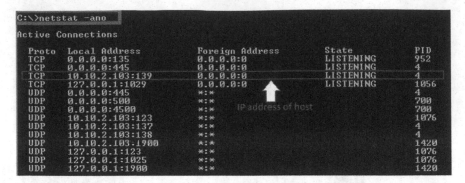

Fig. 4 Network connection before establishing attacking scenario

Fig. 5 No connection established between victim and attacker

Data acquisition and collection of virtual machines is an essential step to investigate cloud system where they contain valuable information. Investigation of virtual machines can be done in two procedures; live forensic and dead forensic. The live forensic approach can be done where the virtual machine is running. In this case, the investigator can collect and analysis volatile data such as open files, network information, and connection status. Volatile information describe the virtual machine system at the point of time. On the other hand, the dead forensic approach can be done after shutdown the virtual machine (i.e., Offline). This can be done by acquiring hard disk of the virtual machine (i.e., VMDK file) then start forensic analysis process of VM's hard disk. The live and dead forensics are done as follows:

1. **Live Forensic**: Live forensics for research purpose will apply in three cases before, during and after hacking activity as:

   - *Before Hacking*: In this research, during the hacking scenario, we opened CMD terminal to write some commands like (*netstat–ano*) to test the network connection between attacker and victim. Before the attacker establishing the connection with victim machine, noticed that no connection as showing in Fig. 4. Another command (*netstat–o*), before the attacker establishing the connection with victim machine, noticed that no connection as showing in Fig. 5. From this step, we noticed that no connection between the victim and attacker before hacking launched.
   - *During Hacking*: Attacking scenario is launched between attacker and victim as shown in Fig. 6 using (*netstat–ano*) command. From that noticed that there is a connection established between the attacker's machine which has IP address 10.10.2.101 that attacked the victim machine which has IP 10.10.2.103. Also,

```
C:\>netstat -ano

Active Connections

  Proto  Local Address          Foreign Address        State           PID
  TCP    0.0.0.0:135            0.0.0.0:0              LISTENING       952
  TCP    0.0.0.0:445            0.0.0.0:0              LISTENING       4
  TCP    10.10.2.103:139        0.0.0.0:0              LISTENING       4
  TCP    10.10.2.103:1036       10.10.2.101:4444      ESTABLISHED     1076
  TCP    127.0.0.1:1029         0.0.0.0:0              LISTENING       1056
  UDP    0.0.0.0:445            *:*                                    4
  UDP    0.0.0.0:500            *:*                                    700
  UDP    0.0.0.0:4500           *:*                                    700
  UDP    10.10.2.103:123        *:*        IP Address of Attacker      1076
  UDP    10.10.2.103:137        *:*                                    4
  UDP    10.10.2.103:138        *:*                                    4
  UDP    10.10.2.103:1900       *:*                                    1420
  UDP    127.0.0.1:123          *:*                                    1076
  UDP    127.0.0.1:1025         *:*                                    1076
  UDP    127.0.0.1:1900         *:*                                    1420
```

Fig. 6 Network connection established between attacker victim machines

```
C:\>netstat -o                    Connection established with attacker

Active Connections

  Proto  Local Address    ↓      Foreign Address        State           PID
  TCP    vm1:1036                10.10.2.101:4444      ESTABLISHED     1076
```

Fig. 7 Connection established between victim and attacker

using (*netstat–o*) as shown in Fig. 7, we noticed there is a connection between the victim and attacker. Another command (*tasklist*) is used to list all running processes on the victim machine, noticed that there are two processes about using CMD; because already two CMD used. The first used by an attacker to control the victim machine remotely and the second used by us to write testing commands as shown in Fig. 8. We had known that the PID of CMD of the attacker is 2020 because we run the CMD before the attacking scenario occurred as shown in Fig. 9.

- *After Hacking*: After the hacking scenario is done, we opened CMD terminal to write commands (*netstat–ano*) and (*netstat–o*) to test the network connection between attacker and victim. After the attacker releasing the connection with victim machine, noticed that no connection between the victim and attackers after the attacker disconnected.

2. **Dead Forensic**: The dead forensic is done after turning off the virtual machine (i.e., Offline). This is done by acquiring the hard disk of the virtual machine (i.e., VMDK file) then start forensic analysis using forensic tools such as Encase, AccessData FTK Imager, and Autopsy. In this research, FTK imager is used for converting VMDF to E01 image format. This is the image of a virtual machine hard disk which is used for testing purpose. Then, Autopsy software is used for analysis of the E01 file to extract traces related to criminal activities.

```
C:\>tasklist

Image Name                       PID Session Name       Session#      Mem Usage
========================= ====== ================== ======== ==============
System Idle Process                0 Console                    0           28 K
System                             4 Console                    0          236 K
smss.exe                         556 Console                    0          388 K
csrss.exe                        620 Console                    0        3,392 K
winlogon.exe                     644 Console                    0          724 K
services.exe                     688 Console                    0        3,716 K
lsass.exe                        700 Console                    0        1,380 K
svchost.exe                      856 Console                    0        4,376 K
svchost.exe                      952 Console                    0        3,820 K
svchost.exe                     1076 Console                    0       22,056 K
svchost.exe                     1248 Console                    0        2,620 K
svchost.exe                     1420 Console                    0        4,168 K
explorer.exe                    1496 Console                    0       11,024 K
spoolsv.exe                     1628 Console                    0        4,236 K
alg.exe                         1056 Console                    0        3,196 K
wscntfy.exe                     1068 Console                    0        1,780 K
wuauclt.exe                     1716 Console                    0        4,872 K
cmd.exe                         1192 Console                    0        2,340 K
taskmgr.exe                      960 Console                    0        1,264 K
cmd.exe                         2020 Console                    0        2,188 K
tasklist.exe                    1668 Console                    0        4,044 K
wmiprvse.exe                    1104 Console                    0        5,296 K
```

Fig. 8 CMD terminal used by attacker and victim machine

```
C:\>tasklist

Image Name                       PID Session Name       Session#      Mem Usage
========================= ====== ================== ======== ==============
System Idle Process                0 Console                    0           28 K
System                             4 Console                    0          236 K
smss.exe                         556 Console                    0          388 K
csrss.exe                        620 Console                    0        1,072 K
winlogon.exe                     644 Console                    0        2,888 K
services.exe                     688 Console                    0        3,708 K
lsass.exe                        700 Console                    0        1,416 K
svchost.exe                      856 Console                    0        4,388 K
svchost.exe                      952 Console                    0        3,820 K
svchost.exe                     1076 Console                    0       22,436 K
svchost.exe                     1248 Console                    0        2,608 K
svchost.exe                     1420 Console                    0        4,168 K
explorer.exe                    1496 Console                    0       11,040 K
spoolsv.exe                     1628 Console                    0        4,236 K
alg.exe                         1056 Console                    0        3,196 K
wscntfy.exe                     1068 Console                    0        1,780 K
wuauclt.exe                     1716 Console                    0        4,872 K
cmd.exe                         1192 Console                    0          764 K
taskmgr.exe                      960 Console                    0        1,728 K
tasklist.exe                    1672 Console                    0        4,044 K
wmiprvse.exe                    1976 Console                    0        5,300 K
```

Fig. 9 CMD terminal used by victim machine only

5   Conclusion

Recently, Cloud computing has become one of the popular computing paradigms. One of the significant models of cloud computing for large companies and organizations is a private cloud. The private cloud is a type of cloud computing infrastructure that created for in-house datacentre user. In this paper, a digital forensic approach for investigation of cybercrimes in private cloud computing is proposed. This approach can help digital investigators and examiners to know how to collect and extract forensic data from the private cloud environment.

References

1. DFRWS Technical Report: A road map for digital forensic research. In: Palmer, G. (ed.) Digital Forensic Research Workshop. Utica, New York (2001)
2. McKemmish, R.: What is Forensic Computing? Australian Institute of Criminology, Canberra (1999)
3. Ruan, K., Carthy, J., Kechadi, T., Baggili, I.: Cloud forensics definitions and critical criteria for cloud forensic capability: an overview of survey results. Digit. Investig. (Elsevier) **10**, 34–43 (2013)
4. Mell, P., Grance, T.: The NIST Definition of Cloud Computing, pp. 20–23 (2011)
5. RightScale 2016 State of the Cloud Report. http://www.rightscale.com/lp/2016-state-of-the-cloud-report?campaign=701700000015euW. Accessed 19 Aug 2016
6. Jeong, D., et al.: Investigation methodology of a virtual desktop infrastructure for IoT. J. Appl. Math. **2015** (2015)
7. Dykstra, J., Sherman, A.: Acquiring forensic evidence from infrastructure-as-a-service cloud computing: exploring and evaluating tools, trust, and techniques. In: DoD Cyber Crime Conference, Jan 2012
8. Rani, D.R., Geethakumari, G.: An efficient approach to forensic investigation in cloud using VM snapshot. In: IEEE International Conference on Pervasive Computing (ICPC) (2015)

Cryptanalysis of a Multifactor Authentication Protocol

Soorea Likitha and R. Saravanan

Abstract Cybercrime analysis reports state that the crime percentage has been doubled from past 3 to 4 years, and most of the crimes are being successful due to the lack of proper identity verification mechanism that is nothing but the authenticity verification mechanism. The authentication of the user can be done by various authentication protocols but still, the attacks are being success; this shows the necessity of improving the efficiency of the authentication algorithms, and this can be done accurately by analysis various existing mechanisms proposed by various researchers. On analyzing the various protocols of authentication, the loop pits of the existing schemes can be recorded, which gives an ideology of filling the loopholes of the same. Therefore, this paper analyzes the most recently proposed multifactor authentication scheme by Mishra et al. Based on various characteristics that are to be considered in designing an authentication protocol, the scheme is been analyzed. A detailed explanation of various vulnerabilities of Mishra et al.'s scheme is given in this paper.

Keywords Cryptanalysis · Smart card · Authentication · Session agreement
Secure channel · Public channel

1 Introduction

Rapid growth in the field of networks has gradually increased the use of networks for various activities in our daily life such as telecasting the news, online banking transactions, etc. The increase in radius of usage circle of network has indirectly or directly increased the need for the security over the data shared or transferred through the network. Not only to protect the data on the network but also to maintain the

S. Likitha (✉) · R. Saravanan
School of Information Technology and Engineering, VIT University,
Vellore, Vellore, India
e-mail: likitha_soorea@yahoo.in

R. Saravanan
e-mail: rsaravanan@vit.ac.in

© Springer Nature Singapore Pte Ltd. 2019   35
P. K. Sa et al. (eds.), *Recent Findings in Intelligent Computing Techniques*,
Advances in Intelligent Systems and Computing 707,
https://doi.org/10.1007/978-981-10-8639-7_4

privacy of the communications made over the network providing network security is required. There are six major security parameters [1] namely, authentication, access control, confidentiality, integrity, non-repudiation, and availability.

There are various mechanisms designed by various researchers for each individual security parameter in order to make the Internet secure, but till date it can be observed that there exists one or the other online theft [2] or breakage happening and most of them are being successful due to lack of proper identity verification of the person (i.e., proper identification of genuine user is not done). This falls under authentication security parameter. Various researchers are currently working in this aspect to improvise the mechanism [3]. Initially, the identity of the user was just verified based upon the single factor, that is, by user password verification. This was easily captured by various attackers, and then researchers came up with two-factor authentication such as password (PW) and also the smart card [4]. The best real-time application of two-factor authentication is ATM machines; it needs card as well as the PIN number to authenticate the user.

Currently, the research is happening on multifactor authentication in finding the effective way of verifying the identity of the user. In 2016, Mishra et al. [5] have proposed a robust and secure smart card-based authentication scheme for session initiation protocol using elliptic curve concepts of cryptography [6]. Mishra et al. have analyzed the robust authentication scheme on sip based on smart card using elliptic curve cryptography [7] introduced by Yeh et al. in the year 2014.

Mishra et al. have proven that the scheme proposed by Yeh et al. is vulnerable to various attacks. It states that the approach of Yeh et al. is prone to offline dictionary attack as the identity of the user is transferred in the plain form in the login request message. Therefore, they have improvised it in their scheme by preventing the transfer of UID directly in the communication network. Similarly, they have proven that Yeh et al.'s scheme is vulnerable to privileged insider attack, user anonymity failure, and does not maintain the perfect forward secrecy. Mishra et al. have also stated that their scheme is secure against privileged insider attack, man-in-middle attack, replay attack, and impersonation attack (on user as well as server). Mishra et al.'s paper also states that their proposed scheme resists all the vulnerabilities identified in Yeh et al.'s scheme. Mishra et al.'s scheme is the multi factor authentication scheme as it uses password, smart card as well as biometric to verify the identity of the user. Still after the detailed analysis their scheme it is found that there exist various vulnerabilities and most of the vulnerabilities have been explained in detailed in the cryptanalysis section of this paper. There are many other researches working on predicting an efficient way of identity verification or authenticity verification of users accessing various servers. There are many smart card-based multi-server authentication mechanisms [8].

Analysis of this scheme is done based on few characteristics that are needed to be followed by an efficient authentication protocol [9]. They are,

- It should resist various network attacks such as impersonation attack, replay attack, etc.
- It should provide mutual authentication between user and the server.

- It should provide session key agreement.
- It should maintain perfect forward secrecy.

Challenger's model

- The information stored in the smart card can be retrieved by applying power analysis.
- Adversary "A" can access all the information passed through the public channel (i.e., can read, modify, tamper, etc.).
- "A" can control the public communication channel as it is insecure and not trustworthy.
- "A" is capable of obtaining certain users identity from public resources.
- "A" can be one of the legal users registered with the server or an insider at server.

The rest of the paper is organized in the following order: Sect. 2 reviews Mishra et al.'s scheme, Sect. 3 provides the detailed cryptanalysis of Mishra et al.'s scheme, and finally Sect. 3 concludes the paper.

2 Review of Mishra et al.'s Scheme

This segment briefly describes the Mishra et al.'s scheme. This scheme consists of four different phases including the password and biometric updating phase. Here, the brief explanation of initialization, registration, and authentication phases as these are the major phases on which the cryptanalysis is given.

2.1 Initialization Phase

1. The server S selects an elliptic curve $E_p(a, b)$ over the Z_p which is the finite field and p is the large prime. It selects P that is a point on E_p.
2. S also defines a collision-resistant hash function h(.).
3. Now, S chooses the master key "s" from Z_n^* in a randomized manner and then computes the public key P_{pub} to be P * s.

After selecting all the parameters required, S closes the initialization phase by publishing $\{E_p(a, b), P, P_{pub}, h(.)\}$ in the smart card SC. Therefore, it can be seen that s is the only secret parameter chosen by the server S.

2.2 Registration Phase

All the new users of the network are supposed to register themselves to the server in order to access the server and its services.

User (U)	Secure channel	Server (S)
Select u, ID_U and PW_U.		
Compute $RPW = h(ID_U \| PW_U \| u)$.	$\langle ID_U, RPW \rangle$	
Imprint biometrics B_U.		Compute $X_U = h(s \| ID_U)$.
Compute $V = h(ID_U \| RPW \| H(B_U))$,	$SC\{Y_U, P_{pub}, h(\cdot)\}$	$Y_U = X_U \oplus RPW$.
$W = u \oplus h(PW_U \| H(B_U))$.		
Store V and W into the smart card.		
Insert $H(\cdot)$ into the smart card.		

Fig. 1 Flow of registration

1. The valid new user selects his/her user id (ID_U), password (PW) and a secret number "u".
2. User U computes RPW =h ($ID_U \| PW \| u$) and sends ID_U along with RPW to the server through a secure channel.
3. Server S verifies the ID_U after receiving the registration request message and asks the user to reselect the ID_U if it already exists else computes $X_U = h(s \| ID_U)$ and $Y_U = h(X_U + RPW)[+ - XOR$ operation] (Fig. 1).
4. After computation, S feeds $\{Y_U, P_{pub}, h(.)\}$ into the smart card and sends it to the user.
5. User U on receiving the smart card imprints its biometric at the sensor and computes V and W as shown in the flow, and finally stores V and W to the card.

Finally, the smart card of the user consists of $\{Y_U, P_{pub}, h(.), V, W, H(.)\}$, where H(.) is the bio hash function(i.e., used for hashing the biometric scanned by the sensor.

2.3 Authentication Phase

Actual authentication or the verification of the registered user is carried out in this phase. In this phase initially the user inserts his/her card into the machine and inputs ID_U, PW, Biometric along with the card. Verification of users identity based on the inputs given is done in several steps as described below.

1. On receiving the input, the user machine computes U^* and the pseudo-random password RPW with U^*, and further verifies V by computing V^* if true then moves to the next step by extracting X_U from the card details.
2. U computes M_U along with the computations C_1, C_U, and DID_U. After completion of calculations, the user sends a login request message to the server S $\{DID_U, M_U, C_1, T_U\}$.
3. Now on receiving the request message from user, S verifies the freshness of the message by ΔT and then verifies the M_U by computing $h(ID_U \| T_U \| X_U \| sC_1 \| C_1)$.
4. On valid verification of M_U the server S sends a challenge message {realm, C_2, M_s, T_s} to the user U. In order to send the challenge message, S com-

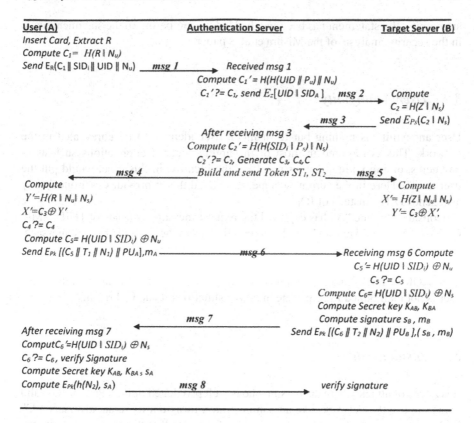

Fig. 2 Flow of authentication phase

puts $C_2 = bP$ (where b is the random number selected by S), $C_3 = bC_1$, sk $=$ h(ID$_U$||X$_U$||T$_U$||C$_3$||sC$_1$) and finally M$_s =$ h(sk||T$_U$||T$_s$||X$_U$||ID$_U$).

5. After receiving the challenge message, SC computes sk' $=$ h(ID$_U$||X$_U$||T$_U$||aC$_2$||C$_U$) and then verifies M$_s$ by comparing it with h(sk'||T$_U$||T$_s$||X$_U$||ID$_U$) if it is valid the sends an response message {realm, Auth} to S, where Auth $=$ h(ID$_U$||X$_U$||realm||sk||T$_s$).

6. "S" on receiving the response message verifies Auth and if verified to be valid then it sends an invite message acknowledgment (Fig. 2).

3 Cryptanalysis of the Mishra et al.

This section of the paper explains the various vulnerabilities of the Mishra et al.'s scheme in detail. This scheme fails to provide perfect forward secrecy and user anonymity; the statement has been justified below comparing to the statement made in the security analysis of the Mishra et al.'s paper.

3.1 User Anonymity

User anonymity is nothing but to maintain user identity to be unrevealed in the network. This is required in most of the highly secured applications such as e-payments, online voting, etc., as revealing the identity, in this case, would put the user at risk. Here in the current scheme, it is stated that it provides user anonymity as it sends DID_U instead of ID_U.

Consider that the "A" has captured the request message consists of $\{DID_U, M_U, C_1, T_U\}$. "A" should know C_U, DID_U to find ID_U from the captured request message. Now assume that the privacy of "s" is been broken, and it is already known that $C_U = sC_1$. Therefore, the adversary "A" on knowing C_U can easily compute ID_U which is an XOR of DID_U and x coordinate of C_U. This shows that the anonymity of the user is not maintained to be private in all the situations it can be broken.

3.2 Insider Attack

It is a type of attack where the person who is well privileged or treated to be the valid person to use the network, like an administrator, turns out to be an adversary "A". In this case, the adversary is aware of most of the data that the user shares with the server because the adversary is an internal person.

Considering an adversary to be an insider, "A" knows ID_U, pseudo-random password RPW and also X_U, Y_U of a particular user. Now, assume that "A" stays in between the user and server and captures the request message and computes C_U as "A" is aware of the secret key (s) of the server. As the adversary knows all the required parameters to compute the challenge message, he/she can impersonate the server and send its own challenge message to the user blocking the servers challenge message. Therefore, through this, it is understood that the proposed scheme does not resist the insider attack.

In another scenario where the adversary is an insider and he/she also stoles the smart card of the user, through power analysis they can access V and W. "A" can also find $H(B_U)$ from the V by trailing on substitution of ID_U and RPW values. He applies the same brute force method to W and finds u. Therefore on knowing all the secret values of the user, "A" can easily impersonate the user. Therefore, an insider can easily break the scheme impersonating both the user and the server.

3.3 Stolen Smart Card Attack

Consider an adversary "A" who have already analyzed or tracked the password observing the users usage patterns or by brute force attack, also gets the biometric traces of the user B_U and then stoles the card of the user then he/she can retrieve the data stored in the card through the power analysis (i.e., V, W, Y_U). As "A" is aware of both PW and B_U he/she can compute RPW. Applying the brute force attack on W or RPW "u" can be retrieved. Now the adversary can change the password and update the values of the card as well as pretend to be the genuine user and access the server. Therefore, this scenario proves that the Mishra et al.'s scheme is vulnerable toward the stolen smart card attack.

3.4 Man-in-Middle and Impersonation Attack

It is the attack where the adversary stays in the communication network between the user and server and access the data communicated between them. Here assuming a scenario where the adversary "A" captures the login request sent by the user to the server and replaces it with values calculated with the random number selected by "A" leaving the ID to be the same as the users ID_U. That is,

- "A" replaces the request message $\{DID_U, M_U, C_1, T_U\}$ to $\{DID_U, M_a, C_{1a}, T_a\}$.
- Now, S generates the challenge message $\{realm, C_2, M_s, T_s\}$ "A" blocks the challenge message of S and replaces it with that of "A" $\{realm, C_{2a}, M_a, T_a\}$ (i.e., A replaces the random numbers of user as well as server).
- The user U verifies freshness and then generates the response which will again be replaced by the adversary "A".

Therefore, the above scenario clearly shows that the Mishra et al.'s scheme is prone to the man-in-middle attack.

4 Conclusion

This paper initially recollects the authentication scheme proposed by Mishra et al. in Sect. 2, and then explains various vulnerabilities of their scheme in detail in Sect. 3. The introductory section of this paper explains all the goals mentioned in Mishra et al.'s scheme but most of them are disproven in the cryptanalysis section. It is a challenge to design an authentication mechanism without using the third party but the use of realm could again give an advantage to the adversary. The analysis of the scheme is been done considering all the four characteristics mentioned in the introductory section. Considering the scenario explained in the proof of stolen smart card attack of the Sect. 3 of the paper it can be observed that the information present in the card can be manipulated. This makes the whole scheme weak. Therefore, Mishra et al.'s scheme would be more secured if it would be improvised in such a way that it resists all the attacks mentioned in the cryptanalysis section of this paper and that would be considered to be the future work of this paper.

References

1. Shunmuganathan, S., Saravanan, R.D., Palanichamy, Y.: Secure and efficient smart-card-based remote user authentication scheme for multiserver environment. Can. J. Electr. Comput. Eng. **38**(1) (2015)
2. Fan, C.-I., Chan, Y.-C., Zhang, Z.-K.: Robust remote authentication scheme with smart cards. Comput. Secur. (2005)
3. Khan, M.K., Kim, S.-K., Alghathbar, K.: Cryptanalysis and security enhancement of a more efficient & secure dynamic ID-based remote user authentication scheme. Comput. Commun. **34**, 305–309 (2011)
4. Leung, K.-C., Cheng, L.M., Fong, A.S., Chan, C.-K.: Cryptanalysis of a modified remote user authentication scheme using smart cards. IEEE Trans. Consum. Electron. **49**(4), 1244 (2003)
5. Wan, T., Jiang, N., Ma, J., Yang, L.: Cryptanalysis of a biometric-based multi-server authentication scheme. Int. J. Secur. Appl. (2016)
6. Mishra, D., Das, A.K., Mukhopadhyay, S., Wazid, M.: A secure and robust smartcard-based authentication scheme for session initiation protocol using elliptic curve cryptography. Wirel. Pers. Commun. (2016)
7. Yeh, H.-L., Chen, T.-H., Shih, W.-K.: Robust smart card secured authentication scheme on SIP using elliptic curve cryptography. Comput. Stand. Interfaces (2014)
8. Ding, W., Chun-guang, M.A.: Cryptanalysis and security enhancement of a remote user authentication scheme using smart cards. J. China Univ. Posts Telecommun. (2012)
9. Chaudhry, S.A., Khan, M.T., Khan, M.K., Shon, T.: A multiserver biometric authentication scheme for TMIS using elliptic curve cryptography. Mob. Wirel. Health. **40**(11), 230 (2016)
10. Wang, X.-M., Zhang, W.-F., Zhang, J.-S., Khan, M.K.: Cryptanalysis and improvement on two efficient remote user authentication scheme using smart cards. Comput. Stand. Interfaces **29**, 507–512 (2007)

Dual-Core Implementation of Right-to-Left Modular Exponentiation

Satyanarayana Vollala, N. Ramasubramanian, B. Shameedha Begum
and Amit D. Joshi

Abstract Modular exponentiation is one of the core operations in most of the public-key cryptosystems. It consists of a sequence of modular multiplications. The performance of public-key cryptographic transformations is strongly influenced by the competent implementation of modular exponentiation and modular multiplication. This paper presents the hardware implementation of modular exponentiation on two processor cores. Montgomery multiplication method is modified according to the needs of dual-core implementation to improve the core utilization. It is implemented with different radices ranging from 2^2 to 2^{32}. The performance of the proposed design is analyzed and compared with the existing techniques in terms of number of clock cycles, throughput, power, and area. The proposed design has been developed using Verilog and synthesized using Xilinx-14.6 ISE for usage in FPGA, and the same has been synthesized using Cadence for ASIC. But here the results are presented based on FPGA.

Keywords Public-key cryptography · Modular exponentiation · Modular
multiplication · Montgomery multiplication · Dual-core architectures

S. Vollala (✉) · N. Ramasubramanian · B. Shameedha Begum · A. D. Joshi
Department of Computer Science and Engineering, National Institute of Technology,
Tiruchirappalli, Tiruchirappalli, India
e-mail: satya4nitt@gmail.com

N. Ramasubramanian
e-mail: nrs@nitt.edu

B. Shameedha Begum
e-mail: shameedha@nitt.edu

A. D. Joshi
e-mail: amitjoshi233@gmail.com

© Springer Nature Singapore Pte Ltd. 2019
P. K. Sa et al. (eds.), *Recent Findings in Intelligent Computing Techniques*,
Advances in Intelligent Systems and Computing 707,
https://doi.org/10.1007/978-981-10-8639-7_5

1 Introduction

Security has become an important part of human life due to the explosive growth in data communication and Internet services. In order to secure a system and transactions over the network, various cryptographic algorithms are required. Based on the usage of keys, cryptographic techniques can be classified into two types, viz., secret key cryptography and public-key cryptography. Secret-key cryptography uses only one key for encryption as well as decryption and public-key cryptography uses two different but related keys, one for encryption and other for decryption. To minimize the complexity of key management between sender and receiver, in 1976, Diffie and Hellman proposed the theory of Public-key cryptography [1] and digital signature scheme to uniquely bind a message to its destination. Public-key cryptographic techniques such as Rivest–Shamir–Adleman (RSA) [2], elliptic curve cryptography (ECC) [3–5], Diffie–Hellman key exchange algorithm (DH), Koblitz scheme, ElGmal [6] public-key scheme, and digital signature scheme (DSS) play an important role for providing security services in the form of secrecy, authentication, integration, and non-repudiation. In contemporary secure electronic communication, RSA is the widely used and one of the most dominant algorithms among all public-key schemes [7].

RSA public-key scheme was proposed by Rivest, Shamir, and Adleman in 1978, based on the idea originally proposed by Diffie and Hellman. The core component of RSA is the modular exponentiation. The security of RSA lies in inability of factoring large integers. For long-term security, the key (exponent) length should be at least 1024-bits [8]. Modular exponentiation consumes a large amount of time due to trial divisions [9]. As a result, the performance of RSA mainly depends on the implementation of modular multiplication and frequency of modular multiplications in modular exponentiation. Montgomery presented a method to speed up the modular multiplication. The basic idea was to replace the complicated arithmetic operations called divisions and multiplications with a simple shift operation.

The remaining content of this paper is structured as follows: Sect. 2 presents the work related to the Montgomery multiplication and modular exponentiation. In Sect. 3, proposed algorithms are elaborated, and Sect. 4 explains the hardware realization and results and analysis. Finally, Sect. 5 concludes this paper.

2 Related Works

In this section, the brief description about Montgomery multiplication and modular exponential algorithms is presented.

2.1 Montgomery Multiplication Method

Many algorithms have been proposed for implementation of modular multiplication in hardware. But the most efficient method was proposed by Montgomery [10]. Let $P = \sum_{i=0}^{n-1} p_i \cdot 2^i$, $Q = \sum_{i=0}^{n-1} q_i \cdot 2^i$ and an n-bit modulus $T = \sum_{i=0}^{n-1} t_i \cdot 2^i$, where $p_i, q_i, t_i \in \{0, 1\}$, and then the Montgomery multiplication of P, Q with respect to the modulus T is given by $P \cdot Q \cdot 2^{-n} mod\ T$.

The Montgomery algorithm has undergone many modifications for improving the performance. Meng Qiang et al. provided a new operand scalable hybrid modular multiplication algorithm [11], which is scalable in key size also. Ming-Der Shieh et al. explained how to relax the data dependency between multiplication, quotient determination, and modular reduction that exists in conventional Montgomery multiplication. Ming-Der Shieh et al. introduced a new modular exponentiation hardware design [12] with unified multiplication and square module that reduce the count of input operands for the CSA to speed up the computation. Yinan Kong et al. presented a new modular multiplication [13] algorithm to speed up the modular exponentiation process in RNS (residue number system) with the help of core function to calculate $A.B mod\ N = X - \lfloor \frac{X}{X} \rfloor \cdot N$, where $X = A \cdot B$. To reduce the critical path and area overhead, Wu Tao et al. proposed a new hardware design for RSA decryption with the help of CRT in [14]. This architecture is optimal in terms of speed, area, and frequency with reference to the literature in [14]. Atsushi Miyamoto et al. presented a systematic design of RSA processor with the help of high-radix Montgomery multipliers [15]. Gavin Xiaoxu Yao et al. presented RNS parameter selection process for computational efficiency [16]. Two methods to select RNS moduli with reduced complexity were also proposed. Shiann-Rong Kuang et al. presented an energy efficient high-speed Montgomery multiplier for RSA [17] with the help of carry save addition. This architecture reduces the energy consumption by avoiding the superfluous CSA. Shiann-Rong Kuang et al. proposed a simple and high-performance Montgomery multiplier [18], where multiplier uses only one-level carry-save adder to attain higher performance and significant area-time product. A brief analysis of different Montgomery multiplications and comparisons between them can be found in [19]. Xiaofeng Chen et al. presented a new algorithm for modular exponentiation with modulo a prime number [20]. They claimed that it is superior in both efficiency and checkability. A uniform $k-$partition method has been presented for Montgomery multiplication [21]; it works in parallel. Residue number system is playing tremendously an important role in hardware cryptography. Dimitrios et al. designed a methodology or converting residue number system (RNS) and polynomial residue number system for Montgomery multiplication [22]. Abdalhossein et al. presented multibit-scan-multibit-shift technique or high-throughput modular exponentiation [23].

2.2 Modular Exponentiation

Modular exponentiation is the core component of many public-key cryptosystems [24] such as RSA. Hence, the performance of RSA is entirely dependent on the efficient implementation of modular exponentiation. Many modular exponential algorithms have been proposed, but binary modular exponential algorithm is extensively used in hardware implementation of public-key cryptography. There are two types of binary modular exponential methods, viz., left-to-right (MSB-to-LSB) and right-to-left (LSB-to-MSB). In left-to-right, the digits of the exponent are scanned from left to right. In right-to-left method, the digits of the key are scanned from right to left. These methods are also called as square and multiply algorithms. The detailed survey about modular exponential algorithms is given in [25].

3 Implementation of Right-to-Left Modular Exponentiation in Dual Core

In RSA, both encryption and decryption operations are accomplished by the same modular exponentiation with different inputs. In this paper, right-to-left binary exponential method is modified to make it adaptable for direct implementation in dual core, and the same is presented in Algorithm 1. A high-radix Montgomery multiplication is tuned according the needs of modified right-to-left method in dual-core implementation.

3.1 The Modified Right-to-Left Binary Exponential Algorithm

Algorithm 1 Modified Right-to-left Modular Exponential Algorithm

---

Input: Base B, Exponent E, modulus T, and the Proposed Constant PC
Output: $Z = B^E \bmod T$
1: $S_q = MHRM(B, PC, T)$;
2: $Z = MHRM(1, PC, T)$;
3: **for** $i = 0$ to $k - 1$ **do**

4:    **if** $(\epsilon_i \neq 0)$ **then**
5:       $Z = MHRM(Z, S_q, T)$
6:    **end if**

7:    $Z = MHRM(S_q, S_q, T)$;
8: **end for**
9: $Z = MHRM(Z, 1, T)$;
10: **return** Z

To evaluate the modular multiplication in Algorithm 1, high-radix Montgomery multiplication is customized according to the needs of the modified right-to-left modular exponential algorithm and named as modified high-radix Montgomery multiplication (MHRM). Every time when MHRM is invoked, an extra factor of $R^{-1} mod N$ is multiplied to the product $P \cdot Q$ with respect the modulus T ($MHRM(P, Q, T) = P \cdot Q \cdot R^{-1} mod T$, where R^{-1} is the inverse of $R = 2^{bs}$ with respect to the modulus T). It is not possible to remove this extra factor explicitly. To avoid this extra factor, first preprocess the message B with a newly proposed constant $(2^b)^{bs} mod T$, by invoking MHRM with the inputs B, PC, T as given in line 1 of Algorithm 1. Take the initial result as Z = MHRM(1, PC, T), for every bit of the exponent square the value of S_q and for every nonzero bit of the exponent multiply the updated result Z with S_q. Repeat the same process for k-times, where k represents the number of bits in the exponent, then postprocess the updated result by invoking MHRM along with the inputs Z, 1, and T to get the desired result $B^E mod T$.

Since the square operation is not dependent on multiply operation in Algorithm 1, these two operations can be performed in parallel with two different cores. Thus, $k_1 \times n$ number of clock cycles can be reduced to compute the modular exponentiation value, where k_1 is the number of nonzero bits in the exponent and n is the number of clock cycles required for one modular multiplication.

The architecture diagram of modified right-to-left modular exponential algorithm is given as Fig. 1. All the registers are initialized with the inputs base B, exponent E, modulus T, and the proposed constant PC, and the counter is initialized with k, where k is number of bits in the exponent E. MHRM is invoked in parallel with B,PC and T to calculate S_q and with 1, PC ant T to calculate Z, which is the initial result. Shift register is assigned with the exponent value E and every time LSB of the exponent is checked for nonzero bit to compute Z value, if found Z is updated by using MHRM with the previous value of Z, S_q and T. At the same instance S_q is also computed in parallel by passing the values of S_q, S_q and T to MHRM. Then, the counter is decremented and for every nonzero value of the count, exponent bit is checked from LSB to MSB. When the counter reaches zero, the final result for Z is obtained by invoking MRM with Z, 1 and the modulus T. The components enclosed within dotted lines numbered as 1, 2, and 3 can perform their operations in parallel thereby reducing execution time by $k_1(n + 3)$ clock cycles resulting in an increased throughput, where k_1 is number of nonzero bits in the exponent E and n + 3 number of clock cycles needed to evaluate MHRM.

3.2 High-Radix Montgomery Multiplication Method

In this section, MHRM has been implemented with high radix to increase the throughput by increasing the clock frequency, and by reducing the clock cycles count. This is achieved because the throughput is directly proportional to the clock frequency and inversely proportional to the number of clock cycles. Let the modulus $T = (t_{s-1}, t_{s-2}, \ldots, t_1, t_0)_{2^b}$ is an odd integer, where 2^b is radix, and the multiplier and

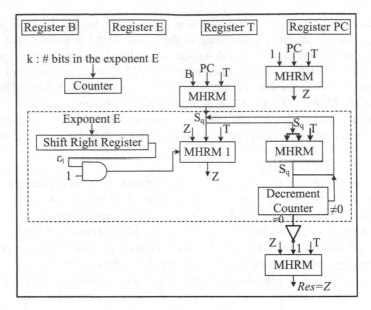

Fig. 1 Architecture of right-to-left modular exponential algorithm

multiplicands are $P = (p_{s-1}, p_{s-2}, \ldots, p_1, p_0)_{2^b}$, $Q = (q_{s-1}, q_{s-2}, ldots, q_1, q_0)_{2^b}$, respectively, where $0 \leq P, Q < T$. Let $T^1 = -T^{-1} mod\ 2^b$, where T^{-1} is the modular inverse of T with respect to the modulus 2^b. Then, the procedure to calculate the Montgomery product $PQR^{-1} mod\ T$, where $R = 2^{bs}$ of two integers P, Q with the modulus T and with radix 2^b is given in Algorithm 2.

Algorithm 2 Modified High-Radix Montgomery multiplication Algorithm

---

Input: P, Q, the modulus T, and T^1
Output: $V = PQR^{-1} mod\ T$, where $R = 2^{bs}$
1: $V = (v_{s-1}, v_{s-2}, ..., v_1, v_0)_{2^b}$;
2: $Z = MHRM(1, PC, T)$;
3: **for** $i = 0$ to $s - 1$ **do**
4:     $U = (v_0 + p_0.q_i).T^1 mod\ 2^b$;
5:     $V = (V + P.q_i + U.T)/2^b$;
6: **end for**

7: **if** $(V \geq T)$ **then**
8:     $V = V - T$
9: **end if**

10: **return** Z

Take the initial result V as zero, for every bit of multiplier Q, update the intermediate result V by using $V = (V + P \cdot q_i((v_0 + p_0.q_i) \cdot T^1 mod\ 2^b) \cdot T)/2^b$. After k iterations, compare the updated result V with the modulus T. If V is greater than or

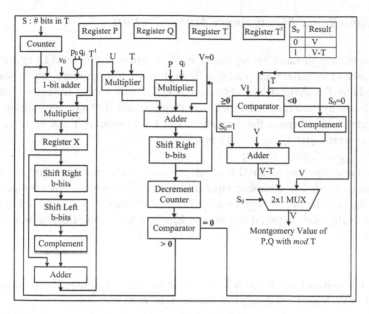

Fig. 2 Architecture diagram MHRM

equal to T, then subtract T from V, to get the desired value $PQR^{-1}mod\ T$, where $R = 2^{bs}$.

The architecture diagram of MHRM is given in Fig. 2. The inputs of MHRM, namely P, Q, T, and T^1, are stored in the respective registers. The counter is initialized with S that represents the number of bits in the modulus T. The value of V is initialized with 0. The computation phase of the algorithm requires adder, multiplier, shift register, and complementing components. Product of p_0, q_i is added with v_0 and then multiplied with T^1, and the result is stored in X. To implement $mod\ 2^b$ of X, i.e., $Xmod\ 2^b$, which is realized with $U = X - ((x >> b) << b)$. To obtain V, two multipliers are used to find $p \cdot q_i$ and $U \cdot T$ which is then added to V. This intermediate result has to be divided by 2^b by shifting $b - bits$ right. This process is repeated till counter reaches zero. After completing all iterations, the final result V is obtained by subtracting T from V only if $V \geq T$. This operation is performed by using a comparator circuit. The 2X1 multiplexer directs V or V-T as the final output based on the status of selection line S_0 which is set 1 for V-T, otherwise zero.

4 Performance Analysis and Comparison

In this section, the results and analysis of proposed hardware designs are discussed in terms of clock cycles, frequency, throughput, power, and area.

4.1 High-Radix Montgomery Multiplication

In this work, a high-radix Montgomery multiplication is proposed to improve the performance of modular multiplication. In high-radix Montgomery multiplication a k-bit modulus T is divided into s blocks, each of length w-bits. Instead of scanning 1-bit of modulus each time, a block of w-bits are scanned at a time to minimize the clock cycles and maximize the frequency. The k-bit modulus T can be represented by $s - bit$ words $t_i (0 \le i \le w - 1)$ as follows:

$$T - \sum_{i=0}^{w-1} t_i (2^s)^i = (t_{s-1}, t_{s-2}, t_{s-3}, \ldots t_1, t_0)_{2^w}$$

As w increases, the required number of clock cycles reduces. The number of clock cycles needed to compute Montgomery multiplication of two integers with k-bit modulus and radix 2^w is given as $\frac{k}{w}(7s) + 2$, where $k = w \cdot s$. Table 1 tabulates the clock cycle count required to compute Montgomery multiplication with respect to the 1024-bit and 2048-bit modulus with various radices ranging from 2^4 to 2^{32}.

The MHRM algorithm is synthesized with various radices 2^2, 2^4, 2^8, 2^{16}, and 2^{32}. Table 2 tabulates the frequency, power area, and throughput of MHRM worked out for different radices ranging from 2^2 to 2^{32} for 1024-bit exponent. It is obvious from the results that higher frequency can be achieved with 2^8.

Table 1 Operation cycles of modular exponentiation

Radix used in MHRM	Number of clock cycles	
	1024-bit modulus	2048-bit modulus
2^2	3672064	7342080
2^4	1837056	3672064
2^8	919552	1837056
2^{16}	460800	919552
2^{32}	231424	460800

Table 2 Power and throughput details of the proposed design

Radix in MHRM	Frequency (MHz)	Power (μW)	Area (μm^2)	Throughput (KBPS)
2^2	385.36	31.81	1256347	104.94
2^4	424.90	35.23	2047845	231.29
2^8	732.45	41.03	3135515	796.52
2^{16}	528.00	52.32	4990460	1145.83
2^{32}	438.86	67.88	7976185	1982.76

Table 3 Performance of 1024-bit proposed modular exponential algorithm

Radix in MHRM	Delay time (ns)	Exponentiation time (ms)	ME time × Area ($\mu m^2 s$)	ME time × Power ($\mu m^2 \mu W$)
2^2	1.81	19.52	24523.893	620.9312
2^4	2.02	09.32	19085.915	328.3436
2^8	2.43	04.05	12698.835	166.1715
2^{16}	2.88	02.58	12875.386	131.8464
2^{32}	3.11	01.96	15633.322	133.0448

4.2 Right-to-Left Binary Exponential Method

To increase the overall throughput, the right-to-left modular exponential algorithm has been implemented on two different cores. One core is used to implement multiply operation, and the second core is used to implement square operation. This dual-core design reduces $k_1(7s + 2)$ number of clock cycles for one exponential operation, where k_1 is the number of nonzero bits in the exponent, and $7s + 2$ is the number of clock cycles needed to implement one modular multiplication with s-bit modulus. Since the throughput is inversely proportional to the clock cycle count, the proposed design increases the throughput. The power consumption, area, maximum frequency, and the throughput of proposed left-to-right modular multiplication design are given in Table 2.

It can be understood from Table 2 that when radix increases, the throughput increases, and also results in increased power consumption. Table 3 shows the performance of proposed design with different radices. Optimized performance with respect to the parameter "*ME time × power*" (ME time: modular exponentiation time) is achieved with radix value of 2^{16}. With radix value of 2^8, the proposed design gives better performance for the parameter "*ME time × Area*". With radix 2^{32}, it is possible to achieve higher throughput.

In comparison with MME42_C2 [17], which is the existing state-of-the-art technology as far as energy efficiency is concerned, the proposed design with a radix value of 2^8 achieves increased throughput of 83.93% with an extra power of 2.2%. With radix value of 2^{16} and 2^{32}, the proposed design achieves better throughput at the rate of 166.60% and 357.86% respectively, with an extra power of 30.7% and 69.57% in comparison with MME42_C2 [17].

5 Conclusion

Modular exponentiation is implemented on two processor cores resulting in reduction of number of clock cycles and increase in throughput without extra power requirements. Modular multiplication is implemented with high-radix Montgomery multiplication. The proposed design with a radix value of 2^8 achieves better throughput at the cost of little extra power. As radix keeps increasing, the complexity of multipliers within each in processing element also increase, resulting in more power consumption but can guarantee to offer better throughput. Optimized performance with respect to the parameter "*ME time* × *power*" is achieved with radix value of 2^{16}. With radix value of 2^8 this proposed design gives better performance for the parameter "*ME time* × *Area*". For applications with a focus on more throughputs, the above-proposed designs seem to be more appropriate.

References

1. Diffie, W., Hellman, M.E.: New directions in cryptography. IEEE Trans. Inf. Theory **22**(6), 644–654 (1976)
2. Rivest, R.L., Shamir, A., Adleman, L.: A method for obtaining digital signatures and public-key cryptosystems. Commun. ACM **21**(2), 120–126 (1978)
3. de Dormale, G.M., Quisquater, J.-J.: High-speed hardware implementations of elliptic curve cryptography: a survey. **53**(2), 72–84 (2007)
4. Tibouchi, M., Kim, T.: Improved elliptic curve hashing and point representation. Des. Codes Cryptogr. 1–17 (2016)
5. Bos, J.W., Costello, C., Longa, P., Naehrig, M.: Selecting elliptic curves for cryptography: an efficiency and security analysis. J. Cryptogr. Eng. 1–28 (2015)
6. ElGamal, T.: A public key cryptosystem and a signature scheme based on discrete logarithms. In: Advances in Cryptology, pp. 10–18. Springer (1985)
7. Kaminaga, M., Yoshikawa, H., Suzuki, T.: Double counting in-ary RSA precomputation reveals the secret exponent. IEEE Trans. Inf. Forensics Secur. **10**(7), 1394–1401 (2015)
8. Huang, X., Wang, W.: A novel and efficient design for an RSA cryptosystem with a very large key size. IEEE Trans. Circuits Syst. II Express Briefs **62**(10), 972–976 (2015)
9. Garg, H.K., Xiao, H.: New residue arithmetic based Barrett algorithms: modular integer computations. IEEE Access **4**, 4882–4890 (2016)
10. Montgomery, P.L.: Modular multiplication without trial division. Math. Comput. **44**(170), 519–521 (1985)
11. Meng, Q., Chen, T., Dai, Z., Chen, Q.: A scalable hybrid modular multiplication algorithm. J. Electron. (China) **25**(3), 378–383 (2008)
12. Shieh, M.-D., Chen, J.-H., Wu, H.-H., Lin, W.-C.: A new modular exponentiation architecture for efficient design of RSA cryptosystem. IEEE Trans. Very Large Scale Integr. (VLSI) Syst. **16**(9), 1151–1161 (2008)
13. Kong, Y., Asif, S., Khan, M.A.U.: Modular multiplication using the core function in the residue number system. Appl. Algebra Eng. Commun. Comput. 1–16 (2015)
14. Wu, T., Li, S.G., Liu, L.T.: Fast RSA decryption through high-radix scalable montgomery modular multipliers. Sci. China Inf. Sci. **58**(6), 1–16 (2015)
15. Miyamoto, A., Homma, N., Aoki, T., Satoh, A.: Systematic design of RSA processors based on high-radix montgomery multipliers. IEEE Trans. Very Large Scale Integr. (VLSI) Syst. **19**(7):1136–1146 (2011)
16. Yao, G.X., Fan, J., Cheung, R.C.C., Verbauwhede, I.: Novel RNS parameter selection for fast modular multiplication. IEEE Trans. Comput. **63**(8), 2099–2105 (2014)
17. Kuang, S.-R., Wang, J.-P., Chang, K.-C., Hsu, H.-W.: Energy-efficient high-throughput montgomery modular multipliers for RSA cryptosystems. IEEE Trans. Very Large Scale Integr. (VLSI) Syst. **21**(11), 1999–2009 (2013)
18. Kuang, S.-R., Wu, K.-Y., Lu, R.-Y.: Low-cost high-performance VLSI architecture for montgomery modular multiplication

19. Koç, C.K., Acar, T., Kaliski, B.S. Jr.: Analyzing and comparing montgomery multiplication algorithms. IEEE Micro 16(3), 26–33 (1996)
20. Chen, X., Li, J., Ma, J., Tang, Q., Lou, W.: New algorithms for secure outsourcing of modular exponentiations. IEEE Trans. Parallel Distrib. Syst. 25(9), 2386–2396 (2014)
21. Néto, J.C., Tenca, A.F., Ruggiero, W.V.: A parallel and uniform-partition method for montgomery multiplication. IEEE Trans. Comput. 63(9), 2122–2133 (2014)
22. Schinianakis, D., Stouraitis, T.: Multifunction residue architectures for cryptography. IEEE Trans. Circuits Syst. I Regul. Pap. 61(4), 1156–1169 (2014)
23. Rezai, A., Keshavarzi, P.: High-throughput modular multiplication and exponentiation algorithms using multibit-scan-multibit-shift technique. IEEE Trans. Very Large Scale Integr. (VLSI) Syst. 23(9), 1710–1719 (2015)
24. Paillier, P.: Public-key cryptosystems based on composite degree residuosity classes. In: Advances in cryptology (UROCRYPT'99), pp. 223–238. Springer (1999)
25. Gordon, D.M.: A survey of fast exponentiation methods. J. Algorithm. 27(1), 129–146 (1998)

Investigation of Privacy Issues in Location-Based Services

Ajaysinh Rathod and Vivaksha Jariwala

Abstract Location-based services (LBS) are one of the emerging technologies in the mobile, networking, and information services (Padmanaban, Int J Adv Remote Sens GIS 2:398–404, 2013 [1]). Users of LBS exchange information based on location like maps and navigation, location-based information, location tracking services, social networking, vehicular direction finding, and location-based advertisement. LBS users' provide highly personalize information to the service providers like location information and personal identity. Location privacy and communication privacy are one of the key issues of LBS. A cryptographic technique is used for avoiding position tracking, secure information sharing, data integrity, and authentication in the mobile network. Location-based services are classified as trusted third party and without trusted third party that uses cryptographic approaches. Trusted third party model guarantees the privacy of the user in LBS. The main issue is that end users are not inevitably gratified by trusting intermediate entities. Trusted third party (TTP) free model can be taken as an excellent approach from all the existing ones for location privacy in LBS. The main aim of the paper is to give a comprehensive survey of location privacy and also proposed effective approach that guarantee the privacy of LBS users with improving scalability, minimum cost.

Keywords Location-based services · Privacy preserving · Cryptography

A. Rathod (✉)
Department of Computer Engineering, Research Development & Innovation Center, C U Shah University, Wadhwan City, Gujarat, India
e-mail: ajay58886@gmail.com

V. Jariwala
Department of Information Technology, Sarvajanik College of Engineering and Technology, Surat, Gujarat, India
e-mail: vivaksha.jariwala@scet.ac.in

© Springer Nature Singapore Pte Ltd. 2019   55
P. K. Sa et al. (eds.), *Recent Findings in Intelligent Computing Techniques*,
Advances in Intelligent Systems and Computing 707,
https://doi.org/10.1007/978-981-10-8639-7_6

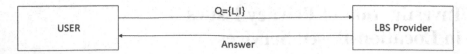

Fig. 1 Simple communication schema single client and server [2]

1 Introduction

In today's growing period of information and communication, location-based services prove to be a prominent technology. User's need some important information and services based on their current position. Emergency services, LBS reminder, map navigation, location-based marketing, location base search, and location-based advertisement are examples of location-based services. LBS are available on a variety of mobile platforms like mobile devices, PDAs, GPS devices, and others devices because they are ubiquitous. Nowadays, the growth of LBS users is very fast. In location-based application, users provide their highly personalized information like their own identity and location to the service provider causing vulnerability to their privacy, e.g., an attacker can also get current location of the user and also track user's daily activities. Due to the tracking capability, it opens many possibilities of computer-based crimes like kidnapping, harassment, car theft, and many more. Varieties of attacks are already possible, so there is a big challenge to protect location privacy with minimum cost.

2 Literature Survey

In this section, we study popular information flow model, privacy requirement, efficiency requirement, and category of crypto-based privacy model for location-based service that has been proposed by many authors.

2.1 Basic Information Flow Model

In LBS, there is a requirement to protect privacy for all users. There are two popular information flow models in LBS.

2.1.1 Client–Server Model

Figure 1 demonstrates that client needs some information based on their location. Client (users) sends a query to the server (service provider) like search for nearby ATM/restaurant. As a result, the server will respond back to the user as required.

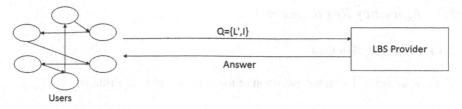

Fig. 2 Communication schema between set of collaborative users and LBS provider [2]

2.1.2 Peer-to-Peer Model

An alternative communication model is the peer-to-peer model where a group of peers would like to cooperatively compute some location without seeing the help of a centralized location-based server [2] (Fig. 2).

2.2 Privacy Requirement

After the thorough study of the proposed schemas in [2–4], we have summarized following properties that must be satisfied to achieve privacy preservation in LBS.

2.2.1 Location Privacy

The protocol does not reveal the (extract) user's location information to the LBS provider [1]. No authorized entity can access or extract user location information.

2.2.2 Identity Privacy

Any unauthorized access for tracing the identity of the user would be blocked.

2.2.3 Tracking Protection

LBS service provider is not able to link two or more successive user's location with time. No unauthorized entity should be able to link user's location from different session. With user's position and time, unauthorized entity is also able to track user's daily routine and activities.

2.3 Efficiency Requirement

2.3.1 Computation Cost

The cost associated with the execution of the cryptographic algorithms.

2.3.2 Communication Cost

It is the cost that is associated with the information transmission and reception.

2.4 Categorization of Crypto-Based Privacy Model for LBS

In the simple form of communication between an LBS user and LBS provider, the former sends a simple query (Q) containing an ID and his location (L) and a request for information (I) that he wants to retrieve from P [5]. The user provides his identity and location to provider, but provider cannot always trustworthy.

There are many solutions that are already proposed by using TTP-based schemas. TTP-based schema is very often used and easy to deploy. This schema has many drawbacks as do not rely on TTP. Many schemas are already proposed as TTP free schemas. Crypto-based privacy model is been categorize into two main parts that are describe below.

2.4.1 TTP-Based Schema

Most schemas within this category adopt a centralized model for privacy [2]. TTP-based schema is easy to understand and easy to develop.

Policy-Based Schema

Policy-based schema is a modest schema in that providers are bound to follow a set of privacy policies known by users. Provider must have to follow this otherwise strict and legal action will be taken against provider. This schema is wildly used on the Internet.

Pseudonym Schema

Pseudonymisers are the intermediate entity between users and provider. Pseudonymisers are protecting the privacy of the user by hiding its identity (Fig. 3).

Fig. 3 Communication schema between users, pseudonymisers, and LBS provider [2]

The main loophole in this schema is that unauthorized access can breach the actual identity of the user by liking the user location in other records.

Anonymisers Schema

Anonymisers can calculate obfuscated positions covering several users [5]. Anonymisers aim to hide user's identity with its location information using k-anonymity property from LBS provider. However, it also has a drawback that an attacker can infer this by using quasi-identifier.

2.4.2 TTP Free Schema

Instead of relying on the third party, users will team up to protect their privacy. It is vigorous against the attack of a vengeful user and service provider and is also scattered.

Obfuscation-Based Schema

It provides a technique in that user's location information is undignified to improve the privacy. This method is based on imprecision. Various algorithms are already proposed based on this schema as negotiation algorithm, space twist algorithms, etc. [3, 5]. The advantage of this approach is that we do not rely on the trusted third party. The disadvantage of this method is not able to achieve k-anonymity and l-diversity and lack of collaboration [5].

PIR-Based Schema

This technique is also TTP- free in that user executes periodic queries at LBS provider. This schema guarantees about privacy in location-based queries. This technique is based on quadratic residues in modulo arithmetic of a large number for the product of two large prime numbers [6, 7]. It is also known as a client–server approach that provides strong privacy using a cryptographic technique. The disadvantage of this method is related to computation and efficiency (Fig. 4).

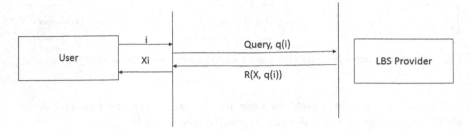

Fig. 4 Communication schema of PIR method between users and LBS provider [2]

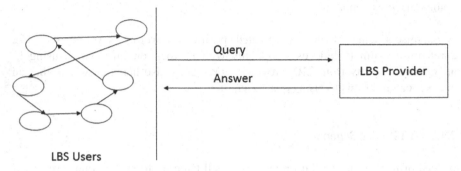

Fig. 5 Communication schema of collaborative method between users and LBS provider [2]

Collaborative-Based Schema

It is fully distributed schema. The trust is scattered among the nodes that form an ad hoc network. All peers work collaboratively to achieve privacy among untrusted entities. Various algorithms are already proposed as Solanas and Balleste, Rebollo-Monedero, Ardagna et al., etc. [8–10]. The advantage of this approach is that it does not reply on TTP, it is distributed, and it also guarantees user's privacy. The main disadvantage of this method is related to the computation and communication cost, and scalability issues (Fig. 5 and Table 1).

3   Motivation of Proposed Approach

3.1   Problem Statement

Location-based services become a cornerstone in the development of the new information society [23]. LBS user provides highly personalized information to the serviced provider. There is always a possibility of threats because LBS provider cannot always trustworthy. Location privacy and communication privacy are one of the key

Table 1 Comparison between different crypto-based privacy models for LBS

No.	Categorization of crypto-based privacy model	Information flow model	TTP-based/TTP free	Various schemas based on category	Advantage	Disadvantage
1	Policy-based	Client–server	TTP-based	Geographic loca-tion/privacy [2]	Easy to implement	No identity privacy, no location privacy
2	Pseudonyms-based	Client–server	TTP-based	Simple pseudonym [11, 12]	Identity privacy	No Location privacy, low accuracy, unlinkability
3	Anonyms-based	Client–server	TTP-based	K-anonymity [13], PrivacyGrid, personalized k-anonymity [14, 12]	Location privacy	No identity privacy, unlinkability
4	Obfuscation-based	Client–server	TTP free	SpaceTwist [15]	High server efficiency, location privacy	No identity privacy, unlinkability
5	PIR-based	Client–server/peer-to-peer	TTP free	PIR protocol, collaborative PIR [2, 8, 16]	Location privacy	No identity privacy, low lbs server efficiency, low accuracy
6	Collaborative-based	Peer-to-peer	TTP free [8]	Distributed TTP-based scheme [9, 10, 17–22]	Robust against collision of malicious users	High cost, unlinkability. Low accuracy, no identity privacy

issues. Location-based service is classified as trusted third party and without trusted third party that uses cryptographic approach. TTP-based model gives a guarantee of privacy of LBS users. The main problem is that user cannot rely on the trusted third party. Collaborative TTP free model is one of the best models that provide location privacy in LBS. The cryptographic approach is secure and achieves highest location privacy. There are many open challenges needed to be solved, such as communication cost, computational cost, and poor scalability.

3.2 Proposed Approach

Our aim is to propose a novel solution that provides location privacy to the LBS users. The main goal of the research is to achieve features as not rely on TTP, hybrid

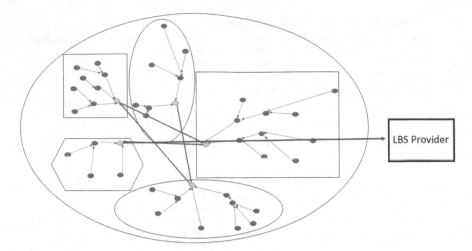

Fig. 6 Proposed communication schema of centralized and decentralized method between users and LBS provider

approach, improve scalability, reduce cost in resource constraint devices, and enhance security and privacy.

Figure 6 represents the system architecture of proposed schema. It contains two main components as LBS users and LBS provider. Each user has their private information on their mobile like user ID, U_{id}, and location information (x_i, y_i). There is need of preserving the privacy of LBS users. Find out the number of users U_i in clocking region who are requesting for location-based information. Next, we generate random region R_i based on the computation of a Voronoi diagram for users in spatial cloaking region, e.g., square, hexagon, circle, and rectangle as shown in Fig. 6. Then, use centralize approach to performing secure data aggregation using privacy homomorphism **PH** [24, 25] in each random region R_i using tree topology that is shown in Fig. 6 with blue edges. Next, we use the decentralized approach to perform random chaining **RC** for all distributed random region R_i to compute the secure centroid **C** as shown in Fig. 6 with red edges. The last user, **U** sends the encrypted sum of location **C** to LBS provider **P** as shown in Fig. 6 with green edge.

The main aim is to hide the user's location within the other users and also give inaccurate location information to the LBS provider. We use a hybrid approach that includes centralized and distributed method to achieve minimum cost and scalability. In this paper, we propose protocol schema for privacy preservation between users and LBS provider. Our proposed protocol schema is as follows (Fig. 7):

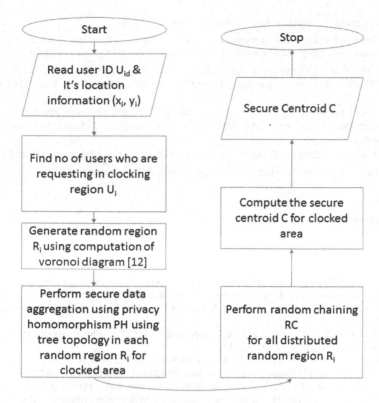

Fig. 7 Flowchart for proposed protocol schema

ALGORITHM 1: Our Proposed Protocol Schema

Input: LBS Users U_i (User ID U_{id}, Location information (x_i, y_i))

Output: Compute Secure Centroid C

4 Conclusion

Privacy preservation is of paramount importance with the rapid growth in the number of LBS users. In this is paper, we discussed various classification method that is proposed by various authors and also covered a comprehensive survey of location privacy. Our proposed approach does not rely on TTP rather uses a hybrid approach that improves scalability and reduces the cost. In addition to that our approach also guarantees the privacy of LBS users.

1	Clocking region for Users requesting U_i.
2	Generate Random Region R_i using computation of Voronoi diagram [26] for users in spatial clocking algorithm.
3	Perform Secure Data Aggregation using privacy homomorphism PH using tree topology in each random region R_i for clocked area.
4	Perform Random Chaining RC for all distributed random region R_i and compute the secure centroid C for clocked area.
5	The Last User, U send encrypted sum C to LBS provider P.
6	LBS provider P decrypts the sum of locations.

References

1. Padmanaban, R.: Location privacy in location based services: unsolved problem and challenge. Int. J. Adv. Remote Sens. GIS **2**(1), 398–404 (2013)
2. Magkos, E.: Cryptographic approaches for privacy preservation in location-based services: a survey. Int. J. Inf. Technol. Syst. Approach (ACM) **4**(2) (2011)
3. Yang, G., Li, J., Zhang, S., Zhou, H.: A survey of location-based privacy preserving. JCIT (2013)
4. Yang, N., Cao, Y., Liu, Q., Zheng, J.: A novel personalized TTP-free location privacy preserving method. Int. J. Secur. Appl. **8**(2) (2014)
5. Solanas, A., Domingo-Ferrer, J., Martínez-Ballesté, A.: Location Privacy in Location-Based Services: Beyond TTP-based Schemes, projects TSI2007-65406-C03-01"E-AEGIS" (2010)
6. Ghinita, G., Kalnis, P., Khoshgozaran, A., Shahabi, C., Tan, K.-L.: Private queries in location based services: anonymizers are not necessary. In: ACM SIGMOD International Conference on Management of Data, pp. 121–132 (2008). ISBN: 978-1-60558-102-6
7. Ghinita, G., Kalnis, P., Kantarcioglu, M., Bertino, E.: A Hybrid Technique for Private Location-Based Queries with Database Protection, Springer Advances in Spatial and Temporal Databases Volume 5644 of the series Lecture Notes in Computer Science, pp. 98–116 (2009), ISBN: 978-3-642-02982-0
8. Solanas, A., Martínez-Ballesté, A.: A TTP-free protocol for location privacy in location-based services. Elsevier Trans. Comput. Commun. (2008)
9. Solanas, A., Martínez-Ballesté, A.: Privacy Protection in Location-Based Services Through a Public-Key Privacy Homomorphism. Springer (2007)
10. Wernke, M., Skvortsov, P., Dürr, F., Rothermel, K.: A Classification of Location Privacy Attacks and Approaches. Springer (2013)
11. Huang, Y., Vishwanathan, R.: Privacy preserving group nearest neighbor queries in location-based services using cryptographic techniques. In: IEEE Global Telecommunications Conference GLOBECOM (2010)
12. Bettini, C., Sean Wang, X., Jajodia, S.: Protecting Privacy Against Location-based Personal Identification (2006)
13. Jariwala, V., Jinwala, D.: Evaluating homomorphice encryption algorithms for privacy in wireless sensor network. Int. J. Adv. Comput. Technol. **3**(6) (2011)
14. Shastry, S.R., Deshmukh, P.K., Bagwan, A.B.: Generating: random regions in Spatial cloaking algorithm for location privacy preservation. IOSR J. Comput. Eng. (IOSR-JCE) **9**(4), 46–49 (2013). e-ISSN: 2278-0661, p-ISSN: 2278-8727
15. Yang, G., Li, J., Zhang, S., Zhou, H.: A survey of location-based privacy preserving. JCIT (2013)
16. Wernke, M., Skvortsov, P., Dürr, F., Rothermel, K.: A classification of location privacy attacks and approaches. In: Personal and Ubiquitous Computing. Springer (2013)

17. Gupta, R., Rao, U.P.: An exploration to privacy issues in location based services. In: 3rd Security and Privacy Symposium, IIIT–Delhi (2015)
18. Amro, B., Saygin, Y., Levi, A.: Enhancing privacy in collaborative traffic-monitoring systems using autonomous location update, The Institution of Engineering and Technology. IEEE (2013)
19. Zhu, X., Lu, Y., Zhu, X., Qiu, S.: A location privacy-preserving protocol based on homomorphic encryption and key agreement. In: International Conference on Information Science and Cloud Computing Companion. IEEE (2014)
20. Ashouri-Talouki, M., Baraani-Dastjerdi, A.: Homomorphic encryption to preserve location privacy. Int. J. Secur. Appl. 6(4) (2012)
21. Tyagi, A.K., Sreenath, N.: Preserving location privacy in location based services against sybil attacks. Int. J. Secur. Appl. 9(12) (2015)
22. Patil, S., Ramayane, S., Jadhav, M., Pachorkar, P.: Hiding user privacy in location base services through mobile collaboration: a review. In: International Conference on Computational Intelligence and Communication Networks. IEEE (2015)
23. Peng, T., Liu, Q., Wang, G.: Enhanced location privacy preserving scheme in location-based services. IEEE Syst. J. (2014)
24. Shokri, R., Theodorakopoulos, G., Papadimitratos, P., Kazemi, E., Hubaux, J.-P.: Hiding in the mobile crowd: location privacy through collaboration. IEEE Trans. Dependable Secure Comput., Special Issue On Security And Privacy In Mobile Platforms (2014)
25. Tyagi, A.K., Sreenath, N.: Future challenging issues in location based services. Int. J. Comput. Appl. (0975 – 8887) 114(5) (2015)
26. Patil, R.J., Joshi, K.K., Raksha, S.: Analysis on preserving location privacy. Int. J. Adv. Res. Comput. Sci. Softw. Eng. 5(3) (2015). ISSN: 2277 128X

Forensic Implications of Cortana Application in Windows 10

Bhupendra Singh and Upasna Singh

Abstract Cortana is one of the new features introduced by Microsoft in its latest version of desktop operating systems, i.e., Windows 10. The feature is identified by "Ask me anything" text box at the Start Menu and can be used for a number of tasks such as setting up reminders based on time, place, and person; searching stuff on local device or web; sending emails and texts; and more. The feature keeps track of reminders when and where they got finalized, as a result, evidentiary artifacts related to reminders are recorded in a back-end database. The forensic examination of Cortana has been largely unexplored in literature as the platform is relatively new. This paper seeks to determine the databases created by Cortana, their format, and the type of information recorded in these databases. As a part of this paper, six custom Python scripts have been developed for decoding and exporting data to aid forensic investigators. Furthermore, several experiments are conducted to extract information related to reminders such as created and last updated timestamps of a reminder, type of reminder, when a reminder got finalized, and where it got finalized. Finally, forensic usefulness of information stored in a Cortana database is demonstrated in terms of a location timeline constructed over a period of time.

Keywords Windows forensics · Cortana forensics · ESE database · Edge forensics

B. Singh (✉) · U. Singh
Department of Computer Science and Engineering, Defence Institute of Advanced
Technology (DU), Pune, India
e-mail: bhupendra_pcse14@diat.ac.in

U. Singh
e-mail: upasnasingh@diat.ac.in

© Springer Nature Singapore Pte Ltd. 2019
P. K. Sa et al. (eds.), *Recent Findings in Intelligent Computing Techniques*,
Advances in Intelligent Systems and Computing 707,
https://doi.org/10.1007/978-981-10-8639-7_7

1 Introduction

Cortana as a voice-activated personal digital assistant, was first introduced by Microsoft in Windows Phone 8.1 and later on integrated with Windows 10 desktop operating systems. The feature collects personalized data of a user such as user's voice, calendar, contacts, location, and Internet history to help and provide suggestions to user. According to Microsoft, Cortana has access to device location information and location history must be on for Cortana to work. Cortana records location of a user's favorite places such as home, work, or grocery store so that it can help user by providing the traffic alerts, suggestions, ideas, reminders, and more. Thus, Cortana contains a goldmine of terrestrial artifacts that can be leveraged by digital analysts to prove or disapprove a case.

The feature is integrated into Microsoft Edge, the default web browser in Windows 10 desktop OS and is powered by Bing. Thus, Bing search engine obtains the search results related to web searches performed by users. Therefore, information related to web searches can also be located in Jump List of Edge browser and can be successfully parsed using available tools. Initially, Cortana was designed to work with Microsoft Edge browser, however, users were able to customize the default web browser that would launch on performing web searches on Cortana. Singh and Singh [4] identified and compared the information recorded in four popular web browsers' Jump Lists when web searches were performed on Cortana, considering each of them as default web browser. In the further study, the authors in [5] gleaned Cortana artifacts from various locations and performed various experiments to facilitate forensic investigations.

The rest of the paper is organized as follows. The next section provides an overview of two databases created by Cortana application including its format and discusses how information can be decoded into readable format. The next section describes the research methodology followed in this paper. For performing experiments, six Python scripts have been developed as a part of this paper and results are highlighted in subsequent section. Finally, the last section concludes the paper.

2 Technical Details: Cortana Artifacts, Format and Their Location

The Windows 10 desktop OS results in creation of numerous artifacts being stored within user's profile when the user performs actions on Cortana application. There are two main databases associated with Cortana that records the information related to users' interaction with Cortana. The file paths to these databases are as follows [3, 5]:

- %Userprofile%\AppData\Local\Packages\Microsoft.Windows.Cortana_cw5n1h2txyewy\AppData\Indexed DB\IndexedDB.edb

- %Userprofile%\AppData\Local\Packages\Microsoft.Windows.Cortana_cw5n1h2txyewy\LocalState\ESEDatabase_CortanaCoreInstance\CortanaCoreDb.dat

These databases are Microsoft-defined ESE (Extensible Storage Engine) database, i.e., frequently used by Microsoft to store data for many applications such as Active Directory, Instant Messaging, Web Browsers (e.g., Internet Explorer and Microsoft Edge), Windows Search [1], Windows Mail, and System Resource Usage Monitor. The tables in these database can be extracted and viewed using available tools such as ESEDatabaseView [6] and libesedb [2].

2.1   Extracting Tables from an ESE Database

For the purpose of analyzing these ESE databases, investigators need to extract both database files from abovementioned locations. For extracting Cortana databases from a test computer, a Live Boot DVD of Ubuntu 14.04 LTS is used to copy out these databases to a forensic server where *libesedb* tool is installed. This tool can parse the tables in an ESE database and export to tab-delimited files, one for each table in the database. A tab-delimited file can be opened with applications such as Microsoft Office, Notepad, or Wordpad. The data in a tab-delimited file is viewable and readable except date and time information, which needs to be decoded into readable format.

2.2   IndexedDB.edb Database

This database consists of 11 flat tables. The first five tables describe table metadata, such as table and column names, data types, and locales. The *HeaderTable* (Table Id = 9) table contains information related to created, last open, and last modified timestamps (in 64 bit OLE format) of the ESE database. It was observed that whenever any operation on Cortana is performed, this database is modified to store the indexed data and timestamp information in HeaderTable table is updated. Thus, this table may provide the meaningful information to forensic investigators regarding the timestamp of first or last time the Cortana application was used on a suspected Windows 10 system. The other tables in this database were reported to contain information merely relevant to forensic analyst.

2.3   CortanaCoreDb.dat Database

The CortanaCoreDb.dat database records the information related to users' interaction with Cortana application. This database also follows the structure similar to ESE database and 24 flat tables were reported when extracted using libesedb tool. Unlike the previous database, CortanaCoreDb.dat database was reported having a goldmine of terrestrial artifacts. The subsequent subsections highlight the artifacts stored in some of the tables, which are of digital analysts' interest.

2.3.1 Geofences

The location information, such as longitude and latitude coordinates for the places where the user had used Cortana are recorded in this table. In addition, this table can provide the timestamp values of Cortana usage at these coordinates. A geofence is a boundary around a place of interest and Cortana uses location for geofencing to see if the users are crossing in and out of one.

2.3.2 Reminders

The information related to all created reminders with their reminder title is recorded in this table. The table were reported having seven columns with first column as reminder Id that acts as the primary key for the table. Second column contains the status value that represents whether a reminder is active (Status = 1) or completed (Status = 2). Column 8 provides the actual reminder title. Thus, this table can provide immensely useful information to forensic investigators to keep track of a user's reminders.

2.3.3 Triggers

Cortana offers feature to create reminders based on time, place, and person in Windows 10 device. The information regarding the types of reminders and their creation timestamp is recorded in this table. This table contains a column with column name *ReminderId* whose values match with the *Id* field in Reminders table. The column with column name *Kind* can provide the information regarding the type of reminders in Reminders table.

2.3.4 LocationTriggers

The latitude and longitude coordinates, as well as the name of the places are recorded in this table. This table can also provide the distance (in KM) of these locations from the place of interest.

2.3.5 TimeTriggers

Forensic analysts can get timestamp information from this table, which triggers time-based reminders. The important information of last time a reminder (time based) was triggered is also stored in this table.

2.3.6 ContactTriggers

This table records the name of the persons based on which reminders are created. Cortana's ability to contact a user's contacts book comes from the synced file at location: \%Userprofile%\AppData\Local\Packages\Microsoft.Windows.Cortana_cw5n 1h2txyewy\LocalState\Cortana\Upload\Contacts\contacts.json.

2.4 Decoding Columns in a Table

The tables exported from CortanaCoreDb.dat database using libesedb's esedbexport tool contains timestamp values in Google Chrome Value format that needs to be decoded for easy reading. We developed six Python scripts for decoding timestamp value and exporting decoded data into a csv file for each forensically relevant table in the database. The scripts are written in Python 3.4 environment and made available publicly on this link.[1] Below is the snippet for decoding timestamp value in column 2 of exported geofences (with file extension as .14) table.

```
import datetime, csv
tsvin = open(''Geofences.14'','r')
data=csv.reader(tsvin,delimiter='\t')
firstline = True
for row in data:
        if firstline:
                firstline = False
                continue
        filetimevalue = int(row[1])/10
        print(datetime.datetime(1601,1,1) +
        datetime.timedelta(microseconds = filetimevalue)
```

3 Research Methodology

The research method used in this paper is based on observations and experiments. This section describes test environment and requirements which is used to create the test environment for conducting experiments. For extracting the Cortana databases from test computer, a Live Boot DVD of Ubuntu 14.04 LTS was used. The tables from the ESE databases were extracted and exported using libesedb tool. For easy reading, the columns in a table were decoded and exported into a csv file using Python custom scripts.

---

[1]https://github.com/Bhupipal/Cortana-Forensics.

Table 1 Test environment
and requirements used for
conducting research

Device type	Device name	Installed OS and version
PC	HP Compaq Pro 4300 SFF PC	Windows 10 Pro (v1511)
Mobile	Motorola Moto G (3rd Gen)	Android v5.1.1 (Lollipop)

3.1 Test Environment and Requirements

The main purpose of this research is to investigate the forensic artifacts of Cortana
usage in a personal computer (PC) running Windows 10 operating system. For this,
a test computer with the specifications, as described in Table 1, is considered and the
Cortana is configured with default settings using a Microsoft account.

Microsoft announced that Cortana can now show notifications from users'
Android device to their Windows 10 PC. This feature brings the two platforms closer
together, therefore, PC may also contain a wealth of forensically relevant information
related to Android device. For this, an Android phone with the specifications shown
in Table 1 is taken into consideration. The latest version of Cortana App (1.5.6937-
enin-preview) available at the time of writing this paper was installed from Google
play store into the considered phone. The App was also configured with default settings using the same Microsoft account as for Windows 10 PC, so that Cortana data
can be synced on both devices.

4 Experiments and Results

Several experiments were conducted to investigate Cortana artifacts left after performing the series of activities. Conducted experiments mainly focus on the information extracted from the CortanaCoreDb.dat database as the other database does
not contain much forensically relevant information. For analysis, the tables from
CortanaCoreDb.dat were parsed and exported using libesedb tool, as illustrated in
Fig. 1. The experiments also demonstrate the forensic usefulness of this database
and how investigators can leverage this database to track user activities, specially
user's reminders and locations. The next subsections describe about the conducted
experiments and their results.

4.1 Tracking Reminders: When Got Finalized?

This experiment was conducted to track reminders such as what were active
reminders, what were finalized reminders, and when they got completed. For this,

Fig. 1 Screenshot of libesedb usage—exporting tables from CortanaCoreDb.dat with esedbexport

Id	Status	SyncStatus	CreationTime	LastUpdateTime	LastAccessTime	CompletionTime	Title
77a9db0f28f94c46b7241be09cb09ed6	1	0	8/10/2016 6:44	8/10/2016 6:44	8/10/2016 6:44	-1	Collect charger
cd31ada7767c044692752c47ebabf490	1	0	7/19/2016 4:04	7/19/2016 4:04	7/19/2016 4:04	-1	Running
ec33053667f4cd47b26441c0c6cfe626	2	0	8/2/2015 6:57	8/10/2016 7:07	8/10/2016 7:07	8/10/2016 7:07	Cycle
9031de40f5f9864aa76a69fdaaa12fe8	1	0	8/9/2016 17:31	8/9/2016 17:31	8/9/2016 17:31	-1	hi
a7c3e1637d23634dba666ca605f69f2c	2	0	8/10/2016 7:18	8/10/2016 7:18	8/10/2016 7:18	8/10/2016 7:18	Missed call notification
cd941f4690f62945b54b7f1c94d853b2	2	0	8/9/2016 10:00	8/9/2016 10:47	8/9/2016 10:47	8/9/2016 10:47	Make a call to home
3d86036ac4381a4d8edc851ccfb0f3e2	1	0	8/10/2016 6:41	8/10/2016 6:41	8/10/2016 6:41	-1	MTech Degree
a0d7c35bb4b6694ca1e4989feb166c00	2	0	8/9/2016 9:59	8/10/2016 6:39	8/10/2016 6:39	8/10/2016 6:39	buy the cake
5f53215cfeb0654896d7e929097f38f8	1	0	8/9/2016 10:46	8/9/2016 10:46	8/9/2016 10:46	-1	Go to temple

Fig. 2 Information stored in Reminders table

the Reminders table was decoded using a Python script[2] and data was exported into a csv file. The table consists of eight columns as shown in Fig. 2. First column is 16 bytes reminder Id that is used as the primary key for the table. Second column describes whether a reminder is active (Status = 1) or got finalized (Status = 2). As the name suggests, columns with header names Creation-Time, LastUpdateTime, LastAccessTime, and CompletionTime describes the timestamps of when reminders were created, last updated, last accessed, and completed, respectively. The last column represents actual reminder to be reminded about by Cortana to the user. In Fig. 2, it can be seen that the reminders with reminder Id ec33053667f4cd47b26441c0c6cfe626, cd941f4690f62945b54b7f1c94d853b2 and a0d7c35bb4b6694ca1e4989feb166c00 has finalized at timestamp 8/10/2016 7:07 AM, 8/9/2016 10:47 AM, and 8/10/2016 6:39 AM, respectively. The table also consists timestamps of missed calls on Android device with reminder Title as "Missed call notification". Thus, the information recorded in this table is immensely useful to keep track of reminders and forensic investigators can leverage this information to know the reminders title and their timestamps of creation and completion.

---

[2]https://github.com/Bhupipal/Cortana-Forensics/blob/master/reminders.py.

Id	ReminderId	Kind	CreationTime	
e75a7935fab0a14383b1147176f87a0b	5f53215cfeb0654896d7e929097f38f8	1	8/9/2016 10:47	Location based reminders
820fc0143a735149b6ae80772284ce4c	a0d7c35bb4b6694ca1e4989feb166c00	1	8/9/2016 9:59	
2e7997615c109b4baa16eef14913ded5	77a9db0f28f94c46b7241be09cb09ed6	1	8/10/2016 6:44	
9119287af25de84e866a3b147535ffad	cd31ada7767c044692752c47ebabf490	2	7/19/2016 4:03	Time based reminders
dca84878f64d2f438ae2a9e86dcaeeff	a7c3e1637d23634dba666ca605f69f2c	2	8/10/2016 7:23	
b938b91b993df449935fab6a02cac600	ec33053667f4cd47b26441c0c6cfe626	2	7/19/2016 4:04	
46e6968107eeb04ab7d92c5af9bb29fa	3d86036ac4381a4d8edc851ccfb0f3e2	3	8/10/2016 6:41	Person based reminders
899a1c521f46234f858d62a45c3f770a	9031de40f5f9864aa76a69fdaaa12fe8	3	8/9/2016 17:32	
b22d984e546baf4493e2a4c5b009bb09	cd941f4690f62945b54b7f1c94d853b2	3	8/9/2016 10:00	

Fig. 3 Information stored in Triggers table

4.2 Tracking Reminders: Identifying Type of Reminders

This experiment was carried out to determine the type of reminders extracted from Reminders table. As discussed earlier, Triggers table in CortanaCoreDb.dat database consists of four columns: Id, ReminderId, Kind, and CreationTime. The column ReminderId is the foreign key referencing Id column in Reminders table. The ReminderId value in this table matches with Id value in Reminders table. The columns with header names Kind and CreationTime describes the type of reminder and the timestamp of creation of that reminder. From Fig. 3, it can be identified that the reminders with reminder Id ec33053667f4cd47b26441c0c6cfe626, cd941f4690f62945b54b7f1c94d853b2, and a0d7c35bb4b6694ca1e4989feb166c00 are time, person, and place-based reminders, respectively. The information from this table can be combined with the information in Reminders table. For example, a reminder with reminder Id a0d7c35bb4b6694ca1e4989feb166c00 was a location-based reminder that reminded user to "buy the cake" at timestamp 8/10/2016 6:39 AM when Cortana got the specified location.

4.3 Tracking Reminders: Where Got Finalized?

The purpose of this experiment was to track the location of reminders where they got finalized. For this, LocationTriggers table was decoded and exported into a csv file using a Python script[3] as shown in Fig. 4. Cortana triggers place-based reminders when she gets the specified location to be stored in this table. From Fig. 4, it can be observed that Id value in this table matches with Id value in LocationTriggers, thus, can be used to get reminder Id. The combined information from Reminders, Triggers, and LocationTriggers tables in CortanaCoreDb.dat database can be utilized to figure out the actual name of place where place-based reminders got finalized. For example, a reminder with reminder Id a0d7c35bb4b6694ca1e4989feb166c00 was a place-based reminder (refer to Fig. 3) that reminded user to "buy the cake" at timestamp 8/10/2016 6:39 AM (refer to Fig. 2) when Cortana got Pune Camp

---

[3]https://github.com/Bhupipal/Cortana-Forensics/blob/master/locationtriggers.py.

Id	Latitude	Longitude	Radius	RuleStatus	Name	EntityId
e75a7935fab0a14383b1147176f87a0b	18.9388	72.8258	200	1	Wankhede Stadium	YN4070x16043416103765192071
820fc0143a735149b6ae80772284ce4c	18.4236	73.9089	200	2	Pune Camp	sid:ed6a2606-82f6-eccc-1cb8-ef0d9fe9fc40
2e7997615c109b4baa16eef14913ded5	18.4175	73.76444	200	1	Haveli, Maharashtra	

Fig. 4 Information stored in LocationTriggers table

Id	RuleStatus	AggregateId	ContactHandle	ContactName
46e6968107eeb04ab7d92c5af9bb29fa	1	{5.70002.116}	{82CE2D73-F102-401F-B4FA-5A5C17BE92E5}	Ashish Tiwary
899a1c521f46234f858d62a45c3f770a	1		{C95B5B44-4460-4C92-BF32-09A49B4E8EBF}	DP Sir
b22d984e546baf4493e2a4c5b009bb09	2	{5.70002.94}	{685F6EB0-BE6F-45CF-BEB2-F9F86244B153}	Mummy2

Fig. 5 Information stored in ContactTriggers table

location (refer to Fig. 4). Thus, it can be inferred that the user in test scenario was created a reminder at timestamp 8/9/2016 9:59 AM to remember "buy the cake" and completed at timestamp 8/10/2016 6:39 AM in *Pune Camp*.

4.4 Tracking Person-Based Reminders

This experiment was performed to track person-based reminders. For the same, ContactTriggers table was taken into consideration and data was exported into a csv file using Python script.[4] The first column in this table is Id whose value matches with Id value in Triggers table. This value can be utilized to determine the reminder Id that can be further used to know the actual reminder. The column with header name Contact-Name in this table describes the name of person that is used by Cortana to trigger the reminder when user contacts to this person. Forensic investigators can combine information from Reminders, Triggers, and ContactTriggers tables to track the complete information about a person-based reminder. For example, a reminder with reminder Id cd941f4690f62945b54b7f1c94d853b2 was a person-based reminder (refer to Fig. 3) that reminded user to "Make a call to home" (refer to Fig. 2) when user contacts to *Mummy2* (refer to Fig. 5).

4.5 Constructing Location Timeline

This experiment was conducted to assimilate a timeline of user location over a period of time using Geofences table in CortanaCoreDb.dat database. For this, a Python script[5] was developed to decode and export data into a csv file. Figure 6 shows the output data containing nine columns. The location of a place is stored in terms of

---

[4]https://github.com/Bhupipal/Cortana-Forensics/blob/master/contacttriggers.py.

[5]https://github.com/Bhupipal/Cortana-Forensics/blob/master/geofences.py.

ID	Time	State	Latitude	Longitude	Accuracy	Speed	Heading	PositionSource
a562f7bada36df46b8ad0b0704cf090f	8/7/2016 16:30	1	18.4177	73.76419	74	1.#QNAN0	1.#QNAN0	2
8b52a9e591cfe2478a3f1107e3d652d9	8/8/2016 4:39	1	18.4223	73.7612	1014	1.#QNAN0	1.#QNAN0	3
75afa40814b1b7489c231218cb995e10	7/19/2016 3:53	1	18.42548	73.75822	88	1.#QNAN0	1.#QNAN0	2
ccde12c9fb7f8246ad2c19e876bf30e0	8/9/2016 4:00	1	18.4223	73.7612	1014	1.#QNAN0	1.#QNAN0	3
8f23d67b13ef964cbf601ef6a3ae6c48	8/2/2016 14:08	1	18.41745	73.76441	105	1.#QNAN0	1.#QNAN0	2
a8faa6c21b86d34594e623ab2c7569a6	7/29/2016 17:03	1	18.41751	73.76436	98	1.#QNAN0	1.#QNAN0	2
0f801d4a211f0846a72d376b31832883	7/26/2016 16:25	1	18.41772	73.76423	74	1.#QNAN0	1.#QNAN0	2
ac5b8fd96acabc4fa036395d3db452d2	8/10/2016 5:32	2	18.42665	73.75964	209	1.#QNAN0	1.#QNAN0	2
9262491e828901448b9b3fbf1e749eac	7/26/2016 9:51	1	18.4254	73.75825	93	1.#QNAN0	1.#QNAN0	2
31fb7d5e51219046b24343e655b43b60	8/3/2016 12:23	1	18.4177	73.76419	74	1.#QNAN0	1.#QNAN0	2

Fig. 6 Information stored in Geofences table

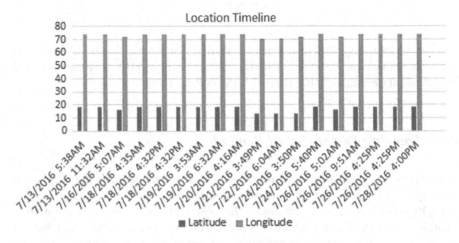

Fig. 7 Constructed location timeline of 15 days using Geofences table

latitude and longitude information. From the figure, it is quite easy to identify the location of the user at a certain date and time.

To evaluate forensic usefulness of Cortana database, a timeline of user location is constructed for a period of 15 days starting from July 13, 2016 to July 27, 2016. The timeline of user location can be helpful to keep track of user location over a period of time. From Fig. 7, it can be deduced that there are no significant changes in the latitude and longitude values, means, and for the considered time period, the user of computer system has been in the boundary of place of interest (home or work). However, on specific timestamps such as *16 July 2016 5:07 AM* user was farther from place of interest than on *13 July 2016 11:32 AM*. The missing days in the constructed timeline such as 14 and 15 July of 2016 refers that the Cortana or the computer system was not started on these days. Thus, Cortana database can provide immensely useful information to investigator in the cases when user's location history is of investigator's interest.

5 Conclusion and Future Research

This paper provides a forensic insight into the databases created by Cortana, their location, format, and type of information recorded in these databases. Cortana stores user data into two back-end databases: IndexedDB.edb and CortanaCoreDb.dat within user's profile. These databases follow the structure similar to Microsoft-defined Extensible Storage Engine (ESE) format and tables from these databases can be extracted using libesedb tool. It was observed that the CortanaCoreDb.dat database contains information which can be of immense importance to a forensic investigator. The extracted tables from the CortanaCoreDb.dat were decoded and exported into csv files using custom Python scripts for easy reading. To understand the forensic usefulness of the information recorded in CortanaCoreDb.dat database, a test case was prepared and several experiments were carried out to extract the expected artifacts. The results of the conducted experiments demonstrate that this database can provide important information related to reminders such as what are active reminders, when they created or last updated, and when and where they got finalized. The location history of a user was also recovered from *Geofences* table in the database. Further, this information was leveraged to construct a timeline of location over a period of 15 days.

References

1. Chivers, H., Hargreaves, C.: Forensic data recovery from the windows search database. Digit. Investig. **7**(3), 114–126 (2011)
2. Metz, J.: libesedb. https://github.com/libyal/libesedb (2012). Accessed 5 July 2016
3. Muir, B.: Windows 10 cortana & notification center forensics. http://bsmuir.kinja.com/windows-10-cortana-notification-center-forenics-1724511442 (2015). Accessed 10 Aug 2016
4. Singh, B., Singh, U.: A forensic insight into windows 10 jump lists. Digit. Investig. **17**, 1–13 (2016)
5. Singh, B., Singh, U.: A forensic insight into windows 10 cortana search. Comput. Secur. **66**, 142–154 (2017)
6. Sofer, N.: Esedatabaseview. http://www.nirsoft.net/utils/ese_database_view.html (2013). Accessed 5 Aug 2016

Secure Data Transfer by Implementing Mixed Algorithms

P. Naga Hemanth, N. Abhinay Raj and Nishi Yadav

Abstract Nowadays, data sharing is a common need for the humans to transfer information. There are different security threats to be overcome while transmitting data in Internet. Cryptography helps to solve all the issues that are generated. The objective of this paper is to enhance the security level by using mixed security algorithms. This work is proposed of three parts—playfair algorithm of 9×6 matrix, RSA algorithm, and an XOR operation. The mixed algorithm enriches the security strength of individual symmetric and asymmetric algorithms. Also, this work yields good results in many parameters like message strength, key strength, encryption, and decryption time. The idea of mixed algorithms in cryptography is helpful for researchers in the nearby future.

Keywords Security algorithms · Playfair cipher · RSA algorithm · XOR operation · Data security

1 Introduction

In current tech-savvy world, securing people information has been a major issue during transmission of various messages. In order to find a solution to the problem cryptography have been introduced. Cryptography is a practice and study of techniques for secure transmission of data. It is divided into two cryptosystems: symmetric cryptosystem and symmetric cryptosystem. For which we have proposed a mixed algorithm for asymmetric cryptosystem. This paper mainly compares the proposed algorithm with the 5 * 5 playfair matrix [1, 2] to overcome the demerits

P. Naga Hemanth (✉) · N. Abhinay Raj · N. Yadav
Department of C.S.E., Central University, Bilaspur, India
e-mail: hemanth.vja@gmail.com

N. Abhinay Raj
e-mail: abraj1995@gmail.com

N. Yadav
e-mail: nishidv@gmail.com

© Springer Nature Singapore Pte Ltd. 2019
P. K. Sa et al. (eds.), *Recent Findings in Intelligent Computing Techniques*,
Advances in Intelligent Systems and Computing 707,
https://doi.org/10.1007/978-981-10-8639-7_8

such as key size and message strength. Now, in this paper, the mixed algorithms consist of playfair algorithm [3] and RSA algorithm, which are described in later sections of this paper.

2 Literature Survey

These are some of the papers which have given basic idea for this work.

In [4], the authors worked on dual level security for the message transfer in playfair cipher. There is no security for the key in existing scheme and can not be able to provide user authentication. Finally, their implementation proved that security and performance of playfair have been increased.

In [5], a brief study on number theory has made by authors, and a new improved RSA (IRSA) algorithm is introduced to improve the security feature in existing RSA algorithm. This work has a constant number of private keys and increases in number of public keys used. This IRSA is only for such a system that needs high-level security.

In [6], this paper discussed the multiple attacks in data security of cloud computing. It concluded that dual security algorithm helps in cloud environment and improves in sharing performances of data. But the existence of those algorithms is still at the research level.

In [7], there is an enhancement in the playfair cipher dropping the restrictions of playfair cipher. As a result, this cipher is stronger than previous ciphers. In [8], playfair cipher represents an improvement in security over substitution ciphers. In order to increase security, the text has been hidden in a digital image. The stegno images are tested by using PSNR. This technique is not susceptible to histogram-based attacks.

In [9], they have modified the original playfair cipher with 5 * 5 matrix to n * m matrix, so that we can encrypt messages using any natural languages. In this piece of work, Urdu was used as a natural language.

3 Proposed Work

3.1 A Brief Description of Proposed Algorithm

In this proposed algorithm, a matrix of 9×6 size is used [10], which consists of characters mentioned in Table 1.

Table 2 represents the notations used in this work.

Table 1 Characters used in the proposed algorithm

a	b	c	d	e	f	g	h	i	j	k	l	m	n	o	p	q	r
s	t	u	v	w	x	y	z	0	1	2	3	4	5	6	7	8	9
!	|	+	−	*	/	<	=	>	%	^	&	{	}	[	]	(	)

Table 2 Notations used in the algorithm

Symbols	Description
pt	Plaintext
ck_1	Key to encrypt
ck_1^I	Encrypted key using RSA
enc_1	Playfair encrypted text
enc_2	Encrypted text after XOR operation
enc_3	Final encrypted text

3.2 Proposed Methodology

Step I: Playfair Cipher

1. First, construct a 9 * 6 matrix [11] with the letters in key by removing the duplicates.
2. Insert alphabets, braces, numbers, and operators, respectively, in the remaining blocks of matrix.
3. Interchange the pair of letters, according to the rules of 5 * 5 playfair matrix [12].
4. Finally, plaintext pt and key ck_1 are encrypted as follows:
   $enc_1 = $ encrypt (pt, ck_1)

Step II: XOR Operation

1. Apply XOR operation between encrypted text [13] enc_1 and the key ck_1

$$enc_2 = enc_1 \oplus ck_1$$

Step III: RSA Encryption

1. Apply RSA algorithm on key ck_1 and store the result in ck_1^I [14].
2. Again, XOR operation is applied on enc_2 and ck_1^I as follows:

$$enc3 = enc2 \oplus ck_1^I$$

3. Receiver receives the final output enc_3 and ck_1^I.

Table 3 Arrangement of 9 × 6 key matrix

u	s	e	r	d		f	i	n	a
b	c	g	h	j		k	l	m	o
q	s	t	u	v		w	y	z	[
]	{	}	(	)		0	1	2	3
4	5	6	7	8		9	*	/	+
–	^	%	<	>		=	&	I	!

4 Example

Assume plaintext pt is "this is the hidden message".

Key ck_1 is "userdefined".

Step I:

Drop duplicates from the given key ck_1, and the output after dropping is—userdfin. Insert the output into the key array at the sender side. With the help of array table, make the matrix of 9 × 6 size for playfair cipher as shown in successive table. First insert these characters into our new matrix table and free blocks with the remaining characters as mentioned above by dropping the duplicates. So, finally, this table is created according to proposed methodology (Table 3).

An even number of characters (20) were present in plaintext; there is no need to add null at last position. Now encrypt this plaintext, with two letters at a time—TH IS TH EH ID DE NM ES SA GE

After encryption, ciphertext is stored in enc_1

$$enc_1 = \text{ug ne ug rg nf fr zm re eu tg}$$

Step II:

In this step, mathematical XOR operation is performed between enc_1 and ck_1. The resulted string enc_2 is in the form of hexadecimal format [15]

1711580756430b42481f0a4d110d45000649430e45420b491002441304505f09525e0f

Step III:

In this final step, RSA encryption is applied on key ck_1 = userdefined.

Take any two prime numbers as a = 3 and b = 5:

Step i: n = a * b = 3 * 5 = 15.
Step ii: s(n) = (a − 1)(b − 1) = 2 * 4 = 8.
Step iii: Calculate any relative prime for s(n). Here we can take it as 7. So e = 7.
Step iv: Apply encryption formula, for ck_1^1, i.e., $ck_1^1 = (ck_1)^e \bmod n$.

For every alphabet, there is a value which is substituted in above encryption formula. So the key array after encrypting with RSA becomes "F M E I M E F C H E M". To decrypt the key at the receiver end, take the value of d = 3 which is the relative prime of e and apply decryption formula, $M = C^d$ mod n. Here, $sk_1 = (sk_1^I)^d$ mod n. By decrypting [16] with above formula the original key is obtained.

5 Simulation Work

The proposed algorithm and original playfair cipher, RSA algorithms [17] have been compared on different data sizes. So this work indicates that the strength of key and message is increased in the following manner:

1. **Key strength**:

By introducing a matrix size of 9×6 compared to 5×5 playfair matrix [18], it increases the complexity of key by many folds. A small diagram (Fig. 1) is represented below to explain the key strength of our algorithm using a number of characters changed in output.

In Fig. 1, the proposed algorithm and 5 * 5 playfair cipher are compared using different key sizes to observe the complexity of the ciphertext. So, the graph shows us that the proposed algorithm is generating more complex ciphertext compared to the 5×5 playfair cipher. As a result, it will be harder for the hackers to crack the original text.

2. **Message strength**:

By introducing XOR operation between the encrypted text and the key which makes original message hidden from the attackers and makes the message strong enough in the aspect of security. Figure 2 is drawn below to show the decryption time of plaintext in the proposed algorithm at different data sizes.

In Fig. 2, the proposed algorithm and 5 * 5 playfair cipher are compared using different plaintext sizes to observe the decryption time of the ciphertext. So, the graph shows us that the proposed algorithm makes the cracker to take huge amount of time to crack the data rather than the 5×5 playfair cipher because of the presence of extra characters in the key and also due to the XOR operation performed between the encrypted text and key.

Fig. 1 Key size representation

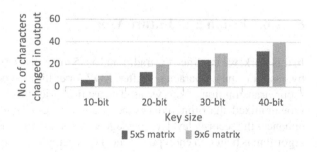

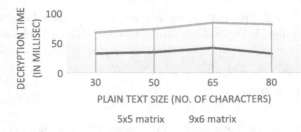

Fig. 2 Message strength representation

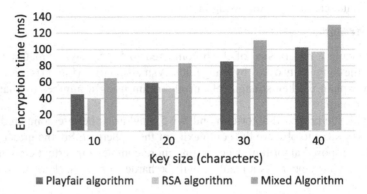

Fig. 3 Comparison of encryption time

3. **Comparison of encryption time**:

Encryption is the process of encoding a message or information in such a way that only authorized parties can access it. The time taken by the program to encrypt the message is called "Encryption Time". Decryption is the process of decoding a message or information in such a way that only authorized parties can access it. The time taken by the program to decrypt the message is called "Decryption Time". In Fig. 3, a comparison is done in between play fair algorithm, RSA algorithm, and mixed algorithm [19].

6   Conclusion and Future Work

In this work, we modified the traditional 5 * 5 matrix to 9 * 6 matrix playfair algorithm by inserting new characters. After XOR operation is performed. Further for the secure transmission of key, we used an asymmetric algorithm named as RSA. So this type of mixed algorithm reduces the disadvantages of traditional playfair cipher and enhances the security level for message transfer. Finally, the performance of proposed algorithm is better, in encryption time, key strength, and message strength, compared to individual symmetric and asymmetric algorithms. Also, many efforts are already been ensured by researchers on these mixed algorithms to increase secureness for data transfer between the sender and receiver.

References

1. Basu, S.: Modified playfair cipher with rectangular matrix (2012)
2. Salam, M., Rashid, N., Khalid, S., Khan, M.R.: A n × m version of 5 × 5 playfair cipher for any natural language. Int. J. Comput. Electr. Autom. Control Inf. Eng. 5(1) (2011)
3. Ravindra Babu, K., Uday Kumar, S., Vinay Babu, A., Aditya, I.V.N.S., Komuraiah, P.: An extension to traditional playfair cryptographic method. Int. J. Comput. Appl. 17(5) (2011)
4. Iqbal, Z., Gola, K.Kr., Gupta, B., Kandpal, M.: Dual level security for key exchange using modified RSA public key encryption in playfair technique. Int. J. Comput. Appl. 111(13) (0975-8887) (2015)
5. Jahan, I., Asif, M., Rozario, L.J.: Improved RSA cryptosystem based on the study of number theory and public key cryptosystems. Am. J. Eng. Res. (AJER) 4(1), 143–149 (2015)
6. Naik, U., Kotak, V.C.: Security issues with implementation of RSA and proposed dual security algorithm for cloud computing. IOSR J. Electron. Commun. Eng. (IOSR-JECE) 9(1), 4347 (2014)
7. Chand, N., Bhattacharyya, S.: A novel approach for encryption of text messages using playfair cipher 6 by 6 matrix with four iteration steps. Int. J. Eng. Sci. Innov. Technol. 3(1) (2014)
8. Obayes, H.K.: Suggest approach to embedded playfair cipher message in digital image. Int. J. Eng. Res. Appl. 3(5) (2013)
9. Murali, P., SenthilKumar, G.: Modified version of playfair cipher using linear feedback shift register. Int. J. Comput. Sci. Netw. Secur. 8(12) (2008)
10. Srivastava, S.S., Gupta, N.: A novel approach to security using extended playfair cipher. Int. J. Comput. Appl. 20(6) (2011)
11. Alam, A.A., Khalid, B.S., Salam, C.M.: A modified version of playfair cipher using 7 * 4 matrix. Int. J. Comput. Theory Eng. 5(4) (2013)
12. Mondal, U.Kr., Mandal, S.N., PalChoudhury, J.: A framework for the development of new approach of playfair cipher. In: Proceedings of the 2nd National Conference (2008)
13. Dar, J.A., Sharma, S.: Implementation of one time pad cipher with rail fence and simple columnar transposition cipher, for achieving data security. Int. J. Sci. Res. 3(11) (2014)
14. Khan, S.S., Tuteja, R.R.: Security in cloud computing using cryptographic algorithms. Int. J. Innov. Res. Comput. Commun. Eng. 3(1) (2015)
15. Iqbal, Z., Gupta, B., Gola, K.Kr., Gupta, P.: Enhanced the security of playfair technique using excess 3 code(xs3) and caesar cipher. Int. J. Comput. Appl. 103(3) (0973-8887) (2014)
16. Dar, J.A., Verma, A.: Enhancing the security of playfair cipher by double substitution and transposition techniques. Int. J. Sci. Res. 4(1) (2015)
17. Preetha, M., Nithya, M.: A study and performance analysis of RSA algorithm. Int. J. Comput. Sci. Mobile Comput. 2(6) (2013)
18. Khan, S.A.: Design and analysis of playfair ciphers with different matrix sizes. Int. J. Comput. Netw. Technol. 3(3) (2015)
19. Durai Raj Vincent, P.M.: RSA encryption algorithm—a survey on its various forms and its security level. Int. J. Pharm. Technol. ISSN: 0975-766X (2016)

Android Malware Detection Mechanism Based on Bayesian Model Averaging

S. Roopak, Tony Thomas and Sabu Emmanuel

Abstract Since Android is the most widely used operating system for mobile devices, it has been a target for widespread malware attacks. During the past years, many new malware detection mechanisms have been introduced for the Android platform. These methods are generally classified as static analysis and dynamic analysis methods. However, none of the existing mechanisms are able to detect the malware applications with reasonable false positive and negative rates. This is a major concern in the field of Android malware detection. In this paper, we propose a novel malware detection mechanism by combining the estimated malicious probability values of three distinct naive Bayes classifiers based on API calls, permissions, and system calls using Bayesian model averaging approach. The majority of the existing Android malwares have signatures in at least one of API calls, permissions, or system call sequences. Hence, the proposed mechanism can overcome the limitations of the existing static and dynamic malware detection mechanism to a good extent. Our experiments have shown that the proposed mechanism is more accurate than the existing static and dynamic malware detection mechanisms.

Keywords Smartphone · Malware applications · Bayesian model averaging

1 Introduction

Nowadays, smartphones have become the principal target of malware attacks. This is because of the widespread usage of third-party applications. Smartphone OS market share is still ruled by Android operating system [1]. It is a Linux-based smartphone

S. Roopak (✉) · T. Thomas
Indian Institute of Information Technology and Management, Kerala,
Thiruvananthapuram, India
e-mail: roopak.res15@iiitmk.ac.in

T. Thomas
e-mail: tony.thomas@iiitmk.ac.in

S. Emmanuel
Kuwait University, Kuwait City, Kuwait
e-mail: sabu@cs.ku.edu.kw

© Springer Nature Singapore Pte Ltd. 2019
P. K. Sa et al. (eds.), *Recent Findings in Intelligent Computing Techniques*,
Advances in Intelligent Systems and Computing 707,
https://doi.org/10.1007/978-981-10-8639-7_9

OS developed by Google. Its source code is released under open source licenses. Android allows third-party developers to develop their own applications and publish in the app stores. This facility can be misused by malware developers. They use API (Application Programming Interface) provided by the application framework in a certain combination for developing the malware. These kinds of malware applications can perform the activities such as stealing information like IMEI (International Mobile Equipment Identity) code, remotely controlling the device, monitoring the user activities, corrupting the device, etc.

Google Play Store is the official app store for Android. The security mechanism called bouncer checks for malware in every new application [2]. It works by analyzing system calls generated by an application in a virtual environment. However, modern-day malwares can easily escape from bouncer verification by system call evasion techniques. Anti-malware products are still relying on static and signature-based analysis. Static analysis is the method of detecting malware application by analyzing its binary code. In signature-based analysis, the hash value of an application is compared to the hash values of known malicious applications for detecting whether it is a malware or not. These mechanisms can be evaded by code transformation attacks [3]. Thus, anti-malware solutions cannot effectively detect transformed malware applications.

According to Xie et al. [4], the signature of a malware application can be identified from its API calls and system call sequences. An API call will execute only when its corresponding permission is granted. While executing an API call, the application switches from user mode to kernel mode. In the kernel mode, the application will carry out the required activities by generating system calls. Therefore, detection of malware applications can be improved by combining the prediction results of API calls, permissions, and system call sequences.

In this paper, we propose a novel mechanism to detect malware application by first estimating the malicious probability values of an application based on its API calls, permissions, and system calls and then combining them by using Bayesian model averaging approach. In Sect. 2, a review of the related works is given. In Sect. 3, the proposed malware detection mechanism is given. The experimental results are given in Sect. 4. Conclusions and future directions for research are given in Sect. 5.

2 Literature Review

In this section, we briefly review the existing static and dynamic analysis-based Android malware detection mechanisms.

2.1 Static Analysis

In [5], Aafer et al. developed a mechanism to detect Android malware applications by analyzing its source code. The mechanism consists of two steps. In the first step, they extract API level features from a set of previously known malware and goodware applications. In the next step, a machine learning classifier such as SVM or KNN is trained with these features. The machine learning classifier will examine whether an unknown application is malware or not based on its API level features. Malware applications can evade this mechanism by using Java reflection and bytecode encryption.

In [6], Enck et al. proposed a method for detecting Android malware application at the time of installation based on some security rules. This method has two steps. In the first step, the apk installer extracts the security configurations from the package manifest file. There is a set of security rules stored in the database. In the second step, the Kirin security service checks the extracted security configurations with all security rules in the database. If the application passes the security rules, then Kirin allows the application to install into the device.

Rovelli et al. [7] proposed a mechanism for detecting Android applications by analyzing its permission patterns. It is based on client–server architecture. In this approach, a machine learning classifier in the server is trained with extracted permissions from a group of malware and goodware applications. A lightweight client application installed in the device sends the permissions requested by an application to the server. In the server, the classifier predicts the malicious behavior of an application based on these permission patterns. Malware applications can evade this mechanism by spoofing the intents and sharing user ID to other goodware apps.

2.2 Dynamic Analysis

Andromaly [8] monitors various system events such as CPU usage, number of packets sent through Wi-Fi, number of running processes, and battery level from the device during its runtime. A previously trained machine learning classifier can predict the malicious behavior based on these features. The main limitation is that it cannot detect short running malicious applications like IMEI stealer.

Wei et al. [9] suggested a behavioral-based approach for detecting malware applications in Android. This approach has two phases. In offline training phase, they execute a set of previously known goodware and malware applications and trace their system calls. Then, they preprocess these system calls and build a dataset with their density values. The density value of a system call is obtained by dividing its count by the total number of invoked system calls. A classifier such as naive Bayes is trained with this dataset. In online testing phase, this classifier will predict whether an unknown application is malware or not based on system call density values.

3 Methodology

In the proposed mechanism, we combine the estimated malicious probability values of an application based on API calls, permissions, and system calls for predicting whether it is malware or not. Malware application usually invokes dangerous API calls with their associated permission along with a sequence of system calls for accessing the resources in a device. Thus, malicious activity can be identified by analyzing API calls or requested permissions or their system call sequences. The proposed method consists of two phases. They are as follows:

- Probability estimation and
- Classifier fusion.

3.1 Probability Estimation

In this phase, the application for malicious functionality is analyzed by examining its source code, permission patterns, and system calls. In the first step, we need to train naive Bayes classifiers for the analysis of API calls, permission, and system call of an application. This framework is shown in Fig. 1. The trained naive Bayes classifier can predict the malicious probability values of an unknown application's source code, permission, and system calls [10]. The multi-feature-based malware probability estimations involve three steps. They are the following:

- API call analysis,
- Permission analysis, and
- System call analysis.

API Call Analysis: In the existing method [5], API level features such as class level information, package level information, and functional level information of an

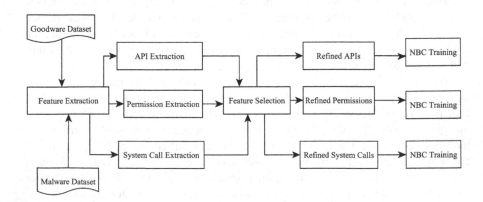

Fig. 1 Training naive Bayes classifiers for different analyses

application are used for detecting a malware application. Among these features, only the functional level information (API calls) is contributing to a malicious activity [4]. Therefore, we will not consider the class level and package level information of an application and instead we analyze only the API calls in the application to determine the probability of the application to be malicious. It is done by decompiling the applications in the collected goodware and malware database. In the decompiling procedure, we use apktool [11] to decompress the application. Then, we use dex2jar tool [12] to convert the bytecode (classes.dex) to jar file. Java source code of the application is recovered from this jar file by using jd-gui tool and APIs are extracted from this source code. A training dataset is constructed with these extracted API calls from malware and goodware applications in the dataset. Most of the features in the dataset are irrelevant and redundant in nature. So we need to select most important API calls from the dataset. We use information gain algorithm for removing this redundant features [13]. In this step, we train a naive Bayes classifier n_1 based on this refined API features from a set of malware and goodware applications.

Let $a = (a_1, a_2, ..., a_m)$ denotes API call-based binary feature vector of an application A. Then, a_i in A is computed as

$$
a_i = \begin{cases} 1, & \text{if feature } i \text{ is found in } A; \\ 0, & \text{otherwise.} \end{cases}
$$

The dependencies among API calls will result in the bugginess of an application as the bug rate is directly proportional to rate of dependency among the API calls [14]. Thus, we can assume that the API calls a_i and a_j are independent of each other for $i \neq j$.

Let M be the event that the application A is malicious. In the training phase we find out the probability values of $p(M)$, $p((a_i = 1)|M)$, $p((a_i = 0)|M)$, $p(a_i = 1)$ and $p(a_i = 0)$ for $i = 1, 2, \ldots, m$ from the dataset. Then, $P(M|a)$ is computed as

$$
p(M|a) = P(M|(a_1, ..., a_m)) = p(M) \cdot \frac{p((a_1, a_2, .., a_m)|M)}{p(a_1, a_2, .., a_m)}
$$
$$
= p(M) \cdot \frac{\prod_{i=1}^{m} p(a_i|M)}{\prod_{i=1}^{m} p(a_i)}.
$$

(1)

Permission Analysis: In this phase, the permissions requested by the application are analyzed for finding the malicious probability of the application. We decompress a set of goodware and malware applications in the database by using apktool [11] for recovering the manifest file. The manifest file contains the necessary permissions required for the working of an application. The permissions are extracted from this manifest file and a dataset is constructed with it. Many irrelevant permissions such as SET_WALLPAPER, SET_TIME are used as features in the existing permission-based malware detection mechanism [7]. So we will eliminate them by using information gain feature selection algorithm [13]. We train a naive Bayes classifier n_2 based on the above dataset.

Let $b = (b_1, b_2, ..., b_l)$ denotes the permission-based binary feature vector of an application A. Then, b_i in A is computed as

$$b_i = \begin{cases} 1, & \text{if feature } i \text{ is found in } A; \\ 0, & \text{otherwise.} \end{cases}$$

The permissions are unique strings for granting access to the service provided by resources in a device. A permission b_i is independent to another particular permission b_j for $i \neq j$ because a programmer can declare any permission b_i in the manifest file independent of other permission b_j for $i \neq j$.

Let M be the event that the application A is malicious. In the training phase we find out the probability values of $p(M)$, $p((b_i = 1)|M)$, $p((b_i = 0)|M)$, $p(b_i = 1)$ and $p(b_i = 0)$ for $i = 1, 2, \ldots, l$ from the dataset. Then, $P(M|b)$ is computed as

$$\begin{aligned} p(M|b) = P(M|(b_1, ..., b_l)) &= p(M). \frac{p((b_1, b_2, .., b_l)|M)}{p(b_1, b_2, .., b_l)} \\ &= p(M). \frac{\prod_{i=1}^{l} p(b_i|M)}{\prod_{i=1}^{l} p(b_i)}. \end{aligned} \tag{2}$$

System Call Analysis: In the existing system call analysis mechanism, the density value of each system call is treated as a feature of a machine learning classifier [9]. However, they have not examined the conditional independence between features while using naive Bayes as a classifier. The density value of a system call depends on the total number of invoked system calls. That means, the density value of a system call changes with the density values of other system calls, and they are dependent on each other. Therefore, we can conclude that their assumptions may not hold. However, counts of individual system calls generated by an application are independent of each other [15] and follow a multinomial distribution. So we have analyzed the count of individual system calls generated by the application at runtime for determining the malicious probability of the application. We run a set of malware and goodware applications in an emulator for collecting the system calls generated by them. These system calls are collected by using strace utility [16]. Some of the system calls such as clock_gettime(),getuid32() are irrelevant. So we remove them by using information gain algorithm [13]. In the next step, we train a naive Bayes classifier n_3 based on these features.

Let $c = (c_1, c_2, \ldots, c_k)$ denotes the system call-based feature vector of an application A and c_i corresponds to the count of the ith system call in a fixed time duration (1 min. in our case). Let s_i denotes the event that the ith system call is present in A. we can assume that the count of a system call c_i is independent of other system call count c_j for $i \neq j$ [15].

Let M be the event that the application A is malicious. In the training phase, we find out the probability values of $p(s_i|M)$, $p(\bar{M})$, $p(s_i|\bar{M})$ and $p(M)$ for $i = 1, 2, \ldots, k$. Then, $p(M|c)$ is computed as

$$p(M|c) = P(M).\frac{p(c|M)}{p(c)} = p(M).\frac{p((c_1, c_2, .., c_k)|M)}{p(c)} = p(M).\frac{\prod_{i=1}^{k} p(c_i|M)}{p(c)}$$

$$= p(M).\frac{\prod_{i=1}^{k} p(s_i|M)^{c_i}}{p(c)},$$

(3)

where $p(c)$ is calculated as

$$p(c) = p(M).p(c|M) + p(\bar{M}).p(c|\bar{M})$$

$$= p(M).p((c_1, c_2, ..., c_k)|M) + p(\bar{M}).p((c_1, c_2, .., c_k)|\bar{M})$$

$$= p(M).\prod_{i=1}^{k} p(c_i|M) + p(\bar{M}).\prod_{i=1}^{k} p(c_i|\bar{M})$$

$$= p(M).\prod_{i=1}^{k} p(s_i|M)^{c_i} + p(\bar{M}).\prod_{i=1}^{k} p(s_i|\bar{M})^{c_i}.$$

3.2 Classifier Fusion

In this phase, we combine the results of above three classifiers by using Bayesian model averaging approach [17]. The procedure is shown in Fig. 2. Let A be an unknown application and a, b, c correspond to the API call-based feature vector, permission-based feature vector, and system call-based feature vector of A, respectively. Let $P(M|n_i, a, b, c)$ for $i = 1, 2, 3$ denotes the probability that the app A is detected as a malware by the classifier n_i by using the feature vector a, b, and c, respectively. Let $p(w_i|n_i)$ for $i = 1, 2, 3$ denotes the belief probabilities of n_i. We assume that $\sum_{i=1}^{3} p(w_i|n_i) = 1$.

The belief probability $p(w_i|n_i)$ denotes the probability with which classifier n_i for $i = 1, 2, 3$ can correctly predict. Then, the approximate malicious probability of an application is given by

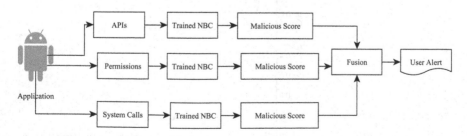

Fig. 2 Malware detection by probabilistic approach

$$p(M) = \sum_{i=1}^{3} p(M|n_i, a, b, c).p(w_i|n_i).$$

Let t be a threshold value between 0 and 1. An application A is treated as a malware when its malicious probability value $p(M)$ is greater than t.

4 Results

We have tested the accuracy of our approach in a dataset of 100 applications containing both malware and goodware applications. The naive Bayes classifiers are trained with 25 malware and 25 goodware samples from the dataset. Further, we have implemented the mechanisms proposed by Rovelli et al. [7], Aefer et al. [5] and Wei et al. [9] using the same dataset and evaluated the performance against our approach. These types of malware applications are downloaded from the Drebin dataset [18]. We have collected the goodware applications from Google Play Store and other trustworthy repositories.

The FPR $(FalsePositiveRate)$ and FNR $(FalseNegativeRate)$ against different threshold values at equal belief probability values of $p(w_i|b_i) = \frac{1}{3}$ for $i = 1$, 2, 3 are shown in Fig. 3. In this figure, we can see that FPR and FNR are lowest at around a threshold value of $t = 0.4$. Hence, we fix the value 0.4 as threshold t. The comparative results include FPR, FNR, $TNR(TrueNegativeRate)$, $Precision$, $Recall$, $Accuracy$, and F_1score are shown in Table 1. In this table, the proposed method is compared with existing mechanisms [5, 7, 9]. From this table, we can see that the accuracy rate of the proposed approach is 0.94 whereas the accuracy rates of existing mechanisms are 0.88, 0.78, and 0.60, respectively. The F_1score

Fig. 3 False positive and negative rates against different threshold values

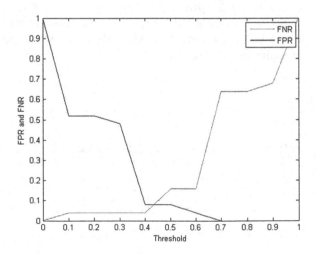

Table 1 Detection rates of proposed approach compared to existing methods

Methodology	FPR	FNR	Recall	Precision	TNR	Accuracy	F₁ score
Proposed approach	0.08	0.04	0.96	0.92	0.92	0.94	0.93
Rovelli et al. [7]	0.12	0.32	0.68	0.88	0.88	0.78	0.77
Aafer et al. [5]	0.12	0.12	0.88	0.88	0.88	0.88	0.88
Wei et al. [9]	0.44	0.36	0.64	0.59	0.56	0.60	0.61

of the proposed approach is 0.93 whereas the $F_1 score$ of existing mechanisms are 0.88, 0.77, and 0.61, respectively. Hence, we conclude that the Bayesian averaging approach of multiple (possibly) mutually exclusive classifiers give superior results in malware detection. The accuracy may be improved by adaptively varying the belief probabilities $p(w_i|b_i)$ and threshold t depending on the category of the application.

5 Conclusion

In this paper, we have introduced a novel approach for detecting Android malware applications by fusing various Bayesian classifiers. Our methodology can effectively detect repacked and obfuscated malwares. This mechanism has an improved accuracy in Android malware detection compared to existing mechanisms.

More stealthy malwares can evade the proposed mechanism by applying source code encryption, intent spoofing, and process hiding applied together at the same time [3, 19, 20]. In those cases, not much evidence is available for detecting malicious nature of an application. This means that API calls, permissions, and system call sequences are not present in the application for the analysis. Therefore, a future direction of research is to detect such kind of stealthy malwares using improved classifier fusion mechanism while minimizing the computational load.

References

1. Worldwide smartphone OS market share. http://www.idc.com
2. A look at Google bouncer. http://blog.trendmicro.com
3. Rastogi, V., Chen, Y., Jiang, X.: DroidChameleon: evaluating android anti-malware against transformation attacks. In: 8th ACM Symposium on Information, Computer and Communications Security (ASIACCS 2013), pp. 329–334. ACM Digital Library, Arizona, USA (2013)
4. Xie, L., Zhang, X., Seifert, J.P., Zhu, S.: PBMDS: a behavior-based malware detection system for cellphone devices. In: Proceedings of the Third ACM Conference on Wireless Network Security, pp. 37–48 (2010)
5. Aafer, Y., Du, W., Yin, H.: DroidAPIMiner: mining API-level features for robust malware detection in android. Security and Privacy in Communication Networks. Springer, pp. 86–103 (2013)

6. Enck, W., Ongtang, M., McDaniel, P.: On lightweight mobile phone application certification. In: Proceedings of the 16th ACM Conference on Computer and Communications Security, pp. 235–245. ACM Digital Library (2009)
7. Rovelli, P., Vigfússon, Y.: PMDS: permission-based malware detection system. In: International Conference on Information Systems Security. Springer, pp. 338–357 (2014)
8. Shabtai, A., Shabtai, A., Kanonov, U., Elovici, Y., Glezer, C., Weiss, Y.: Andromaly: a behavioral malware detection framework for android devices. J. Intell. Inf. Syst. **38**(1), 161–190 (2012)
9. Wei, Y., Zhang, H., Ge, L., Hardy, R.: On behavior-based detection of malware on android platform. In: Global Communications Conference (GLOBECOM), pp. 814–819. IEEE (2013)
10. Lewis, D.D.: Naive (Bayes) at forty: the independence assumption in information retrieval. In: European Conference on Machine Learning. Springer, Berlin, Heidelberg (1998)
11. Apktool. http://ibotpeaches.github.io/Apktool/
12. dex2jar. http://sourceforge.net/projects/dex2jar
13. Azhagusundari, B., Thanamani, A.S.: Feature selection based on information gain. Int. J. Innov. Technol. Explor. Eng. (IJITEE) 2278–3075 (2013). ISSN
14. Cataldo, M., Mockus, A., Roberts, J.A., Herbsleb, J.D.: Software dependencies, work dependencies, and their impact on failures. IEEE Trans. Softw. Eng. **35**(6), 864–878 (2009)
15. Xiao, X., Jiang, Y., Liu, X., Ye, R.: Identifying Android malware with system call co-occurrence matrices. Trans. Emerg. Telecommun. Technol. (2016)
16. Strace. https://sourceforge.net/projects/strace
17. Madigan, D., Raftery, A.E., Volinsky, C., Hoeting, J.: Bayesian model averaging. In: Proceedings of the AAAI Workshop on Integrating Multiple Learned Models, pp. 77–83, Portland (1996)
18. Arp, D., Spreitzenbarth, M., Hubner, M., Gascon, H., Rieck, K.: DREBIN: effective and explainable detection of android malware in your pocket. In: Proceedings of the 20th Annual Network & Distributed System Security Symposium (NDSS), Siemens (2014)
19. Sbîrlea, D., Burke, M.G., Guarnieri, S., Pistoia, M., Sarkar, V.: Automatic detection of inter-application permission leaks in android applications. IBM Journal of Research and Development **57**(6), 10–1 (2013)
20. Butler, J.: DKOM (direct Kernel Object Manipulation). Black Hat USA (2004)

A Provably Secure Re-encryption-Based Access Control in Hierarchy

Gaurav Pareek and B. R. Purushothama

Abstract A hierarchical access control mechanism is required when some users of a system have more access privileges than others. A set of users in the system can be divided into subsets called classes which are hierarchically organized in a way that any class can access data meant for the classes lower in the hierarchy but not vice versa. In this paper, we propose a re-encryption-based access control scheme for a hierarchy of security classes in which storage overhead on the users belonging to every security class is constant with public storage requirements being linear in the number of classes in the hierarchy. Direct re-encryption key from a class lower down the hierarchy to the one higher up can be derived using intermediate re-encryption keys available between each pair of adjacent classes. The re-encryption key derivation procedure requires steps of the order of the depth of the class for which the data was initially encrypted. As a result of re-encryption key derivation, just one re-encryption is needed.

Keywords Hierarchical access control · Proxy re-encryption

1 Introduction

In a hierarchical organizational structure, users at different levels in the hierarchy are given different access privileges. Users higher up in the hierarchy should be able to access information that is accessible to users lower in the hierarchy, in addition to information available at their level.

G. Pareek (✉) · B. R. Purushothama
National Institute of Technology Goa, Ponda, India
e-mail: gpareek@nitgoa.ac.in

B. R. Purushothama
e-mail: puru@nitgoa.ac.in

© Springer Nature Singapore Pte Ltd. 2019
P. K. Sa et al. (eds.), *Recent Findings in Intelligent Computing Techniques*,
Advances in Intelligent Systems and Computing 707,
https://doi.org/10.1007/978-981-10-8639-7_10

Proxy re-encryption [1] is a cryptographic tool used for transforming a data item encrypted under public key of any party into ciphertext under public key of another party without letting anyone learn anything about the underlying message or secret keys of either parties. A re-encryption scheme which is both unidirectional and transitive may be useful in this scenario. This is because using such a re-encryption scheme, the re-encryption keys are not required to be stored but are derived as and when required. But unfortunately, a re-encryption scheme that exhibits transitivity and unidirectionality in key derivation is not present. We propose a re-encryption scheme that is unidirectional and supports re-encryption key derivation from a lower to higher class using the intermediate re-encryption keys between adjacent classes. In our proposed scheme, the event of class addition and deletion from the access hierarchy is handled with minimal overhead on the users of the classes.

Contents of the rest of the paper are as follows. Related research in the area of hierarchical access control in hierarchy is discussed in Sect. 2. In Sect. 3, we present the background for proposed re-encryption scheme and its concrete construction. Section 4 contains description of our concrete method for hierarchical access control and the paper concludes in Sect. 5.

2 Related Work

Akl et al. [2] in their seminal work performed the initial work in the area of access control in hierarchical structure. Chang et al. [3] proposed a scheme that supports dynamic access management with inefficient key derivation procedure. Another scheme due to Hui et al. [4] supports dynamics but requires updation of every class secret key. He et al. [5] proposed a scheme based on the Chinese Remainder Theorem and the Rabin public key cryptosystem. The schemes due to Sun et al. [6] have a high rekeying overhead during the event of a class addition or a class deletion from the access hierarchy.

3 Proposed Unidirectional Re-encryption Scheme with Rekey Derivation

3.1 Preliminaries

Bilinear Maps: If $\mathbb{G}_1$ and $\mathbb{G}_2$ are two cyclic groups each one of which is of prime order p and $g \in \mathbb{G}_1$ its generator, a *bilinear map* $e : \mathbb{G}_1 \times \mathbb{G}_1 \to \mathbb{G}_2$ has the following properties:

1. Bilinearity: $e(g^l, g^k) = e(g, g)^{lk}, \forall\, g \in \mathbb{G}_1, l, k \in \mathbb{Z}_p^*$.
2. Non-degeneracy: $e(g, g) \neq 1$.

3. Symmetric: $e(g^l, g^k) = e(g^k, g^l) = e(g, g)^{lk}, \forall g \in \mathbb{G}_1, l, k \in \mathbb{Z}_p^*$.
4. Computability: An efficient algorithm exists for computing the bilinear map.

Complexity Assumptions

Definition 1 *(Discrete Logarithm (DLP) Assumption)* Given the distribution (g, g^a) where $a \in_R \mathbb{Z}_p$ and $g \in \mathbb{G}_1$ is the generator of $\mathbb{G}_1$, the **DLP** assumption states that it is computationally infeasible for a PPT algorithm $\mathcal{A}$ with non-negligible advantage to guess the value of a.

Definition 2 *(Decisional Bilinear Diffie–Hellman (DBDH) Assumption)* Given $(g, g^u, g^v, g^w, e(g, g)^{uvw})$ and $(g, g^u, g^v, g^w, e(g, g)^Q)$ where $u, v, w, Q \in_R \mathbb{Z}_p^*$, the **DBDH** assumption means that any PPT adversary $\mathcal{A}$ cannot distinguish these two distributions with non-negligible advantage λ. So,

$$|Pr[\mathcal{A}(g, g^u, g^v, g^w, e(g, g)^{uvw}) = 0] - Pr[\mathcal{A}(g, g^u, g^v, g^w, e(g, g)^Q) = 0]| \geq \lambda.$$

3.2 Construction for the Proposed Proxy Re-encryption Scheme

Setup(1^k): Taking security parameter as input, it determines public parameters: $param = (\mathbb{G}_1, \mathbb{G}_2, g_1, g_2, e, p)$ where $\mathbb{G}_1$ and $\mathbb{G}_2$ are distinct cyclic groups each of prime order p, $g_1, g_2 \in \mathbb{G}_1$ are the generators of $\mathbb{G}_1$ and a bilinear mapping $e : \mathbb{G}_1 \times \mathbb{G}_1 \rightarrow \mathbb{G}_2$.

KeyGen($i, param$): This algorithm uses identifier and the global parameters generated in the previous step to produce (pk_i, sk_i) as under:
$sk_i = (s_i, r_i, u_i) \in_R (\mathbb{Z}_p^*)^3$ and $pk_i = (\alpha_i, \beta_i, \gamma_i, \delta_i) = (g_1^{r_i s_i}, g_1^{s_i}, g_1^{r_i}, g_2^{u_i})$.

Enc(pk_i, m): Any message $m \in \mathbb{G}_2$ can be encrypted under the public key pk_i to produce CT_i as under:
choose $t \in_R \mathbb{Z}_p$ and compute $CT_i = (A_i = e(\alpha_i^t, \delta_i)m, B_i = \beta_i^t)$.

Dec$_1$(sk_i, CT_i): This procedure obtains the underlying message m from CT_i using sk_i:
compute: $A_i e(B_i, g_2^{-r_i u_i}) = m$.

ReKeyGen(sk_i, sk_j): Re-encryption key $rk_{i \rightarrow j}$ is generated using an interactive rekey generation algorithm as:
$rk_{i \rightarrow j} = g_2^{-(r_i + r_j s_j)u_i}$.

ReEnc($rk_{i \rightarrow j}, CT_i$): This procedure takes as input ciphertext containing the message encrypted under pk_i and re-encryption key $rk_{i \rightarrow j}$ and computes re-encrypted ciphertext under pk_j as:
$C'_j = (A'_j = A_i e(B_i, rk_{i \rightarrow j}), B'_j = B_i)$.

Dec$_2$ (sk_i, C_i'): Decrypts ciphertext re-encrypted under pk_i using sk_i as input:
compute: $A_i'e(B_i', g_2^{-r_iu_is_i}) = m$.
Derive$(rk_{i\to j}, rk_{j\to k}, sk_j, sk_k)$: Takes as input two one-step re-encryption keys and generates interactively, a two-step transitive re-encryption key. The algorithm works as under:
derive-j(sk_j).
    Output: $\zeta = (\delta_i)^{s_jr_j}rk_{i\to j}\beta^{r_js_j}$.
derive-k(sk_k, ζ).
    Output: $rk_{i\to k} = \zeta\alpha_j^{s_k}\delta_i^{-r_ks_k}$.

3.3 Correctness of Decryption

Ciphertext produced directly under a public key is of the form:
$CT_i = (A_i = e(\alpha_i^s, \delta_i)m, B_i = \beta_i^s)$.
Here, $A_i = e(g_1^{r_is_it}, g_2^{u_i})m = e(g_1, g_2)^{r_is_iu_it}m$, and $B_i = g_1^{s_it}$.
Decryption step gives the following:
$A_ie(B_i, g_2^{-r_iu_i}) = e(g_1, g_2)^{r_is_iu_it}e(g_1^{s_it}, g_2^{-u_ir_i})m = m$, meaning **Dec$_1$** is correct.

3.4 Re-encryption

The output of the procedure **ReEnc** using $rk_{i\to j}$ is:
$C_j' = (A_j' = A_ie(B_i, rk_{i\to j}), B_j' = B_i)$, where $A_j' = e(g_1^{r_is_it}, g_2^{u_i})e(g_1^{s_it}, g_2^{-(r_i+r_js_j)u_i})m$
$= e(g_1, g_2)^{s_is_jr_ju_it}m$ and $B_j' = g_1^{s_it}$, which is decrypted using **Dec$_2$** as:
$e(g_1, g_2)^{s_is_jr_ju_it}e(g_1^{s_it}, g_2^{r_js_ju_j})m = m$, where (s_j, r_j, u_j) is the secret key.

3.5 Re-encryption Key Derivation

Derive has two sub-procedures, namely, **derive-j** and **derive-k**. Sub-procedure **derive-j** produces $\zeta = (\delta_i)^{s_jr_j}rk_{i\to j}\beta^{r_js_j} = g_2^{-u_ir_i}g_1^{-r_ks_ku_j}$.
derive-k obtains $rk_{i\to k} = \zeta\alpha_j^{s_k}\delta_i^{r_ks_k} = g_2^{-r_iu_i-r_ks_ku_i}$, which is indeed $rk_{i\to k}$.

3.6 Security Analysis

Security Definitions We denote the target user as u_{i*}, set of corrupted users as $\{u_x\}$ and $\{u_h\}$ is the set of honest users.

Standard Security
Standard Security captures the inability of anyone to distinguish the two ciphertexts

encrypted under the same public key if the attacker does not have the secret key to decrypt it.

$$Pr[\{(pk_{i^*}, sk_{i^*}) \leftarrow \textbf{KeyGen}(i^*, param)\}, \{(pk_x, sk_x) \leftarrow \textbf{KeyGen}(x, param)\},$$
$$\{(pk_h, sk_h) \leftarrow \textbf{KeyGen}(h, param)\}, \{rk_{x \to i^*} \leftarrow \textbf{ReKeyGen}(sk_x, sk_{i^*})\},$$
$$\{rk_{i^* \to h} \leftarrow \textbf{ReKeyGen}(sk_{i^*}, sk_h)\}, \{rk_{h \to i^*} \leftarrow \textbf{ReKeyGen}(sk_h, sk_{i^*})\},$$
$$(m_0, m_1, \alpha) \leftarrow \mathcal{A}(\{pk_{i^*}\}, \{pk_h\}, \{(pk_x, sk_x)\}, \{rk_{x \to i^*}\}, \{rk_{i^* \to h}\}, \{rk_{h \to i^*}\},$$
$$b \leftarrow \{0, 1\}, b' \leftarrow \mathcal{A}_k(\alpha, \textbf{Enc}(pk_{i^*}, m_b)) : b = b'] < \frac{1}{2} + negl(k)$$

Master Secret Security
The definition captures inability of guessing the secret key of the delegator with non-negligible advantage.

$$Pr[(pk_{i^*}, sk_{i^*}) \leftarrow \textbf{KeyGen}(i^*, param), (pk_x, sk_x) \leftarrow \textbf{KeyGen}(x, param),$$
$$(pk_h, sk_h) \leftarrow \textbf{KeyGen}(h, param), \{rk_{x \to i^*} \leftarrow \textbf{ReKeyGen}(sk_x, sk_{i^*})\},$$
$$\{rk_{i^* \to x} \leftarrow \textbf{ReKeyGen}(sk_{i^*}, sk_x)\}, \{rk_{x \to h} \leftarrow \textbf{ReKeyGen}(sk_x, sk_h)\},$$
$$\alpha \leftarrow \mathcal{A}_k(\{pk_{i^*}\}, \{pk_h\}, \{(pk_x, sk_x)\}, \{rk_{x \to i^*}\}, \{rk_{i^* \to x}\}, \{rk_{x \to h}\}) : \alpha = sk_{i^*},$$
$$] < negl(k)$$

Theorem 1 *The scheme proposed in Sect. 3.2 is secure under the Decisional Bilinear Diffie–Hellman (DBDH) and Discrete Logarithm (DLP) Assumption.*

4 Dynamic Access Control in Hierarchy

System Model: A set of users U in the system are divided into disjoint classes, $C_1, C_2, \ldots, C_n$. Let $S = \{C_1, C_2, \ldots, C_n\}$ that has a class secret key, CSK_i and a class encryption key, CEK_i associated with it. Let $S = \{C_1, C_2, \ldots, C_n\}$ be ordered by partially ordered binary relation $\preceq$. That is if $C_i \preceq C_j$, users assigned to the class C_j are authorized to access information destined to users in class C_i. Also, there is a re-encryption key corresponding to each directed edge in the access graph (hierarchy). At any point in time, a security class addition or deletion from the hierarchy may take place. There exists a central trusted Key Generation Center (KGC) which is responsible for generating the class secret keys, class encryption keys, and re-encryption keys.

4.1 Proposed Approach for Access Control in Hierarchy

Setup: This procedure is executed by KGC to produce the global system parameters. This procedure uses the **Setup** defined in the construction in Sect. 3.2.

Gen (key generation and re-encryption key generation): Executed by the KGC, this procedure uses the **KeyGen** procedure defined in Sect. 3.2 to generate secret keys for all the security classes in the system and publishes the corresponding public keys (the (CEK, CSK) key pair). The procedure also uses the **ReKeyGen** procedure defined in Sect. 3.2 to produce re-encryption keys for each directed edge in the hierarchy, i.e., $rk_{i \rightarrow j}$ for all $C_i \preceq C_j$.

Derive (re-encryption key derivation and data access): When a data item encrypted under public key of class C_i arrives and if class C_k is the immediate predecessor of C_i, $rk_{i \rightarrow k}$ is directly available as a result of the **Gen** phase defined above. If C_k is an antecedent of C_i separated by more than one levels, C_k requests execution of the procedure **Derive** defined in Sect. 3.2 recursively starting with the class C_i through all the intermediate classes between C_i and C_k to the class C_k deriving one direct re-encryption key at a time as shown in Fig. 1. Now, **ReEnc** procedure of the re-encryption scheme defined in Sect. 3.2 is used with $rk_{i \rightarrow k}$ and CT_i as input to produce CT'_k. The re-encrypted ciphertext can now be decrypted using **Dec₂** (sk_k, CT'_k).

4.2 Handling Dynamics

If any new class is introduced in the hierarchy, the KGC generates the secret key $(sk_i \in_R (\mathbb{Z}_p^*)^3)$ for the joining class and publishes its public key (pk_i) using the procedure **KeyGen**. Based on the security clearance of the incoming class (say C_i), the GC generates re-encryption keys $rk_{j \rightarrow i} \ \forall C_j \preceq C_i$ and $rk_{i \rightarrow k} \ \forall C_i \preceq C_k$. In case a security class is to be removed, the descendants of the leaving class C_i need to update one of their three secret key parameters (u_j) and then the re-encryption keys for delegations in which they are the delegators can also be updated by them individually by computing $(rk_{j \rightarrow i})^{u'_j / u_j}$.

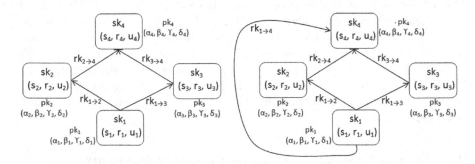

Fig. 1 Example of rekey derivation in hierarchy

Table 1 Comparison with existing hierarchical access control schemes. Here, $|V|$ is the number of security classes, $|E|$ the number of directed edges in the access graph, L the relative depth of the target class, and w the width of the access graph

Scheme	Private storage	Public storage	Key derivation cost
Lin [7]	$O(1)$	$O(|E|)$	$O(L)$
Atallah et al. [8]	$O(1)$	$O(|E| + |V|)$	$O(L)$
D'Arco et al. [9]	$O(1)$	$O(|V|)$	$O(L)$
Freire et al. [10]	$O(w)$	$O(1)$	$O(L)$
Tang et al. [11]	$O(1)$	$O(|V|^2)$	$O(L)$
Proposed scheme	$O(1)$	$O(|E|)$	$O(L)$

4.3 Security of the Proposed Approach

The output of the procedure **Derive** defined above must not succumb to the key distinguishability attacks whose aim is to distinguish outputs of different instances of the **Derive** procedure after collusion with various parties in the system. Security of the proposed hierarchical key assignment scheme depends on security of the proxy re-encryption scheme defined in Theorem 1.

4.4 Performance Analysis

A comparison of the proposed re-encryption-based method for hierarchical access control against various existing schemes is given in Table 1.

5 Conclusion

Cryptographic access control is important to realize in hierarchy of security classes with minimum storage and computation overhead on the users in the security classes. We proposed a re-encryption-based access control management in hierarchy which is both secure and efficient. The underlying re-encryption scheme is interactive and requires a trusted Group Controller to generate the re-encryption keys. Proposing a noninteractive scheme remains as ongoing future work.

References

1. Ateniese, G., Fu, K., Green, M., Hohenberger, S.: Improved proxy re-encryption schemes with applications to secure distributed storage. ACM Trans. Inf. Syst. Secur. (TISSEC) **9**(1), 1–30 (2006)
2. Akl, S.G., Taylor, P.D.: Cryptographic solution to a problem of access control in a hierarchy. ACM Trans. Comput. Syst. (TOCS) **1**(3), 239–248 (1983)

3. Chang, C.C., Buehrer, D.J.: Access control in a hierarchy using a one-way trap door function. Comput. Math. Appl. **26**(5), 71–6 (1993)
4. Hui-Min, T., Chin-Chen, C.: A cryptographic implementation for dynamic access control in a user hierarchy. Comput. Secur. **14**(2), 159–66 (1995)
5. He, M., Fan, P., Kaderali, F., Yuan, D.: Access key distribution scheme for level-based hierarchy. In: Fourth International Conference on Parallel and Distributed Computing, Applications and Technologies, pp. 942–945. IEEE (2003)
6. Sun, Y., Liu, K.R.: Scalable hierarchical access control in secure group communications. In: Twenty-third Annual Joint Conference of the IEEE Computer and Communications Societies, pp. 1296–1306. IEEE (2004)
7. Lin, C.H.: Hierarchical key assignment without public-key cryptography. Comput. Secur. **20**(7), 612–619 (2001)
8. Atallah, M.J., Blanton, M., Fazio, N., Frikken, K.B.: Dynamic and efficient key management for access hierarchies. ACM Trans. Inf. Syst. Secur. (TISSEC). **12**(3), 1–43 (2009)
9. D'Arco, P., De, Santis, A., Ferrara, A.L., Masucci, B., : Variations on a theme by Akl and taylor: security and tradeoffs. Theor. Comput. Sci. **411**(1), 213–27 (2010)
10. Freire, E.S., Paterson, K.G., Poettering, B.: Simple, efficient and strongly KI-secure hierarchical key assignment schemes. In: Cryptographers' Track at the RSA Conference, pp. 101–114. Springer, Berlin, Heidelberg (2013)
11. Tang, S., Li, X., Huang, X., Xiang, Y., Xu, L.: Achieving simple, secure and efficient hierarchical access control in cloud computing. IEEE Trans. Comput. **65**(7), 2325–2331 (2016)

Prevention of Hardware Trojan by Reducing Unused Pins and AES in FPGA

Navneet Kaur Brar, Anaahat Dhindsa and Sunil Agrawal

Abstract With the advent of globalization in the fabrication of IC (integrated circuits) industry, the threat of Hardware Trojan (HT) has increased. The malevolent HT inserted at unused resources and the user is unaware. We proposed a technique, i.e., the combination of modified BISA (Built In Self Authentication) and AES (Advance Encryption Standard) to protect the system from HT hazard. In modified BISA, the unused resources pins are filled with the dummy functional logic. Then the encryption is done by AES algorithm, the code is decrypted with the key that is known only to user. The proposed technique is implemented on Xilinx 10.1_ISE in family Automatic Sparton3 and device XA3S50. The scheme reduces the probability of HT insertion and performance remains same as of original circuitry.

Keywords AES (Advance encryption Standard) · FPGA (Field programmable gate Array) · Encryption · Hardware trojan · Xilinx ISE tool

1 Introduction

Integrated circuit (IC), i.e., the collection of electronic components at one semiconductor chip has become a poignant part of every stream like military, government, commercialization, etc. The entire fabrication of IC at one-place costs high and to reduce cost ICs fabrication is done at different places. The third party can conveniently alter RTL (Resistor Transistor Logic), add vicious rare conditions or codes at any level of abstraction. Hardware Trojan (HT) is a malignant program that enters off the record in the system and performs abominable tasks such as leaks confidential

N. K. Brar (✉) · A. Dhindsa · S. Agrawal
UIET, Panjab University, Chandigarh, India
e-mail: navubrar1992@gmail.com

A. Dhindsa
e-mail: anaahat.dhindsa85@gmail.com

S. Agrawal
e-mail: s.agrawal@hotmail.com

© Springer Nature Singapore Pte Ltd. 2019
P. K. Sa et al. (eds.), *Recent Findings in Intelligent Computing Techniques*,
Advances in Intelligent Systems and Computing 707,
https://doi.org/10.1007/978-981-10-8639-7_11

data, abrupt the functioning, degrade the performance of system, annihilates data, etc. The Trojan insertion is havoc to any circuitry. The Trojan insertion in voting machine can block the counter of any candidate even when the candidate is getting votes. The HT can annihilate the data in the breach of the transmitter and receiver [1].

The researchers have been focusing on this malignant Trojan detection and prevention. Still the Trojan enters the circuitry with use of some loopholes due to small size, rare activation and detection and many instances to enter in the circuit. The Trojan detection methods previously used had some limitations like side channel analysis faced problem of process noise, many techniques required golden chip, i.e., free of any fault or error for reference. The propose scheme tries to fix the menace.

In the proposed scheme the unused resources are filled with dummy logic functional cells in order to leave no vacant space for Trojan insertion. However, there was still a probability that Trojan may replace the dummy logic as in BISA [2]. The user is not aware of Trojan replacement, as there is no change in circuit functionality. To prevent such state, the encryption of code by AES algorithm is done. The scheme reduces the chance of Trojan insertion not at the cost of system performance.

The remainder of this paper is organizing as follows: Sect. 2 introduces to the recent work done for prevention and detection of HT. Section 3 gives the overview of HT structure, types, and testing approach. Section 4 demonstrates the proposed scheme in the paper. Section 5 manifests the experimented results of the scheme. Finally, the conclusion is drawn in Sect. 6.

2 Related Work

Since long, researchers have been performing many experiments for detection and prevention of HT. The Trojan enters the circuitry with use of some open-loop holes. Yier Jin et al. [3] detected Trojan by using Path Delay Fingerprinting technique. The technique-analyzed finger prints delay for all path of the circuitry and compared delay parameters. Hence the technique was able to detect explicit payload Trojan effectively. However, this technique was not effective to detect implicit payload Trojan. Rithesh M et al. [4] analyzed and detected Trojan via Scan Chain method. In this method, the power is analyzed and the presence of malicious Trojan circuitry changes the original power of circuitry. Hence, the Trojan is detected.

Sheng Wei et al. [5] proposed a technique for detection and diagnosis of HT, i.e., self-Consistency. The technique-analyzed gate level characterization by applying segmentation and consistency. Any change in scaling factor depicted the presence of HT along with Trojans location. M Sahib Zamani et al. [6] experimented with clock period to improve detection of HT through Trust Driven Retiming (TDR). In retiming the delay elements, e.g., flip flops, etc., were replaced in order to minimize the clock period. The author experimentally showed TDR boosted the Hardware Detection Probability (HDP) up to 90%.

The vulnerability of Trojan attacks on FPGA had increased tremendously. The researchers performed many experiments to prevent FPGA devices from Trojan.

Behnam Khaleghi et al. [7] proposed a technique based on protecting HT's insertion in FPGA using dummy logic. The adversary placed the Trojan at unused spaces. So to prevent the Trojan insertion unused resources in the FPGA were filled with dummy logic cells. The technique showed no effect on performance and power. Sanchita Mal-Sarkar et al. [8] used approach that worked on the combined approach of side channel and logic testing analysis. This scheme used Adapted Triple Modular Redundancy (ATMR) that firmly protected FPGA devices against Trojan. This scheme gave more power overhead at the cost of utilizing more time-sharing of resources.

Nandeesha Veeranna et al. [9] detected Trojan using Property Checking Techniques at Behavioral level. The method required no golden chips, i.e., chip with 0% error reference as in path delay fingerprinting, etc. The Trojan detection by third party encrypted IP is done through High Level Synthesis (HLS). The C code reformed to do 3PIPs attestation in order to detect Trojan. M Tarek Ibn Ziad et al. [10] used Homomorphic Data Isolation to protect from Trojan. The encryption/decryption has shown admirable results in security. The author used partial Homomorphic design of ElGamal encryption/decryption. The two homomorphism used are additive and multiplicative homomorphism which is done at minimum cost. The technique showed area and power savage with no resource sharing.

3 Overview of Hardware Trojan

Hardware Trojan is an unwanted program or software inserted in the circuitry to degrade the system performance, steal or manipulate data, malfunctioning of system, tamper the circuit design or Denial of Service. HT can be inserted at any phase or level, i.e., gate level, design fabrication, RTL (Register Transfer Level), Testing level, etc. In any automatic machine Trojan insertion is disastrous to human lives, e.g., in airplane automated machinery if Trojan is inserted can over ride the plane speed, can land in populate areas, etc. HT occupies negligible area and power when intruded in circuitry. Because of the small circuitry of Trojan, the detection is quite difficult in large ICs (Integrated Circuits).

The Trojan structure consists of trigger and payload. The trigger part contains the information of way of activation of Trojan. The payload depicts the functionality of Trojan [4]. The Trojan triggers by either analog or digital means. The Trojan triggered by analog conditions such as temperature, power failure, climatic changes, etc. The digital condition may be any Boolean expression for Trojan activation [1]. The mechanism of Trojan trigger and payload explained in Fig. 1 [9]. Based on payload functioning Trojan is further classified as implicit and explicit payload Trojan. The implicit payload Trojan takes signals as stimulus of the trigger. The implicit Trojan leaks information or may collapse the system through radio signals. The implicit payload Trojan is difficult to detect due to masking by process variations [3]. In explicit payload Trojan, the payload part changes the internal signal when Trojan triggers and creates delay in some paths [3].

| TRIGGER | PAYLOAD |

Fig. 1 Structure of HT

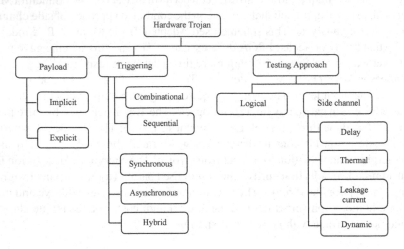

Fig. 2 Classification of HT

The Trojan triggers through rare conditions that can be combinational or sequential. The Trojan triggered via rare signal, payload reverse the bus selection signal to distort the system working is combinational triggering [3]. The Trojan triggered through rare events and k bit counter is sequential triggering of Trojan [1, 3]. The adversary set the rare condition according to the system where user could not understand the cause of abruption in system. The Trojan activates though internal signals as well as number of clock edges. The Trojan that activates by counting number of clock edges is synchronous Trojan [4]. The Trojan activated through edge of internal signal, *i.e.,* based on counter value is asynchronous Trojan [4]. The combination of both synchronous and asynchronous is hybrid Trojan, i.e., activation by edge of internal signal and clock signal [4].

The havoc of Trojan is spreading among all streams, i.e., in medical, government, communication, commercialization, defense, etc. The testing of Trojan is done by two methods: side channel and logical testing. In logical testing, the arbitrary input waveforms are given to activate Trojan. In complex circuit, so many nodes are there and to check each point is quite cumbersome and makes logical testing less effective. In side channel analysis, the changes in parameters like power, delay, thermal, current, etc, are compared with golden reference values. Any deviation between calculated and golden value depict presence of Trojan. Side channel analysis faces the menace of process variations [11].The HT can be generated by using if else or switch case statement [9]. The classification of Hardware Trojan explained in Fig. 2.

Fig. 3 Flowchart of
proposed scheme

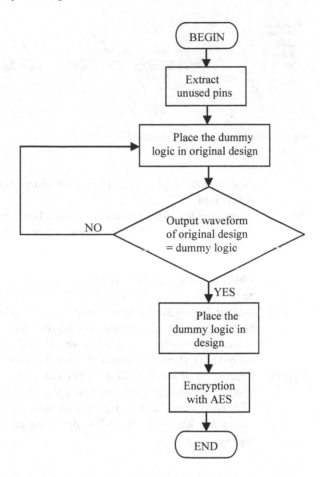

4 Proposed Technique

The HT is a malicious intruder that consumes negligible area and power. The user remains unaware when HT occupies the unused spaces. The HT presence is known when Trojan triggers. This technique provides solution to the problem. The proposed technique is implemented on Xilinx Automatic Sparton3 device XA3S50, Package VQG100 and speed-4 in Xilinx 10.1 ISE. The outlay of the proposed method is shown in Fig. 3. The objective for prevention of HT includes two steps

1. To utilize the unused pins by adding dummy functional logic that don't affects the functionality of the system.
2. Encrypt the modified design with AES (Advance Encryption Standard) 128 bits.

These steps are achieved via using following procedure:

Step I: *Coding*: The code of desired system is implemented on the Xilinx ISE tool. The information of standard cells for cells geometry, macro cells used, pin

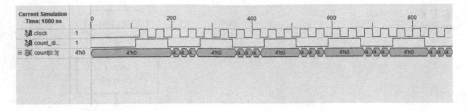

Fig. 4 Circuit waveform with inserted HT

used or unused pins as in [7] are obtained through design summary in Xilinx tool.

Step II: *Identifying the unused resources (pins)*: The used pins in Xilinx FPGA are obtained by *name.ncd* (native circuit description). The ncd file is extracted through the Xilinx process window in View/Edit Routed Design (FPGA Editor) under Generate Post-Place and Route Static Timing. The pins which are not used in ncd file are unused resources where Trojan can be placed.

Step III: *Adding dummy logic*: The dummy logic, i.e., a logic which does not affect the functionality of the circuit, the output remains the same. But the number of used pins has increased as shown in Fig. 6 due to which unused spaces are reduced. Hence the vulnerability to HT is reduced at unused resources.

Step IV: *Encryption of modified design*: HT can replace the dummy logic in the system. To prevent such condition the modified design encrypts using AES (Advance Encrypted standard) algorithm of length 128 bits using symmetric key [12]. The AES algorithm increase throughput provides high security.

5   Results and Discussion

This segment shows the result of proposed scheme explained in previous section. The proposed scheme is implemented on Automatic Sparton3 device XA3S50, Package VQG100 and speed −4 in Xilinx 10.1_ISE. The results are calculated on Xilinx tool 14.5 ISE. The benchmarks of HT are taken from [13, 14]. In the up–down counter when Trojan is inserted, the output is changed arbitrarily as shown in Fig. 4. The changed outputs create ambiguity for user, as user is not able to find the reason for waveform distortion. The adversary takes benefit of such situation to degrade the system performance.

The original design used few pins due to which the Trojan could easily insert without changing system functionality. The original used pins layout is shown in Fig. 5.

	Name	Site	Type	#Pins
1	CLOCK_BUFG	BUFGMUX	BUFGMUX	3
2	COUNT_OUT	P92	IOB	1
3	DIRECTION	P91	IOB	1
4	CLOCK	P90	IOB	1
5	COUNT_OUT	P89	IOB	1
6	COUNT_OUT	P88	IOB	1
7	COUNT_OUT	P87	IOB	1
8	count_int<3>	SLICE_X6	SLICEL	10
9	count_int<2>	SLICE_X6	SLICEL	10
10	count_int<0>	SLICE_X7	SLICEL	4

Fig. 5 Used pins of original circuit

	Name	Site	Type	#Pins
1	CLOCK	P90	IOB	1
2	CLOCK_BUFG	BUFGMUX	BUFGMUX	3
3	COUNT_OUT	P92	IOB	1
4	COUNT_OUT	P87	IOB	1
5	COUNT_OUT	P88	IOB	1
6	COUNT_OUT	P89	IOB	1
7	DIRECTION	P91	IOB	1
8	count_int<0>	SLICE_X7	SLICEL	4
9	count_int<2>	SLICE_X6	SLICEL	10
10	count_int<3>	SLICE_X6	SLICEL	10
11	dummy<0>	P23	IOB	1
12	dummy<1>	P22	IOB	1
13	dummy<2>	P21	IOB	1
14	dummy<3>	P17	IOB	1
15	dummy<4>	P16	IOB	1
16	dummy<5>	P15	IOB	1

Fig. 6 Used pins of modified circuits

The design is modified by inserting dummy logic in the circuit so that the number of used pins is increased. The dummy logic has no affect on the output waveform as well on power consumption. The increased number of used pins is increased in Fig. 6. The dummy logic is used in the technique looks like following:

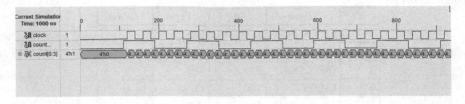

Fig. 7 Output waveform after modification

9RjCmqMFlCJzq7CYYDY4qZZt/4g0EDb1J4MCıQkГaEn32VDywqqpKXTgqGRKMgV+SDdr8kRYPoXE73qyl
ZgwKHIVCGX8PpR3PrOMBWtGPEjBqjsHURtwwndQ90EF59sAlcFvbqxNBtIPo1Rn+lgil++1TZSJyIU9w==

Fig. 8 Encrypted code

```
if clock ='1' and clock'event then
if direction = '1' then
 for i in 0 to 19 loop
 tmp (i) <= tmp (i+1) ;
 end loop;
else
 for j in 0 to 5 loop
 smp (j) <= smp (j + 1);
 end loop;
end if;
  end if;
```

After the dummy logic insertion the output waveform remains same as of original circuit. The output waveform is shown in Fig. 7 is similar to the original design which shows there is no change in functionality after insertion of dummy logic.

Hence, at last the encryption of this modified design is done in order to prevent form replacement of dummy logic. The encrypted code is shown in Fig. 8.

Therefore, the system is prevented from Trojan insertion. The throughput and security of system is increased not at cost of extra circuitry requirement.

6   Conclusion and Future Scope

In this paper, focus on menace of HT placement at unused resources is resolved by reducing unused pins in FPGA. The encryption of modified design is done using AES encryption algorithm. In this way possibility of Trojan placement at unused resources is reduced. The AES algorithm will further prevent the code from replacement of dummy logic with Trojan. The proposed method used does not require extra circuitry.

Experimental results show that proposed method increased the throughput and gave high security to the system. Hence this method gives a better approach of prevention from HT. This work can further be expanded to explore impact of encryption on HT prevention.

References

1. Li, H., Liu, Q., Zhang, J.: A survey of hardware Trojan threat and defence. Int. VLSI J. **55**, 426–437 (2016)
2. Xiao, K., Tehranipoor, M.: BISA: Built-in self-authentication for preventing hardware Trojan insertion. In: 2013 IEEE International Symposium on Hardware-Oriented Security and Trust (HOST), Austin, TX pp. 45–50 (2013)
3. Lamech, C., Rad, R.M., Tehranipoor, M., Plus quellic, J.: An experimental analysis of power and delay signal-to-noise requirements for detecting Trojans and methods for achieving the required detection sensitivities. IEEE Trans. Inf. Forensics Secur. **3**, 1170–1179 (2011)
4. Ridhesh, M., et al. Detection and analysis of hardware trojan using scan chain method. In: 19th International Symposium on VLSI Design and Test (VDAT). IEEE (2015)
5. Wei, S., Potkonjak, M.: Self-consistency and consistency-based detection and diagnosis of malicious circuitry. IEEE Trans. VLSI Systems **22**, 1845–1853 (2014)
6. Seyed Mohammad Hossein, S., Morteza Saheb Zamani. Improving hardware Trojan detection by retiming. Microprocess. Microsyst. **39** 145–156 (2015)
7. Khaleghi, B., et al.: Fpga-based protection scheme against hardware trojan horse insertion using dummy logic. IEEE Embed. Syst. Lett. **7** 46–50 (2015)
8. Sanchita, M.-S., et al.: Hardware trojan attacks in fpga devices: threat analysis and effective counter measures. In: Proceedings of the 24th Edition of the Great Lakes Symposium on VLSI ACM, pp. 287–292 (2014)
9. Nandeesha, V., Schafer, B.: Hardware trojan detection in behavioral intellectual properties (ips) using property checking techniques. IEEE Trans. Emerg. Top. Comput. (2016)
10. Ziad, S., Tarek Ibn, M., et al.: Homomorphic data isolation for hardware trojan protection. In: 2015 IEEE Computer Society Annual Symposium on VLSI. IEEE (2015)
11. Samer, M., et al.: Classification of hardware trojan detection techniques. In: Tenth International Conference on Computer Engineering and Systems (ICCES). IEEE (2015)
12. Stallings, W.: Cryptography Netw. Secur., 3rd edn. Pearson, Boston (2017)
13. Tehranipoor, M., Karri, R., Koushanfar, F., Potkonjak, M.: TrustHub. https://www.trust-hub
14. Albrecht, C.: Iwls 2005 benchmarks. In: International on Workshop for Logic Synthesis (IWLS) (2005). http://www.iwls.org

Scrabble-O-Graphy: An Encryption Technique for Security Enhancement

Vaishnavi Ketan Kamat

Abstract The advancements of high end technology have enabled the people to connect and communicate around any corner of the world through the usage of Internet. Communication through such mass media includes exchange of information which can be personal, official, or confidential in nature. To ensure the secured transmission, the ability of the encryption algorithms have to be enhanced. This paper illustrates a novel technique "Scrabble-O-graphy" which showcases a different route for generating the secured cipher text, without the use of intensive mathematical computation. This paper describes how a simple game of scrabble can be of great importance for the generation of the complex cipher text.

Keywords Scrabble · Cryptography · Encryption · Decryption symmetric key
Variable · Dictionary · Cipher

1 Introduction

"Necessity is the Mother of Invention". This proverb suits best for the current scenario of information transmission involving Internet and its users. The number of Internet users currently have risen to a billion since its conception and so is the level of the security for such transmissions. This paper describes the scrabble game as a tool for generating a strong cipher text.

V. K. Kamat (✉)
Department of Computer Engineering, Agnel Institute of Technology and Design,
Assagao, Bardez, Goa, India
e-mail: vaishnavikunkolikerkamat@gmail.com

© Springer Nature Singapore Pte Ltd. 2019
P. K. Sa et al. (eds.), *Recent Findings in Intelligent Computing Techniques*,
Advances in Intelligent Systems and Computing 707,
https://doi.org/10.1007/978-981-10-8639-7_12

1.1 Scrabble Game

The scrabble game involves two to four players to place their set of tiles on the scrabble board which is divided into a grid of 15 × 15. Tiles are placed based on the word that a player wishes to form on the board. Words can be formed horizontally or vertically but not diagonally. A total of 100 tiles are available out of which 98 tiles are marked with letters and their respective points and 02 tile are blank without any points, which can be substituted for any letter. In the sequence of play, a player will form a word on the board only if the word exists in the dictionary [1].

1.2 Cryptography

"Kryptos" the Greek word for cryptography means "Secret Writing". Cryptography is a tool which encapsulates the original message and displays an unintelligible (cipher) message during message transmission and cryptanalysis is the art of analyzing and decoding the original message. Depending upon the nature of the algorithm symmetric (secret) key or asymmetric (public and private) keys can be used in encryption and decryption process [2].

1.3 Cryptanalysis Attacks

A cryptanalyst depending on what information is available, can lead to following set of attacks:

(a) Cipher Text Only Attack (b) Known Plain Text Attack
(c) Chosen Plain Text Attack (d) Chosen Cipher Text Attack

In this paper, it is illustrated how the designed scrabble cipher holds strong enough against the two main types of attacks, i.e., Cipher Text Only Attack and Known Plain Text Attack.

2 Scrabble-O-Graphy Ideology

Scrabble a game with 225 squares on the board, has lot of potential for generating different combinations of words horizontally as well as vertically. In this paper, some rules of scrabble are modified to make it convenient for the process of cipher text generation. This scrabble game will have number of tiles equal to the number of characters in the given plain text. The game will be played between two users. Any character from the plain text can be selected in the set of chosen letters. Game will finish only when there is no scope for any word formation on the board.

The question that needs to be answered as to why the game of scrabble for cipher text generation? Letters from the plain text will be selected randomly and altogether a different word existing in the dictionary will be formed, which does not have any relation directly to the plain text, i.e., there is no established pattern in selecting and forming the words. Two basic techniques are successfully utilized in this game "Substitution and Transposition". Substitution because any letter can be replaced randomly by any other letter from the set, depending on what word is formed. Transposition because words can be placed anywhere on the board horizontally or vertically.

3 Scrabble-O-Graphy Algorithm and Illustration

Consider the text given in Fig. 1 to be the plain text to be encrypted and transmitted. Figure 2 displays the information regarding letters and their corresponding values. Having a good vocabulary is a pre-requisite for this algorithm.

Step 1: Convert the accepted plain text into lowercase and mark the line numbers and location of each character in each line.
Illustration: There are 05 lines in the given plain text. Considering the first line, 'a' will be marked at location 1, 's' will be marked at location 2 and so on till 'y' marked at location 102. Every new line location will start from 1 till all letters in that line are covered.

Step 2 Each user/player will randomly select 08 set of letters each from the available plain text and form a valid word, which should be present in the dictionary used. Users will play sequentially. As the word is formed, first and second level keys should be generated simultaneously.
Illustration: Referring to Figs. 3 and 4, User1 has selected 08 letters namely "S A E F U H L I" and User 2 has selected letters namely "A C P Q T L I E". Starting with User1, the word "SHELF" is formed and simultaneously the key is formed with the letters that are utilized. Next User2 forms the

As I look around I see the crumbling ruins of a proud civilization strewn like a vast heap of futility. And yet I shall not commit the grievous sin of losing faith in Man. I would rather look forward to the opening of a new chapter in his history after the cataclysm is over and the atmosphere rendered clean with the spirit of service and sacrifice. Perhaps that dawn will come from this horizon, from the East where the sun rises. A day will come when unvanquished Man will retrace his path of conquest, to win back his lost human heritage.

Fig. 1 Input plain text

Letter	A	B	C	D	E	F	G	H	I	J	K	L	M	N	O	P	Q	R	S	T	U	V	W	X	Y	Z
Value	1	3	3	2	1	4	2	4	1	8	5	1	3	1	1	3	10	1	1	1	1	4	4	8	4	10

Fig. 2 Letters and their value

Sr. No.	Letter Set	Player 1 Word & Score	Key Generated	Letter Set	Player 2 Word & Score	Key Generated
01	SAEFUH LI	SHELF = 15	(1,20)(1,87)(3,131)(2,48)(4,47)	ACPQTL IE	PIETA = 15	(3,114)(3,40)(5,102)(1,1)
02	AUIPVD PO	AVOID = 13	(5,24)(3,17)(1,56)(3,127)	CEQLLF IN	FIELD = 17	(4,29)(3,94)(5,19)(1,6)
03	AUPPIW HR	SHARP = 13	(5,92)(4,57)(3,147)(3,56)	CQLNSD LC	ILLS = 8	(1,33)(1,59)(5,30)
04	UPIWE OZA	ZOEAS = 24	(1,61)(2,45)(4,78)(5,36)	CQNDCM IE	CINDER = 18	(3,84)(1,76)(5,96)(4,14)(2,35)
05	UPIWQ ARE	PIQUE = 32	(5,54)(5,65)(1,46)(4,2)	CQMFAE IU	CAFÉ = 18	(2,21)(2,13)(3,45)
06	IWARN AUV	URIC = 07	(1,52)(1,12)(3,68)	QMEIUO HA	EQUIP = 18	(3,58)(5,27)(1,14)(2,42)
07	WANAV FDN	DEAVE = 13	(3,123)(3,13)(3,75)(3,98)	MOHAES VT	HASTED = 18	(5,98)(1,47)(2,39)(4,34)(1,88)
08	WANNC EPH	CHEAP = 39	(3,129)(1,71)(5,1)(4,6)	MOVOYT DW	STOW = 14	(3,77)(4,31)(5,74)
09	WNNHF AOR	DWARF = 14	(5,7)(3,101)(4,75)(1,95)	MVOYDG WR	FOGY = 33	(3,151)(2,32)(2,5)
10	NNHOO CIS	COIN + PI = 10	(1,28)(3,71)(4,20)	MVDWR OYR	ARMORY = 14	(3,14)(3,92)(4,25)(3,169)(1,102)
11	NNHOS YAB	BOY = 16	(5,78)(5,88)	VDWSGB NF	VOWS = 20	(3,157)(3,51)(2,11)

Fig. 3 Game sequence 1

Sr. No.	Letter Set	Player 1 Word & Score	Key Generated	Letter Set	Player 2 Word & Score	Key Generated
12	NNHSY AWO	SWAM = 27	(1,68)(2,24)(2,1)	DGBN FHWI	OW = 10	(3,135)
13	NNHY OSLK	KIN = 12	(3,19)(1,41)	DGBN FHIU	DISH = 08	(3,27)(2,61)(4,62)
14	NHYOS LVM	AVOIDS + SHY = 23	(1,42)(1,25)(3,90)	GBNF UIRI	SHYING = 39	(2,25)(3,163)(2,53)
15	NOLV MNOY	LONG = 06	(3,6)(1,92)(4,17)	BFUIR KIF	DUB = 18	(5,28)(1,32)
16	NVMO YDHC	BODY = 14	(5,13)(5,3)(3,73)	FIRKI FIC	OF = 13	(1,45)
17	NVMH CAOR	HAVOC = 42	(4,39)(5,47)(1,82)(3,30)	FIRKI ICE	CRAKE = 19	(5,62)(5,46)(5,81)(3,118)
18	NMCR NCOT	SON = 03	(4,44)(3,41)	FIIIF OEAT	IT = 02	(3,64)
19	MCRN CTOS	BOYS+SO = 11	(3,144)(2,18)	FIIFO EAT	OATH = 08	(5,103)(3,81)
20	MCRN CTIE	REQUIP + RECOIN = 25	(4,3)(3,155)	FIIFO EUM	ELONG = 06	(4,56)
21	MCNC TIAP	ASON = 04	(2,56)	FIIFO UMD	AM = 04	(2,24)
22	MCNC TIPE	IN + NAM = 07	(3,62)	FIIFO UDE	AD = 05	(1,53)

Fig. 4 Game sequence 2

word "PIETA". **This is the strong point of the algorithm. For example, the word "SHELF" has no relation to the words present in the plain text, so establishing pattern based on cipher text knowledge to trace the plain text is not possible because all the letters are selected randomly across the plain text.** Now 08 letters in a set are to be maintained, some more letters are fetched from the plain text by both the players. In a similar manner the rest of the game is played till no more words can be formed. In the illustrated example, each player has played maximum 22 times.

Step 3: As words are formed by each player, score is recorded and also the second-level key depending upon scrabble board rows and columns is formed.

Fig. 5 Complete scrabble board game

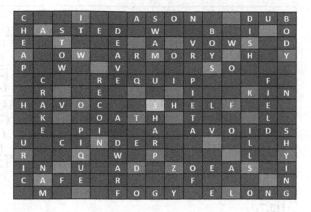

Fig. 6 Primary and final cipher text generated from the scrabble board

CSSISSASONSSDUB-
HASTEDSWSSBSISO-
ESISSESASVOWSSD-
ASOWSARMORYSHSY-
PSWSSVSSSSOSSS-
SCSSREQUIPSSSFS-
SRSSESSSSISSKIN-
HAVOCSSHELFSES-
SKSSOATHSTSSSLS-
SESPISSASAVOIDS-
USCINDERSSSSLSH-
RSSQSWSPSSSSLSY-
INSUSADSZOEASSI-
CAFÉSRSSSFSSSSN-
SMSSSFOGYSELONG

Box 1 : Primary Cipher Text

NSSDUBHASTEDSWS
SBSISOESTSSESAS
VOWSSDASOWSARMO
RYSHSYPSWSSVSSS
SSOSSSSCSSREQUI
PSSSFSSRSSESSSS
ISSKINHAVOCSSSH
ELFSESSKSSOATHS
TSSSLSSESPISSAS
AVOIDSUSCINDERS
SSSLSHRSSQSWSPS
SSSLSYINSUSADSZ
OEASSICAFÉSRSSS
FSSSSNSMSSSFOGY
SELONGCSSISSASO

Box 2 : Final Cipher Text

Illustration: Once the game is complete, as displayed in Fig. 5, to generate the cipher text read the text either row wise or column wise. The output cipher text generated in shown in the first box of Fig. 6.

Step 4: Once primary cipher text is generated, then to add a little bit of fuzziness to the cipher text, any basic encryption technique can be applied. In this paper, simple Ceaser cipher technique is applied.

Illustration: A key value equal to 6 is selected randomly between 1 to 225. Analyzing the second box in Fig. 6, where the cipher text is shifted by 6 places, the cipher text is completely different from the words present in the plain text, that to without any pattern visible between plain text and the cipher text. The second box in Fig. 6 shows the final cipher text.

(8,8)-	(6,10)-	(10,10)-	(6,14)-	(8,8)-	(10,13)-	(13,9)-	(11,3)-
(8,12)	(10,10)	(10,14)	(10,14)	(12,8)	(13,13)	(13,13)	(11,8)
(10,4)-	(14,1)-	(11,1)-	(6,6)-	(2,6)-	(2,1)-	(1,1)-	(2,3)-
(14,4)	(14,4)	(14,1)	(6,10)	(6,6)	(2,6)	(5,1)	(5,3)
(11,6)-	(15,6)-	(8,5)-	(10,4)-	(4,6)-	(2,11)-	(3,10)-	(1,8)-
(15,6)	(15,9)	(11,5)	(10,5)	(4,11)	(4,11)	(3,13)	(4,8)
(4,3)-	(7,13)-	(1,13)-	(10,10)-	(10,15)-	(10,15)-	(15,12)-	(1,13)-
(4,4)	(7,15)	(4,13)	(10,15)	(12,15)	(15,15)	(15,15)	(1,15)
(1,15)-	(13,10)-	(8,1)-	(6,2)-	(1,8)-	(1,4)-	(5,11)-	(2,11)-
(4,15)	(14,10)	(8,5)	(10,2)	(1,10)	(2,4)	(5,12)	(5,11)
(9,5)-	(6,5)-	(6,5)-	(15,11)-	(1,7)-	(14,2)-	(13,1)-	(13,2)-
(9,8)	(6,10)	(11,5)	(15,15)	(1,10)	(15,2)	(13,2)	(15,2)
(13,6)-							
(13,7)							

Fig. 7   s level key

In Fig. 5 green color tiles are normal tiles, red colour tiles are triple word score tiles, pink color double word score tiles, light blue color tiles are double letter score tiles and dark blue colored tiles are triple letter score tiles.

4   Multilevel and Variable Key Generation

This algorithm makes use of **multilevel keys**. At the first level key is generated for each word. Key size depends upon how many letters from the letter set are utilized. Length of key will vary depending on the utility of the letters. Each new letter used will have the key in following form (**Line no.,Letter Location**). For example, referring to Fig. 3 serial number 11, the word "BOY" is formed by User1. Now letter "Y" already existing on the scrabble board due to the previous word "ARMORY". So, when the key is formed for the word "**BOY**" its **length** will be **02**, i.e., [(5, 78) (5, 88)]. The key value for letter "Y" will be formed for the word "ARMORY".

The first-level key is given in the pair form and the total length of the key will be equal to the number of letters present on the completed scrabble game board.

The second-level key as shown in Fig. 7, is used for tracking the position and the length of the word on the scrabble board. This key is also in a double pair form [{**starting row no.,col no.**} - {**ending row no., col no.**}].

The third-level key is a simple key, used for Ceaser Cipher, i.e., any number between 1 to 225.

In this algorithm these keys play an important role, as the relationship between the plain and cipher text can be established only with the help of these keys. Since the letters are randomly selected and words are also randomly formed, that forms a cipher whose interpretation merely by analyzing it will be difficult.

Important characteristics of key generation are as follows:

15	15	13	17	13	8	24	18
32	18	07	18	13	18	39	14
14	33	10	14	16	20	27	10
12	08	23	39	06	18	14	13
42	19	03	02	11	08	25	06
04	04	07	05				

Fig. 8 Word score matrix by User1 and User2 sequentially

(1) Keys are used at multiple levels during encryption and decryption.

(2) The first-level key depends upon how many letters are there on the board.

(3) The second-level key depends upon how many words are there on the board.

(4) Both the keys cannot be formed by an attacker by merely guessing, as each player has a different way to execute the word formation.

(5) The third-level key may look like a simple key, but its value can range from 1 to 225. Decoding this key without the proper channel will consume a great amount of time.

For example, consider any cell from the given second-level key. Say **[(1, 1) – (5, 1)]**. The interpretation of the key is as follows. (1, 1) means (row no. 1, column no. 1) to (row no. 5, column no. (1). From this we can state that the word is vertically placed with a length of 05.

The third key only substitutes one character by other character depending upon the key value. The second level key provides information regarding the placement of words on the board along with their length. The first level key will tell exactly where is the actual position of the letter in the plain text.

Figure 8 displays the score for each word by each user sequentially.

5 The Decryption Process

(1) The user has to receive the key to decrypt the final cipher text to primary cipher text.

(2) Once the primary cipher text is obtained, then obtain second level key. Examine each cell of the key and compute the total score of the word. This computed score should match the cell in the Word Score Matrix as given in Fig. 8. If the score is matching than full word is retrieved.

Checking score is important because a user can create 2 words at the same time. For example, in Fig. 3, there are 02 words formed by User1, **COIN + PI** whose **total score** is **10**. Now, when key cell {**(8, 5) to (11, 5)**} is retrieved, and we do not compare it with the **Score Matrix**, then calculated value at the time of decryption will be **06**, whereas in the Score Matrix it is 10. If the scores calculated are not matching then use the next consecutive key cell, calculate the value of that cell and add it to the previous cell's total. Continue doing this till

Fig. 9 Partially decrypted
plain text

As I look around I see the crumbling ruins
of a proud civilization strewn like a vast
heap of futility And yet I shall not commit
the grievous sin of losing faith in Man I
would rather look forward to the opening
of a new chapter in his history after the
cataclysm is over and the atmosphere
rendered clean with the spirit of service
and sacrifice. Perhaps that dawn will come
from this horizon, from the East where the
sun rises. A day will come when
unvanquished Man will retrace his path of
conquest, to win back his lost human
heritage.

the calculated score does not match the Score Matrix Value. Score Matrix will have same number of cell as the total number of moves by each user.

(3) Next, step after retrieving the word is to find out its exact location in the plain text. For this part of decryption, the first level key will be required to map words to their respective letters.

The decrypted plain text is displayed in Fig. 9. Every scrabble game will encrypt certain set of letters. For the remaining set of letters, again the same strategy needs to be applied. The advantage of this approach is that, even after first phase of decryption, the plain text is not fully visible, as seen in Fig. 9 only bold black colored letters are visible as the plain text output.

6 Complexity and Analysis

The complexity of the Scrabble-O-graphy algorithm can be best analyzed with the help of the concept of Permutations and Combinations. To measure the game complexity, two types of complexity characterize board games. Computational complexity is based on the size of the problem space in a given game and includes measures such as state space complexity and game tree complexity, average branching factor and average game length [3].

Consider the two cryptanalyst attacks namely, Cipher Text Only attack, where the intruder has a part of cipher text and based on assumptions and algorithm knowledge it tries to decipher it and the secondly the Known Plain Text Only attack where intruder has some knowledge about the relation between cipher and plain text.

Scrabble-O-graphy algorithm is able to foil both types of attacks. In Cipher Text Only attack, even if the intruder gets a piece of cipher text, it is very difficult to relate what it is in the plain text, without the key. For example, the word "SHYING", such word is not at all available in the plain text and nor it is substituted for some other letters and words directly. Since the cipher text is not formed based on a particular pattern, establishing plain – cipher text pair is impossible, so it takes care of Known Plain Text attack also.

If the intruder wishes to crack the cipher text, it would require 225 attempts in the first place and still it will not be sufficient to tell whether what key is applied is the correct one or not because there is nothing fixed on the scrabble board with respect to formation of words.

Retrieving the scrabble board words will only convey that these letters are present in the plain text, but no information regarding their location is provided. If depending upon the letters retrieved, if combinations or permutations of letters are applied to recreate the original word, it will take great amount of time. Recreation of word will require to map words from the dictionary. The second edition of 20 volume Oxford English Dictionary contains 171,476 words in current use [4].

For the given example, scrabble board contains a total of 129 letters on the board. So, a combination of these words is $2^{129} = 680564733841876926926749214863536422912$ words, say fifty percent are rejected, as some words may not be present as part of vocabulary, it will consume ample amount of time to crack the cipher.

7 Conclusion

The Scrabble-O-graphy algorithm devised in this paper showcases a unique way to generate the encrypted text. Several permutations and combinations are involved in this algorithm but on letters in order to form words. Any word can be formed anywhere on the scrabble board by maintaining the connectivity of words. Substitutions for letters are not fixed. Even after having the letters from the plain text, an intruder will find it difficult and time-consuming to crack the word as all the letters are in a scrambled manner, in form of different words. This algorithm does not require the knowledge of complex mathematical functions, but knowledge of good vocabulary is required.

The first- and second-level symmetric keys are only generated during encryption, not used for any operation on the text, only the third key is used on the primary cipher text. Keys play a vital role in the process of decryption. Deciphering the cipher text without proper key values is very difficult.

In the scope for future work, improvements are to be done on increasing the security of the first and second level paired key, which comes under the area of Key Exchange.

This algorithm aims at providing secured and reliable encrypted text.

References

1. https://en.wikipedia.org/wiki/Scrabble
2. Bruce, K.B., Cardelli, L., Pierce, B.C.: Comparing Object Encodings. In: Abadi, M., Ito, T. (eds.) Theoretical Aspects of Computer Software. Lecture Notes in Computer Science, vol. 1281. Springer, Berlin, Heidelberg, New York pp. 415–438 (1997)
3. Gobet, F., Retschitzki, J., de Voogt, A.: Moves in Mind: The Psychology of Board Games. Psychology Press, pp. 26–30 (2004)
4. https://en.oxforddictionaries.com/explore/how-many-words-are-there-in-the-english-language

Vertex Magic Total Labelling and Its Application in Cryptography

Rahul Chawla, Sagar Deshpande, M. N. Manas, Saahil Chhabria
and H. K. Krishnappa

Abstract Vertex magic total labelling, VMTL, of a graph G is a labelling concept wherein for each vertex, the sum of all the weights of incident edges on a vertex and the weight of the vertex is a constant, and this constant is known as magic constant. VMTL is a labelling which generates constant, unique number patterns for a particular graph. Such labelling of a graph can be leveraged to develop a cryptosystem by mapping sensitive data on VMTL and encrypting the data. For every word in the input text, different labellings are achieved based on the characteristic of the text so that two same words in different messages have different encoded forms. A hash function is used for the same which takes in the date of text and size of the word to generate a hash value which is the order of the graph, for which VMTL is achieved, and then the word is mapped on it. Efficiency for generating labelling for the graph plays an important role; thus, low, odd ordered complete graphs are considered.

Keywords Vertex magic total labelling · Cryptosystem · Complete graphs
Hashing

R. Chawla (✉) · S. Deshpande · M. N. Manas · S. Chhabria · H. K. Krishnappa
Department of Computer Science and Engineering, Rashtreeya Vidyalaya College
of Engineering, Bengaluru 560059, Karnataka, India
e-mail: rahulchawla2801@gmail.com

S. Deshpande
e-mail: sagardespande.76@gmail.com

M. N. Manas
e-mail: manasmn@rvce.edu.in

S. Chhabria
e-mail: saahil.work@gmail.com

H. K. Krishnappa
e-mail: krishnappahk@rvce.edu.in

© Springer Nature Singapore Pte Ltd. 2019
P. K. Sa et al. (eds.), *Recent Findings in Intelligent Computing Techniques*,
Advances in Intelligent Systems and Computing 707,
https://doi.org/10.1007/978-981-10-8639-7_13

1 Introduction

It is a known fact that vertex magic total labelling, VMTL, is exhibited by all complete graphs, among other graphs. Labelling obtained through the process of VTML on a graph has applications in the domain of data mining and network security. We refer algorithm by Krishnappa, Srinath, Ramakanth from [1] to understand how a VMTL table is generated for an odd complete graph. Each column in the table stores the weights of all the incident edges on a vertex and weight of that vertex. The sum of all the weights in a column is a constant that is magic constant. The pattern of numbers generated is unique and constant with respect to a particular graph, and thus, mapping and unmapping of text on such stream of numbers are easily achievable. Sender and receiver use the same algorithm to encrypt and decrypt the messages. The computation speed of VMTL table for odd ordered complete graphs is evaluated on a regular system and a graph of number of vertices versus time is plotted, and it is quite evident that computation speed grows in an unproportionate manner. Only odd ordered graphs are considered and used for mapping as achieving labelling for it is simpler than even ordered graphs.

Depending on the characteristic of the input text example word length, date of creation the stream of numbers is generated, which is nothing but labellings of a particular order of complete graph. The cryptosystem considered here involves a hash function which for each word in the input text returns the order of the graph and then that word is mapped on the labellings of the graph fetched. As these characteristics are constant even after encoding, the receiver at the receiving end can use a similar hash function to decode the message.

2 Related Work

The research on applications of vertex magic total labelling in the field of computer science is relatively new. Few algorithms exist to achieve labelling on different types of graphs, although our research is limited to dealing with labelling of only complete graphs and applying in cryptography. Cryptography is a domain in computer science now, but it existed way before the onset of Internet and was facilitated by mathematicians working on mathematical functions to form ciphers. New algorithms have been introduced in recent years which contribute to the efficiency and throughput of the cryptosystem, like algorithm explained by Nilesh and Nagle in [2]. Cryptography can be broadly classified in two ways: first is to convert the sensitive data to cipher text using a cipher which both sender and receiver side share, and the second is to hide sensitive data in audio-video files as explained by Praveen and Arun in [3], a technique called as steganography. Our technique falls into the first category which deals with conversion of sensitive text to an encoded form.

The algorithms to compute vertex magic total labelling are well known which include the method proposed by authors in [1] that deals with labelling of complete graphs. They propose two algorithms to construct the VMTL table, one for each odd and even number of vertices in the graph. Security has always been a concern since the onset of Internet till today as full assurance is not achieved by conventionally used RSA algorithm to obtain perfect secrecy. An extra mapping that will make the cryptosystem more robust as explained by authors in [4] is necessary as double-layered encryption is hard to decipher without a key if data packet is sniffed in between the transmission. Multithreading is used for dividing the encrypting process among threads also providing them with locks while dealing with shared variables in order to make a significant impact on computation speed as described by Balasubramani and Subba Rao in [5].

3 Methodology

The process involves encrypting a message by mapping the text on the content of the vertex magic total labelling of different orders of graphs retrieved by feeding a key to a hash function, H1 as shown in Fig. 1, to generate hash values. The structure of the key and the hash function can be decided beforehand by both the communicating parties. In this case, let us assume key that includes the date of the creation of the message and word length of different words in the message. The hash function is

$$H(D, L) = D * 10 + (2 * L + 1) = \text{hashvalue} \tag{1}$$

where L is the word length, and D is the day of the month ranging from 1 to 31. The hash value returned in Eq. (1) will always be odd, and it is nothing but the order of the graph whose labelling is to be considered for mapping. For instance, the message is sent on 01/02/2015 and the word length is 4; the hash value becomes 19; thus, vertex magic total labelling of graph of order 19 is considered using the algorithm referred from [1]. The adjacency matrix has 19 * 19 = 361 different numeric values which can be used for encoding by mapping 3 61 different characters on it.

The ASCII code of the character in the input word acts as index in the adjacency matrix of the labelled graph to retrieve a code for the character. The boundary con-

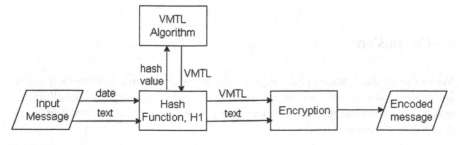

Fig. 1 Encryption

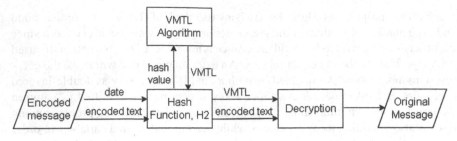

Fig. 2 Decryption

ditions are taken care of as the minimum hash value is 13, which contains 13 * 13 = 169, that is, 169 different characters can be mapped on the labelling which is greater than 128 different ASCII characters, that is, the range of characters in the input text.

As shown in the figure the sender side generates hash values and sends it to the script running the algorithm to compute the vertex magic total labelling. For each word, a thread is created so that the execution can take place asynchronously, and the time complexity is reduced. The labellings are returned to the sender side, encryption takes place by mapping the characters on labelling and then sent to the receiver. The receiver on the other side has a similar hash function, H2 as shown in Fig. 2 to decrypt the code.

4 Experimental Analysis

Execution time for algorithm expressed by authors in [1] to compute VMTL for odd ordered complete graph is recorded for different orders. This is done to check whether dynamically computing VMTL for every word while encrypting the message is feasible or not. The graph as shown in Fig. 3 is formulated for the computation time in seconds of labelling for different vertices. It can be clearly observed that the time taken for computing labelling for low orders is less and increases unproportionally for higher orders. It is quite evident that low order labellings can be easily computed dynamically while encrypting and will not hinder the efficiency.

5 Conclusion

Vertex magic total labelling of a complete graph produces a stream of numbers which are unique and constant for each graph. The efficiency of computing labelling for odd ordered complete grows unproportional thus using low odd ordered graphs for encryption is a feasible solution.

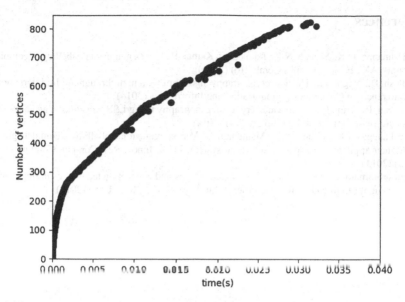

Fig. 3 Order versus time

This property helps in mapping sensitive text to the labelling and form encoded text which can be transferred to the receiver side with secrecy even if security breach takes place and packets get sniffed in between the transmission. Selection of a specific VMTL of a complete graph is determined by the size of the word which acts as input key to the hash function which returns order of the graph as hash value. Data is mapped on the labellings achieved on the ordered graph which is in the form of a stream of numbers.

6 Future Scope

The encryption of information in a way that same words have different encoded forms in same message depending on other factors such as length of the document, placement of word in the document, serial number of the original message, occurrence of data around specific keywords, date of the message transmission all together in considering the selection of ordered graph labelling thus making the cryptosystem robust.

Encryption of audio-video files can be achieved once the computation of labelling for high ordered complete graph is efficient as the size of such files is big and will require high ordered graph mapping. Caching of the stream of numbers for graph labelling at the sender and receiver side for a message rather than computing the labelling again and again for every word will be a huge step up in the complexity.

References

1. Krishnappa, H.K., Srinath, N.K., Ramakanth Kumar, P.: Vertex magic total labelling of complete graphs. AKCE Int. J. Graphs Comb. **1**(6) (2009)
2. Nilesh, D., Nagle, M.: The new cryptography algorithm with high throughput. In: International Conference on Computer Communication and Informatics (2014)
3. Praveen, P., Arun, R.: Audio-video crypto steganography using LSB substitution and advanced chaotic algorithm. Int. J. Eng. Invent. **2**(4) (2014)
4. Krishnappa, H.K., Srinath, N.K., Manjunath, S.: Vertex magic total labelling of complete graphs and their application for public-key cryptosystem. Int. J. Innov. Res. Comput. Commun. Eng. **1**(2) (2013)
5. Balasubramani, A., Subba Rao, Chdv.: Sliced images and encryption techniques in steganography using multi threading for fast retrieval. Int. J. Appl. Eng. Res. **11**(9) (2016)

Hybrid Traceback Scheme for DDoS Attacks

Vipul and Virender Ranga

Abstract In the modernized era, numerous types of attacks are espied on the Internet, along with the utmost destructive attacks, Distributed Denial of Service (DDoS) Attacks. After these types of attacks, legitimate users are not able to access the authorized services. IP Traceback scheme is the only way to trace the original source of the attack. Researchers have been implemented various traceback schemes in the past, but none are able to provide comprehensive efficient results in terms of marking rate and traceback rate. In this paper, we proposed a hybrid approach based on star coloring with autonomous systems, which depicts improved results in terms of marking rate and traceback rates. The simulation results are shown and compared with previous dataset CAIDA.

Keywords DDoS · IP traceback · ASN · Packet marking · Star coloring algorithm · Hybrid traceback scheme

1 Introduction

Many unlawful activities are committed on the Internet. DDoS attack is one of such activities. The originator IP address attached to the packet can be used by the attacker in order to mask himself from discerning in this type of attacks. The process of seeking the attackers is not that much easy because the attacker may be at any place in the world it may be thousands of kilometer away from the victim. In this paper, we have proposed a hybrid IP traceback scheme which will provide the better result as compared to other existing techniques in terms of marking rate. Autonomous system-based traceback schemes are used nowadays for improvement of traceback results.

Vipul · V. Ranga (✉)
Department of Computer Engineering, National Institute of Technology
Kurukshetra, Kurukshetra, Haryana, India
e-mail: virender.ranga@nitkkr.ac.in

Vipul
e-mail: Vipulmandhar130793@gmail.com

© Springer Nature Singapore Pte Ltd. 2019
P. K. Sa et al. (eds.), *Recent Findings in Intelligent Computing Techniques*,
Advances in Intelligent Systems and Computing 707,
https://doi.org/10.1007/978-981-10-8639-7_14

131

This technique is enhanced in our proposed model by coupling with Star coloring technique on the Autonomous system. Here we use pre-assigned Color of the router and ASN (Autonomous System Number) for healthier traceback.

2 Similar Work

Burch and Cheswick, earliest literature survey can be outlined to the concept of Network traceback [1]. Local Topology marking proposed by Ihab et al. [2], which is basically a hybrid traceback scheme, i.e., a combination of packet marking and packet logging schemes. Autonomous system-based marking is discussed by Paruchuri et al. [3]. DPM (deterministic packet marking) scheme was proposed by Belenky and Ansari [4], it has the problem with tracing beyond corporate firewall while PPM (probabilistic packet marking) was being proposed by Savage et al. in [5] it has low scalability and very low entropy variation, packet marking and packet logging. Various other improvements to PPM schemes have been proposed in [6, 7]. To improve the performance of PPM schemes various encoding techniques have been used, as Tabu marking Ma et al. [8], Huffman Codes is used by Choi et al. [9], and hybrid traceback is discussed by Yang et al. [10] which is basically a combination of more than one traceback schemes.

3 Proposed Approach

In our approach, instead of marking the traffic in the autonomous system with the IP address of a router, we mark the traffic with the AS Number and preassigned color of the router. An Autonomous System (AS) is a group of Internet Protocol networks managed by one or more than one network operators, which has solo and plainly defined independent routing guidance [1]. An Autonomous System Number (ASN) is a globally solitary number on the Internet to classify an Autonomous System. AS numbers are 32-bit integers [11], assigned and managed by Internet Assigned Numbers Authority (IANA). Autonomous System Border Routers (ASBRs) are connected to more than one AS, to interchange the information with routers i.e. routing information in other Autonomous Systems. The traffic to/from an AS is guarded by its ASBRs. These ASBRs has preferred color and ASN that is used as a mark and attached to the en-router IP packet. Finally, the attack path is constructed using these marks. Any graph G of the proposed scheme is a vertex coloring of G so no path of length 3 is bi-colored. Most of the existing packet marking execution requires a large count of packets to pool on the attack path but in this style, it reduces the aerial (overhead). Star chromatic no. of the graph G, denoted as $Xs(G)$ which is minimum no of colors required for star color G. Colors are pre-mentioned to each router and (worn) used as marks for attack path construction [12]. To trim the IP-header the colors are reused but at the same time, we can distinguish between routers by using

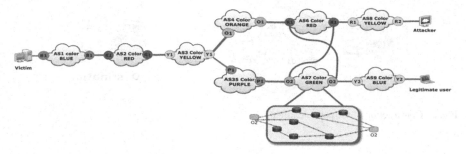

Fig. 1 Star coloring on border routers

some action of its unique ASN and IP address to trace back the actual source of the attack. The actual attack path built up by the legitimate user is thus a flow of colors and their Autonomous system number for better results. Given a flow of colors, Star coloring satisfies that from any distinct node the attack path can be variously traced as shown in Fig. 1.

3.1 Attack Path Construction

In this approach, each packet on their path to the destination is marked by the Autonomous System Border Routers (ASBR) with probability p. All the information collected in the packets from the traversed node of the attack path is stored and set by the victim. If any border router is d hop distance away from the victim then the point of victim swing the border router will be inclined with the probability $p(1 - p)^{d-1}$, where d is maximum node degree which is less than n, $(d \ll n)$. The range for the total number of colors is $[\Omega(d); O(d2)([13])]$. In star coloring algorithm, the single color can be repeated in an attack path and hence truthful separate the packets placed on the basis of their color and their count. If we see the actual attack path, red–yellow–orange–red–yellow (as shown in Fig. 1), the victim will accept the packets from border router Red 1 and border router Red 2 and from border router Yellow 1 and border router Yellow 2 pose same color data. To build up the attack path there must be the least number of packets received from the border router, i.e., closest to the attacker. The actual attack path will be constructed by ASN and distinct marks sorting in decreasing order from the victim's distance. The TTL field is used during reformation of the attack path [14]. If two nodes lie on two different paths then their TTL value must be same, otherwise different. The main problem with existing packet marking technique is successfully addressed by imprinting the fingerprint of the path in each packet as shown in Fig. 2. In this process, a path identifier referred to as *(path ID)* is embedded in the packet by the en-routers along with their color. To keep the path ID field size compatible with the color field we can either XOR or can use one's complement to compute the sum at each en-router. Victim when finally

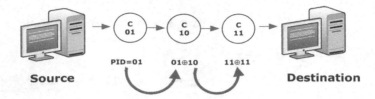

Fig. 2 Construction of the path identifier

receives the packet, it receives XOR summation as a mark with that packet. All this process is done on Autonomous Systems which are used for marking the packets for the traceback process, we knew that there are fixed 51,459 autonomous systems on the Internet as compared to millions of router [1]. The coloring scheme is done only on the autonomous system border routers but inside these autonomous Systems there are hundreds or thousands of routers, i.e., used for intradomain processes or internal packet passing in this proposed scheme we used one method on the basis of hop distance between the routers which are inside the same autonomous system to find the three shortest paths between the autonomous system border routers which are connected to each other so processing inside the AS will be fast and traceback time will also decrease.

4   Results and Discussion

Three cases are considered for the analysis of simulated network environment while a different number of attackers and benign users are created. Omnet++ network simulator is used for the simulation of these different cases as it provides a flexible and powerful tool to develop these scenarios. We have generated our results, i.e., traceback rate on the basis of marking rate. The simulation environment has poles apart components, which are connected to each other and depends upon each other functionality. Our proposed simulation environment has 24 routers and each router has pre-assigned color and no path of length 3 is bi-colored as shown in Fig. 3. We have a setup of approximately 100 users including benign users and attackers. The following parameters are considered for analysis of proposed approach.

Traceback rate: Traceback rate is basically the ratio of Attack packet with a total number of packets.

Marking rate: Marking rate is basically the ratio of Attack packets with all packets.

$$Traceback\ Rate = \frac{Traceback\ Packets}{Total\ attack\ Packets} \quad Marking\ Rate = \frac{Attack\ Packets}{All\ Packets}$$

Case 1: We have analyzed the traffic coming in our proposed simulation environment, where total number of users are approximately hundred from these there are Four malicious users that are connected to Router 22, which are sending attack

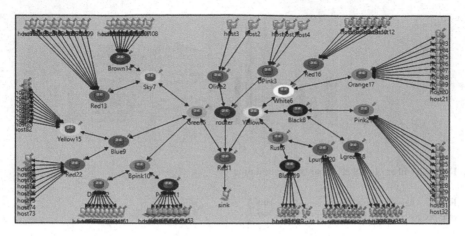

Fig. 3 Proposed simulation environment

Table 1 Details of attack packets coming from a single source

Router color	Time					
	10 s		30 s		60 s	
	Total packets	Attack packets	Total packets	Attack packets	Total packets	Attack packets
Red 22	545	255	1562	778	3130	1583
Blue 9	1098	255	3257	778	6571	1582
Green 5	3352	255	10,008	778	19,987	1582
Red 1	6484	255	19,429	778	38,973	1582

traffic to the overall simulated network environment, all the other users are benign and sending normal traffic. We are capturing a different kind of packets for 10 s, 30 s, and 60 s to analyze the simulated environment behavior during the attack as shown in Table 1. Figure 4 shows the plot between received packets and simulation time (s). It has been observed that the number of attack packets is increased by the system as simulation time is increased and the number of traced attack packet is more in the proposed scheme so high accuracy traceback rate of 99.87% is achieved.

$$Marking\ Rate = 0.080\%\ Traceback\ Rate = 99.87\%$$

Case 2: Similarly, we have analyzed the traffic with 8 malicious users that are connected to Router 13 and Router 21 and the results were marking rate = 0.178% and Traceback rate is 99.72% further information described in Fig. 4.

Case 3: In this case, we have analyzed the traffic with 16 malicious users that are connected to R13, R17, R21, and R22 and the results were marking rate = 0.35% and traceback rate is 99.44% further information described in Fig. 4.

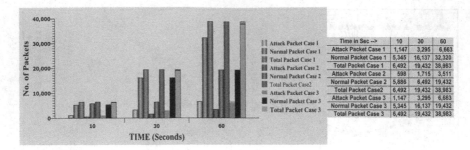

Time in Sec -->	10	30	60
Attack Packet Case 1	1,147	3,295	6,663
Normal Packet Case 1	5,345	16,137	32,320
Total Packet Case 1	6,492	19,432	38,983
Attack Packet Case 2	598	1,715	3,511
Normal Packet Case 2	5,886	6,492	19,432
Total Packet Case2	6,492	19,432	38,983
Attack Packet Case 3	1,147	3,295	6,663
Normal Packet Case 3	5,345	16,137	19,432
Total Packet Case 3	6,492	19,432	38,983

Fig. 4 Number of packets versus simulation time

Fig. 5 Analysis of traceback rate proposed of CAIDA data set

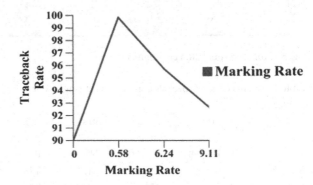

Fig. 6 Analysis of traceback rate of our proposed approach

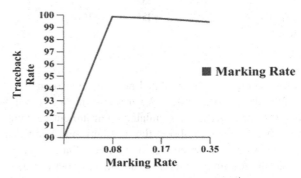

In our proposed scheme, we have considered the data set of CAIDA for comparison of results and it is shown in Figs. 5 and 6. It has been observed from Figs. 5 and 6 that in our proposed approach, the packet Traceback rate improves as Marking rate decreases. This improvement is better in our proposed approach because marking rate is less due to the marking of less number of attack packets in the system. However, marking rate is high in CAIDA dataset as compared to our proposed approach and consequently shows less accuracy in the system.

5 Conclusion

IP traceback is an impressive approach to find the actual path of network packets from where it is actually provoked. Different IP Traceback schemes like link testing, messaging, packet marking, packet logging have been used to trace back the attack. Some advanced and hybrid traceback schemes are also developed in the literature. But nonetheless is given an efficient solution in terms of computational overhead, bandwidth overhead, memory overhead, low traceback rate and false positive rate. Our proposed approach is proposed to trace the source of attack packet with high efficiency and low overhead, which is the limitation of the previously proposed approaches. Through experimental setup and evaluation, we have derived the results where marking rate is low, which further helps to achieve the high traceback rate.

References

1. Burch, H., Cheswick, B.: Tracing anonymous packets to their approximate source. In: Proceedings of USENIX LISA, Dec 2000
2. Ihab, F.H., Kesidis, G.: Performance of IP address fragmentation strategies for DDoS traceback. In: 3rd IEEE Workshop on IP Operations and Management, 2003 (IPOM 2003), pp. 1–7. IEEE (2003)
3. Paruchuri, V., Durresi, A., Barolli, L.: FAST: fast autonomous system traceback. In: 21st International Conference on Advanced Information Networking and Applications (AINA2007), pp. 498–505, May 2007
4. Belenky, A., Ansari, N.: IP traceback with deterministic packet marking. IEEE Commun. Lett. 7(4), 162–164 (2003)
5. Savage, S., Wetherall, D., Karlin, A., Anderson, T.: Network support for IP traceback. IEEE/ACM Trans. Netw. (TON) 9(3), 226–237 (2001)
6. Peng, T., Leckie, C., Ramamohanarao, K.: Adjusted probabilistic packet marking for IP traceback. In: International Conference on Research in Networking, pp. 697–708. Springer, Berlin, Heidelberg (2002)
7. Adler, M.: Tradeoffs in probabilistic packet marking for IP traceback. In: Proceedings of 34th ACM Symposium on Theory of Computing (STOC 2001)
8. Ma, M.: Tabu marking schemes for traceback. In: IPDPS (2005)
9. Choi, K.H., Dai, H.K.: A marking scheme using Huffman codes for IP traceback. In: 7th International Symposium on Parallel Architectures, Algorithms and Networks, 2004. Proceedings, pp. 421–428. IEEE (2004)
10. Yang, M.-H., Yang, M.-C.: RIHT: a novel hybrid IP traceback scheme. IEEE Trans. Inf. Forensics Secur. 7(2), 789–797 (2012)
11. Aghaei-Foroushani, V., Nur Zincir-Heywood, A.: Autonomous system based flow marking scheme for IP-Traceback. In: Network Operations and Management Symposium (NOMS), 2016 IEEE/IFIP, pp. 121–128. IEEE (2016)

12. Roy, S., Singh, A., Sairam, A.S.: IP traceback in star colored networks. In: 2013 Fifth International Conference on Communication Systems and Networks (COMSNETS), pp. 1–9. IEEE (2013)
13. Muthuprasanna, M., Manimaran, G., Alicherry, M., Kumar, V.: Coloring the Internet: IP traceback. In: 12th International Conference on Parallel and Distributed Systems-(ICPADS'06), vol. 1, 8 pp. IEEE (2006)
14. Yaar, A., Perrig, A., Song, D.: FIT: fast Internet traceback. In: Proceedings IEEE 24th Annual Joint Conference of the IEEE Computer and Communications Societies, vol. 2, pp. 1395–1406. IEEE (2005)

Dynamic Detection and Prevention of Denial of Service and Peer Attacks with IP Address Processing and Scrutinizing

Aditya Vikram Agarwal, Navneet Verma, Sandeep Saha and Sanjeev Kumar

Abstract Security has undeniably been a major concern for people in all aspects and implementations of computer science. Be it a website, database, or any other software, every application needs security to prevent any possible attack on them. One of the many popular attacks is Denial of Service (DoS), which is widely used to slow down, and even stop, many applications from processing any request. In this paper, we present a method, an algorithm, which can essentially prevent DoS and peer attacks like Distributed Denial of Service (DDoS). We use storage Data Structures like tries and process IP Addresses as they make any new request to the server. In multiple stages of IP processing and IP scrutinizing, any possible malicious request made by potential DoS attackers will be detected and, essentially, all normal requests will be separated, allowing them normal access, whereas, the attackers will be denied access to the application.

Keywords Cyber security · Denial of service · Distributed denial of service
UDP flood · ICMP flood · Algorithms · Searching techniques · Parallel
computing · Tries

A. V. Agarwal (✉) · N. Verma · S. Saha · S. Kumar
Department of Computer Science, School of Engineering and Technology,
Jaipur National University, Jaipur, India
e-mail: adityavagarwal.research@gmail.com

N. Verma
e-mail: navneetverma.cs@gmail.com

S. Saha
e-mail: sandeepsahajnu@gmail.com

S. Kumar
e-mail: sanjeev.network.bs@gmail.com

© Springer Nature Singapore Pte Ltd. 2019
P. K. Sa et al. (eds.), *Recent Findings in Intelligent Computing Techniques*,
Advances in Intelligent Systems and Computing 707,
https://doi.org/10.1007/978-981-10-8639-7_15

1 Introduction

Attacks are a severe threat to any machine, website, or any other application. With the advent of numerous technologies and with ever-growing size of applications, the number of loopholes in any software is bound to increase. This, of course, increases the vulnerability of any software, even if it is programmed with care and consistency. To aggravate the issue, there are even many tools primarily designed for attacking websites. As the sophistication of software increases, so does the number of possibilities of attacking it. In the computer age, with ever-increasing concern for cyber security, and now more than ever before, with increasing numbers of attacks on servers and applications, methods to counter those attacks are not only a need and necessity of servers but also improves their overall reliability, robustness, and quality.

There are various attacks that can, depending on the server, application, and the attack, be done. One of the most popular ways to attack any website or application running on a server is sending so many requests to the server that it cannot process them, due to the limit of processing requests per unit time being reached. When the server capacity to process the number of requests is reached, it either crashes or does not process any of the excess requests. This incorporates Denial of Service (DoS) attack, where any request beyond the server capacity is not processed. The problem, however, is that even requests made by normal users are not processed. The server denies processing any of the requests which it receives beyond its capacity [1].

With increasing popularity and easiness to attack websites, servers, applications, and other software, due to many applications being developed for this, undoubtedly, the number of attacks and the ways to attack any service application has increased. Even the number of ways to attack via DoS has increased [2]. There are many ways to do DoS attack, including ICMP (Internet Control Message Protocol) Flood, SYN/TCP (Synchronize/Transmission Control Protocol) Flood [3–5], UDP (User Datagram Protocol) Flood, and DDOS (Distributed Denial of Service) [2, 6].

There is a fixed bandwidth and server capacity to process requests, due to which only a limited amount of people can be served at a time. However, due to artificial bots made by hackers, in order to engage websites, and their servers, lots of traffic is produced on those websites. Due to this, actual people who want to be served cannot be served and website server goes down. We present an algorithm that will essentially help the Internet society (World Wide Web) by stopping DoS and such peer attacks. In this algorithm, we have many phases in the presence of which, if any attacker wants to attack websites, and in turn their servers, then their IP Addresses pass through these phases, and will be detected if are being requested by any artificial bot.

1.1 Internet Control Message Protocol Flood

In ICMP (Internet Control Message Protocol) Flood attack, a large number of IP packets with the source address appearing as the address of the victim are sent. Due to this, all the bandwidth is used by those IP packets and so, the actual person may not get any benefit from that service.

1.2 Synchronize/Transmission Control Protocol Flood

In SYN/TCP (Synchronize/Transmission Control Protocol) Flood attack, the host sends a flood of SYN/TCP packets, simultaneously, to the victim. However, their corresponding ACK (Acknowledgement) packets wait for a packet to get a response from the sender's address whereas the sender does not send any response [3–5].

1.3 User Datagram Protocol Flood

In UDP (User Datagram Protocol) Flood attack, a large number of UDP (User Datagram Protocol) packets is sent to random ports on a remote host. When so many UDP packets are sent, the victimized system will be forced into sending many ICMP packets in response to those UDP requests. Hence, most ICMP responses to actual people who make normal requests are not processed.

1.4 Distributed Denial of Service

In DDOS (Distributed Denial of Service) attack, multiple systems perform a DoS attack on a single target. This even makes a forceful shutdown of a targeted system. It is one of the most powerful peer DoS attack, as due to the presence of more than one system, the strength of the attack is considerably increased [6].

2 Overview

A lot of websites, search engines, and their servers are dealing with DoS and its peer problems. So, we come to the decision that we must take this step forward and essentially prevent these types of attacks. We use a comprehensive algorithm to process every IP Address that makes a request to the server and analyze the number of times the same IP Address makes requests in a unit time. Using efficient storage Data Structures to insert, search, and delete these IP Addresses makes our work easier. One such Data Structure is Tries. It is, in fact, a tree in which each individual node stores only one element of a string. The elements, we store in our algorithm, are the digit (0-9) of a number which represents the IP Address of any system that makes a request to the server. This is in contrast to the Dictionary Sorting Algorithm where the alphabets (A–Z) of a word are stored in the nodes instead of the digits (0–9) of the incoming IP Addresses. As every node has some value and we want to know if the incoming IP Address is already stored in the Trie, in order to check if the number of requests made by the current IP Address, we trace the path of the IP Address in the Trie from the root to the leaf node. The count, representing the number of times the IP Address has come, for each IP Address and the time in which that count is determined, are used to determine the count rate (count/time) of that IP Address. This count rate is used with a standard count rate (CR), such that if the count rate generated by the incoming IP Address is more than CR, we detect a possible, if not definite, DoS attack. However, we are yet not sure if the requests made by that IP Address are normal or if they are a part of a DoS attack on the server. So, we have multiple phases of IP processing and scrutinizing where we check if the IP Address is malicious. Each of these phases is stricter than the previous in terms of checking the lag generated by the incoming IP Address and if it can pass the deliberate lag generated by the server. If the IP Address is normal and one time unit has passed with the count rate generated by that IP Address less than the CR of that phase, then the IP Address is removed from the Trie.

This algorithm has an impressive time complexity seeing the number of possible IP Addresses, which in the case of Internet Protocol Version 6 (IPv6) is 2^{128} and in the case of Internet Protocol Version 4 (IPv4) is 2^{32}. Every operation including searching, insertion, and deletion is considerably fast and cheap and irrespective of the number of incoming IP Address, our algorithm will robustly and reliably detect each IP Address with a possible DoS Attack.

3　Algorithm

In order to effectively detect any possible intent of DoS attack while separating normal benign users, we have a discerning implementation of our algorithm. We also apply a stricter IP processing and scanning for dubious IP Addresses and come to a conclusion if an IP Address is really malicious only after rigorous processing of possible DoS attack IP Addresses. Thus, we have multiple phases for processing IP Addresses with each successive phase have an even more discerning way of processing the IP Addresses that have been filtered out by the preceding phases and are possible suspects of a DoS attack.

Every IP Address is first converted to its decimal form its corresponding binary forms. So, for an IPv4 Address (similarly for an IPv6 Address) of the form XXXX.XXXX.XXXX.XXXX in decimal, first it is converted to its binary form by taking each of the four decimal digits, separated by periods, in sets. These four binary numbers are then merged (appended), representing a single binary number, and, then the decimal form of the resulting binary number is computed. For our algorithm, however, we use IPv6 into consideration due to its tougher implementation. So, if the number of digits in the decimal form is less than 39 (the maximum number of decimal digits possible in case of an IPv6 Address), then additional zeroes are appended to the left so as to make the number of digits to be 39 (or 10 in the case of IPv6 Addresses).

For example, consider an IPv4 Address 192.168.15.254, which is converted into its binary form by taking sets of decimal numbers separated by periods. The resulting binary form of the IP Address is 11000000.10101000.00001111.11111110, which is appended thereby resulting in 110000001010100000001111111111110. Now, this binary number is converted into its decimal form, the effective IP Address, which is 3232239614. This form of the IP Address is used for every processing in the algorithm instead of its binary form. However, for the rest of our paper, we consider IPv6 Address by default instead. This is in order to chalk out the worst possible scenario that could be dealt with by the algorithm. For an IPv4 Addressing mode, the algorithm can be modified a bit just by changing the length of the IP Address to 10 instead of 39 (Fig. 1).

Every resulting IP Address is processed in a Trie that we maintain for each of our phases. The Tries present in every phase have similar structures. In each of the phases, the Tries have their internal node, other than their head node, representation of an array (as shown in Fig. 2) with the Index '0' of the array containing a digit of the IP Address and Indices '1' to '10' of the array containing references of their next nodes if they store the digits 0 through 9 and contain the reference of NULL otherwise. However, the head nodes of the Tries have an array representation where the Indices '0' to '10' of the array contains references to their next nodes if they store the digits 0 through 9 and NULL otherwise. Since the head nodes themselves do not store any digit of any IP Address, hence there is no need of an extra array index for the head nodes. Finally, every leaf node of the Tries in each phase has its node representation of an array (as shown in Fig. 3) with Index '0' containing the last digit of the IP Address, Index '1' containing the count of the number of times the IP Address has made a request, Index '2' containing the time in which the count has been determined, and Index '3' containing the reference to its previous node. Whenever an IP Address makes a request to the server implementing this algorithm, it is processed in the Phase 1 of the algorithm.

3.1 Phase 1

Every IP Address that makes a request to a software or an application, and in turn their servers, is processed in this phase, irrespective of it being a possible DoS suspect or not. We have a standard CR, denoted by α for Phase 1, above which the IP Address may be malicious and below which the IP Address is normal. In this phase, first it is checked if the incoming IP Address is already present in T1 (the Trie of Phase 1). If it is not already present, the IP Address is inserted in T1 with its leaf node containing the count value (C1) initialized to 1 and the time value (ψ1) initialized to 0. Whenever an IP Address which is already present in T1 makes a request, then T1 is traversed to its leaf node corresponding to the incoming IP Address and its C1 is increased by 1. Thereby, the requests are processed by the server. However, ψ1,

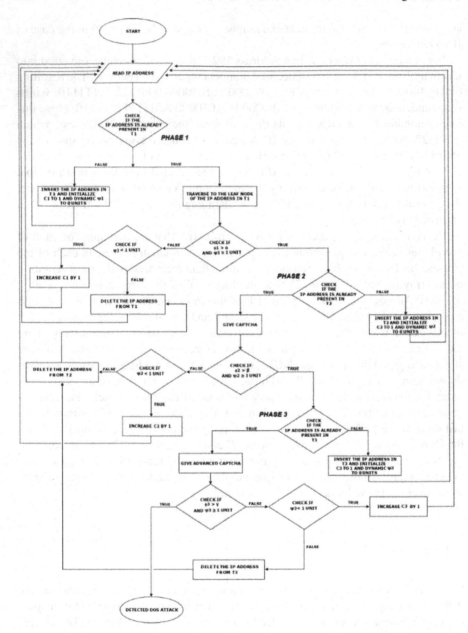

Fig. 1 Control Flow for the DoS and peer attacks prevention Algorithm where incoming IP Addresses are scrutinized for a potential DoS attack and if they are detected to be definitively malicious then they are processed accordingly and prevented access to the application and, in turn, the server

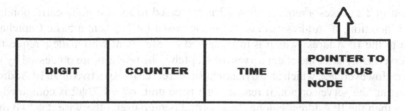

Fig. 2 An internal node's, other than the head node's and the leaf node's, array representation in the Tries of every phase

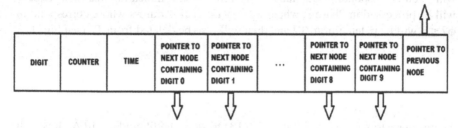

Fig. 3 A leaf node's array representation in the Tries of every phase

which dynamically increases irrespective of an IP Address requesting the server or not, if reaches one time unit, then $\phi 1 = C1/\psi 1$ is computed. If $\phi 1 \le \alpha$ then the IP Address belongs to a normal benign user, otherwise if $\phi 1 > \alpha$ then that IP Address is doubtful and is to be further processed. So, in this phase, the IP Addresses having their corresponding $\phi 1 > \alpha$ detected, will only go to Phase 2 with their corresponding $\psi 1$ values frozen to one unit indicating that their requests will be processed in Phase 2, whereas every other IP Address whose corresponding $\phi 1 \le \alpha$ will be deleted from T1 (see Fig. 1).

3.2 Phase 2

Since the Trie we use in this phase (T2) to store the filtered out IP Addresses in Phase 1 is similar to T1; it is also efficient in complexity, cheap, reliable, and robust. This Trie is used to process IP Addresses which are possible suspects and need further processing. Each such IP Address whose corresponding $\phi 1 > \alpha$ is processed in this Phase instead of Phase 1. In this phase, the CR corresponding to Captcha insertion, denoted by β, is such that above which the IP Address is malicious and below which the IP Address is normal. If an IP Address making a request is not already present in T2, then it is inserted in T2 with its corresponding count value (C2), the number of times the IP Addresses has made a request to the server, stored in the leaf node initialized to 1 and the time value ($\psi 2$), denoting the time in which the requests were made, initialized to 0. Similar to Phase 1, whenever an IP Address which is already present in T2 makes a request, then T2 is traversed to its leaf node corresponding to the incoming IP Address and its C2 is increased by 1. Then, a basic Captcha is given to the IP Address which is to be served by the client who made a request to the server. Only after the client serves the Captcha, the requests are processed by the server. However, $\psi2$, which also dynamically increases irrespective of an IP Address requesting the server or not, if reaches one time unit, $\phi2 = C2/\psi2$ is computed. If $\phi2 \leq \beta$ then the IP Address belongs to a normal benign user, otherwise if $\phi2 > \beta$ then that IP Address is even more doubtful and is to be further processed. So, in this phase, the IP Addresses having their corresponding $\phi2 > \beta$ detected, will only go to Phase 3 with their corresponding $\psi2$ values frozen to one unit indicating that their requests will be processed in Phase 3, whereas every other IP Address whose corresponding $\phi2 \leq \beta$ will be deleted from T2 and then will also be deleted from T1 (see Fig. 1).

3.3 Phase 3

In this phase too, we use a similar Trie (T3) to store the filtered out IP Addresses in Phase 2, which is similar to both T1 and T2. This Trie is used to process IP Addresses which are even more possible suspects and need further processing and scrutinizing. Each such IP Address whose corresponding $\phi2 > \beta$ is processed in this Phase instead of Phase 1 and Phase 2. In this phase, the CR corresponding to Advanced Captcha insertion, denoted by γ, is such that above which the IP Address is of an actual DoS attack and below which the IP Address is normal. If an IP Address making a request is not already present in T3, then it is inserted in T3 with its corresponding count value (C3), the number of times the IP Addresses has made a request to the server, stored in the leaf node initialized to 1 and the time value ($\psi3$), denoting the time in which the requests were made, initialized to 0. Whenever an IP Address which is already present in T3 makes a request, then T3 is traversed to its leaf node corresponding to the incoming IP Address and its C3 is increased by 1. Then, an Advanced Captcha is given to the IP Address which is to be served by the client who made a request to the server. Only after the client serves the Advanced Captcha, the requests are processed by the server. However, $\psi3$, which also dynamically increases irrespective of an IP Address requesting the server or not, if reaches one time unit, $\phi3 = C3/\psi3$ is computed. If $\phi3 \leq \gamma$ then the IP Address belongs to a normal user, otherwise if $\phi3 > \gamma$ then that IP Address is malicious and is of a DoS attack, which is reported and strict actions are taken against it. It can also be prevented access to the application or software, and in turn the server. So, in this phase, the IP Addresses having their corresponding $\phi3 > \gamma$ detected, will be prevented access and reported as a DoS attack with it being deleted from all T3, T2, and T1, in order. Every other IP Address, whose corresponding $\phi3 \leq \gamma$, will also be deleted from T3, and then from T2, and then also from T1 (see Fig. 1).

4   Analysis and Complexity

The design of our algorithm keeps time and space complexity primarily in check along with working reliably, robustly, and consistently. This is impressive considering the possible number of IP Addresses making a request to the server implementing this algorithm. Every operation performed on every IP Address in this algorithm is considerably cheap considering the importance of server time and is designed to work in parallel and in cohesion with the other operations of the algorithm. For each IP Address, only three operations are performed in addition to extra processing of Catptcha insertion in Phase 2 and Advanced Captcha insertion in Phase 3. These three primary operations are Inserting an IP Address into the Tries, Searching for the IP Address in the Tries, and Deleting the IP Address from the Tries (see Fig. 1).

4.1   Inserting an IP Address

In every Trie, there is only one root node from which there are multiple branches. Due to the constant sizes of IP Addresses, the height of the Trie is same as the number of digits in the IP Address, which we denote as η (39 in the case of IPv6 Addresses) for generality. Insertion of an IP Address is just the checking of the branch nodes of the current node. If a branch node has the next digit of the IP Address, then the processing goes to that node. Otherwise, it makes a new node with containing the next digit of the IP Address [7, 8]. Hence, the time complexity of insertion of the IP Address in a Trie is the number of digits in the IP Address, η, which is $O(\eta)$. Since η is 39 for IPv6 Addressing and 10 for IPv4 Addressing, so insertion of an IP Address actually takes constant time.

4.2   Searching an IP Address

Searching of any IP Address in the Trie starts from the root node. First, the immediate branches of the current node are searched. If the immediate next digit of the IP Address is found, traverses to it, otherwise detects that the IP Address is not present in the Trie. As the number of possible digits occupying a position of the IP Address is 10 (0–9), denoted by δ, so the number of possible checks made on the immediate next level is also δ. This makes the search complexity to a considerable minimum. Hence, the search complexity of an IP Address in the Trie is at most $O(\delta * \eta)$, which is if the IP Address is present in the Trie. Due to δ and η being very small, with δ being 10 and η being 39, the search complexity of an IP Address can instead be treated to be of a constant time. This is, in fact, a Breadth First Search (BFS) Algorithm with searching time complexity being dependent upon the breadth and the height of the tree [8].

4.3 Deleting an IP Address

IP Addresses whose $\psi 1$ reaches one and $\phi 1 \leq \alpha$, or $\psi 2$ reaches one and $\phi 2 \leq \beta$, or $\psi 3$ reaches one, are removed from T1; T2 and T1; T3, T2, and T1, respectively. Hence, during the removal of an IP Address from a Trie, a deletion operation is done on the Trie. This is in order to keep the space occupied by the algorithm in check. If a deletion operation is not performed, then IP Addresses are kept on inserting into the Trie, thereby increasing the number of nodes in the Trie, and, thus increasing the space complexity of the algorithm. Hence, between a minimum of one unit time and a maximum of three unit times, depending upon their $\phi 1$, $\phi 2$, and $\phi 3$ values, the IP Addresses are deleted. Deletion of an IP Address from a Trie starts from its leaf node in the Trie. So, first, the Trie is traversed to reach the leaf node of the Trie, and then the deletion operation of nodes of the IP Address one-by-one in the Trie is done. The time taken for deleting an IP Address is, in fact, reaching the leaf node of the IP Address in the Trie and then deleting the nodes of the IP Address, which is same as $O(\delta * \eta + \eta)$ or $O(\delta * \eta)$. Since both, δ and η, are very small, just like the searching operation, the deletion operation is instead treated to be of a constant time [7, 8].

5 Future Scope

Increasing numbers of cyber security breaches and increasing vulnerability of applications, softwares, and their servers, and now more than ever before, desperately needs some techniques to prevent any type of cyber threat. Techniques to prevent DoS and peer attacks like ICMP Flood, SYN/TCP Flood [3, 4], UDP Flood, DDoS attack [9, 10], and the like is definitively a boon to cyber security. The algorithm chalked out in this paper essentially frees servers from all such attacks. Due to the correctness and consistency of this algorithm and its learning of CR values α, in Phase 1, β, in Phase 2, and γ, in Phase 3, it can well be used in Deep-learning and Machine learning as components of Artificial Intelligence, in which instead of those α, β, and γ values, some other values can be learnt by taking incoming IP Addresses as inputs and generating desired outputs depending on those α, β, and γ values.

6 Conclusion

In the computer age, any algorithm, method, or technique that fortifies cyber security is a boon. The reliable, robust, consistent, cheap, and discerning algorithm presented in this paper, also belonging to the category of fortifying cyber security, is a definitive savior of the Internet, and in turn the servers, from any type of DoS and peer attacks like ICMP Flood, SYN/TCP Flood, UDP Flood, and DDoS. It is not only almost error-free but also can be implemented by any type of software and any type of server. The generalization of this algorithm not only for IPv4 and IPv6 Addressing modes but also for any other form of IP Addressing makes it even more dependable. So, the addition of this new algorithm to fortify cyber security will be a definitive good move as this algorithm is, in itself, an asset to cyber security.

References

1. Wang, X., Reiter, M.K.: Mitigating bandwidth-exhaustion attacks using congestion puzzles. In: Proceedings of the 11th ACM Conference on Computer and Communications Security (CCS), pp. 257–267 (2004)
2. Kumar, M., Kumar, N.: Detection and prevention of ddos attack in manet's using disable ip broadcast technique. Int. J. Appl. Innov. Eng. Manag. 2(7), 29–36 (2013)
3. Kalra, P., Pandey, K., Varshney, A.: Comparative analysis of SYN flooding attacks on TCP connections. Int. J. Inf. Comput. Technol. 4(3), 279–284 (2014)
4. Wang, H., Zhang, D., Shin, K.G.: Detecting SYN flooding attacks. In: Proceedings of the 21st Annual Joint Conference of the IEEE Computer and Communications Societies (2002)
5. Kavisankar L., Chellappan C.: A mitigation model for TCP SYN flooding with IP spoofing. Proceedings of the 2011 IEEE International Conference on Recent Trends in Information Technology (2011)
6. Specht, S.M., Lee, R.B.: Distributed denial of service: taxonomies of attacks, tools, and countermeasures. In: Proceedings of the ICSA 17th International Conference on Parallel and Distributed Computing Systems (PDCS 2004), 2004 International Workshop on Security in Parallel and Distributed Systems, pp. 543–550 (2004)
7. Askitis, N., Sinha, R.: HAT-trie: A Cache-conscious trie-based data Structure for strings. In: Proceedings of the 30th Australasian Conference on Computer, vol. 62, pp. 97–105 (2007)
8. Rao, J., Ross, K.A..: Making b+-trees cache conscious in main memory. In: Proceedings of 2000 ACM SIGMOD International Conference on Management of Data, vol. 29, no. 2, pp. 475–486 (2000)
9. Akiwate, B., Desai, M., Surpurmath, S., Khot, R., Power, D.: Detection and prevention of DoS attack. Int. J. Adv. Res. Comput. Sci. Softw. Eng. 6(5), 639–642 (2016)
10. Bhuyan, M.H., Kashyap, H.J., Bhattacharyya, D.K., Kalita, J.K.: Detecting Distributed Denial of Service Attacks: Methods. Tools Future Dir. Comput. J. 57(4), 537–556 (2014)
11. Bodon, F.: A Trie-based APRIORI implementation for mining frequent item sequences. In: Proceedings of the 1st International Workshop on Open Source Data Mining: Frequent Pattern Mining Implementations, pp. 56–65 (2005)

Comparative Study of ECG-Based Key Agreement Schemes in Wireless Body Sensor Networks

Saroj Kumar Panigrahy, Bibhu Prasad Dash, Sathya Babu Korra, Ashok Kumar Turuk and Sanjay Kumar Jena

Abstract Securing an individual's privacy in a proficient manner is an important task for critical infrastructures like Wireless Body Sensor Networks (WBSN). WBSNs provide real-time monitoring of patients. By protecting the WBSN, lives of patients or soldiers on battleground is secured. This paper gives a comparative study of electrocardiogram (ECG)-based key agreement schemes in WBSNs. The usage of ECG aims to bring plug-and-play capability in WBSNs, i.e., the deployment of sensors on the human body will enable wireless and secure communication for exchanging information. The schemes use Fast Fourier Transform (FFT) and Discrete Wavelet Transform (DWT) for feature extraction. The process is made secure by adding watermark. The watermarking technique does the locking and unlocking technique at the sender and receiver's side improving the security while not affecting the plug-and-play paradigm. The simulation and analysis is done by using ECG signal taken from MIT PhysioBank database.

Keywords Electrocardiogram (ECG) · Body Sensor Network (BSN)
Watermarking · Feature Extraction · Fast Fourier Transform (FFT) · Discrete
Wavelet Transform (DWT)

S. Kumar Panigrahy (✉)
Department of Computer Science and Engineering, Vellore Institute
of Technology Andhra Pradesh, Amaravati 522237, Andhra Pradesh, India
e-mail: skp.nitrkl@gmail.com

B. Prasad Dash · S. Babu Korra · A. Kumar Turuk · S. Kumar Jena
Department of Computer Science and Engineering, National Institute
of Technology Rourkela, Rourkela 769008, Odisha, India
e-mail: bibhudas96@gmail.com

S. Babu Korra
e-mail: ksathyababu@nitrkl.ac.in

A. Kumar Turuk
e-mail: akturuk@nitrkl.ac.in

S. Kumar Jena
e-mail: skjena@nitrkl.ac.in

© Springer Nature Singapore Pte Ltd. 2019   151
P. K. Sa et al. (eds.), *Recent Findings in Intelligent Computing Techniques*,
Advances in Intelligent Systems and Computing 707,
https://doi.org/10.1007/978-981-10-8639-7_16

1   Introduction

A Body Area Network (BAN) also called a wireless body area network (WBAN) or also referred to as Wireless Body Sensor Network (WBSN), is a remote or wireless network of wearable devices having computational ability. WBSN can be embedded inside the body, or implanted on the surface of the body. Development of this wearable technology started around 1995 on utilizing wireless personal area network (WPAN) technologies to communicate around the human body [1]. In the year 1996, Zimmerman [1] was the first person to conceive and implement the idea of WBAN. After 6 years, the term WBAN was used to refer such systems where communication is usually around the immediate proximity of human body [2]. Recently, interests in the use of WBAN have become impressively increased. Various numbers of small wireless sensors, placed on body create a network of wireless connection providing vital signs, real-time data to the client and medical authorities. These sensor networks ceaselessly monitor the health and well-being of patients to prevent and early hazard detection by offering the data to caretakers and doctors [3, 4]. The capability to monitor the health of a person in real-time environment is very important in case of emergency situation such as calamities, battlegrounds, and illness (diabetes, heart attacks, etc.). Recent advancements in electronics which are low powered have led to designing sensors in such a way that it can perform monitoring of patients. Absence of satisfactory security measures may not just prompt a rupture of patient protection privacy, additionally enable adversaries to alter real information bringing about wrong determination and treatment [5, 6]. Sensors are dependent on cryptographic keys to secure the communication of information. The electrocardiogram (ECG)-based scheme of Key Agreement uses ECG signals for generation of cryptographic keys [7]. The scheme allows the sensors to handle the data of sensors securely using cryptographic keys which are long, time invariant, and random. There is an ample amount of research work to be done and lot of scope available with respect to the challenges of the WBSN, i.e., security issues, management of data, data storage, etc. [8].

Challenges: As mentioned in [8], some of the key challenges concerning a WBSN has been mentioned as follows.

1. *Data Privacy and Confidentiality*: Since WBAN are being highly utilized and implemented in healthcare conditions, they are relied to be able to communicate securely with the patients data and privacy being intact.
2. *Data Integrity*: It should be mandatory or highly essential that the data should be protected from any type of changes or modification by application of required protocols.
3. *Data Freshness*: It assures the fact that old information is not recycled. So, it helps in avoiding replaying the previous data by the intruder and deceive the WBSN coordinator.

4. *Availability of the Network*: Network availability is essential for having the access to the information of a patient by the physician, which is very sensitive and has the potential of saving life.
5. *Data Authentication*: Data authentication is required for both medical and non-medical applications in a WBAN.

The rest of the paper has been organized as follows. Section 2 gives the necessary literature survey for the related works. Section 3 presents the implementaion of ECG-based key agreement schemes. Section 4 presents the results obtained from the schemes and their comparative performance analysis. Finally, Sect. 5 gives the conclusion.

2 Related Work

WBSN has been a important research area recently and a good amount of research is being done to make the security of the sensor data more efficient and secure. In this section, we discuss various techniques available in the literature.

Kumar et al. [9] client verification for medicinal services application is presented in which two factor (password and smart card) confirmation plans has been proposed. Keoh et al. [10] proposed a secure connection of biosensors with the management of key and the patients. The methodology is designed to cater to the security, confidentiality, and authenticity and is a highly trusted association among the biosensors and patients. For key administration, a novel plan is discussed which utilized key chains for gathering key establishment for the WBSNs. There have been hardware encryption, use of elliptic curve cryptography, and use of biometric methods proposed by Dimitriou and Athens [11]. Çamtepe and Yener [12] proposed a DWSN (Distributed wireless sensor network) scheme. In the given methodology, three methodologies have been received to convey key, which are probabilistic, deterministic or hybrid.

A biometric key agreement was proposed for WBSN by Yao et al. [13], in which ECG signal used to generate key is produced before the transmission of information for secure communication. The ECG that is used to generate key ensures authentication, data integrity, and confidentiality. The main drawback is being able to produce the same random signal from the sensors [14]. In [15], Mehmood et al. proposed an efficient key agreement scheme, which guarantees privacy and validation. The proposed scheme depends on symmetric cryptography and RSA cryptography. Communication among sensors for Physiological signals was first devised by Cherukuri et al. [16]. InterPulse-Interval (IPI) was being worked on the proposed idea for generation of cryptographic keys. Venkatasubramanian et al. [7] has proposed EKG-based key agreement schemes, which use FFT for feature generation. An enhanced version of the above scheme has been proposed by Ali and Khan [17] which uses DWT and other biometric features such as fingerprint or iris.

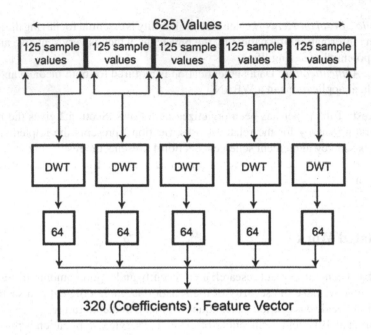

Fig. 1 Feature generation of ECG signal

3 Implementation of ECG-Based Key Agreement Schemes

EKG-based Key Agreement schemes have been compared that enables two sensors to agree upon a common key that is being generated. The schemes use DWT and FFT for feature generation. The Key agreement schemes have two main steps: *Feature Generation* and *Key Agreement*.

Feature Generation In a WBSN, secure and protected communication requires feature extraction. A frequency domain analysis is being done of the EKG signals for feature generation. It is done because the components of frequency of the signals obtained at a any specified time will have similar values irrespective of the measured place of the human body. It involves two steps: Extraction of Features and Quantization of blocks. The ECG signal from the MIT PhysioBank[1] is sampled at a certain frequency for a specified time (125 Hz sampling rate and 5 s duration). Unnecessary components of frequency has removed by passing the signal through filters. The obtained sample which produces 625 samples. It is further divided into 125 samples of 5 parts. The DWT is applied on each of the parts after filtering. The 64 coefficients are chosen and concatenated to make a 320 coefficients feature vector, as shown in Fig. 1.

The next step is quantization of the binary blocks of the concatenated key block. To obtain a key from the binary stream the block is divided to form 20 blocks each having

---

[1]https://physionet.org/physiobank/.

16 coefficients. Exponential quantization was used to represent the coefficients in 4-bit binary values. It results in 64-bit block in each of the 20 blocks.

Key Agreement: Right after the quantization process, feature vectors are created and blocks are formed and these blocks are exchanged for the communication of sensors. These communicating sensors are used to generate a common key. The whole process is mentioned in Fig. 2. There are three steps in the key agreement stage (phase) which have been explained as follows.

1. *Watermarked Commitment Phase*: To embed machine-readable information within digital media content, the technology of digital watermarking is used. This encoding is done by making some changes to the blocks (which are the signals here). The phase of watermarked commitment is one where blocks are exchanged by the sensors. For example, at sensor B, blocks $B_{s1} = b_{1,1}, b_{1,2}, ..., b_{1,20}$ are created. Since these blocks are what builds the key, it becomes imperative to implement security measures to exchange these blocks among other sensors, securely. For security, hash functions (MD5, SHA-256 and SHA-512) have been used to hash the blocks. The commitment phase is performed in two steps:

Step 1: $s_1 \rightarrow s_2 :< ID, hash(b_{1,1}, N), ..., hash(b_{1,20}, N),$
$MAC(Key'_R, ID, N, hash(b_{1,1}, N), ..., hash(b_{1,20}, N)) >$
Step 2: $s_2 \rightarrow s_1 :< ID, N', hash(b_{2,1}, N'), ..., hash(b_{2,20}, N'),$
$MAC(Key'_R, ID, N', hash(b_{2,1}, N'), ..., hash(b_{2,20}, N)) >$
where node ids are ID and ID', and the nonce are N and N' for sustaining freshness of transaction. MAC refers to the Message Authentication Code.

2. *Processing Phase*: After exchange of the blocks, the hash values that have been received of block size 20×64. Let the matrices be named U and V, respectively. A 20×20 dimension of matrix W is computed from the given matrices such that each of the element of $W(i, j)$ is equal to the Hamming distance between the jth row of V and ith row of U and $i > 1$ and $j < 20$. For derivation of the Key, KeyGen Algorithm [7] has been used which is given follows. The generated common Key, obtained from the hashing of the KeyMat Block elements are identical at both the communicating sensors.

Algorithm KeyGen(W)
1. $Key = \phi$
2. $KeyMat = \phi$
3. while(all $W(i, j) \neq 1$) do
4. $(i, j) == min(W)$
5. $if(min(W) == 0)$
6. $KeyMat = KeyMat + (b_i)^1$
7. $W(i, k) = 1 \; \forall 1 \leq k \leq 20$
8. $W(u, v) = 1 \; \forall 1 \leq k \leq 20$
9.   else
10.   return error

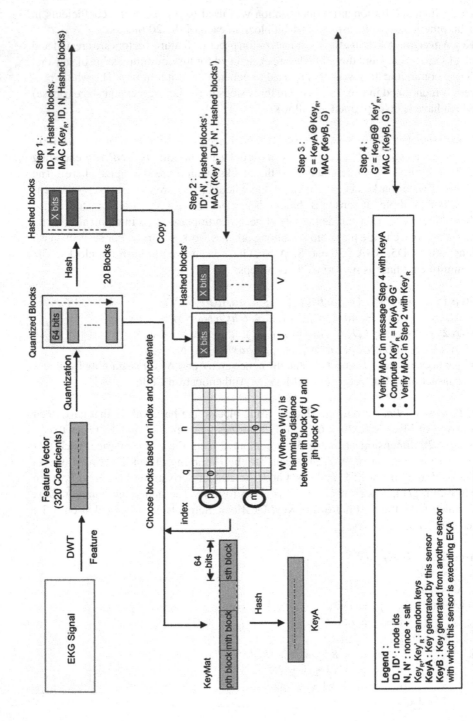

Fig. 2 Key Agreement at an arbitrary sensor in WBSN

11.   end if
12. end while
13. $Key = hash(KeyMat)$
14. return Key

3. *Decommitment Phase*: The authenticity of the keys is verified in this phase after the keys have been generated in the commitment phase. For such verification the below-mentioned exchange takes place. After the processing phase, the keys which are generated are Key_A and Key_B at the sensors s_1 and s_2 which are both located on the same person on different parts of the body. Both the keys generated should be identical. The verification of Message Authentication Code (MAC) are received in the decommitment phase using Key_A and Key_B respectively.

Step 3: $s1 \rightarrow s2 :< G = Key_R \oplus Key_A, MAC(Key_A, G) >$
Step 4: $s2 \rightarrow s1 :< G' = Key_R \oplus Key_B, MAC(Key_B, G') >$

The random keys generated at the sensors A and B can be extracted by XOR-ing with G and G' if the verification is found successful. If both the extracted keys match in the evaluation, the keys generated are accepted. The values of Key_A and Key_B, generate temporary keys $K_{temp} = hash(Key_A, l) = hash(Key_B, l)$ for performing actual communication, where l is a random number. Thus, the actual key is secured.

4 Results and Performance Analysis

The implementation has been done in Python and analysis has been done using MATLAB. We have implemented the key agreement scheme in [17] using MD5 hash, SHA-256, and SHA-512 and computational time is measured for performance of the scheme. Changing the block size after the quantization process also affects the Key generation time. As we increase the block size, the key generation time also increases. Plots of different frequencies and block sizes with respect to time is shown in Figs. 3 and 4. Figure 5 represents the comparison of time taken by FFT and DWT for data with various frequencies and Fig. 6 represents the comparison of MD5, SHA-256 and SHA-512 hash functions.

Security Analysis: The benefit of usage of hash function is that correct values are unknown to the adversaries who get hold of the packet. Besides this, it also takes care of the randomness, distinctiveness, temporal variance. It is very difficult to guess the key from the blocks due to the randomness of the generated key. Security against the replay attacks is prevented by the nonce. It was seen that it does not not produce the same keys for new set of ECG signals.

Computational Cost: The FFT-based feature extraction in [7] has a time complexity of $O(n \log(n))$, where as the scheme in [17] has a time complexity of $O(n)$.

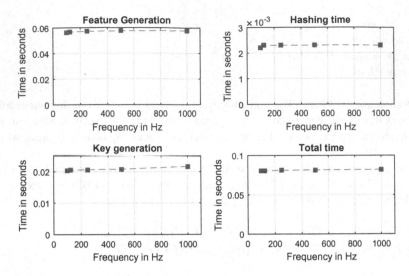

Fig. 3 Time taken for different frequencies

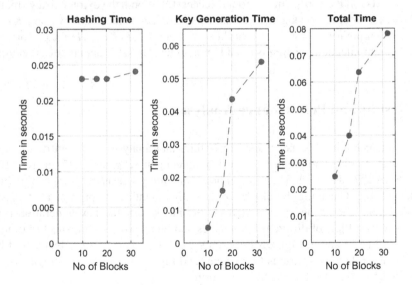

Fig. 4 Time taken for different block sizes

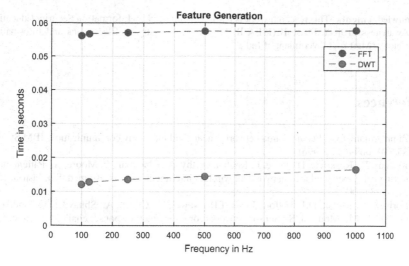

Fig. 5 Comparison between FFT and DWT

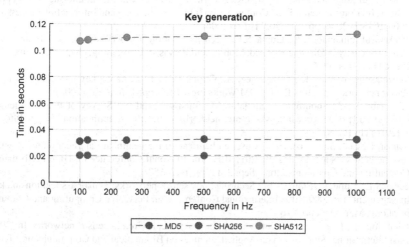

Fig. 6 Comparison between MD5, SHA-256, SHA-512

5 Conclusion

In this paper, we have presented the comparative analysis of the ECG-based Key Agreement schemes using DWT and FFT and performance evaluation of the generated key and the scheme has been done. As the sensors are very low powered and are not able to do heavy calculation, the scheme has a linear complexity which is advantageous to the sensors. Moreover the watermarking scheme during locking and unlocking makes the schemes more secure.

Acknowledgements This research work is partially supported by Information Security Education and Awareness Project, Phase-II (ISEA-II) funded by Ministry of Electronics and Information Technology (MeitY), Government of India.

References

1. Zimmerman, T.G.: Personal area networks: near-field intrabody communication. IBM Syst. J. **35**(3.4), 609–617 (1996)
2. Jovanov, E., Raskovic, D., Price, J., Krishnamurthy, A., Chapman, J., Moore, A.: Patient monitoring using personal area networks of wireless intelligent sensors. Biomed. Sci. Instrum. **37**, 373–378 (2001)
3. Lorincz, K., Malan, D.J., Fulford-Jones, T.R., Nawoj, A., Clavel, A., Shnayder, V., Mainland, G., Welsh, M., Moulton, S.: Sensor networks for emergency response: challenges and opportunities. IEEE Pervasive Comput. **3**(4), 16–23 (2004)
4. Schwiebert, L., Gupta, S.K., Weinmann, J.: Research challenges in wireless networks of biomedical sensors. In: Proceedings of the 7th annual international conference on Mobile computing and networking, pp. 151–165. ACM (2001)
5. Venkatasubramanian, K.K., Gupta, S.K.: Security for pervasive health monitoring sensor applications. In: Fourth International Conference on Intelligent Sensing and Information Processing, 2006. ICISIP 2006, pp. 197–202. IEEE (2006)
6. Venkatasubramanian, K.K., Gupta, S.K.: Security solutions for pervasive healthcare. In: Xiao, Y. (ed.) Security in Distributed, Grid, Mobile, and Pervasive Computing, chap. 15, p. 349. CRC Press: NY, UK (2007)
7. Venkatasubramanian, K.K., Banerjee, A., Gupta, S.K.: EKG-based key agreement in body sensor networks. In: 2008 INFOCOM Workshops IEEE. pp. 1–6. IEEE (2008)
8. Al-Janabi, S., Al-Shourbaji, I., Shojafar, M., Shamshirband, S.: Survey of main challenges (security and privacy) in wireless body area networks for healthcare applications. Egypt. Inform. J. (2016) (In Press)
9. Kumar, P., Lee, S.G., Lee, H.J.: A user authentication for healthcare application using wireless medical sensor networks. In: 2011 IEEE International Conference on High Performance Computing and Communications, Sept 2011, pp. 647–652
10. Keoh, S.L., Lupu, E., Sloman, M.: Securing body sensor networks: Sensor association and key management. In: 2009 IEEE International Conference on Pervasive Computing and Communications, Mar 2009, pp. 1–6
11. Dimitriou, T., Ioannis, K.: Security issues in biomedical wireless sensor networks. In: 2008 First International Symposium on Applied Sciences on Biomedical and Communication Technologies, Oct 2008, pp. 1–5
12. Çamtepe, S.A., Yener, B.: Key distribution mechanisms for wireless sensor networks: a survey. Rensselaer Polytechnic Institute, Troy, New York, Technical Report pp. 05–07 (2005)
13. Yao, L., Liu, B., Wu, G., Yao, K., Wang, J.: A biometric key establishment protocol for body area networks. Int. J. Distrib. Sens. Netw. (2011)
14. Le, X.H., Khalid, M., Sankar, R., Lee, S.: An efficient mutual authentication and access control scheme for wireless sensor networks in healthcare. J. Netw. **6**(3), 355–364 (2011)
15. Mehmood, Z., Nizamuddin, N., Ch, S., Nasar, W., Ghani, A.: An efficient key agreement with rekeying for secured body sensor networks. In: 2012 Second International Conference on Digital Information Processing and Communications (ICDIPC), pp. 164–167. IEEE (2012)
16. Cherukuri, S., Venkatasubramanian, K.K., Gupta, S.K.S.: BioSec: a biometric based approach for securing communication in wireless networks of biosensors implanted in the human body. In: International Conference on Parallel Processing Workshops (ICPPW'03), Oct 2003, pp. 432–439

17. Ali, A., Khan, F.A.: An improved EKG-based key agreement scheme for body area networks. In: Bandyopadhyay, S.K., Adi, W., Kim, T.h., Xiao, Y. (eds.) Proceedings: Information Security and Assurance: 4th International Conference (ISA'10), Miyazaki, Japan, Jun 2010, pp. 298–308. Springer, Berlin, Heidelberg

Wireless Device Authentication Using Fingerprinting Technique

Asish Kumar Dalai, Kaushal Kumar and Sanjay Kumar Jena

Abstract Wireless Device fingerprinting is a technique used to uniquely identify a wireless device. Device fingerprinting plays an important role in authenticating devices in a network. In this paper, a technique of wireless devices fingerprinting has been presented. We have taken both frame Inter-arrival Time (IAT) and Transmission Time (TT) as the features for device fingerprinting. The proposed model has been evaluated using traffic traces from Sigcomm2008 datasets. The performance analysis of two different training function, i.e., Scaled Conjugate Gradient (SCG) and Bayesian Regularization (BR) has been done. The result shows that BR performs better than SCG in identifying unique devices. The proposed method gives better accuracy and identifies more number of devices than the existing techniques.

Keywords Wireless network security · Device fingerprinting · Device
authentication · Artificial neural network

1 Introduction

Wireless devices are becoming extremely popular due to ease of use and cost-effectiveness, leading to large-scale migration of wired infrastructures to wireless ones. One major problem that can occur in wireless setups is that it has no specific boundary. The wireless network can easily extend outside the walls, opening up the internal network to attackers. If the attacker can crack the security key, he can gain access to the network. Also, a legitimate user of the network may try to connect their devices to the malicious network, which lures the victims for speed and open access, exposing it to possible threats. To prevent such scenarios, there is a need for

A. Kumar Dalai (✉) · K. Kumar · S. Kumar Jena
National Institute of Technology Rourkela, Rourkela, India
e-mail: dalai.asish@gmail.com

K. Kumar
e-mail: kumarkaushal400@gmail.com

S. Kumar Jena
e-mail: skjena@nitrkl.ac.in

© Springer Nature Singapore Pte Ltd. 2019
P. K. Sa et al. (eds.), *Recent Findings in Intelligent Computing Techniques*,
Advances in Intelligent Systems and Computing 707,
https://doi.org/10.1007/978-981-10-8639-7_17

Fig. 1 Overview of a Device
Fingerprinting System

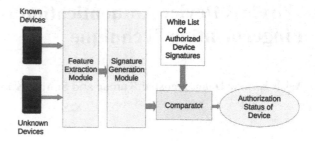

an efficient device authentication system that allows only the authenticated device to access the network.

Several methods have been proposed to identify a device and to decide if the device is a legitimate one that should be granted access or not. In this work, a device authentication system has been proposed using device fingerprinting. The basic idea behind fingerprinting is to passively or actively extract unique patterns of a device from the network traffic. A variety of features can be extracted and utilized including physical layer features, medium access control layer features, and upper layer features. Effective device fingerprints must satisfy two properties that include: i) they are difficult or impossible to forge, and ii) the features should be stable in the presence of environment changes and device mobility. The first requirement renders identifiers such as MAC addresses, IP addresses are unsuitable candidates as all these identifiers have been shown to be easily modifiable via software [1]. In contrast, location-dependent features such as the popular Radio Signal Strength (RSS) cannot be used on their own as fingerprints, as they are susceptible to mobility and environmental changes. Therefore, there is a need of unique, reliable, and reproducible feature/ features that can identify a device.

Device fingerprinting is a robust solution to the above-discussed problem of unique device identification. It produces a signature, also called a fingerprint that is unique for each device. The overview of a device fingerprinting system is given in Fig. 1. The model is designed to incorporate the identification of rogue or unauthorized devices.

The rest of the paper is organized as follows. Section 2 describes the related works. The proposed approach has been presented in Sect. 3. Evaluation of the proposed approach is done in Sect. 4. Finally, the concluding remarks are given Sect. 5.

2 Related Work

Several research has been proposed for device identification/authentication using fingerprinting techniques [2, 3, 5–8, 8, 9]. The fingerprinting approaches can be classified into active or passive in the context of the traffic analysis of the network packets done by the fingerprinter for generating the device signature. For active analysis, the fingerprinter sends data packets, and the responses to those packets are examined to fingerprint the target device. Whereas, The passive fingerprinting approach, generates the device signature from the traffic traces collected passively, may be using any traffic capturing tool without the knowledge of the target device.

Bratus et al. [2] have proposed an active fingerprinting technique to identify access points and can classify client nodes by their drivers and chipsets used. Their method relies on the concept that the every device may respond differently, to the malformed and nonstandard packets. They have used a frame generator named BAFFLE to inject the malformed and nonstandard frames to the target device. As the connection establishment process of AP and the station are different, they have used two different approaches, one for fingerprinting of the access point and the other one for client stations. For classifying the devices, a decision list-based learning algorithm has been used. The results show that their method is more suitable to classify APs than the client stations.

Ureten et al. [3] showed another type of a radio-frequency-based mechanism for the identification of the wireless devices. It uses a challenge-response mechanism to verify the client if the fingerprint is not valid or suspicious. The fingerprint is generated by analyzing the transient characteristics of the antenna at the beginning of the transmission. Their work shows an entire identification system consisting of data extraction, transience detection mechanism, fingerprint generation, and classification. However, it suffers from the drawbacks and gives good results when coupled with traditional IDS solutions.

Cache et al. [4] proposed a couple of different approaches that look at the 802.11 protocol's association redirection mechanism. This mechanism is a loosely implemented one in most 802.11 deployments, and this creates the possibility for fingerprinting the devices using the widely different ways in which the protocol handles its redirection. This method performs analysis on the duration field values on the various frames of the 802.11 protocol like the data and management frames. This mechanism works well for the detection of the drivers being used by the AP as different vendors have their devices compute their duration field values in slightly different ways.

Gao et. al. [5] has proposed a passive fingerprinting approach to identify the APs type. A packet train that emulates normal traffic is sent to the AP. They have observed that due to the heterogeneity in the device architecture, APs respond differently to the packet train. The resultant time difference in the arrival of frames and the corresponding bin size are used as the feature tuple for each AP. By applying discrete Haar wavelet transformation to the feature vector, a distinct signature is generated, which can uniquely classify the APs. Even though they claim a 100 percent accuracy, but the number of devices used in their experiment is very less. Franklin et al. [6] have shown a passive model to extract the fingerprint from devices by using the frequency of certain management frames of the 802.11 protocol. They used the probe request frames that the device sends at fixed intervals to analyze the device identity. They used a binning-based technique on the data collected from the device and generated various statistical parameters from it such as the mean difference and percentage distributions between the bins to create the signatures. This method had varying degrees of success depending on the bin size. To effectively detect MAC address spoofing Jana et al. [7] calculated the clock skew of an AP from the IEEE 802.11

Time Synchronization Function (TSF) timestamps sent in the beacon/probe response frames and use it as device feature. They then analyzed whether the devices were authorized or not using LSF and LPM methods and analyzed the clock skews. The signature was derived using the variations in the clocks of the fingerprinter and the target device by performing analysis on the same. Corbett et al. [8] also designed a fingerprinting approach that looks at the rate switching mechanism employed in the 802.11 protocol and differentiates between different devices using the spectral profiles of the NIC. Implementation of this rate switching mechanism exerts an influence on the transmission patterns of the wireless stream and can be observed in traffic analysis. This approach uses signal processing to analyze the periodic pattern induced in communication due to this rate switching and creates a fingerprint for the devices using that as a feature. A stable profile is created using the spectral analysis of the traffic components to determine the identity of the device.

3 Proposed Work

In this work, a technique for wireless device authentication has been proposed based on device fingerprinting. The proposed technique deals with device authentication using the information manifested by a device through its network traffic. This is accomplished by utilizing the heterogeneity found in the timing behavior of the devices. We took the network parameters from the wireless traffic to find out the most suitable features. From the experimental analysis, we found that the combination of IAT and TT as the feature vector gives better performance. Therefore, those features are utilized to generate the device signature. To generate the signature, it uses data binning approach to the resulted histogram of the features. A multilayer feed-forward ANN has been used to train the model. The model has been configured to use BR backpropagation as the training function. The similarity score produced from the trained module is used to authenticate the device. The proposed method has four major components: feature selection, signature generation, enrollment, and device authentication.

3.1 Feature Selection

Device fingerprinting can be done by utilizing a variety of features. But the selected feature/features must be unique, reproducible and immune to forgery attacks. Considering this requirement, a set of network parameters from the traffic has been calculated, which can be utilized as a feature for device identification. The objective is to select feature/features that represent the device and guarantees its uniqueness. By experimental analysis and the survey of the related literature, it has been found that the combination of IAT and TT is the most appropriate feature for device fingerprinting. The successive delay between the consecutive packets of a device is measured to calculate the IAT and TT is the amount of time taken by the frame to transmit. The feature vector which combines both IAT and TT gives a time-series data. The subsequent feature vector with time-variant behavior is passed through signature construction process for time-series inspection. The feature vector can be represented as

$$F = (\delta t_1, \delta t_2, \delta t_3, \ldots, \delta t_n, \alpha t_1, \alpha t_2, \alpha t_3, \ldots, \alpha t_n) \tag{1}$$

where, $\delta t_n = t_n - t_{n-1}$ represents the IAT and $\alpha t_n = \frac{size_n}{rate_n}$ represents the TT.

3.2 Signature Construction

The signature construction method uses the frequency density measurement of the feature vector. The frequency counts of the IAT and TT values are calculated using the histogram. The histogram depends on the number of bins and the bin width. In this work, based on empirical analysis, we have taken N equally spaced time bins to generate the signature. The signature vector contains the frequency counts of the IAT and TT values that fall in these bins. We generate two histograms of each device d_i (one for IAT and one for TT) present in the traffic traces. We formulate it in range of bins $[b_1, \ldots, b_N]$.

We denote f_k^i (where $1 \leq k \leq N$) the frequency count of IAT and TT in bin b_k. Where the percentage frequency of the bin b_k is $P_k^i = f_k^i / |P^i(S)|$. So the resultant histogram for a given device d_i is represented as:

$$hist^i(s) = \{P_k^i \forall j \varepsilon 1 \leq k \leq N\}. \tag{2}$$

The signature of the device is sensitive to bin ranges, and different bin ranges may reveal different information about the device feature vector. If we consider smaller bin ranges, then less number of IAT values fall within a bin. Larger bin ranges may not contain important information about the devices. From the experimental analysis, it is found that N = 20 would be an ideal choice for all traffic type tested in this paper. We use the value of N to find the bin ranges for each traffic type. The range between starting and ending points of the histogram are split into 20 equal bins. Finally, the wireless device signature contains 42 values, which includes the deviceID, device weight and 20 bin values each for IAT and TT respectively, which is represented as: $< DeviceID_i, Weight(Device_i), hist_i(S_{iat}), hist_i(S_{tt}) >$.

3.3 Enrollment

To enroll the device, we have trained the signatures using Bayesian regularized ANN. ANN is a computing system made up of some highly interconnected processing elements, which process information by their dynamic state response to external inputs. This highly interconnected processing element is called neurons which accept input and produce an output according to an activation function. BR training function updates the weight and bias values according to Levenbergp–Marquardt optimization [10]. It minimizes a combination of squared errors and weights and then determines the correct combination so as to produce a network that generalizes well.

The multilayer feed-forward neural network takes an input of size N (X_1 to X_N) and generates a result as M different devices (Y_1 to Y_M). Input taken by input layer is a signature, which is probability distribution with N bins and the output produces by output layer is the similarity measure between input and the output signature that has been already trained on. We consider the hidden layer size (p), which provides an optimum result at p $=$ 100. Neural systems of this kind are utilized for each device that has to be examined. The ANN (θ) are put into the master database after training.

3.4 Device Authentication

At the device authentication step, for each device, signatures are produced and then matched with previously stored master signatures for authentication. We assign a unique ID for each device and calculate their signature. The signatures are training using ANN, and the result is stored in master database. Further, every device that needs to access the network has to be authenticated. Therefore, we generate the signature for the device and measure the similarity between unknown signature (ω) and stored signature. This similarity method gives the closeness value matched with the previously observed closeness value, if closeness value is matched it authenticates the device and allows the access otherwise denies. The steps are as given in Algorithm 1.

4 Evaluation

We evaluate the method using publicly available umd/sigcomm2008 dataset (v. 2009-03-02) [11] collected and distributed by A Community Resource for Archiving Wireless Data (CRAWDAD).

- **umd/sigcomm2008 dataset**: This dataset contain wireless traces captured at SIGCOMM 2008 conference. The wireless traffics collected by inviting people to join the traced SSID. The data capturing process continued for 5 long days, and a large number of people joined the traced network. The dataset contains the traffic traces from a variety of devices. The data packets include information about different layers depending on the packet type. Therefore, using this dataset enable us to extract the network parameters (IAT and TT) from a wide range of devices.

Data: Intialize the feature vector for all devices.

1 **for** *all device_i* $\in S$ **do**
2 | $hist^i(S) = f_j^i\ /\ |P^i(s)|$;
3 **end**
4 Store $\{\theta_{ID}, weight^i(S), hist^i(S)\} \leftarrow sig(S)$ in database
5 $\{\theta_{ID}, Dev_{list}\} \leftarrow Master\_Database$
6 $U \leftarrow Unknown\_Samples()$
7 $\omega \leftarrow Generate\_Signature(U)$
8 $Out_{ID} \leftarrow sim(\theta_{ID}, \omega)$
9 $index, closeness \leftarrow max(Out_{ID})$
10 $M \leftarrow previous\_observed\_closeness\_value$
11 **if** *(closeness > M)* **then**
12 | return Device Authenticated;
13 **else**
14 | return Unauthorized Device Denied;
15 **end**

Algorithm 1: Device Authentication

4.1 Impelmentation

In sigcomm2008 dataset each packet incorporates physical layer data such as IAT, TT, MAT, frame length, etc. We considered only the IAT and TT values of the frame along with its MAC address. We provide a unique ID correspond to each MAC address. Similarly, we repeat this for all the traces and combine all the traces. Then, we divide the dataset into twelve blocks and find the devices, which are common in all twelve blocks. We generate signatures for those unique devices. We consider sigcomm2008 dataset as we got both the desired network parameters (TT and IAT) in all traffic traces.

The next step is to generate the signatures from each device from their corresponding feature vector. The signatures are generated and are enrolled using ANN and stored in the master database. Whenever any device wants to access the network, the features are collected followed by signature construction. Then, the newly generated signature is matched with the signature stored in master database. This similarity measure technique gives a value between 0 to 1. If the value is close to 1, then the new device is legitimate and otherwise unauthorized.

4.2 Results

Experiments have been conducted, and the results are presented. Table 1 represents the accuracies of our technique by combining IAT and TT as the features for the Sigcomm2008 dataset. Accuracies for device identification using IAT and TT as features have been presented for varying thresholds (frame counts of unique devices) starting from 200,000 to 400,000 with an increment of 50,000. In all these cases, BR training function performs better than that of SCG.

Table 1 Performance analysis of scg and br training function with varying threshold

Threshold (frame count)	Total no. of devices	Devices common in all blocks	Accuracy (IAT + TT)	
			scg (%)	br (%)
200,000	41	27	48.5	93.9
250,000	34	21	42.9	95.6
300,000	25	14	75	97
350,000	20	11	84.8	97
400,000	17	10	99.2	99.2

Fig. 2 ROC curves (TPR vs FPR) for classification of 14 devices

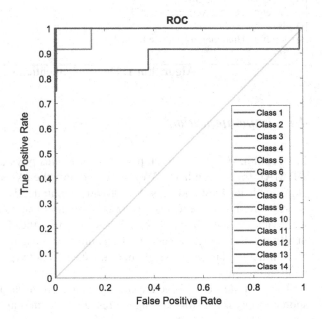

From the result given in Table 1, it is clear that the threshold value of 300,000 frame counts gives comparatively better accuracy and classifies more number of devices. Figure 2 represent the ROC plot for the threshold value 300,000 and 14 unique devices considering IAT and TT as the features using BR. The confusion matrix is given in Fig. 3. It is clear from these figures that the proposed method can classify 14 different devices with an accuracy of 97%.

From the results given in Table 1, one can observe that the accuracy increases as the threshold value increases and the device count decreases. Therefore, for finding an optimal threshold value, which gives better accuracy and classifies more number of devices, we plotted the graphs of normalized threshold versus accuracy and normalized device count vs. accuracy as shown in Fig. 4. Both threshold and device count have been normalized in [0, 1]. From these figures, it is calculated that the optimal threshold value is 281246, which gives an accuracy of 96.6% for classifying 17 unique devices by combining IAT and TT as the feature.

Fig. 3 Confusion matrix for classification of 14 devices

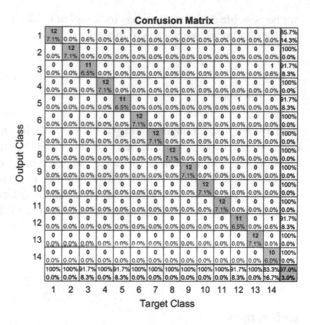

Fig. 4 Trade-off between accuracy, threshold, and device count

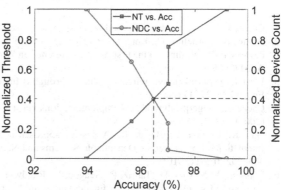

4.3 Attacks

The proposed technique relies on IAT and TT, which contains the transmission time and the emitting client station's idle time. Both these parameters have an impact on the signature value. Therefore, tampering these parameters may create a distorted signature. An attacker can introduce delays to packets, vary the packet size, fluctuate the data rate and can use different protocols to tunnel the packets to bypass the security mechanism. A highly skilled attacker can emulate the authorized device's traffic pattern. Therefore, to mitigate such kind of attack, we have to harden the authentication process by combining the conventional procedures with the proposed approach. Also, the accurate emulation of the authorized device's traffic pattern could

not be possible as the attacker has to completely hide his details and make changes
to a lot of network parameters to meet the requirement.

5 Conclusion

We have proposed a fingerprinting technique, which can uniquely identify wireless
devices for the purpose of device authentication. We have utilized both IAT and
TT from the wireless traces for device fingerprinting using ANN. The proposed
technique achieves better accuracy and identifies more number of devices. With its
promising results, this technique can complement existing security solutions such
as those that provide authentication and access control. In our future work, we will
focus on selecting and or combine other network parameters to reduce the frame
count for signature generation, which shall eventually reduce the time to generate a
stable signature.

References

1. Cache, J., Liu, V.: Hacking exposed wireless: wireless security secrets & solutions (2007)
2. Bratus, S., Cornelius, C., Kotz, D., Peebles, D.: Active behavioral fingerprinting of wireless
   devices. In:Proceedings of the first ACM Conference on Wireless Network Security, pp. 56–61.
   ACM (2008)
3. Ureten, O., Serinken, N.: Wireless security through rf fingerprinting. Can. J. Electr. Comput.
   Eng. **32**(1), 27–33 (2007)
4. Cache, J.: Fingerprinting 802.11 implementations via statistical analysis of the duration field
   (2006)
5. Gao, K., Corbett, C., Beyah, R.: A passive approach to wireless device fingerprinting. In:
   International Conference on Dependable Systems and Networks (DSN), 2010 IEEE/IFIP, pp.
   383–392. IEEE (2010)
6. Franklin, J., McCoy, D., Tabriz, P., Neagoe, V., Randwyk, J.V., Sicker, D.: Passive data link
   layer 802.11 wireless device driver fingerprinting. In: Usenix Security, vol. 6 (2006)
7. Jana, S., Kasera, S.K.: On fast and accurate detection of unauthorized wireless access points
   using clock skews. IEEE Trans. Mob. Comput. **9**(3), 449–462 (2010)
8. Corbett, C.L., Beyah, R.A., Copeland, J.A.: Passive classification of wireless nics during active
   scanning. Int. J. Inf. Secur. **7**(5), 335–348 (2008)
9. Kumar, K., Dalai, A.K., Panigrahy, S.K., Jena, S.K.: An ANN based approach for wireless
   device fingerprinting. In: IEEE International Conference on Recent Trends in Electronics Information Communication Technology, 19–20 May2017, India
10. Gopalakrishnan, Kasthurirangan: Effect of training algorithms on neural networks aided pave-
    ment diagnosis. Int. J. Eng. Sci. Technol. **2**(2), 83–92 (2010)
11. Aaron Schulman, Dave Levin, and Neil Spring. CRAWDAD dataset umd/sigcomm2008 (v.
    2009-03-02). http://crawdad.org/umd/sigcomm2008/20090302. Accessed from Mar 2009

Part II
Identity Management in Digital Systems

An Improved Framework for Human Face Recognition

Nasir Fareed Shah and Priyanka

Abstract In recent years considerable progress has been made by the researchers in the field of pattern recognition in general and face recognition in particular. Computers can now outperform human brain in face recognition and verification tasks. While most of the methods related to face recognition perform well under specific conditions, some show anomalous behavior when the degree of accuracy is concerned. In this paper, we have divided the face recognition task into three sub-parts as Segmentation, Feature Extraction, and Classification. Information from face image is extracted and modelled using Eigenvectors. The weights calculated from Eigenvectors are classified by the statistical classifier using distance metric specification. The system is capable of recognition to an accuracy of 96%, having a standard deviation of 0.662 for facial expression variations.

Keywords Face recognition · Feature vector · Eigevalues · Eigevectors
Pattern recognition · Biometrics

1 Introduction

Face being the primary focus of attraction and attention is often described as the index of the mind. Through various facial postures and poses, we can convey meaningful information without using any other sensory organ(s). Pain, anxiety, temptation, happiness, and sorrow can be directly guessed from the person's facial expression. The best example in the virtual world is smiley, which conveys exclamation to other

N. F. Shah (✉)
Department of Computer Science and Engineering, Birla Institute of Technology,
Mesra, Ranchi 835215, India
e-mail: saednasir@gmail.com

Priyanka
Bhagalpur College of Engineering, Sabour, Bhagalpur 813210, Bihar, India
e-mail: priya.pal05@gmail.com

© Springer Nature Singapore Pte Ltd. 2019
P. K. Sa et al. (eds.), *Recent Findings in Intelligent Computing Techniques*,
Advances in Intelligent Systems and Computing 707,
https://doi.org/10.1007/978-981-10-8639-7_18

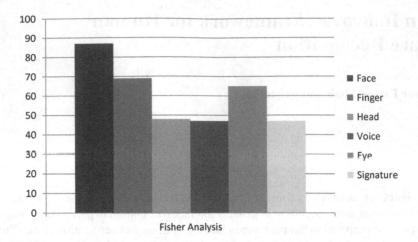

Fig. 1 Fishers analysis

user(s) online. Facial expressions are the effective biometrics to detect person's identity, mood, behaviors, etc.

Nowadays various biometric applications are available and the majority are based on facial scan. Because the accuracy and easiness in facial sensing are high. Human face recognition framework has many applications ranging from crowd surveillance to human–computer interaction.

The major goal of our research is to develop a framework which effectively and efficiently scans human face and then accurately detects its virtual impressions, example a photograph.

In recent years Face recognition has shown a boom in information security field. Figure 1 shows the Fisher analysis in the field of pattern recognition. Human–computer interaction has now made machines to understand, learn, and adapt as per the input data, user commands, or sensory signals. Human beings can perceive the information matter easily from the given picture, but for the computer, it is sometimes difficult to differentiate between similar backgrounds and foregrounds. When compared to human brain, computers are sophisticated machines with the capability of recognizing complex patterns and memorizing them.

2 Literature Survey

The information in facial postures has become more conspicuous, since information captured is useful for decision-making, system building, and authentication. The world is generating an enormous amount of information in the field of Information Security particularly in the field of pattern recognition. One amongst the challenging fields in the world of computation is Pattern recognition and evaluation.

The field of pattern recognition has attracted researchers from every section of the learning societies. Human face recognition makes use of various types of stimuli obtained from many sensory organs including visual, auditory, olfactory, tactile, etc. The data collected from these stimuli is either used individually or collectively for decision-making. In many cases, contextual knowledge also plays an important role in building the decision-making system. Many scientists and researchers in the field of neuroscience and psychology have shown their relevance in designing algorithms or systems for machine recognition of faces. Pattern recognition field, especially face recognition has attracted many researches since late nineteenth century as reported by Benson and Pratt.

Sir Francis Galton, An English scientist studied the case of French prisoners, where prisoners were identified on the basis of four features; Head length, Head breadth, foot length, and middle finger length of hand and foot. The idea of comparing introduced by Galton was used by scientists in face recognition field.

Turk and Pentland [1] put forwarded the approach of face profiling based on geometrics. He used 12 subjects with three or more training images and one test image with each subject being represented using 17-element feature vector.

Pentland and Starner [2] introduced the neural network approach for face detection. They successfully trained the system which identifies the face of the person from the dataset available.

Bellman [3] used Principal Component Analysis (PCA) to efficiently represent face profiles. They further argued the possibility of reconstruction of face image from small collection of weights of each face including standard face image. Turk and Pentland [1] Used the method of [4] for construction of efficient face recognition system. They used the modular space composed of facial features like eye nose and mouth. Their system achieved the identification rate of 95% on FERET database of numerous images. Recently researchers in the field of face recognition have developed an architecture based on PCA. They used the Eigenface method based on a probabilistic measure of accuracy, which was initially based on simple subspace—restricted norms.

Fisher [4] Introduce a Linear Discriminate Analysis (LDA) method for face verification and identification. The method searches for those vectors in the underlying space that best discriminates the vectors under that class.

Zhao et al. [5] prepared a face recognition architecture based on PCA and LDA. It is a dual-step process. In the first step, the face image is projected from original vector space to a face subspace using PCA. After the first step, the results are treated to obtain best linear classification using LDA. The main idea to ensemble PCA and LDA was to improve the capability of generalization of LDA if small numbers of samples are used per class.

Liy and Lu [6] proposed a novel classification idea called the Nearest Feature Line for face recognition. Any two features from the same class were categorized using the feature line passing through two points. The feature classification is based on the nearest distance from the query feature points to each FL. The combined error rate of NFL with the face database is 44–65% approximately.

Wiskott et al. [7] presented a system classification of facial feature from simple singular images from the collected set of large databases. The task of classification is difficult in terms of expression, pose, size, and position since images have different or irregular variations across the plane. In facial feature extraction, Active Shape Model (ASM) has been used successfully to capture facial feature under various positions, e.g., frontal view. The accuracy of ASM degrades when the face has irregular variations across the shape.

3   Research Methodology

Almost two-thirds of the previous work based on face recognition has ignored the issue that face stimulus is as important as any other feature corresponding to face recognition. The facial stimuli during agony, anxiety, happiness, or excitement are necessary for building robust face recognition architecture.

Facial image can be considered as a vector of different rows. If the image has height h and width w, then the number of component vectors can be w * h. Here, every pixel is a coded vector.

Similar to face vector, the image or face can be categorized into image space, if having different image space vectors corresponding to pixels. Consider a black and white image of size N * N I(x, y), where I(x, y) is simply a matrix of 8 bit matrix with each bit represents the intensity of that particular pixel.

Calculating Eigenface:

Let a face image I(x, y) be a two dimensional N * N array of 8-bit intensity value. So a simple image of 256 * 256 size will become a vector of size 65,536. An $N \times N$ matrix has 'A' Eigenvectors, and corresponding Eigenvalue (λ) as

$$AX = \lambda. \tag{1}$$

This can be calculated as

$$\text{Det}|A - \lambda I| = 0. \tag{2}$$

After some insight of mathematical formulae, let us denote the training set of face image as 1, 2, 3, …, M then their average is calculated as

$$\Psi = \frac{1}{M} \sum_{n=1}^{M} \Gamma_n \tag{3}$$

Functional units involved in the library formation phase:

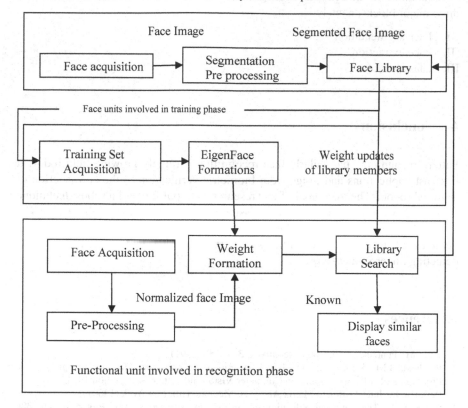

Fig. 2 Functional block diagram of the proposed face recognition system

Training Phase:

The training library contains a huge set of face images. We calculated the Eigenvalue and Eigenvector of each face image. After successful calculation of face images, we calculated values of only those face images that have high Eigenvalues.

Recognition Phase:

After finalizing training set and calculation of Eigenvalues, the system is ready to perform recognition task. The user starts the recognition task by choosing the face image. The selected image is preprocessed, normalized based on the user request. Once the normalization process is done its weights (weight vectors) are calculated similarly to that in training phase. The weight vectors calculated in the previous step are compared with the weight vectors of every face image present in the database. If there exists at least one face member similar to selected face image in the database then the face image is called known face image, otherwise unknown (Fig. 2).

To summarize the above face recognition system, it can be broadly classified into three major blocks, namely:

 I. Face segmentation
 II. Face extraction.
 III. Classification.

4 Conclusion

Pattern recognition, particularly face recognition is rapidly growing field today. It has vast applications and usage from biometric verification to the construction of artificial robots. The main goal of our research is to put forward a robust technique for face recognition and investigate the issue that hinders in getting right results. The approach put forward by us in this paper is applied to face images only, but it can be modified and applied to other images as well. The idea can be effectively used in detecting fingerprints at large.

References

1. Turk, M., Pentland, A.: J. Cogn. Neourosci. **3**, 72–86 (1991)
2. Pentland, B.M., Starner, T.: View-based and modular Eigen spaces for face recognition. In: Proceeding of IEEE Conference on Computer Vision and Pattern Recognition 1994
3. Bellman, R.: Introduction to Matrix Analysis. McGraw-Hill, New York (1960)
4. Fisher, R.A.: The statistical utilization of multiple measurements. Ann. Eugen. **8**, 376–386 (1938)
5. Zhao, W., Chellapa, R., Krishnoswammy, A.: Discriminant Analysis of Principal Components to Face Recognition. Centre for Automation Research University of Maryland (2001)
6. Liy, S.Z., Lu, J.: Face recognition using the nearest feature line method. IEEE Trans. Neural Netw. **10**, 439–443
7. Wiskott, L., Fellous, J.M., von der Malsburg, C.: Face recognition by elastic bunch graph matching. IREEE Trans. Pattern Anal. Mach. Intell. **19**, 775–779 (1997)

A Semantic Approach to Text Steganography in Sanskrit Using Numerical Encoding

K. Vaishakh, A. Pravalika, D. V. Abhishek, N. P. Meghana
and Gaurav Prasad

Abstract Steganography is the art of hiding a message within another so that the presence of the hidden message is indiscernible. People who are not intended to be the recipients of the message should not even suspect that a hidden message exists. Text steganography is challenging as it is difficult to hide data in text without affecting the semantics. Retention of the semantics in the generated stego-text is crucial to minimize suspicion.This paper proposes a technique for text steganography using classical language Sanskrit. As Sanskrit is morphologically rich with a very large vocabulary, it is possible to modify the cover text without affecting the semantics. In addition numerical encoding is used to map a Sanskrit character to a numerical value. This helps in hiding the message effectively. Moreover, in this technique, a key is used for additional security. The key is generated dynamically and is appended to the final message to further add security to the proposed method. The proposed method generated stego-texts with syntactic correctness of 96.7%, semantic correctness of 86.6%, and with a suspicion factor of just 23.4% upon evaluation.

Keywords Text steganography · Morphology · Semantics

K. Vaishakh (✉) · A. Pravalika · D. V. Abhishek · N. P. Meghana
Department of Information Technology, National Institute
of Technology Karnataka, Surathkal, Manglore, India
e-mail: kvaishakhnambiar@gmail.com

A. Pravalika
e-mail: avvarupravalika@gmail.com

D. V. Abhishek
e-mail: dv.abhishek@gmail.com

N. P. Meghana
e-mail: meghananpmegha@gmail.com

G. Prasad
Department of Information Technology, Manipal University, Jaipur, India
e-mail: nitkitgauravprasad@gmail.com

© Springer Nature Singapore Pte Ltd. 2019   181
P. K. Sa et al. (eds.), *Recent Findings in Intelligent Computing Techniques*,
Advances in Intelligent Systems and Computing 707,
https://doi.org/10.1007/978-981-10-8639-7_19

1 Introduction

The multiple methods of modern communication need to be secured especially on the computer network, hence network security has become one of the most crucial research areas, while exchanging data on the Internet. Information hiding has been used to establish confidentiality and data integrity to protect against unauthorized access and to provide a secure communication channel between sender and recipient. Research in steganography has steeply increased as it has attracted more attention than the other methods of Information Hiding due to its advantages.

Redundant information in audio and pictures are exploited to perform steganography, which is not suitable for text steganography due to its relative lack of redundant information hence making it a difficult process. But its small memory occupation and simple communication make it more advantageous over other methods. Also, it is not prone to high data alteration whereas audio and image steganography can be dampened by noise during communication, hence altering the message.

Different languages can be used as cover text to hide secret message. In this paper, we explore an innovative method to perform text steganography using Sanskrit. Text in Sanskrit is used as the cover text as it is morphologically rich and very structured. Also, the words in this language are not limited since each word has many inflections and new words can be generated using Paninian rules. This paper presents an algorithm for hiding information by exploiting the properties of Sanskrit Language.

The paper is organized as follows. Section 2 presents the existing work done in text steganography and about Sanskrit. Section 3 describes the proposed methodology which includes parsing, information hiding and key generation. Section 4 describes the implementation details with example. The results are discussed in Section 5. This is followed by conclusion and future work.

2 Literature Review

A good amount of research has been done in text steganography in the recent times. One of the initial works used the syntactic structure of the text to hide the information [1]. Context Free Grammar was used to build correct sentences so that the suspicion factor is minimal. Works to perform text steganography by exploiting the syntactic properties was successful in generating syntactically correct sentences, but it failed at producing an output that is grammatically and semantically correct [2]. Using this disadvantage of NICETEXT, steganalysis algorithms have been developed to identify the presence of hidden information in the cover file which was generated by NICETEXT [3].

Text Steganography for information hiding by using *Specific Character of Words* approach where certain characters from particular words are identified to hide the information is discussed in [4]. In text steganography by Line Shifting Method [5], the lines are shifted vertically to a certain degree α. For example, lines are shifted

अ	आ	इ	ई	उ	ऊ	ऋ	ॠ	ऌ	ए	ऐ	ओ	औ	अं	अः	क्
a	A	i	I	u	U	q	Q	L	e	E	o	O	M	H	k
ख	ग	घ	ङ	च	छ	ज	झ	ञ	ट	ठ	ड	ढ	ण	त	थ
K	g	G	f	c	C	j	J	F	t	T	d	D	N	w	W
द	ध	न	प	फ	ब	भ	म	य	र	ल	व	श	ष	स	ह
x	X	n	p	P	b	B	m	y	r	l	v	S	R	s	h

Fig. 1 WX notation

vertically to degree say α or $-\alpha$. The information is 1 and 0 for α and $-\alpha$ respectively. This method is appropriate for printed text. *Random Character and Word Sequences* method hides information by generating a random sequence of characters or words to hide the secret information [6]. Changder et al. [7] and Kim et al. [8] describes text steganography by *Feature Coding* method. The features of the text are changed to hide data. This method allows hiding a huge amount of private data in the text securely. Features like elongation or shortening of the end portion of some characters, or vertical displacement of points of characters like 'i', 'j' are used to hide data. Hiding information using the *Open Spaces* method [9] was another method, where extra spaces were added to the text to hide data. But these techniques are not robust. Another method of text steganography is by creating Spam Texts in a HTML file [6]. It uses the concept of case sensitiveness of the HTML tags to hide the data.

Regarding the work done in text steganography through Indian Languages, there are algorithms that have used numerical coding to hide the secret text [10]. An approach for text steganography in Bengali text is discussed in [11]. Changder et al. [12] describes a greedy approach to text steganography using properties of sentences. The main disadvantages with these methods are that they aren't very dynamic and robust. Also, these methods produce semantically ambiguous sentences.

Sanskrit is a classical language with an ancient history. Unlike various other languages, it can be characterized by a rich system of inflections, derivations, and compound formations. Sanskrit grammar (called *Vyakarana*) started evolving in the late Vedic period in India and culminated in Panini's monumental work 'Ashtadhyayi' ([13]), which consists of 3990 sutras (circa. 5th century BC). The phonology and morphology of Sanskrit make it very special when compared to other languages. A large number of words are derived from the root word by specific orthographic rules in Sanskrit [13]. To represent Sanskrit in a machine processable form, a transliteration scheme called the WX notation [14] is used. Figure 1 illustrates the notations used in the WX scheme for representing vowels, phonetic sounds, and consonants for the Devanagari script (used to write Sanskrit).

In this paper, we have developed a new approach for text steganography using Sanskrit to handle all the disadvantages discussed in earlier methods. The algorithm is robust, dynamic so that data integrity, confidentiality are achieved by exploiting the properties of Sanskrit. The method is robust because there is no loss or alteration of data during transmission. The proposed system hides the secret message as well as the *Key*. The *Key* is generated dynamically according to the generated stego-text. The algorithm tries to achieve syntactic and semantic correctness to a large extent.

3   Methodology

The proposed technique ensures the final message is dynamic, where all words do not contribute to hiding the secret message. This technique is suitable for hiding messages which can be represented in ASCII format.

3.1   System Architecture

Figure 2 depicts the architecture of the proposed system.

1. *Secret Message*: This is the secret message to be conveyed.
2. *Cover Text*: The Sanskrit text which is used to hide the secret message. The choice of the cover text depends on the secret message. The cover text needs to be long enough to be able to hide the message. This text can comprise of *sandhied* or *unsandhied* words (The words can be simple or compound).
3. *Stego-Text Generator*: This component generates the stego-text by finding suitable words in the cover text, where secret message can be hidden. This is integrated with other components like Sanskrit Parser, Sanskrit WordNet, and Morphological Generator.

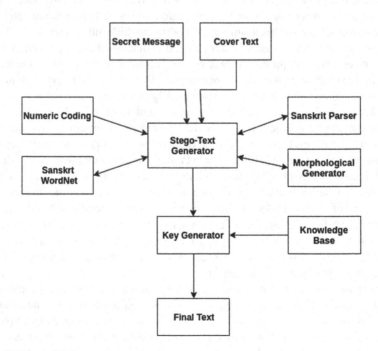

Fig. 2 System architecture

4. *Numeric Coding*: All the Sanskrit vowels and consonants are mapped to a number between 0 to 15 inclusive. The mapping is based on the frequency of occurrence of a particular value and also the frequency of occurrence of Sanskrit words beginning with that character.
5. *Sanskrit Parser*: Sanskrit parser is used to obtain the morphological details and also to resolve real-time ambiguities in the cover text.
6. *Sanskrit WordNet*: Sanskrit WordNet [15] is a linked lexical knowledge base which is used to find synonyms of a word.
7. *Morphological Generator*: Morphological generation refers to generating the appropriate inflection for the given root according to the required properties.
8. *Key Generator*: It generates the key which represents the relevance of each word in the stego-text. It is generated with the help of a knowledge base.
9. *Knowledge Base*: The knowledge base consists of nouns and verbs with proper semantic mapping. This is used to generate semantically correct short texts of size 2–4 words which form the key.
10. *Final Text*: The final text which contains the embedded secret message is sent through the medium of communication.

3.2 Numeric Coding

A huge cover text which is syntactically and semantically correct is taken. The secret message from the user which consists of characters 0–9, A–Z, and a–z is converted into a binary scheme.

An analysis on the number of synsets beginning with each character is made using Sanskrit WordNet and is described in Table 1. This gives an idea of the words occurring or can be replaced with each character. Based on the above analysis, we propose a numeric coding which is described in Table 2.

Table 1 WordNet synset count

a	A	i	I	u	U	q	Q	L	e	E	o
4613	1836	353	121	1541	106	101	0	0	238	56	61
O	M	k	K	g	G	f	c	C	j	J	F
66	47	4725	480	1576	235	2	1162	179	1180	1180	1
t	T	d	D	N	w	W	x	X	n	p	P
75	6	66	18	4	1549	9	1654	564	2152	5024	165
b	B	m	y	r	l	v	h	S	R	s	
1458	1247	2586	562	1256	758	4220	1048	2550	130	4827	

Table 2 Numeric coding

0	1	2	3	4	5	6	7
m q	e E	u U	Q i o l	a M k w	A O p y	I W g r	P x b
8	9	10	11	12	13	14	15
G X v	f n C	c R	K T s	j d	B J D	t h F	S N

3.3 Encoding

Initially, the sentences in the cover text are parsed to obtain the morphological properties of each word. Once this is done, the algorithm takes the binary scheme obtained from conversion of secret text and divides it into groups of 4 bits. If the number of bits is not divisible by 4, then pad the binary string with zeros appropriately. Each group is then converted to its decimal equivalent whose value lies between 0 and 15 inclusive. Next, the decimal equivalent of first group is taken and then the corresponding letters that have been given that score are identified from Table 2. All those letters are stored in a list. The system then checks if first character of the word in the cover text matches with any of the letters in the list. If it matches, then that word is given a score of 1. If it does not match, then the synsets of that word are found by using the Sanskrit WordNet and checked if any of those synsets start with any of the letters in the list. If so, then original word is replaced with its synonym and then the word is given a score of 1. The morphological properties obtained during parsing is used by the morphological generator to ensure that the replaced synonym has the same property as the original word. If this fails then the word is given a score 0 and system will start to process the next word in the cover text.

This process continues till the algorithm finds a word matching for all the groups that were created initially using the secret message. After hiding the secret message into the cover text, a media text is obtained which needs to be transferred. At the end of the process a sequence of 0's and 1's is generated which denote if a word in the cover text contributes to the process of hiding or not. This sequence is termed as the *Key* and is dynamically created based on the stego-text and the secret message to be hidden. This key is again divided into groups each containing 4 bits. The decimal equivalent of the groups are calculated and are mapped to a set of characters (Table 2) with which a word can possibly start.

A knowledge base is used to generate the short sentences whose words start with the characters which have been mapped to the groups of the *Key*. The thematic relations between the possible words in the context of the cover text are present in the knowledge base. If the key needs to be two words long then an agent and a verb is picked from the knowledge base. If the key is to be of size 3, then a meaningful sentence of 3 words is formed from the knowledge base. Hence, the algorithm generates a text containing the hidden message which can be transferred without any suspicion.

3.4 Decoding

In the stego-text, the receiver is unaware of the point which divides the message into the part which contains the hidden message and the *Key*. During encoding, as sentences are appended at the end, it is known that a full stop would separate the stego-text into two parts. The algorithm starts checking from the last sentence and iterates backwards. If the number of words before a particular full stop is almost 4 times the number of words after it, that full stop is identified as the point of partition. The system then takes the second part and converts the starting letter of each word into its equivalent integer between 0 and 15 hence creating a sequence of 0's and 1's. After this, the starting letter of all the words before the partition point with a score of 1 are considered. These characters are converted to its integer mapping and then to its binary equivalent by grouping them into bits of 8. Each group of 8 bits represents an ASCII value which in turn represents a character. The combination of these characters represents the hidden message.

4 Implementation

Each sentence in the cover text is parsed to get the morphological properties which also resolves the real time ambiguities. Once parsing is done, the encoding process starts. During the encoding process, a searchKey is needed to search for synonyms from WordNet. The searchKey is the base form for nouns and the root with optional prefix for verbs. The morphological generator is used to get the base form in case of a noun. For verbs, the root (and prefix) is obtained while parsing. After obtaining the searchKey, WordNet is queried to search for a synset beginning with the required character. If the query is successful, morphological generator is used to get the required inflected form of the word. The word is scored 1. If no match is found then it will be scored 0.

Consider the word in cover text *vanasya*. Upon parsing the properties of the word *vanasya* as vana (root) neutral (gender), genetive (case), singular (number) are obtained. Suppose a word beginning with 'k' is required to hide the message. Clearly *vanasya* does not begin with 'k'. The base form of *vanasya* i.e. *vanam* is obtained using the Morphological generator. Then, the WordNet is queried for *vanam*. The search-results obtained contain words *araNyam, atavI, vipinam, kAnanam, gahanam* etc. Among these search-results, the word *kAnanam* begins with 'k'. The morphological generator is then used to get the inflected form of *kAnanam* in genetive case, singular, which is *kAnanasya*. Thus *vanasya* is replaced by *kAnanasya* and marked as 1.

In the fusional language Sanskrit, compound words are likely to occur in the text. It is possible to split a compound word into its individual morphemes and then perform our encoding operation. This will probably reduce the size of the cover text required to hide the secret message, but splitting the compound word is likely to affect the orthographic properties of the language. Additionally, the stego-text will be more suspicious if one finds Sandhi split at unusual places. Hence this method refrains from performing Sandhi splitting during encoding.

For example consider a scenario where the word *prawyupakAraH* is in the cover text and the encoding algorithm needs a word beginning with 's'. In our proposed algorithm the word *prawyupakAraH* is marked 0 and the next word is processed. This is because there exists no synonym for *prawyupakAraH* in the WordNet. On the other hand if we perform a sandhi split, $prawyupakAraH = prawi + upakAraH$. *sahAyam* is a synonym of *upakAraH*. So now $prawyupakAraH = prawi + sahAyam$. *prawi* will be given a 0 and *sahAyam* a 1, hence hiding becomes more efficient. Normally *prawyupakAraH* is extensively used and when a person encounters *prawi sahAyam* in the text, he is most likely to notice this usage and can arouse some suspicion.

4.1 Morphological Analyzer and Generator

The Sanskrit tools available at *The Sanskrit Heritage Site* (http://sanskrit.inria.fr/) [16] are used for performing Morphological analysis and parsing. These tools can resolve the real time ambiguities as well.

4.2 Knowledge Base

For each cover text there are a set of possible thematic relations which forms a knowledge base. Semantically meaningful short sentences related to the context of the cover text are generated using this knowledge base. Table 3 gives the structure of the knowledge base.

Table 3 Thematic relations in knowledge base

Property	Description
Text	Cover text
Nominative	Subject
Accusative	Direct object
Instrumental	Object used to perform action
Dative	Indirect object
Ablative	Movement from something
Genitive	Possession
Locative	Location
Verb	Action or event
Other words	Other words

4.3 Example—Encoding

Message to Hide: IAS

Cover Text: purA BArawe aSoka nAma rAjA AsIw. eRaH prajA priyaH ca AsIw.

Internal Computations

$ASCII - I = 73, A = 65, S = 83$

$Binary - I = 0100\ 1001, A = 0100\ 0001, S = 0101\ 0011$

$Grouping - 0100(4)\ 1001(9)\ 0100(4)\ 0001(1)\ 0101(5)\ 0011(3)$

Applying the proposed method

purA BArawe aSoka(*) nAma(*) rAjA(*) AsIw. eRaH(*) prajA(*) priyaH(*) ca AsIw.
where * indicates the possibility to hide in that word on replacement by a synonym.

Stego-Text—*purA BArawe aSoka nAma aXipaH AsIw. eRaH prajA iRtaH ca AsIw.*

Bit Sequence of the text 0011(3) 1011(11) 1000(8)

Key Generation

From the bit sequence a sentence with 3 words needs to be generated as the *Key*. From the knowledge base a sentence is formed *ixAnIM saH viKyAwaH*. This is appended to the stego-text to get the final message.

Final Message

purA BArawe aSoka nAma aXipaH AsIw. eRaH prajA iRtaH ca AsIw. ixAnIM saH viKyAwaH.

4.4 Example—Decoding

Start dividing the sentence into 2 parts. Start this in the backward direction.

Message

purA BArawe aSoka nAma aXipaH AsIw. eRaH prajA iRtaH ca AsIw. ixAnIM saH viKyAwaH.

Initially it is assumed that the last sentence is the point of separation.
ixAnIM saH viKyAwaH − no of words = 3

Also 3 * 4 = 12 is almost equal to number of words in rest of the text. Thus this is identified as the point of separation. The last sentence is the *Key* and the sentences before it contain the hidden message.

Using the defined Numeric Encoding, the numeric equivalent of the *Key* is,
i = 3(0011), s = 11(1011), v = 8(1000)

Thus Contribution Array is 0011 1011 1000

Such that all the words in position of 0 (zero) are ignored.

Decoding

purA BArawe aSoka nAma aXipaH AsIw. eRaH prajA iRtaH ca AsIw.
0 0 1 1 1 0 1 1 1 0 0
a = 4(0100), n = 9(1001), a = 4(0100), e = 1(0001), p = 5(0101), i = 3(0011)

Hidden Message in Bits – 01001001 01000001 01010011

Grouping the bits into groups of 8

Binary – I = 0100 1001, A = 0100 0001, S = 0101 0011

ASCII – I = 73, A = 65, S = 83

Thus secret message is IAS.

5　Results

A set of 15 stego-texts were given to 10 individuals who are well versed in Sanskrit and they were asked to score (range 0–10) on the syntactic and semantic correctness of the stego-text along with their views on if the text is suspicious. Table 4 describes the average score given by each of the 10 evaluators for the 15 texts. The system generated sentences with syntactic correctness of 96.7%, semantic correctness of 86.6% and a suspicion factor of only 23.4%.

Table 4 Results

Syntactic correctness	Semantic correctness	Suspicion score
9.8	9.2	1.6
9.5	8.8	2.4
9.7	9.1	2
9.4	8.5	2.6
9.6	8.3	3
9.8	8.7	2.5
9.9	9.1	1.2
9.5	8.6	2.2
9.7	8.4	2.3
9.8	7.9	3.6
9.67	**8.66**	**2.34**

6 Conclusion

Text Steganography can be used for sharing information securely. In this work, a unique technique is presented which uses the morphological and grammatical properties of Sanskrit to make the hiding efficient. A knowledge base is used to generate a key dynamically and the key is also sent as a part of the message. Overall, the properties and flexibility of Sanskrit language make this method dynamic and logical. As the stego-text is syntactically and semantically correct to a large extent, the goal to generate the least suspicious text is achieved. Further, the dynamically generated key is hidden within the message, hence making it difficult for a middleman to figure out what words are to be considered in decoding even if he finds the message suspicious. So this method can securely transmit data across the sender and receiver.

As a part of our future work, we wish to make the key contextually more relevant to the cover text. Better exploitation of the structure and properties of Sanskrit and a better knowledge base will surely help improve the semantic correctness of the message.

References

1. Wayner, P.: Mimic functions. Cryptologia **XVI**, 193–214 (1992)
2. Chapman, M.T.: Hiding the hidden: a software system for concealing ciphertext as innocuous text. In: Information and Communications Security Volume 1334 of the series Lecture Notes in Computer Science, pp. 335–345 (1997)
3. Meng, P., Huang, Chen, L.Z., Yang, W., Li, D.: Linguistic steganography detection based on perplexity. In: International Conference on MultiMedia and Information Technology, pp. 217–220 (2008)
4. Moerland, T.: Steganography and Steganalysis (2003)
5. Low, S.H., Maxemchuk, N.F., Brassil, J.T., O'gorman, L.: Document marking and identification using both line and word shifting. In: Proceedings of the Fourteenth Annual Joint Conference of the IEEE Computer and Communications Societies (INFOCOM), vol. 2, pp. 853–860 (1995)
6. Bennett, K.: Linguistic Steganography: Survey, Analysis, and Robustness Concerns for Hiding Information in Text. Purdue University, CERIAS Tech
7. Changder, S., Das, S., Ghosh, D.: Text steganography through Indian Languages using feature coding method. In: Eighth International Conference on Information Technology: New Generations (2011)
8. Kim, Y., Moon, K., Oh, L.: A text watermarking algorithm based on word classification and inter-word space statistics. In: Proceedings of the Seventh International Conference on Document Analysis and Recognition (LCDAR), pp. 775–779 (2003)
9. Huang, D., Van, H.: Interword distance changes represented by sine waves for watermarking text images. IEEE Trans. Circuits Syst. Video Technol. **II**(12), 1237–1245 (2001)
10. Pathak, M.: A new approach for text steganography using Hindi Numerical Code. Int. J. Comput. Appl. **1**(8) (2010)
11. Changder, S., Debnath, N.C.: A new approach for steganography in Bengali text. In: Proceedings of the International Conference on Software Engineering and Data Engineering, pp. 74–78 (2009)
12. Changder, S., Ghosh, D., Debnath, N.C.: A Greedy approach to text steganography using properties of sentences. In: Eighth International Conference on Information Technology: New Generations (2011)

13. Vasu, S.C. (ed.): The Ashtadhyayi of Panini, vol. 2. Motilal Banarsidass Publishers (1962)
14. Gupta, R., Goyal, P., Diwakar, S.: Transliteration among Indian Languages using WX notation. In: Proceedings of KONVENS, Saarbrucken, Germany (2010)
15. Kulkarni, M., Dangarikar, C., Kulkarni, I., Nanda, A., Bhattacharyya, P.: Introducing Sanskrit Wordnet (2010)
16. Huet, G.: Formal structure of Sanskrit text: requirements analysis for a mechanical Sanskrit processor. In: Sanskrit Computational Linguistics 1 & 2. Springer, LNAI 5402 (2009)

A Novel Real-Time Face Detection System Using Modified Affine Transformation and Haar Cascades

Rohit Sharma, T. S. Ashwin and Ram Mohana Reddy Guddeti

Abstract Human Face Detection is an important problem in the area of Computer Vision. Several approaches are used to detect the face for a given frame of an image but most of them fail to detect the faces which are tilted, occluded, or with different illuminations. In this paper, we propose a novel real-time face detection system which detects the faces that are tilted, occluded, or with different illuminations, any difficult pose. The proposed system is a desktop application with a user interface that not only collects the images from web camera but also detects the faces in the image using a Haar-cascaded classifier consisting of Modified Census Transform features. The problem with cascaded classifier is that it does not detect the tilted or occluded faces with different illuminations. Hence to overcome this problem, we proposed a system using Modified Affine Transformation with Viola Jones. Experimental results demonstrate that proposed face detection system outperforms Viola–Jones method by 6% (99.7% accuracy for the proposed system when compare to 93.5% for Voila Jones) with respect to three different datasets namely FDDB, YALE and "Google top 25 'tilted face'" image datasets.

Keywords Face detection · Face tracking · Affine transformation
Haar cascades · AdaBoost · OpenCV

R. Sharma (✉) · T. S. Ashwin · R. M. R. Guddeti
Information Technology Department, National Institute of Technology Karnataka,
Surathkal, Mangalore, Karnataka, India
e-mail: sharmarohit0013@gmail.com

T. S. Ashwin
e-mail: ashwindixit9@gmail.com

R. M. R. Guddeti
e-mail: profgrmreddy@gmail.com

© Springer Nature Singapore Pte Ltd. 2019   193
P. K. Sa et al. (eds.), *Recent Findings in Intelligent Computing Techniques*,
Advances in Intelligent Systems and Computing 707,
https://doi.org/10.1007/978-981-10-8639-7_20

1   Introduction

Face recognition systems are used in many applications such as remote surveillance for E-learning, home, industry, national security, traffic monitoring, etc. But these systems require robust method to recognize the face even if it is tilted, occluded or with a very difficult recognizable pose.

There are many approaches to detect the face in a given frame of an image but it is very difficult when the face has affine distortions, Clutter, environment and unwanted noise, different pose, orientation (tilted), scaling, illumination, etc. The most generic limitation of face recognition by several states of the art algorithms like Viola Jones, etc. [1] is when the face is tilted. There are methods [2–4] like Local Binary pattern (LBP), Linear Discriminant Analysis, Scale Invariant Feature Transform (SIFT), Independent Component Analysis (ICA), Eigen signatures, etc., available to detect the frontal, tilted and occluded faces but with 62–88% accuracy.

Linear Analysis Methods [3] for face detection cannot address the variations in illumination, pose, expression whereas techniques like LDA, Principle Component Analysis (PCA), ICA, Laplacian methods recognize the entire face as a raw input for processing, i.e., it requires recognition of eyes mouth and nose to recognize it as a face. Hence these approaches fail to recognize the face even if it is tilted or occluded. On the other hand, we have dense local descriptors and sparse local descriptors like SIFT and LBP which address the issues related to the location, scale and also it is robust to affine transformation and changes in illuminations. Although the feature based techniques are more tolerant to missing parts, but fail to give better results with respect to emotion recognition from facial expressions.

The majority of existing approaches will not address the accurate recognition of frontal faces, tilted faces ($-80°$ to $+80°$), occluded and with different illuminations. Hence our proposed algorithm combines the state of the art technique, i.e., Voila Jones and Modified Affine transformation for the real-time face recognition system that not only detects the frontal face accurately but also it detects the tilted face (even if the face is completely rotated in $80°$), occluded and different illuminated face present in the given image. Further, our proposed system tracks the face in any streaming data.

The core idea of our proposed algorithm is to combine the Modified Affine Transformation along with Viola–Jones Haar-cascade in order to detect the face in the unknown image or given image even in the real-time streaming environment.

The key contributions of this paper are as follows:

- Combined approach of both Modified Affine Transformation and Viola–Jones Haar-cascade in order to detect face in the unknown image or given image even in the real-time streaming environment.
- The proposed system accurately detects the frontal, Different Orientation or Tilted, Occluded faces with Various Illumination Conditions and with Various Poses for a given frame of an image.

The rest of this paper is organized as follows. Section 2 describes the Related Work for the paper. Section 3 explains the Proposed Methodology and the algorithms

Table 1 Summary of existing works

Authors	Approaches	Merits	Limitations
Paul et al. [1]	Integral image, AdaBoost, cascade	Detect face with high accuracy and very less computation or processing power	Cannot detect a face if it is tilted
Wheeler et al. [8]	Biometric recognition	Biometric recognition at a significant range	The system fails when person's face is tilted and not aligned in an upright direction
Lee et al. [9]	PTZ camera is used with automatic operation	Cost-effective solution in surveillance systems	Fails when the face is tilted or not aligned in up right direction
Lang [10]	AdaBoost	Detect face with high accuracy	Cannot detect face which is tilted to some degrees
Zhu et al. [11]	AdaBoost and Haar-like feature	High performance in both accuracy and speed of the developed system	Fails to detect face which its tilted to some degree

used in the detection and tracking of faces for a given frame in an image. Section 4 discusses the Experimental Setup. Section 5 describes the Results and Analysis for the face detection and tracking. Finally, Sect. 6 Concludes with Future directions.

2 Related Works

Face detection is itself a big problem with differences in poses, an orientation of the face, expression, and lighting. The processing image becomes difficult because of the presence or absence of glasses and facial hair [5]. Knowledge-based, template matching, feature invariant, and appearances based are some current methods which are used to detect or find the face in a given image. Table 1 shows the summary of existing face detection techniques their merits and limitations.

Hence, in this paper, we propose a robust system using Modified Affine transformation and Haar cascades which recognize the face which is tilted, occluded or with any difficult pose with different illuminations. The proposed algorithm can detect the face irrespective of its pose in the image. Further, proposed system can detect faces even if they are tilted up to |80|° in the image.

3 Proposed Methodology

Figure 1 shows the flow of how the face is detected, which takes input as the image then try to detect the face in that image via Viola–Jones Haar Cascade. If the face is not detected then we use modified Affine Transformation to rotate the image to a degree and detect the face again, increment the degree by 3 for every step till face is detected in that image from $-80°$ to $+80°$. If loop completes with no face detected then we can conclude that there is no face in that image. For real-time face detection, frames in the video stream are used and detected independently.

The following steps are performed to recognize and track the face in a streaming data of an image:

i. Recognize face in image,
ii. Follow the face in subsequent frames.

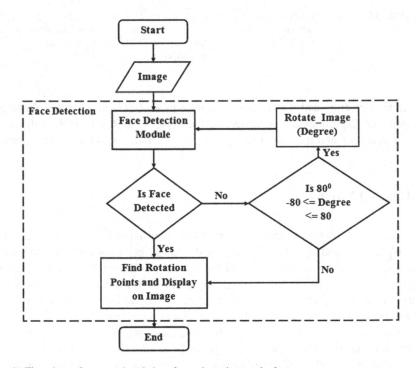

Fig. 1 Flowchart of proposed real-time faces detection method

3.1 Affine Transformation

Linear 2-D geometric transformations consist of affine transformation which is one of the important class which maps variables or coordinates (e.g., in an input image values of the pixel intensity at position (x_1, y_1) into a new variables or coordinates (e.g., (x_2, y_2) in the output image), this is done by applying a linear combination of scaling, translation, shearing, and rotation operations is applied to achieve this.

The affine transformation is generally written in homogeneous coordinates as shown in the Eq. (1).

$$\begin{bmatrix} X_2 \\ X_2 \end{bmatrix} = A \times \begin{bmatrix} X_1 \\ Y_1 \end{bmatrix} + B \tag{1}$$

Matrix A is used for pure rotation and it is defined as (Eq. (2)):

$$A = \begin{bmatrix} \cos\theta & -\sin\theta \\ \sin\theta & \cos\theta \end{bmatrix}, \quad B = \begin{bmatrix} 0 \\ 0 \end{bmatrix}$$

(For clockwise rotations) (2)

Now to get the new coordinates (x', y') for the image using the old coordinates (x, y), let us consider $u = $ initial angle, $i = $ angle of rotation.

$$x = r \cos u$$
$$y = r \sin u$$
$$x' = r \cos(u + i) = r \cos u \cdot \cos i - r \sin u \sin i$$
$$y' = r \sin(u + w) = r \sin u \cdot \cos i + r \cos u \sin i$$

hence:

$$x' = x \cos(i) - y \sin(i)$$
$$y' = y \cos(i) + x \sin(i)$$

Thus, when we rotate the original point (x, y) about another (origin) point (t, q) counterclockwise by an angle θ, we can calculate the new point's coordinates by

1. translating the entire plane so that (t, q) goes to the origin and
2. perform the rotation and translate the entire plain back.

Therefore, the new point's coordinates are given as (Eqs. (3) and (4))

$$x' = (x - t)\cos(\theta) - (y - q)\sin(\theta) + t, \tag{3}$$
$$y' = (x - t)\sin(\theta) + (y - q)\cos(\theta) + q. \tag{4}$$

3.2 Proposed Algorithm

The main goal of the proposed algorithm is to detect the face in an image at real time with high accuracy. If face in the image is not detected by viola Jones then the proposed algorithm rotates the image sequentially to detect the face, if an image is rotated in all possible way and still not able to detect then there is no face in that image, which is not even in the image train dataset. We have used Affine Transformation for rotating the image in a given plain. The flowchart of the proposed real-time faces detection method is shown in Fig. 1.

Algorithm 1: Image Rotation

Start algorithm
1: Degrees=[-80 to 80]
2:     For images in IMAGES{
3:         For angles in Degrees{
4:             Rotate_Image=rotate_image(image,angle)
5:             DetectFace=face.detectMultiscale(rotateImage)
6:             If(detectFace){
7:                 //Face is detected at Angle angle
8:                 Break
9:             }
10:            //Draw Rectangle over the Image
11:            Rectangle(image,pt1,pt2,pt3,pt4)
12:            Image.show()
13: }
End algorithm

Algorithm 1 is used to find the face in the given image, if it is not able to detect any face then it calls Algorithm 2 to rotate the image and angle it in a sequential manner and sends the new image back to the Algorithm 1 and then the detected method again finds the face in that image. But the problem with image rotation is to obtain the coordinates of the face from where we will be recognizing the face using the rectangle to indicate that face has been detected. Here, by using Algorithm 3, one can easily get the coordinates of the face but these face coordinates are detected in that rotated image but not in the original image. Even if we have the coordinates for the face and even if we try to map these coordinates of the original image then we may get wrong results.

Algorithm 2: Affine Transformation

Start algorithm
1://rotate image by degree angle
2://(Using Affine Transformation)
3: rotate_image(image,angle){
4:          if(angle==0 or -1 or 1){
5:              return image
6:          }
7:          //get the height & width of image
8:          Height,width=image.shape()
9:          //image_matrix is rotated by using affine transformation
10:        Rotate_matrix = gctRotationMatrix (height, width, image, angle)
11:        //warp the image
12:        Result_Image=warpAffine(Image,Rotate_matrix)
13:        Return Result_Image
14: }
End algorithm

Algorithm 3: Coordinate Detection

Start algorithm
1: //find the Position or points (x,y) to draw rectangle
2:    //get   the   points   and   rotate   them   in   angle   around   the origin
3: Rotate_Point(Points[],image,angle){
4:        x,y,h,w = Points[] //x,y coordinate and height and width from that x,y
5:        P=image.shape(x)*0.4 //origin x coordinate
6:        q=image.shape(y)*0.4 //origin y coordinate
7:        new_x_point1=(x-p)cos(angle)-(y-q)sin(angle)+p
8:        new_y_point1=(x-p)sin(angle)+(y-q)cos(angle)+q
9:        new_x_point2=(x+h-p)cos(angle)-(y-q)sin(angle)+p
10:      new_y_point2=(x+h-p)sin(angle)+(y-   q)cos(angle)+q
11:      new_x_point3=(x-p)cos(angle)-(y+wq)
            sin(angle)+p
12:      new_y_point3=(x-p)sin(angle)+(y+wq)   cos(angle) + q
13:      new_x_point4=(x+h-p)cos(angle)-(y+wq)   sin(angle)+p
14:      new_4_point4=(x+h-p)sin(angle)+(y+wq)   cos(angle)+q
15:      return
16:      [(new_x_point1,new_y_point1),
            (new_x_point2,new_y_point2),
            (new_x_point3,new_y_point3),   (new_x_point4,new_y_point4)]
16: }*End algorithm*

In some cases, we will get accurate results even if the face is tilted with very small degree (around -150 to 150), but if the face is tilted to greater degree, then we will miss the face coordinate in the original image. To overcome this problem we have used modified version of affine transformation which will rotate image in its origin and then to get the coordinates for the original image. As image is it made up of pixels it can be represents in matrix form and then we apply affine transformation on the image matrix M as given by Eq. (5).

(a)   **(b)**

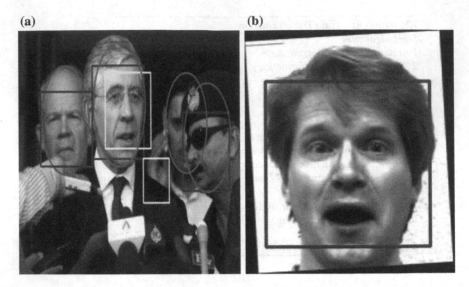

Fig. 2 Sample images from FDDB and YALE dataset

$$M = \begin{bmatrix} \cos q & \sin q & 0 \\ -\sin q & \cos q & 0 \\ 0 & 0 & 1 \end{bmatrix}, \quad B = \begin{bmatrix} 0 \\ 0 \end{bmatrix} \tag{5}$$

4 Experimental Setup

4.1 Datasets Used

Yale [6] database is used for testing which is in. PGM file extension. The images in the dataset are all frontal face views of male and female. Some faces included even glasses.

    FDDB [7] dataset consists of wide variety of 2845 gray scale and color images with a total of 5171 different faces which consist of occluded, difficult poses, low resolution, and out of face images. Sample Image of the dataset is shown in the Fig. 2.

4.2 Webcam and System Requirement

Final testing was performed via Webcam for real-time face detection. Frames of images are taken the video streaming data and faces are recognized for each frame. An IBM Intel core i7 system 3.4 GHz with 320 GB hard drive and 8 GB ram with a

Table 2 Results of face detection for various datasets

Technique	Dataset	Number of images	Number of faces detected	Runtime (s)
Viola Jones	YALE	165	165	1.12
	Google top 25 "tilted face"	25	2	0.02
	FDDB	3539	3312	52.12
Proposed method	YALE	165	165	1.12
	Google top 25 "tilted face"	25	25	1.28
	FDDB	3539	3529	55.88

webcam (resolution 640×360) were used in the experiment. Ubuntu 14.04 operating system is used with the CV2 library for working with Viola–Jones algorithms and some the basic operation on the image.

5 Results and Analysis

We used different datasets for the testing of our proposed algorithm. Following are the results with their runtime. It is evident that out proposed algorithm takes more time than Viola Jones but if we compare it to the number of faces detected our algorithm generates a better result than Voila Jones.

To compare the algorithms with different approaches we used dataset like YALE (Fig. 2b) and for tilted, low resolution, occluded faces with difficult poses, and out of face images, we used FDDB dataset (Fig. 2a) and "top 25 Google's searched 'tilted face'" images (Fig. 4). We obtained 100% accuracy for YALE dataset using our proposed method. Also, we obtained 100% accuracy for those 25 images of "top 25 Google's searched 'tilted face'" whereas Viola Jones was able to detect only 2 images with an accuracy of 8%. For FDDB dataset we ran our algorithm on 3539 images and we obtained 99.7% accuracy rate whereas that of Viola–Jones was 93.5%.

Results are shown in Table 2. The corresponding Number of faces detected for each dataset is compared with the total number of faces present in that dataset and also the result of Voila Jones method. Summary of this is shown in Fig. 3a. Similarly, the runtime for each dataset with respect to three different datasets using Viola–Jones and Our proposed method is shown in Fig. 3b (Fig. 4).

It is clearly observed from Fig. 3a that YALE database has same accuracy and runtime when compared to Voila Jones Method. This is because it contains only Frontal Face but there is a significant difference in other datasets since those datasets contain Tilted Face, Occlusion, different Pose and different Illuminations. Out of 3539 images of FDDB dataset, our proposed algorithm has detected 3529 faces whereas Voila Jones method detected only 3312 (Fig. 3a). But it is also evident that

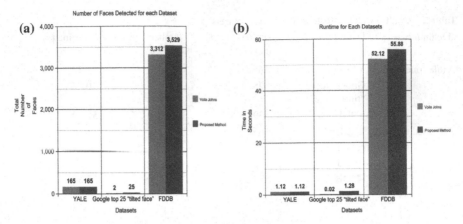

Fig. 3 Number of faces detected (**a**) and runtime for three different datasets (**b**)

Fig. 4 Detection results from Googles Top 25 "tilted face" search

our proposed algorithm took 55.88 s runtime when compared to 52.12 s of Voila Jones Algorithm (Fig. 3b). For real-time face detection, we used the webcam to capture the video stream data and then ran on the system. In real time we were able to detect and also track the face as shown in Fig. 5a, b. Figure 5a shows the detected face when it is horizontally tilted and also occluded to some extent. Further, Fig. 5b shows the detected face when it is vertically tilted.

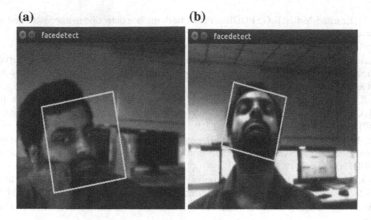

Fig. 5 Results of real-time face detection

6 Conclusion and Future Work

Our proposed algorithms based on Modified Affine Transformation and Viola–Jones-based Haar Cascades accurately detects the face in a given frame for real-time face detection and tracking. Experimental results demonstrate that our proposed algorithm outperforms the existing Voila Jones algorithm by 6% for YALE, FDDB and "top 25 Google's searched 'tilted face'" datasets. These datasets include frontal, occluded, and tilted faces with different illuminations. And also we have tested on real-time face detection and tracking using the webcam which gave better results with very good accuracy. The future work of the proposed method is to reduce the runtime by decreasing the time complexity by working with GPUs and parallelizing.

Declaration Authors have obtained all ethical approvals from appropriate ethical committee and approval from subjects involved in this study.

References

1. Viola, P., Jones, M.: Rapid object detection using a boosted cascade of simple features. In: Proceedings of the 2001 IEEE Computer Society Conference on Computer Vision and Pattern Recognition, 2001. CVPR 2001, vol. 1. IEEE (2001)
2. Kisku, D.R., et al.: Robust multi-camera view face recognition. Int. J. Comput. Appl. **33**(3), 211–219 (2011)
3. Cinque, L., Iovane, G., Sangineto, E.: Comparing SIFT and LDA-based face recognition approach. J. Discrete Math. Sci. Cryptogr. **11**(6), 685–704 (2008)
4. Mawloud, G., Djame, M.: Modified local binary pattern for human face recognition based on sparse representation. Int. J. Comput. Appl. **36**(2), 64–71 (2014)
5. Lee, M.-C., Chen, W.: Image compression and affine transformation for image motion compensation. U.S. Patent No. 5,970,173, 19 Oct 1999

6. Jain, V., Learned-Miller, E.G.: FDDB: a benchmark for face detection in unconstrained settings. UMass Amherst Technical Report (2010)
7. http://vision.ucsd.edu/content/yale-face-database
8. Wheeler, F.W., Weiss, R.L., Tu, P.H.: Face recognition at a distance system for surveillance applications. In: 2010 Fourth IEEE International Conference on Biometrics: Theory Applications and Systems (BTAS). IEEE (2010)
9. Lee, S., Xiong, Z.: A real-time face tracking system based on a single PTZ camera. In: 2015 IEEE China Summit and International Conference on Signal and Information Processing (ChinaSIP). IEEE (2015)
10. Lang, L., Gu, W.: Study of face detection algorithm for real-time face detection system. In: Second International Symposium on Electronic Commerce and Security, 2009. ISECS'09, vol. 2. IEEE (2009)
11. Zhu, J., Chen, Z.: Real time face detection system using Adaboost and Haar-like features. In: 2015 2nd International Conference on Information Science and Control Engineering (ICISCE). IEEE, 2015. Initial Experiment Results of Real-Time Variant Pose Face Detection and Tracking System

Palmprint Identification and Verification System Based on Euclidean Distance and 2D Locality Preserving Projection Method

Mouad M. H. Ali, A. T. Gaikwad and Pravin L. Yannawar

Abstract Biometrics authentication system using palmprint is played nowadays as a good research work on selected modalities, palm print has more features compare to fingerprint and it is hard to be copied and easy to acquire. The piece of work is primarily addressing the mechanism of preprocessing, feature extraction and matching of palmprint data, The Region of Interest (ROI) extracted by using *Euclidean distance*. The appearance-based approach like *Two-Dimension locality preserving projection* (*2DLPP*) is used for feature extraction technique. The matching conducting in two cases Identification and Verification with help of distance measure. The experiments conducted over CASIA Multi-spectral database v1.0 and the results shown the identification was giving the result 97.33% with error rate 5.33%, while the verification result is 94.67% with error rate 2.67% of the palmprint system.

Keywords Biometrics · Palmprint · ROI · 2DLPP · Identification · Verification

1 Introduction

Conventional authentication methods such as passwords, PINs, tokens, and smart card are not relevant for application on systems that require security high. The biometrics system replacing conventional methods by utilizing physical characteristics or behavior characteristics of human that actually represent a person's identity and advantages that are difficult to duplicate, stolen, and falsified [1]. There are three main challenges facing the biometrics system [2], namely accuracy, scale, and usability.

M. M. H. Ali (✉) · P. L. Yannawar
Department of CS&IT, Dr. Babasaheb Ambedkar Marathwada University, Aurangabad, India
e-mail: mouad198080@gmail.com

P. L. Yannawar
e-mail: pravinyannawarr@gmail.com

A. T. Gaikwad
Institute of Management Studies and Information Technology, Aurangabad, India
e-mail: drashokgaikwad@gmail.com

© Springer Nature Singapore Pte Ltd. 2019
P. K. Sa et al. (eds.), *Recent Findings in Intelligent Computing Techniques*,
Advances in Intelligent Systems and Computing 707,
https://doi.org/10.1007/978-981-10-8639-7_21

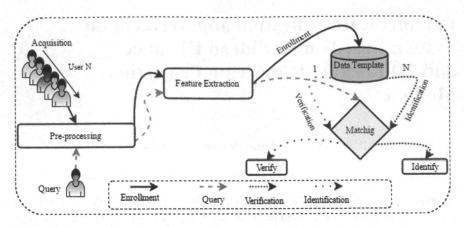

Fig. 1 Biometrics systems mode (Enrolment, Identification and Verification) process

Various proposed ways to improve the accuracy of biometrics systems such as by combining more than one biometric characteristic for the introduction or referred to as multimodal biometrics system [3]. In addition, various techniques of feature extraction is proposed to improve the performance of biometric systems among which the LDA, PCA, ICA, LBP, and the LDP [4–6]. In this study explained about palmprint and Two-Dimensional Locality Preserving Projection (2DLPP) for use in identity recognition proven successful with high accuracy [7, 8]. The LPP techniques is one of appearance based approach for biometrics system. And the main objective of LPP is to preserve the local structure of the image space for this Hu et al. [9] they are working on 2DLPP by extract the feature directly from images matrices, while the He et al. [10] they are used LPP for image feature extraction and dimension reduction and apply to face recognition and called first one implemented this technique and they get good results. In general the 2DLPP technique use to solve the generalized Eigen values problem [9], also in another hand the 2DLPP required more coefficients which are one of the disadvantages of 2DLPP technique for image representation and recognition [11]. 2DLPP can applied in column direction with help of 2DPCA projection in row direction [11]. Any biometrics system passing through above steps is mentioned in (Fig. 1). The main steps are acquisition, preprocessing, feature extraction, matching. There are two modes in the process of introducing biometric systems are made that verification and identification. Verification is the process of matching a person's biometric data with the template data in the database one-to-one matching. Identification is the process of matching a biometric data with all the data in the database one-to-many matching. There are many techniques reviews in [17].

This paper is arranged in four sections, the remaining sections will be explained the methodology given in Sect. 2. In Sect. 3 the Experiment and result analysis are given. Finally the conclusion and future work are given in Sect. 4.

2 Methodology

This piece of research work primarily focused on 2DLPP technique based on palmprint data for Identification and verification propose. This technique applied over CASIA palmprint dataset. The work divided to two scenarios which are identification and verification. The paper addressing many steps likes preprocessing, Feature Extraction, and Matching of palmprint data. The goal of preprocessing is to enhance and extract the ROI from palmprint data by using competitive hand valley detection method. The feature extraction used in this work is Two-Dimensional locality preserving projection (2DLPP) methods which take the ROI and extract the features and passing the features to next stages. The matching done between training and testing, here there are two concepts of matching called as Identification and verification depend on distance and threshold values. The identification process done by (1:N) comparisons between query and all templates the decision either "Known" or "Unknown". The verification is done by (1:1) comparison and the decision either "Accepted" or "Rejected". The Fig. 2 shows the methodology of this work.

2.1 Preprocessing Stage

The palmprint preprocessing stage covers many steps which are used to enhance, remove the noise and clarity of palmprint image which is useful to extract the real part on palmprint image which is called Region of Interest (ROI), it is small area and very important part and contains more information used to recognize the persons. Before the extract of ROI, there are many steps like acquisition the palmprint image and do in this image Binarization process, determine the boundary, Extreme point

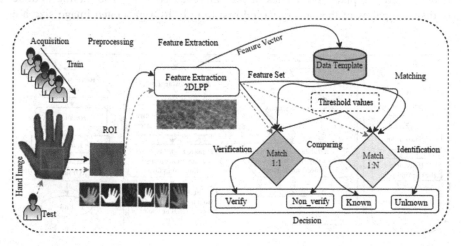

Fig. 2 Methodology of the system

on palmprint surface and determine the valley point, and put the reference point on it. The last steps are cutting the ROI region which is ready to pass to the next stage (Feature Extraction). For extract the ROI there are many methods like 1st is cropping ROI directly from palmprint image without using any enhancement [12]. The 2nd one is used the Competitive Hand Valley Detection (CHVD) [13]. The third used is the Euclidean distance (ED) [14]. In this paper we used Euclidean distance (ED) derived from [14] to extract ROI from palmprint images. The steps of algorithm are shown in flowchart in Fig. 3 started by calculating the Euclidean distance for all pixels on boundary to centroid point, then plotting the ED values to Curve, then plotting the first derivative of distance values from previous steps. After that eliminate the high-frequency components of the curve by using smoothing filter, then get the position of reference point of ROI by finding the point on the curve that passes through the zero value or changes from negative to positive—Fig. 4 shows all the steps. The visualizations of the ED methods applied on samples for extract the ROIs are shown in Fig. 5.

Valley point can be identified from the point of passing a value of 0 on the axis y and experiencing changes sign from negative to the positive. A reference point is used only points in the valley between the index finger with the middle finger (V_1) and between the ring finger and the little finger (V_3). Figure 6 shows the detection of ROI.

The rotation is done to cope with the variation of the position or orientation of the palm. Rotation is done by rotating the image at a certain angle (θ) is obtained from trigonometric Eq. (1)

$$\theta = \tan^{-1}\left(\frac{y_2 - y_1}{x_2 - x_1}\right) \tag{1}$$

Point x_1 and y_1 are the pixel coordinates of the reference point P_1 while the point x_2 and y_2 are the pixel coordinates of the reference point P_2. The scaling is done by distance palms of the camera varied resulting in ROI generated image size are

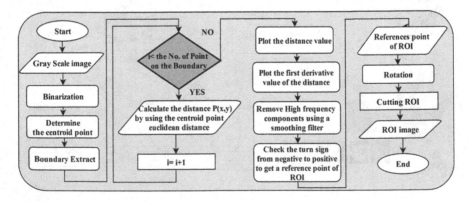

Fig. 3 Flowcharts of Euclidean Distance Algorithm of extract ROI of palmprint

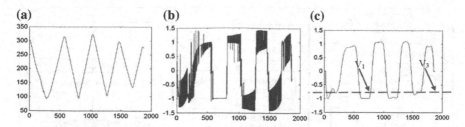

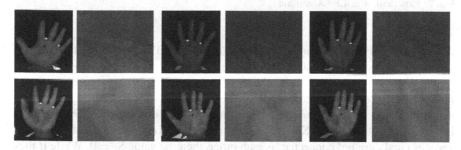

Fig. 4 **a** Euclidean distance of each pixel within the boundary with the centroid, **b** First derivative value, **c** First derivative graph smoothing value of Euclidean distance

Fig. 5 Samples of palm images and ROIs extract by Euclidean Distance Algorithm

Fig. 6 Detection ROI of an image of the palm of the hand

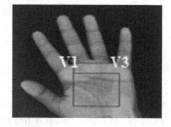

not consistent. Therefore, normalization dimension needs to be done in order to produce the image ROI with fixed size considering several feature extraction algorithms require image data with a consistent and uniform size, one 2DLPP. Normalization is done by scaling using Bicubic interpolation techniques.

2.2 Feature Extraction Stage Using 2DLPP

The hallmark of the (Characteristic) palm print is obtained by projecting the image using a transformation matrix obtained from the 2DLPP algorithm. Representations characteristics 2DLPP results referred to Laplacianpalm. The transformation matrix obtained from 2DLPP process that is applied to a set of training images $X = \{x_1, x_2,$

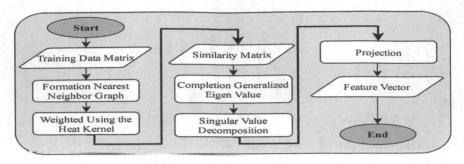

Fig. 7 Flowchart of 2DLPP Algorithm

..., x_N} with N is the number of samples in the dataset. The algorithm of 2DLPP [7, 15] are described in flowchart as show in Fig. 7.

The first step is establishment of nearest neighbor graph (S) is represented as a matrix in which each element S_{ij-ij} to indicate the closeness between images i and j in a dataset. Vertex i and j are connected if the image k_i is close to the neighboring point j using k-nearest neighbor (KNN). The second step is giving weight to the graph: If vertices i and j are not connected to each other, then the value $S_{ij} = 0$. While if they are connected, the weights can be calculated in two ways: first is simple-minded, giving a value of 1 S_{ij} if i and j are connected and the second way is a heat-kernel gives weight using the Eq. (2):

$$S_{ij} = e\frac{\|x_i - x_j\|^2}{t}, \tag{2}$$

where x_i and x_j are two images of observation and t is a constant that has been determined output of this phase is the similarity matrix is symmetric (S) dimension N × N. The third step is the resolution on the problem of generalized at this stage, eigenvector and eigenvalue calculated from the following Eqs. (3), (4) and (5)

$$XLX^T w = \lambda XDX^T w \tag{3}$$

$$D_{ij} = \sum_{ij} S_{ij} \tag{4}$$

$$L = D - S, \tag{5}$$

where X is a collection of images in a dataset matrix, D is the vector sum of the column or row of S given S is symmetric. L is the Laplacian matrix reduction results matrix with the vector S. Equation (3) Will produce a solution matrix w = { $w_1, w_2, ...,$ w_d} which is an eigenvector corresponding to Eigen values in ascending sequences. Finally the Projections are done by performing matrix multiplication between the matrix image X_i and Eigen vector W to get feature vector.

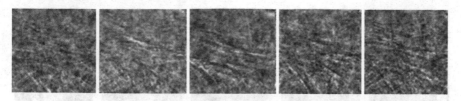

Fig. 8 Representation Laplacianpalm from some samples

$$x_i \rightarrow Y_i = X_i W, \quad i = 1, 2, \ldots, N \tag{6}$$

Vector line dimensionless from (Y_i) is the feature vector that represents the image of X_i. Figure 8 shows Laplacianpalm of some samples.

2.3 Matching Stage

There are two matching can be done by the system biometric verification and identification [1]. Verification is also called positive recognition or one-to-one matching is a matching done between the data belonging to a person's biometric data with the data in the database whose identity is claimed by the person. Identification which is also called a negative recognition or one-to-many matching is a matching done between the data belonging to a person's biometric data with all the data is in the database. We are use Euclidean distance method to calculate the similarity measure between the two images (input and template). The Euclidean distance Eq. (7) Use to measure the distance between the feature vector f_1 and f_2:

$$d(f_1, f_2) = \sqrt{\sum_{n=1}^{N} (f_{n1} - f_{n2})^2}, \tag{7}$$

where, N is the number of attributes or the long dimension of the feature vector f_1 and f_2. In the case of verification, the result is "accepted" or "rejected". While at the identification, the result is "known" or "unidentified" depend on threshold value. *The threshold* (θ) used in our system obtained from an array [16] by Eq. (8).

$$\Delta = \frac{\max(TA) - \min(TA)}{\beta} \tag{8}$$

Threshold Array (*TA*) Euclidean distance obtained from the distance between all the feature vector samples in a dataset. If a dataset there are N individuals with each M samples then there are N pieces of (TA), each TA has M pieces threshold. β is a constant predetermined threshold value to divide into N parts. Then, N values threshold will be tested to obtain optimal values of FAR and FRR by Eq. (9).

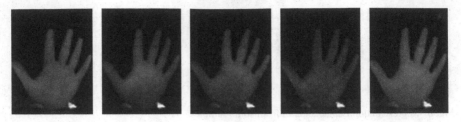

Fig. 9 Samples taken from CASIA dataset

$$\theta_1 = \min(TA) + \Delta$$
$$\theta_2 = \min(TA) + 2\Delta$$
$$\theta_N = \min(TA) + N\Delta \tag{9}$$

The threshold value of all i (i = 1, 2, …, N) or at θ_i then selected when the value approaches the FRR or FAR value is very small [16] depending on the specifications required.

3 Experiment and Result Analysis

The experiment is test on CASIA Multi-Spectral Database v1.0 palmprint comprising palmprint image (illuminator 460 nm) measuring 768×576 pixels belonging to 100 individuals with ID 001-100, 6 samples each individually labeled 01-06, and using the left hand alone ($100 \times 6 \times 2 = 1200$). Figure 9 shows some samples taken from dataset. And run on a computer with an Intel (R) Core (TM) i3 2.4 GHz and 2 GB of RAM with Windows 7, 64-bit and simulation tools MATLAB R2013a.

In this study we test our system performance by using two factors which are the basic types of errors on biometrics system that FAR and FRR [1]. System performance analysis is done with a metric measurement demonstrated by the ROC curve. The performance obtained by the intersection between the curves of the FAR and FRR called the Equal Error Rate (EER). The lower the EER value, the better the resulting performance. This Performance done in two ways: using the accuracy, FAR and FRR. Accuracy, False Accept Rate (FAR) and False Reject Rate (FRR) is calculated using the formulas (10), (11) and (12) respectively.

$$\text{Accuracy} = 100 - \frac{FAR + FRR}{2} \tag{10}$$

$$FAR = \frac{\text{imposter score}}{\text{All imposter score}} \tag{11}$$

$$FRR = \frac{\text{Genuine score failing below thershold}}{\text{All Genuine score}} \tag{12}$$

Table 1 Palmprint system accuracy

(Data test with data training)	Accuracy (%)
Sample [01] with [02, 03, 04, 05, 06]	92
Sample [02] with [01, 03, 04, 05, 06]	95
Sample [03] with [01, 02, 04, 05, 06]	92
Sample [04] with [01, 02, 03, 05, 06]	96
Sample [05] with [01, 02, 03, 04, 06]	98
Sample [06] with [01, 02, 03, 04, 05]	96
Average accuracy	94.83

Table 2 Palmprint accuracy with several feature dimensions

Feature dimension (%)	Feature point	Accuracy (%)
10	60	88.17
30	180	93.67
40	240	94.50
60	**360**	**94.83**
80	480	94.33
90	540	94.67

Table 3 FAR and FRR generated from testing five β values along its threshold value

β	FAR	FRR	Threshold value
25	0.033	0.020	304.3551
50	0.030	0.037	286.0596
75	0.033	0.020	304.3551
100	0.030	0.023	**295.2073**

Tests were conducted first to find the baseline accuracy of palmprint using all of the attributes on which some 600 feature vector attributes (length dimension (n) = 600). This test uses the data to the configuration model to calculate the accuracy. Table 1 shows the baseline accuracy of palmprint.

The number of feature point from feature vector length was tested with multiple values to find the optimal length for palm print feature vector. Best accuracy results from several long dimension of the feature vector tested right on palmprint shown in Table 2.

The value of accuracy if plotted into a graph would yield curve is as Fig. 10 shows the characteristic effect of the number of attributes of the feature vector accuracy. Dimension reduction has the advantage that it can shorten the processing time and eliminate irrelevant attributes which can degrade accuracy. To obtain optimal threshold value we use some values β Table 3 show the FAR and FRR with β value we see that threshold value that generates the value of the FAR and FRR is the smallest of the four values of β.

Fig. 10 ROC accuracy with
different feature vector
length of palmprint system

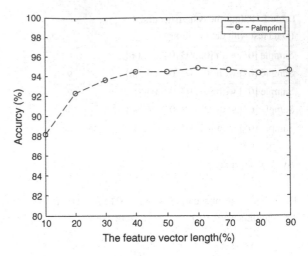

Fig. 11 Relations between
FAR and the FRR depend on
the value of threshold

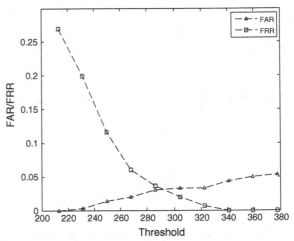

The optimal threshold value is taken based on the value of the FAR and FRR lowest that can be achieved is 0.030 and the 0.023 respectively. The relation between FAR and FRR based on the threshold values are shown in Fig. 11 and the intersection of FAR and FRR is called EER which achieved 0.035 seen from the curve. Figure 12 depicts the performance of the system for all threshold values. The threshold value used for the test is 295.2073 of the 300 trials for each verification and identification of 100 individuals, the result is 16 error occurs when the process of verification and 15 errors (eight errors in reception and seven errors in denial) occurs during the identification process. The system performance for identification and verification purpose is shown in Table 4. The execution times of the steps in the system are shown in Table 5.

Fig. 12 ROC curve of FAR versus GAR

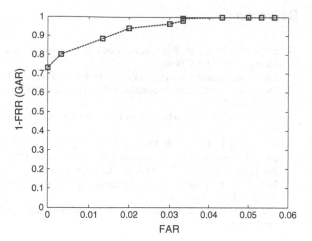

Table 4 Performance of biometrics mode (verification, identification) system

Biometrics mode	Error rate (%)	Recognition rate (%)
Verification	5.33	94.76
Identification	2.67	97.33

Table 5 Execution time in second of the system

Stages	Execution time (s)
Preprocessing	0.102
Feature extraction	0.202
Identification	0.502
Verification	0.305

4 Conclusion and Future Work

This study has been discussed on palm print recognition system Using 2DLPP feature extraction algorithms, which give high accuracy. The advantages of 2DLPP is capable of taking a characteristic topology or information on the local structure of an image with precision. The disadvantage of this algorithm is not resistant to variations in lighting and orientation. Lighting variation overcome by dividing the sample training data and test data with image composition with dark lighting and light evenly while variations in position or orientation overcome by normalizing the image of ROI by performing rotation and scaling. In this system we did the identification and verification separately and the result for identification show the less error rate 2.67% with the recognition rate 97.33%, while the verification system the error rate is very high 5.33% and the recognition rate is 94.76%. This work may extend for combining 2DLPP and LBP with neural network to improve our performance for palmprint recognition system.

References

1. Jain, A.K., Flynn, P., Ross, A.A.: Handbook of Biometrics. Springer Science and Business Media, LLC, USA (2008)
2. Jain, A.K., Pankanti, S., Prabhakar, S., Hong, L., Ross, A.: Biometrics: a grand challenge. In: Proceedings of the 17th International Conference on Pattern Recognition, ICPR 2004, vol. 2, pp. 935–942 (2004)
3. Mishra, A.: Multimodal biometrics it is: need for future systems. Int. J. Comput. Appl. **4**, 28–33 (2010)
4. Connie, T., Teoh, A., Goh, M , Ngo, D.: Palmprint recognition with PCA and ICA. In: Conference of Image and Vision Computing New Zealand (IVCNZ'03), pp. 227–232 (2003)
5. Mirmohamadsadeghi, L., Drygajlo, A.: Palm vein recognition with local binary patterns and local derivative patterns. In: 2011 International Joint Conference on Biometrics (IJCB), Washington, DC, pp. 1–6 (2011)
6. Wu, X.-Q., Wang, K.-Q., Zhang, D.: Palmprint recognition using Fisher's linear discriminant. In: Proceedings of the International Conference on Machine Learning and Cybernetics (IEEE Cat. No.03EX693), vol. 5, pp. 3150–3154 (2003)
7. Chen, S., Zhao, H., Kong, M., Luo, B.: 2DLPP: a two-dimensional extension of locality preserving projections. Neurocomputing **70**(4–6), 912–921 (2007)
8. Wang, J.G., Yau, W.Y., Suwandy, A.: Fusion of palmprint and palm vein images for person recognition based on Laplacian palm feature. Pattern Recogn. **41**(5), 1514–1527 (2008)
9. Hu, D., Feng, G., Zhou, Z.: Two-dimensional locality preserving projections (2DLPP) with its application to palmprint recognition. Pattern Recogn. **40**, 339–342 (2007)
10. He, X., Yan, S., Hu, Y., Niyogi, P., Zhang, H.: Face recognition using Laplacianfaces. IEEE Trans. Pattern Anal. Mach. Intell. **27**(3), 328–340 (2005)
11. Pan, X., Ruan, Q.-Q., Wang, Y.-X.: An improved 2DLPP method on Gabor features for palmprint recognition. In: Proceeding in ICIP 2007, pp. 413–416 (2007)
12. Ali, M.M.H., Yannawar, P., Gaikwad, A.T.: Study of edge detection methods based on palmprint lines. In: 2016 International Conference on Electrical, Electronics, and Optimization Techniques (ICEEOT), Chennai, India, pp. 1344–1350 (2016)
13. Wirayuda, T.A.B., Adhi, H.A., Kuswanto, D.H., Dayawati, R.N.: Real-time hand-tracking on video image based on palm geometry. In: International Conference of Information and Communication Technology, pp. 241–246 (2013)
14. Kekre, H.B., Sarode, T., Vig, R.: An effectual method for extraction of ROI of palmprints. In: International Conference on Communication, Information and Computing Technology (ICCICT), pp. 19–20 (2012)
15. He, X.: Locality preserving projections. In the document dissertation for a degree of Doctor of Philosophy Chicago, Illinois (2005)
16. Malik, J., Girdhar, D.: Reference threshold calculation for biometrics authentication. Int. J. Graph. Signal Process. 246–253 (2014)
17. Ali, M.M.H., Gaikwad, A.T.: Multimodal biometrics enhancement recognition system based on fusion of fingerprint and PalmPrint: a review. Global J. Comput. Sci. Technol. **16**(2-F), 13–26 (2016)

An Improved Remote User Authentication Scheme for Multiserver Environment Using Smart Cards

Shreeya Swagatika Sahoo, Sujata Mohanty, Saurabh Kumar
Sunny and Banshidhar Majhi

Abstract The smart card-based user authentication scheme is a simple and user friendly scheme, which establishes the secure and authorized communication over the open channel. In this paper, we have intimated an efficient remote user authentication scheme for the multiserver environment. Many password-based authentication schemes have been suggested for the multiserver environment using the smart card. We have analyzed Lee et al.'s scheme and observed that this scheme could not withstand online password guessing attack and offline password guessing attack. And also, it cannot achieve session key security and mutual authentication. The suggested scheme overcomes the weaknesses of the Lee et al.'s scheme and provides more security.

Keywords Mutual authentication · Smart card · Session key security · Attack

1 Introduction

Due to the rapid growth of Internet and telecommunication technology, network security has become a major issue in communication. Hence, authentication between user and server can provide secure data exchange over the insecure network. During the

S. S. Sahoo (✉) · S. Mohanty · S. K. Sunny · B. Majhi
Department of Computer Science and Engineering, NIT Rourkela, Rourkela, India
e-mail: shreeya.swagatika@gmail.com

S. Mohanty
e-mail: sujatam@nitrkl.ac.in

S. K. Sunny
e-mail: saurabh.kumar259@gmail.com

B. Majhi
e-mail: bmajhi@nitrkl.ac.in

© Springer Nature Singapore Pte Ltd. 2019
P. K. Sa et al. (eds.), *Recent Findings in Intelligent Computing Techniques*,
Advances in Intelligent Systems and Computing 707,
https://doi.org/10.1007/978-981-10-8639-7_22

217

past decade, many secure password-based authentication schemes using smart card have been proposed. The smart card-based user authentication scheme is a simple and user-friendly scheme, which establishes the secure and authorized communication over the open channel.

Lamport introduced the password-based authentication scheme in 1981, which uses server-based password storing procedure to validate the authentic user [1]. Later, many password-based authentication protocols have been designed to improve security for the single-server environment.

In 2000, Lee et al. [2] suggested a new protocol for the multiserver environment. Unlike single environment, the multiserver environment includes three participants: the user (U_i), the server (S_j), and the registration center (RS). The user registers at RS once and can utilize all the services on a remote server (S_j). In 2001, Li et al. [3] came up with a new scheme in which the user can select their password freely. Lin et al. [4] enhanced the Li et al. scheme and pointed out that the registered user need not remember the login passwords for several servers. In 2004, Juang [5] proposed an advanced scheme which improves the efficiency of Lin et al. scheme. In 2009, Liao et al. [6] suggested a scheme which can update the password dynamically.

Hsiang et al. [7] intimated an advance of Liao et al.'s scheme in which they have pointed out that, the scheme is susceptible to some passive and active attacks. Lee et al. [8] suggested an improved scheme and claim that Hsiang et al.s scheme is still insecure for several attacks. Moreover, the mutual authentication is also not achieved. However, we observed that Lee et al. scheme still can not withstand to offline password guessing attack, online password guessing attack. Furthermore, Lee et al.'s scheme cannot achieve mutual authentication and session key security. In this paper, we have proposed an enhanced scheme to overcome the weaknesses.

2 Cryptanalysis of Lee et al.'s Scheme

In this section, we have analyzed Lee et al.'s scheme. We demonstrated that this scheme is susceptible to some attacks. The scheme has four phases such as registration phase, login phase, authentication phase, and password change phase. In a multiserver environment, three participants that are the user (U_i), the server (S), and the registration center (RS) participate. The RS selects the master key p and the secret number q to calculate $h(p \parallel q)$ and $h(q)$ assuming that RS is the trusted party. Following analysis highlights some of the vulnerabilities of the scheme.

2.1 Offline Password Guessing Attack

An adversary can obtain the login message from the channel and also get the information from the smart card. By using login message, an adversary can retrieve T_i and then compute $h(b \oplus PW_i)$. He guesses only password PW_i' because the random number is already embedded with SC, to compute $h(b \oplus PW_i')$. If both are equal, then adversary successfully guess the user password.

2.2 Online Password-Guessing Attack

Assume that the adversary knows the information stored in the SC and login message. Then, he tries to send the login message to server. He generates a random number N_1^* and calculate $T_i^* = P_{ij} \oplus h(h(q) \parallel N_1^* \parallel SID_j)$, $A_i^* = h(T_i \parallel h(q) \parallel N_1^*)$ and $Q_i^* = h(B_i \parallel A_i^* \parallel N_1^*)$. Then send it to the server. Upon getting the login message, the server will calculate T_i^* and Q_i^* and verifies the message. So, an adversary can easily change the random number N_1^* and calculate the login message using N_1^*. Hence, this scheme can not withstand the online password guessing attacks.

2.3 Mutual Authentication

The suggested scheme cannot achieve mutual authentication. When adversary gets success to achieve online password guessing attack, then he tries to generate another random number in server side and calculate $M_{ij} = h(B_i \parallel N_1^* \parallel A_i \parallel SID_j)$. He sends it to the user, and the user will verify it and sends another message M_{ij}' for authentication of the server. As adversary generates its random number, so mutual authentication is not properly achieved.

2.4 Session Key Attack

Let the smart card has been lost or stolen. Then an adversary gets the information, and also he can eavesdrop the login message. He can get $h(b \oplus PW_i)$ using offline password guess and then try to have B_i of a legitimate user. Using B_i an adversary can steal the session key of that user. The session key $SK = h(M_2 \parallel SID_j \parallel N_i \parallel N_j)$ where N_i and N_j are two random numbers and the adversary can calculate A_i. So, all the information encrypted with SK will be compromised.

3 The Proposed Scheme

In this section, we suggested a secure and enhanced smart card based remote user authentication scheme for the multiserver environment. There are three participants in the proposed scheme, that is the user (U_i), the server (S_j), and the registration center (RS). The RS selects the master secret key p and secret key q. The RS computes

Table 1 Notation used

Notation	Description
U_i	ith user
S_j	jth server
SC	Smart card
RS	Registration center
ID_i	User (ith) identity
PW_i	User's password
p	Master key generated by RS
q	Secret key generated by RS
b	Random number
$h(\cdot)$	Cryptographic one way hash function
$\parallel$	Concatenation operation
$\oplus$	Bitwise XOR operator
SK	Session key for user

$h(p \parallel q), h(q)$ and shares with server through a secure channel. The proposed scheme includes four phases as explained below. The notation used in the proposed scheme are given in Table 1.

3.1 Registration Phase

First, the user U_i will register itself to the RS. Then, the user U_i and RS compute the following steps to finish the registration phase.

Step 1: First the user selects a user ID_i and password PW_i, and chooses a random number b. Then, PW_a is calculated as follows.
$$PW_a = h(PW_i \parallel b) \quad G_i = b \oplus h(ID_i \parallel PW_i)$$
Then the user sends $<ID_i, PW_a>$ to the server.
Step 2: After getting registration message, RS will generate a random number N_1 and calculate the complying parameters.
$$B_i = h(ID_i \parallel PW_a) \oplus h(p \parallel q)$$
$$C_i = B_i \oplus h(ID_i \parallel PW_a \parallel N_1)$$
$$D_i = h(ID_i \parallel h(p \parallel q))$$
$$E_i = h(ID_i \parallel h(q) \parallel PW_a)$$
Then sends the message $<B_i, C_i, D_i, E_i, h(q), h(.)>$ to the user and embedded with the SC. Then the user will embedded G_i with SC.

3.2 Login Phase

In this phase user, U_i computes the login message and transmit it to S_j.

Step 1: U_i inserts his SC into card reader and enters his ID_i and PW_i. Then, the SC computes the random number b from G_i.

$G_i = b \oplus h(ID_i \parallel PW_i)$

Then calculate $PW'_a = h(b \parallel PW_i)$
$E_i^* = h(ID_i \parallel h(q) \parallel PW'_a)$

Now checks $E_i \overset{?}{=} E_i^*$. If it does not satisfy, then the SC will reject the session. Otherwise, it goes for further computations.

Step 2: The smart card generates a random number N_2 and calculates
$h(p \parallel q) = B_i \oplus h(ID_i \parallel PW_a)$
$CID_i = h(D_i \parallel h(q) \parallel PW'_a \parallel N_2 \parallel C_i)$
$L_i = h(SID_j \parallel h(p \parallel q) \parallel h(q))$
$P_{ij} = L_i \oplus N_2$
$M_1 = h(P_{ij} \parallel CID_i \parallel N_2 \parallel L_i)$
Step 3: After calculating, the SC will send the login message $<P_{ij}, CID_i, M_1>$ to the S_j through the public channel.

3.3 Authentication Phase

After obtaining the login message $<P_{ij}, CID_i, M_1>$ from U_i, S_j does the complying steps to verify user and complete mutual authentication.

Step 1: Server computes L_i^* for verification of the user.
$L_i^* = h(SID_j \parallel h(p \parallel q) \parallel h(q))$
$N_2^* = P_{ij} \oplus L_i^*$
Step 2: After calculating L_i^* and N_2^*, the server will verify the user. Then, the user calculates the following equation.

$M_1 \overset{?}{=} h(P_{ij} \parallel CID_i \parallel N_2^* \parallel L_i^*)$

If both conditions are not match, server rejects the session. Otherwise, generates a random number N_r and calculate the following parameters:
$M_2 = h(SID_j \parallel L_i^* \parallel N_r)$
$M_3 = M_2 \oplus N_2^*$
$M_4 = h(SID_j \parallel h(q) \parallel M_2 \parallel N_2^*)$
Step 3: Then, the server will send $<M_3, M_4>$ to the user through the public channel for authentication.
Step 4: Upon getting the message, the user will authenticate the server by computing following parameters.
$M_2^* = M_3 \oplus N_2$

If it satisfies the condition then, the user will calculate
$M_4^* \stackrel{?}{=} h(SID_j \| h(q) \| M_2 \| N_2)$. If yes, then calculate
$M_5 = h(M_2^* \| N_2 \| M_3 \| SID_j)$

And now the user sends $<M_5>$ to the server.

Step 5: After getting the message server will authenticate the user by checking $M_5^* \stackrel{?}{=}$
$h(M_2 \| N_2^* \| M_3 \| SID_j)$

If it satisfies the condition then, the session key will calculate $SK = h(M_2 \|$
$SID_j \| N_2 \| L_i^*)$ for future communication between user and server.

3.4 Password Change Phase

In this phase, the U_i can freely change his/her new password by replacing the old
password.

Step 1: User inserts his/her smart card into card reader and enters user ID_i and
password PW_i. Then SC will calculate the random number and computes PW_a'
using the calculated random number b.

Step 2: Then, it computes $E_i^* = h(ID_i \| h(q) \| PW_a')$ and check with the received
E_i. If calculated E_i^* is the same as E_i, user asked to input the new password PW_n
and the new random number b_n.

Step 3: Now, user sends PW_n and b_n to the RS. Now RS will calculate $B_n = h(h(b_n \oplus$
$PW_n) \| h(p \| q))$ and send it to the user. Finally, the smart card replaces the new
password with the old password.

4 Security Analysis of the Proposed Scheme

In this section, we have described various security analysis of the proposed scheme.

4.1 User Anonymity

The proposed scheme achieved user anonymity entirely throughout the communication. The login message and the smart card does not contain the user ID_i in plain
text. So it is a tough task to find the ID_i for the adversary.

4.2 Offline Password Guessing Attack

Let the smart card be stolen and the adversary retrieves the stored data $<B_i$, C_i, D_i, E_i, G_i, $h(q)$, $h(.)>$ from smart card. And also the assumption is the adversary eavesdropped one valid login message $<P_{ij}$, CID_i, $M_1>$ from the channel. No parameter has not been sent as plain text in login message. The attacker tries to retrieve the password related information PW_a from B_i. But he can not get it because the registration center generates p and q. Hence, the proposed scheme resists the offline guessing attack.

4.3 Stolen Smart Card Attack

Let the smart card has been stolen or lost. The adversary will get the information stored in the SC and also the assumption is that he intercepts the login message $<P_{ij}, CID_i, M_1>$. Now, the adversary attempt to use this information and send a valid login message to the server.

To generate a valid login message $<P_{ij}, CID_i, M_1>$, he has to calculate $M_1 = h(P_{ij} \parallel CID_i \parallel N_2 \parallel L_i)$ where N_2 is a random number generated by user. And also to compute $L_i = h(SID_j \parallel h(p \parallel q) \parallel h(q))$ the adversary has to know the p and q which is generated by the RS. So he cannot guess the secret key and the proposed scheme can withstand the stolen smart card attack.

4.4 Mutual Authentication

The suggested scheme attains proper mutual authentication. User U_i verifies the authenticity of the server with condition $M_4 = h(SID_j \parallel h(q) \parallel M_2 \parallel N_2^*)$ where $M_2 = h(SID_j \parallel L_i^* \parallel N_r)$. After verification of the user, the server will verify the authenticity of the user with condition $M_5 = h(M_2 \parallel N_2^* \parallel M_3 \parallel SID_j)$. When both user and server will authenticated by each other successfully, the session key $SK = h(M_2 \parallel SID_j \parallel N_2 \parallel L_i^*)$ will computed for later communication.

4.5 Replay Attack

The proposed scheme can withstand the replay attack. The adversary eavesdropped the login message and tried to pretend legal user to login the server. But in this scheme, the user and server will generate random number N_2 and N_r for computation. So for each session, the random number will be different, and the adversary cannot launch the replay attack by repeating login message.

4.6 Session Key Security

The proposed scheme achieves session key security. Session key will calculate by using four parameters that is $SK = h(M_2 \parallel SID_j \parallel N_2 \parallel L_i^*)$, where N_2 is a random numbers generated by user. To calculate M_2, an adversary has to know about L_i where $L_i = h(SID_j \parallel h(p \parallel q) \parallel h(q))$. Both p and q are generated by the RS. So, it is an infeasible work to get the secret key for an adversary.

5 Conclusion

In this paper, we have shown an enhanced remote user authentication scheme for the multiserver environment using a smart card. We have demonstrated that Lee et al.'s scheme has some weaknesses such as online and offline password guessing attack, and also could not achieve mutual authentication and session key security. We have analyzed that the suggested scheme can withstand active, passive attacks and provides mutual authentication and user anonymity. And also, the security and the efficiency of the proposed scheme is better than the Lee et al.'s scheme.

References

1. Lamport, L.: Password authentication with insecure communication. Commun. ACM **24**(11), 770–772 (1981)
2. Lee, W.-B., Chang, C.-C.: User identification and key distribution maintaining anonymity for distributed computer networks. Comput. Syst. Sci. Eng. **15**(4), 211–214 (2000)
3. Li, L.-H., Lin, L.-C., Hwang, M.-S.: A remote password authentication scheme for multiserver architecture using neural networks. IEEE Trans. Neural Netw. **12**(6), 1498–1504 (2001)
4. Lin, I.-C., Hwang, M.-S., Li, L.-H.: A new remote user authentication scheme for multi-server architecture. Future Gener. Comput. Syst. **19**(1), 13–22 (2003)
5. Juang, W.-S.: Efficient multi-server password authenticated key agreement using smart cards. IEEE Trans. Consumer Electron. **50**(1), 251–255 (2004)
6. Liao, Y.-P., Wang, S.-S.: A secure dynamic id based remote user authentication scheme for multi-server environment. Comput. Stand. Interfaces **31**(1), 24–29 (2009)
7. Hsiang, H.-C., Shih, W.-K.: Improvement of the secure dynamic id based remote user authentication scheme for multi-server environment. Comput. Stand. Interfaces **31**(6), 1118–1123 (2009)
8. Lee, C.-C., Lin, T.-H., Chang, R.-X.: A secure dynamic id based remote user authentication scheme for multi-server environment using smart cards. Expert Syst. Appl. **38**(11), 13863–13870 (2011)

A Bibliometric Analysis of Recent Research on Machine Learning for Medical Science

Jaina Bhoiwala and Rutvij H. Jhaveri

Abstract Machine learning is a system capable of the independent acquisition and integration of knowledge. Machine learning is a chosen approach to speech recognition, natural language processing computer vision, medical outcome analysis, and computational biology. In this paper we carry out bibliometric analysis of 150 papers from January 2015 to September 2016 in order to recognize various aspects of machine learning when used for medical science. We have considered large number of objectives and top rated publishers for analyzing the papers. For carrying out further research in the machine learning for medical science, our analysis would assist students, researchers, publishers, and experts to study the recent trends.

Keywords Wireless sensor networks · Trend analysis · Graphical interpretation

1 Introduction

Machine learning is a system capable of the independent acquisition and integration of knowledge. Machine learning is a chosen approach to speech recognition, natural language processing computer vision, medical outcome analysis, and computational biology Medical informatics is the subdivision of science concerned with the use of computers and communication technology to obtain, store, examine, communicate, and display medical information and knowledge [1].

Trend analysis is an aspect of technical analysis to predict movement of particular research area in future, based on past data. Since machine learning is an emerging field, our objective behind this trend analysis is to seek in which medical field what and how much of work has been done. In this paper, we propose a bibliometric

J. Bhoiwala (✉) · R. H. Jhaveri
Department of Computer Engineering, Shri S'ad Vidya Mandal Institute
of Technology, Bharuch, India
e-mail: jainabhoiwala@gmail.com

R. H. Jhaveri
e-mail: rhj_svmit@yahoo.com

© Springer Nature Singapore Pte Ltd. 2019
P. K. Sa et al. (eds.), *Recent Findings in Intelligent Computing Techniques*,
Advances in Intelligent Systems and Computing 707,
https://doi.org/10.1007/978-981-10-8639-7_23

analysis on machine learning for medical science showing the current trend of year 2015 till now. We have analyzed 150 research papers of different well-known journals and reputed publishers. Observation of these papers leads us to classification, segmentation, and detection of brain tumor.

The reminder of this paper is organized as follows: in Sect. 2 we include our graphical representation of this trend analysis, Sect. 3 consists of the final conclusion of paper.

2 Graph Interpretation

2.1 Percentage of Papers Published by Different Countries

Recent days, machine learning is an emerging field. There are large number authors from different countries participate in the research work of machine learning for medical science. For the sake of this trend analysis we have analyzed 150 journal papers and conference papers from different publishers, i.e., ACM, Elsevier, IEEE, Inderscience, Springer, Taylor and Francis and Wiley from January 2015 to September 2016. Figures 1 and 2 show that 32% of authors in our data set are from India, 17% of authors are from China, 8% of authors are from USA followed by other countries.

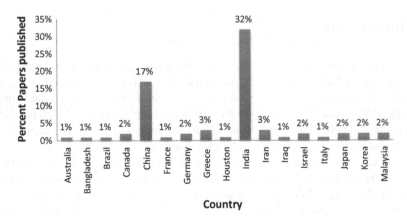

Fig. 1 Percentage of papers published by countries-I

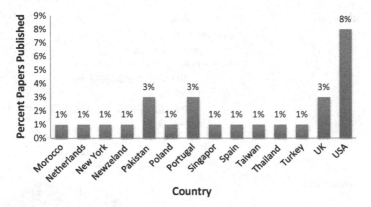

Fig. 2 Percentage of papers published by countries-II

Fig. 3 Percentage of publishers

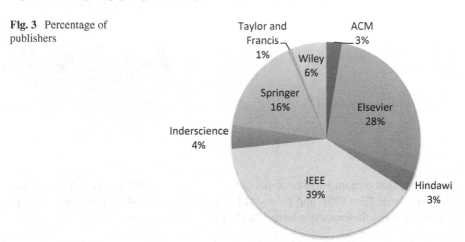

2.2 Percentage of Publishers

In the direction to get accurate analysis, high-quality literature as resource is necessary. As shown in the Fig. 3, IEEE is the most well recognized publisher having 39% of chart covering followed by Elsevier (27%) followed by Springer (16%). Other publishers Wiley, Inderscience, ACM, Hindawi and Taylor and Francis are having 6%, 4%, 3%, 3%, and 1%, respectively.

2.3 Citation per Publisher

This analysis shows number of citations per different publishers. As the number of citation increases, the quality of paper increases. As shown in Fig. 4, Elsevier

Fig. 4 Citations per
publishers

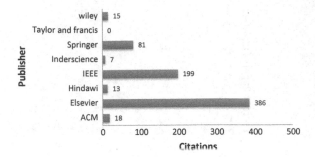

Fig. 5 Conference papers
versus jour

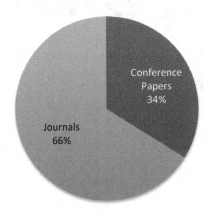

has the maximum number of citations (386) from January 2015 to September 2016, followed by IEEE (199), and followed by Springer (81). Other publishers ACM, Wiley, Hindawi, Inderscience have 18, 15, 13, 7 citations respectively.

2.4 Conference Papers Versus Journals

The place where the researchers can present their research to community is called conference. The article that is published in an issue of journal is journal papers. Figure 5 shows that maximum of our dataset consists of Journal papers (66%), followed by Conference papers (34%).

2.5 Frequency of Number of Authors

The number of authors involved in covering the whole topic in co-authorized manner is analyzed here. Figure 6 shows that maximum of the papers (38) were published by co-authorship of 2 authors, followed by papers (30) written by 3 authors, followed

Fig. 6 Frequency of number of authors

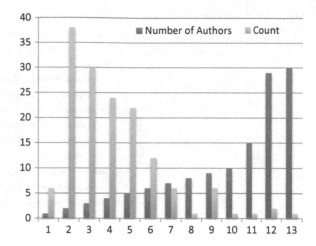

by papers (24) written by 4 authors and followed by papers (22) written by 5 authors. Only 2 papers are having a total 29 number of authors and one paper is having 30 numbers of authors.

2.6 Objective Wise Frequency

Objective wise frequency represents the most trending objective from January 2015 to September 2016. As shown in Fig. 7, Segmentation and Classification have maximum of the chart covering (12%) each. Detection of brain tumor is having 10% of chart covering followed by classification of EEG signal data (9%), Diagnosis of thyroid (8%), Diagnosis of EEG (6%). Least addressed objective here is diagnosis of cancer.

2.7 Objective-Wise Percentage of Citations

This analysis shows the percentage of citations per objective. As shown in Fig. 8, segmentation of brain tumor (27%), classification of brain tumor (11%), classification of EEG signal data (11%) were the most trending objective from January 2015 to September 2016 and were followed by detection of brain tumor (10%), detection of Breast cancer (6%) and diagnosis of EEG (6%). Other objectives are detection of lung cancer, detection of cancer tumor, detection of colon cancer, diagnosis of pancreatic cancer and diagnosis of hepatitis.

Fig. 7 Objective wise
frequency

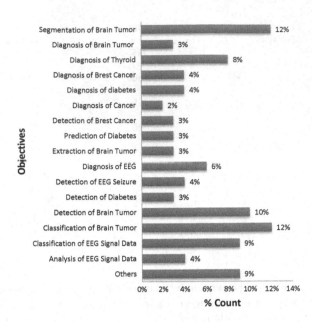

Fig. 8 Percentage of
citations

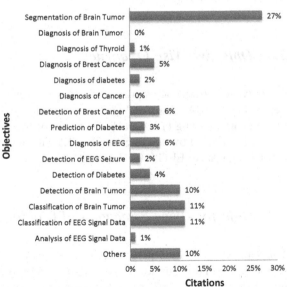

2.8   Methods or Algorithms Used for Research

The methods or algorithms used for research for our data set include classification,
clustering and regression. The most acknowledged algorithm is SVM (Support Vector Machine) with 46% of chart covering. Neural Network algorithms, i.e., ANN

Fig. 9 Methods or algorithms used for research

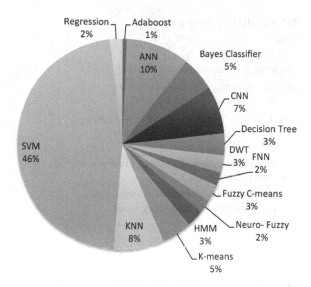

(Artificial Neural Network), CNN (Convolutional Neural Network) are having 10% and 7% of papers respectively. Adaboost is the least addressed method. For this analysis, we first need to standardize the efficiency evaluation for all the papers of our dataset from January 2015 to September 2016. The equation shown below is proposed to standardize the ranking/efficiency for all articles in our dataset [2].

2.9 Paper Efficiency/Ranking

Efficiency of the paper shows how resourceful the paper is. For this analysis, we first need to normalize the efficiency evaluation for all the papers of our dataset from January 2015 to September 2016 needs to normalize the efficiency. For all articles in our dataset, the equation shown below is proposed to standardize the ranking/efficiency [3] (Fig. 9).

$$E = C/(21 - PM)$$

where, E = Efficiency of the Paper, C = Citation of the Paper, PMN = Published Months.

For locating efficiency of each paper, the above equation is used for each paper. The citation of the paper is divided by the value obtained by excluding published months from total number of months till the paper published (Till month of September of year 2016). Total 12 months of year 2015 and month of January to September that leads us to total of 21 months. According to our dataset we have categorized the

Table 1 High efficiency papers

Title	C	PMN	E
The Multimodal Brain Tumor Image Segmentation Benchmark (BRATS) [4]	136	12	15.11
Breast cancer diagnosis using Genetically Optimized Neural Network model [5]	13	20	13
An efficient and effective ensemble of support vector machines for anti-diabetic drug failure prediction [2]	12	8	12
Image based computer aided diagnosis system for cancer detection [6]	10	20	10
Automated classification of brain images using wavelet-energy and biogeography-based optimization [7]	29	17	7.25
Classification of epileptic seizures in EEG signals based on phase space representation of intrinsic mode functions [8]	45	13	5.625
Machine learning applications in cancer prognosis and prediction [9]	42	11	4.2
A rapid automatic brain tumor detection method for MRI images using modified minimum error thresholding technique [10]	42	11	4.2
High performance EEG signal classification using classifiability and the Twin SVM [11]	4	20	4

Note Rest of the papers are having efficiencies less than 4 so they are not shown

results in four categories: (1) High Efficiency Papers and (2) Low Efficiency Papers. Table 1 shows the High Efficiency Papers.

3   Conclusion

In this paper, we thoroughly analyze various bibliometric data of recent papers on machine learning for medical science.. It provides justification and suggestion of the most recent findings for researchers, educators and authors. We conclude that most trending topics from our dataset are classification and segmentation of brain tumor from January 2015 to June 2016. We also concluded the top countries involved in the research in the field of machine learning for medical science are *China*, *India,* and *USA* across the globe. The publisher having maximum research papers published is *IEEE* followed by *Elsevier* followed by other publishers. Along with the maximum number of research papers published, *Elsevier* is having maximum number of citations among any other publisher. The segmentation of brain tumor is having the maximum citations. Furthermore, we conclude that maximum amount of research works are implemented using SVM algorithm. Looking forward to refinement of various performance metrics in machine learning for medical science, the field shows a good potential for further research in this area.

References

1. Jain, R.: Recent Machine Learning Applications to Internet of Things (IoT) Abstract: Table of Contents, pp. 1–19
2. Kang, S., Kang, P., Ko, T., Cho, S., Rhee, S., Yu, K.-S.: An efficient and effective ensemble of support vector machines for anti-diabetic drug failure prediction. Expert Syst. Appl. **42**(9), 4265–4273 (2015)
3. Uddin, S., Hossain, L., Abbasi, A., Rasmussen, K.: Trend and efficiency analysis of co-authorship network. Scientometrics **90**(2), 687–699 (2012)
4. Menze, B.H., Jakab, A., Bauer, S., Kalpathy-Cramer, J., et al.: The multimodal brain tumor image segmentation benchmark (BRATS). IEEE Trans. Med. Imaging **34**(10), 1993–2024 (2015)
5. Bhardwaj, A., Tiwari, A.: Breast cancer diagnosis using genetically optimized neural network model. Expert Syst. Appl. (2015)
6. Lee, H., Chen, Y.P.: Image based computer aided diagnosis system for cancer detection. Expert Syst. Appl. (2015)
7. Yang, G., Zhang, Y., Yang, J., Ji, G., Dong, Z., Wang, S., Feng, C., Wang, Q.: Automated classification of brain images using wavelet-energy and biogeography-based optimization (2015)
8. Sharma, R., Pachori, R.B.: Classification of epileptic seizures in EEG signals based on phase space representation of intrinsic mode functions. Expert Syst. Appl. **42**(3), 1106–1117 (2015)
9. Kourou, K., Exarchos, T.P., Exarchos, K.P., Karamouzis, M.V., Fotiadis, D.I.: Machine learning applications in cancer prognosis and prediction. CSBJ (2014)
10. Kalaiselvi, T., Nagaraja, P.: A rapid automatic brain tumor detection method for MRI images using modified minimum error thresholding technique (2015)
11. Soman, S.: High performance EEG signal classification using classifiability and the Twin SVM. Appl. Soft Comput. J. **30**, 305–318 (2015)

Linear Prediction Model for Joint Movement of Lower Extremity

Chandra Prakash, A. Sujil, Rajesh Kumar and Namita Mittal

Abstract Human gait analysis is an emerging area that has a wide application in medical science specially exoskeleton-based rehabilitation robots. In this paper, a linear time-series-based prediction models have been proposed for joint movement for the lower extremity. The joint movement data is collected at RAMAN Lab, MNIT Jaipur. Experimental results indicate that this approach is better than feedforward neural network in the case of linearly correlated data, considering mean absolute percentage error as an evaluation measure. The proposed prediction model could be used for efficient control of lower extremity robot-assisted device for a smooth gait for the patients.

Keywords Time-series-based prediction · Lower extremities kinematics · Neural network · Gait analysis · Joint angles

1 Introduction

Gait can be considered as stereotyped activity in healthy human. Gait is known as the way of locomotion from one position to another. Gait analysis is the research associated with human walking, and it discloses the pattern of human movement

C. Prakash (✉) · A. Sujil · R. Kumar · N. Mittal
Malaviya National Institute of Technology Jaipur, Jaipur 302017, Rajasthan, India
e-mail: cse.cprakash@gmail.com

A. Sujil
e-mail: sujilavijayan@gmail.com

R. Kumar
e-mail: rkumar.ee@gmail.com

N. Mittal
e-mail: mittalnamita@gmail.com

© Springer Nature Singapore Pte Ltd. 2019   235
P. K. Sa et al. (eds.), *Recent Findings in Intelligent Computing Techniques*,
Advances in Intelligent Systems and Computing 707,
https://doi.org/10.1007/978-981-10-8639-7_24

by quantifying factors responsible for the functionality of the legs. Gait analysis is not only applicable in medical, geriatrics care, biometrics, sports, but also has been widely used in rehabilitation sector [1].

In the past decade, there is a remarkable growth in the both lower and upper extremity-based rehabilitation-assistive devices. Gait rehabilitation- based robots can be classified into end effector and exoskeleton-based devices [2]. Haptic walker and Lokmat® are the examples of each class, respectively. Other devices includes powered prosthetic (C-leg®), human performance augmenting exoskeletons (HAL), assistive exoskeletons (ReWalk®), etc., [1–4]. There is an acute need for more accurate prediction model for online control in case of powered-based rehabilitation-assistive devices.

Dysfunctional gait can be either because of improper biomechanics or because of chronic injury. Orthopedists can analyze and monitor gait movement variables. The literature survey suggests that anthropometric, spatiotemporal, kinematic, kinetic, EMG, and combined parameters are six categories of possible parameters used in gait analysis [1]. In this study, kinematic parameters are worked upon, especially the segment Trajectory and angles. Kinematic parameters includes joint angles by considering the motion of body landmark selected for analysis. Along with angle of joints (hip, trunk, knee angle, ankle angle, etc.), it also comprises angular motion, acceleration, and segment trajectory.

Prediction model directly influences the efficiency of gait rehabilitation robots. Many factors affect gait coordinates trajectory and joint angle prediction, including age, gender, factor, history of injury of gait. Thus, there is need of more accurate and reliable techniques for joint coordinates trajectory and angle prediction.

Several prediction modules using hard and soft-computing techniques have been suggested for hip, knee, and ankle trajectories and joints angles in sagittal plane [5, 6]. Model-based prediction, Hidden Markov models, Gaussian Mixture Models, Gaussian Processes Latent, probabilistic principle component analysis, Neural Network are some of the developed prediction models [7–9].

Kutilek and Viteckova use Artificial Neural Network (ANN) for human gait using angle–angle diagram [10]. Yun et al. suggests Gaussian process regression (GPR) for the prediction of gait kinematics [11]. Some researchers also explore Kalman filter and genetic algorithms for the lower limbs parameters prediction [12]. Salavka et al. use the fuzzy system for gait modeling for lower limb joints angles [13]. But these models require a significant amount of training data.

This paper presents a very simple, new prediction model for gait joint movements based on very less no. of past gait cycle data. The proposed models are compared with persistent models. The generated results indicate the effectiveness accuracy of the proposed prediction model for joint movements.

Section 2 report the experimental setup, followed by the proposed model in Sect. 3. Generated results are reported and analyzed in Sect. 4, followed by conclusions.

2 Experimental Setup

In this section, the lab setup and the subject used for the experiment have been presented. Figure 1 shows the RAMAN Lab setup at MNIT, Jaipur for gait analysis. The walking path, which has been used in this experiment, has a total length of 4.8 m, which gives the subject enough room to take at least two steps. Data was recorded considering both free and plain green background as shown in Fig. 1.

Ten healthy volunteers (seven male and three female) from MNIT participated in this research work. The anthropomorphic parameters of the subject participated in the study; (age: 23.5 ± 3.50 years old; height: 1.7019 ± 0.11 m; weight: 62.99 ± 7.92 kg and BMI: 21.80 ± 2.52). All subjects have neither any history of injuries and surgeries of the lower limb nor did they underwent any rehabilitation. Subjects were aware of the purpose and methodology of the experiment before written consent. In this study, the subject walks at a self-selected normal speed (SSNS) from right to left on ground level gait path at least five times so that natural gait could be captured.

A conventional video recorder and Biopac camera (Zmodo video camera) are used to acquire the motion of the markers in the sagittal plane. Five passive markers are attached on subjects lateral metatarsal, malleolus knee, hip, and shoulder. All these markers are placed on the left portion of the subject's body, in accordance to the previous work [14]. The coordinate of the makers and kinematic parameter (joint trajectories and joints) were collected and used in this study.

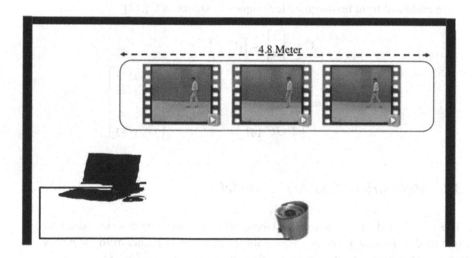

Fig. 1 RAMAN LAB setup for gait analysis

3 Model Development

Prediction is finding the future value for new event based on previously known or observed event data. Two model were developed and analyzed in this paper. These techniques are based on a simple linear time-series-based model using the least-square method, using previous one and three historical gait joint data.

3.1 Univariate-Based Time Series Model

The first model, an univariate-based time series model, is based on the n historical data values from the previous gait cycle (GC_1) and predict the coordinates for the next gait cycle (GC_2) for the same time interval. Thus, by previous n frames of (GC_1), next (GC_2) is predicted. In this study, one gait cycle is composed of 35 frames. The univariate model can be expressed as Eq. (1)

$$Y_{GC_2}(i) = a + bX_{GC_1}(i) \quad where \quad i = 1, 2, \ldots, n \tag{1}$$

where $Y_{GC2}(i)$ is the predicted points for the GC_2 coordinates and angles data values, $X_{GC_1}(i)$ represents the corresponding actual data-points for the GC_1 Joint movement (coordinates and joint) data-points, i is the number of frames. Model variables a and b are evaluated from least-square techniques as shown in Eq. (2)

$$C = \begin{bmatrix} a \\ b \end{bmatrix} = [\beta^T \beta]^{-1} \cdot \beta^T \cdot Y$$

$$where \quad \beta = \begin{bmatrix} 1 & GC_1(1) \\ 1 & GC_1(2) \\ \vdots & \vdots \\ 1 & GC_1(n) \end{bmatrix} \quad X_{GC1} = \begin{bmatrix} GC_1(1) \\ GC_1(2) \\ \vdots \\ GC_1(n) \end{bmatrix} \tag{2}$$

3.2 Multivariate Time Series Model

The second model is the generalized form of the univariate time-series-based prediction model. This model is then used with the next n data-points from the p previous series to predict the next n points for the current series using Eq. (3)

$$Y_{GC_{p+1}}(i) = a + b_1 X_{GC_p}(i) + b_2 X_{GC_{p-1}}(i) + \cdots + b_k X_{GC_1}(i)$$
$$where \quad i = 1, 2, \ldots, n \tag{3}$$

where $Y_{GC2}(i)$ is the predicted points for the GC_2 coordinates and angles data values, $X_{GC_1}(i)$ represents the corresponding actual data-points for the GC_1 Joint movement (coordinates and joint) data-points, i is the number of frames, k is taken as $p + 1$. Model variables a and b are evaluated from least-square techniques as shown in expressed in Eq. (4)

$$C = \begin{bmatrix} a \\ b_1 \\ b_2 \\ \vdots \\ b_{p+1} \end{bmatrix} = [\beta^T \beta]^{-1} \cdot \beta^T \cdot Y$$

(4)

$$where \quad \beta = \begin{bmatrix} 1 & X_{GC_p}(1) & X_{GC_{p-1}}(1) & \dots & X_{GC_1}(1) \\ 1 & X_{GC_p}(2) & X_{GC_{p-1}}(2) & \dots & X_{GC_1}(2) \\ \vdots & \vdots & \vdots & \vdots & \vdots \\ 1 & X_{GC_p}(n) & X_{GC_{p-1}}(n) & \dots & X_{GC_1}(n) \end{bmatrix}$$

4 Results and Discussion

In this section, the result generated using the proposed models has been presented. For the analysis purpose, a random subject is chosen, and his joint movement is analyzed and presented. Mean Absolute Error (MAE), Root Mean Square Error (RMSE), and Mean Absolute Percentage Error (MAPE) are considered as evaluation for the prediction models as presented in Eq. (5)

$$MAE = \frac{1}{n} \sum_{i=1}^{n} |y_i - \widehat{y}| \quad RMSE = \sqrt[2]{\frac{1}{n} \sum_{i=1}^{M} (y_i - \widehat{y})^2},$$

$$MAPE = \frac{100}{n} \sum_{i=1}^{n} \frac{|y_i - \widehat{y}|}{y_i}$$

(5)

where y_t and $\widehat{y}_t$ are the actual and predicted gait movements at frame i and and n is the no. of frames (Fig. 2).

Tables 1 and 2 presents the performance plot for knee angle and coordinates using evaluation parameters, as described earlier. Multivariate time series with the $p = 3$ model has the lowest MAE, RMSE, and MAPE when compared against the other model which is Backpropagation Neural network (BP-NN) model. Figure 2 illustrates the performance plot for knee angle using the proposed and BP-NN-based model.

Trail and error method chooses the number of hidden layer neurons. The training set consists of one and three past patterns. All other network and training parameters are carefully chosen to reduce the sum squared error. Hidden layer selected is 1 with 2 neurons, and the presented evaluation measure are the average of 30 iterations. The

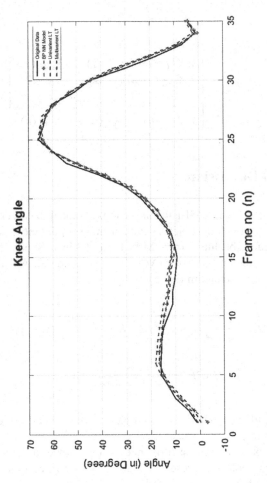

Fig. 2 Performance plot for Knee angle Prediction using various model

Table 1 MAE, RMSE, MAPE measures for the predicted knee angle

Prediction models	MAE	RMSE	MAPE
BP-NN	2.4878	2.1391	16.9561
Univariate time series	1.3442	1.6673	12.4336
Multivariate time series (p = 3)	0.7110	0.8621	11.9363

Table 2 MAE, RMSE, MAPE measures for the predicted knee coordinates

Prediction models	MAE	RMSE	MAPE
BP-NN	9.8920	10.5530045	4.8220
Univariate time series	3.2568	4.0667	1.7071
Multivariate time series (p = 3)	36.2076	36.7316	19.1632

multivariate time series model predicts the knee angle more precisely when compared with other models. BP-NN, MAPE's value is 16.9561 while that of the proposed MTS with P = 3 is 11.9363. There is an improvement of around 5% in MAPE. The possible reason is higher correlation coefficient among the dataset as presented in Fig. 3a. Similarly for the knee coordinates, univariate time series prediction model outperforms than other models. MTS with p = 3 shown higher MAPE. The reason is that the historical data is not linearly correlated as shown in Fig. 3b.

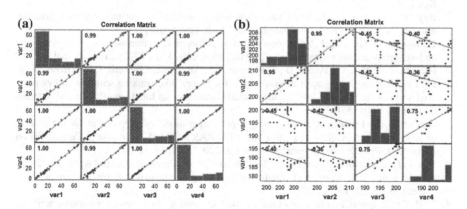

Fig. 3 **a** Correlation coefficient of the knee angle dataset. **b** Correlation coefficient of the knee coordinates dataset

5 Conclusion

This paper presents a very simple, new prediction models; Multivariate and univariate-based linear time series model, for gait joint movements based on very less no. of past gait cycle data. The results reveal the higher accuracy of the proposed models for gait movement prediction for lower extremities. After rigorous analysis of the result and justifying the better accuracy obtained, it is found out the previous data used in the study are highly co-related. The correlation coefficient is very high in the selected data. Thus, it can be concluded that the proposed model works better with linearly co-related data. Less number of training data is needed, as compared to the BP-NN, where weight are adjusted to map the input to the output. Thus, there is need of larger dataset. Accurate prediction model can be of great use for the power-assistive gait rehabilitation robots. As a future scope, one can use these models for the other fields.

Acknowledgements This work was supported by Department of Science and Technology, India; project under grant SR/S3/MERC/0101/2012.

Declaration The authors have obtained all ethical approvals from appropriate ethical committee and written approval from the subjects involved in this study.

References

1. Prakash, C., Kumar, R., Mittal, N.: Recent developments in human gait research: parameters, approaches, applications, machine learning techniques, datasets and challenges. Artif. Intell. Rev. (2016)
2. Beyl, P.: Design and control of a knee exoskeleton powered by pleated pneumatic artificial muscles for robot-assisted gait rehabilitation. Mech. Eng. (2010)
3. Gouwanda, D., Gopalai, A.A.: A robust real-time gait event detection using wireless gyroscope and its application on normal and altered gaits. Med. Eng. Phys. **37**(2), 219–225 (2015)
4. Glackin, C., Salge, C., Polani, D., Tüttemann, M., Vogel, C., Guerrero, C.R., Grosu, V., et al.: Learning gait by therapist demonstration for natural-like walking with the corbys powered orthosis. In: 2015 IEEE/RSJ International Conference on Intelligent Robots and Systems (IROS), pp. 5605–5610. IEEE (2015)
5. Aertbeliën, E., De Schutter, J.: Learning a predictive model of human gait for the control of a lower-limb exoskeleton. In: 2014 5th IEEE RAS & EMBS International Conference on Biomedical Robotics and Biomechatronics, pp. 520–525. IEEE (2014)
6. Sun, J.: Dynamic Modeling of Human Gait Using a Model Predictive Control Approach (2015)
7. Ren, L., Jones, R.K., Howard, D.: Predictive modelling of human walking over a complete gait cycle. J. Biomech. **40**(7) (2007)
8. Beyl, P.: Design and control of a knee exoskeleton powered by pleated pneumatic artificial muscles for robot-assisted gait rehabilitation. Mech. Eng. (2010)
9. Ackermann, M., van den Bogert, A.J.: Optimality principles for model-based prediction of human gait. J. Biomech. **43**(6) (2010)
10. Kutilek, P., Viteckova, S.: Prediction of lower extremity movement by cyclograms. Acta Polytechnica **52**(1) (2012)
11. Yun, Y., Kim, H.-C., Shin, S.Y., Lee, J., Deshpande, A.D., Kim, C.: Statistical method for prediction of gait kinematics with Gaussian process regression. J. Biomech. **47**(1) (2014)

12. Nogueira, S.L., Inoue, R.S., Terra, M.H., Siqueira, A.A.G.: Estimation of lower limbs angular positions using Kalman filter and genetic algorithm. In: Biosignals and Biorobotics Conference (BRC), pp. 1–6. IEEE (2013)
13. Viteckova, S., Kutilek, P., Svoboda, Z., Jirina, M.: Fuzzy inference system for lower limbs angles prediction. In: 2012 35th International Conference on Telecommunications and Signal Processing (TSP), pp. 517–520. IEEE (2012)
14. Prakash, C., Mittal, A., Kumar, R., Mittal, N.: Identification of spatio-temporal and kinematics parameters for 2-D optical gait analysis system using passive markers. In: 2015 International Conference on Advances in Computer Engineering and Applications (ICACEA), pp. 143–149. IEEE (2015)

Multimodal Group Activity State Detection for Classroom Response System Using Convolutional Neural Networks

Abraham Gerard Sebastian, Shreya Singh, P. B. T. Manikanta, T. S. Ashwin and G. Ram Mohana Reddy

Abstract Human–Computer Interaction is a crucial and emerging field in computer science. This is because computers are replacing humans in many jobs to provide services. This has resulted in the computer being needed to interact with the human in the same way as the human does with another. When humans talk to each other, they gain feedback based on how the other person responds non-verbally. Since computers are now interacting with humans, they need to be able to detect these facial cues and accordingly adjust their services based on this feedback. Our proposed method aims at building a Multimodal Group Activity State Detection for Classroom Response System which tries to recognize the learning behavior of a classroom for providing effective feedback and inputs to the teacher. The key challenges dealt here are to detect and analyze as many students as possible for a non-biased evaluation of the mood of the students and classify them into three activity states defined: active, passive, and inactive.

Keywords Emotion detection · OpenCV · Convolutional neural network
Video analytics · Feedback mechanism · Activity states

1 Introduction

The detection of the level of activity or attention in a group based on the facial expressions as detected from a video is a very interesting and seldom attempted problem. Despite of many computer vision techniques developed, recognition of facial activity in video sequences is still considered a very challenging conundrum. To us humans, a person's expression is one of the first and key indicators of said person's mood or level of attention. Similarly in a class environment the attentiveness of the students can be determined by their facial expression and physical gestures. Manual

A. G. Sebastian (✉) · S. Singh · P. B. T. Manikanta · T. S. Ashwin · G. R. M. Reddy
Department of Information Technology, National Institute of Technology Karnataka,
Surathkal, Mangalore, India
e-mail: abraham.g.sebastian@gmail.com

© Springer Nature Singapore Pte Ltd. 2019
P. K. Sa et al. (eds.), *Recent Findings in Intelligent Computing Techniques*,
Advances in Intelligent Systems and Computing 707,
https://doi.org/10.1007/978-981-10-8639-7_25

feedback of a class taken by a pedagogy would be mostly biased and inaccurate because of some reasons like, the student might give false feedback or might feel shy/lazy. But an automated system generates unbiased and accurate feedback based on the states of facial expressions of a student.

In our proposed work we have divided the attentiveness of a student into three states namely: active, inactive, and passive. *Active* is when a student is completely focusing on the lectures. *Inactive* is when a student is distracted away from the lectures and *passive* is when a student is focusing partially on the lectures and is distracted sometimes. The application is trained to be intelligent enough to understand the facial expressions with respect to their emotions; whether the student is active, inactive or passive. Now upon doing the aggregate study of states of every student in the entire class, our application will be able to send the report to pedagogy based on the percentage of attentiveness of the class wholly.

The rest of this paper is organized as follows. Section 2 deals with Literature Survey; Sect. 3 describes Proposed Methodology; Sect. 4 discusses the Experimental Results and Analysis and finally Sect. 5 Concludes the paper with Future Directions.

2 Literature Survey

2.1 Background

The formulation of the of our proposed method required a detailed literature survey of about 5–6 papers in the field of Video Analytics, Deep Learning and Machine Learning. Each paper's description has been tabulated in Table 1.

3 Proposed Methodology

For the achievement of our end goal we have divided the proposed method into three major steps. In this section we will discuss these different steps and also describe the techniques used. The workflow of the entire proposed methodology has been shown in Fig. 1.

3.1 Multiple Face Detection

For the multiple face detection in classroom, we will be using the Haar Cascades detection method. This approach is a very effective approach in which a "cascade" function is trained using a number of positive as well as negative face images. The classifier is then tested by using it to detect the same objects in frames or images.

Table 1 Summary of existing works

Works	Merits	Demerits
Jeon et al. [1]	Method allows for real-time emotion detection	CNN will need very powerful hardware; low testing accuracy when data is skewed
Zhang et al. [2]	This allows simultaneous audio and video emotion detection	DCNN requires very powerful hardware for training purposes, more levels of training required and not currently real time
Mou et al. [3]	Framework to detect the emotional state of a group along valence and arousal measures	High level feature extraction techniques need to be used and data needs to be collected separately
Sun et al. [4]	Evaluates various ML algorithms and also has created an international database for emotions	Classification performed in controlled environment using data generated by authors
Soleymani et al. [5]	Uses text, audio and video features for the emotion detection	Lot of features like pitch, frequency and energy have not been considered
KalaiSelvi et al. [6]	Presents a system for recognizing emotions through facial expression displayed in the video	A current weakness in this area of facial study is still the lack of comparable databases

Fig. 1 Features used in haar classifier

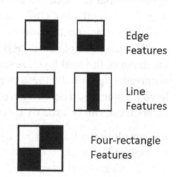

Edge Features

Line Features

Four-rectangle Features

For training the detector, there is a requirement for a large number of positive images (face images) as well as negative images (non-face images). Further, the "haar features" as shown in Fig. 1 are used to extract certain features from these images. All of these features are comprised of a single value that is obtained the sums of pixels under the white box and black box respectively and subtracting the white box's sum from the black box sum.

Kernels of all kinds of sizes and even locations are used to generate a lot large number of features. The most relevant features are selected using AdaBoost. All the features are first applied to the images allotted for training, and for each feature the optimal thresholds that enables them to classify the face to positive or negative are found. Out of all these features, the features with the least error rates are selected.

Finally, the classifier is obtained by taking a weighted sum of all the previously generated "weak" classifiers and is used to effectively detect faces in the images.

All the smaller classifiers are called "weak" because they are unable to classify any images effectively on their own. They are however able to be used in tandem with other such classifiers to create a very robust and accurate classifier.

3.2   Convolutional Neural Network

Convolutional Neural Networks have a high similarity to regular Neural Networks; they also consist of neurons which have weights and biases that can be learned using data. The final layer in CNN generally has a loss function which is essential during the backpropagation step of the training phase. One of the main assumptions of most CNN architectures is that the inputs to these networks are raw images; this aids them in encoding specific properties into the network structure. Another benefit of this structure is that the forward functions are made more efficient by reducing the number of attributes in between layers. This happens because nodes between layers of CNN's are only partially connected unlike regular Neural Nets, in which all nodes in a layer are connected to all other nodes in adjacent layers. Another property of CNN's is that the nodes of each layer are oriented in three dimensions, i.e., height, width, and depth. The CNN transforms the dimensions of the image layer by layer while the various activation functions in the different layers, until it finally reduces to a single vector of scores assigned to each class. These scores are arranged along the depth of the final layer. A simple CNN is a collection of different kinds of layers arranged sequentially, these may be partially or fully connected and may even not have parameters. The CNN architecture consists of a number of different kinds of layers these can be broadly categorized into the below three kinds:

- Convolutional Layer
- Pooling Layer
- Fully Connected Layer.

For our proposed method we used the following architecture to build a model that is able to perform our state detection:

- Three convolutional layers with MaxPooling layers after each of them.
- Each of the Max pooling layers have a pool size of 2×2 and use the relu activation function.
- The last two layers are both fully connected layers of 64 nodes and 3 nodes (no of classes) each.
- The input is a 150×150 RGB image.
- The loss function used is the Maximum Squared Error Loss function.
- The final layer also uses the SoftMax function as its activation function.

The above-mentioned architecture was decided upon through trial and error to maximize accuracy while minimizing complexity of architecture.

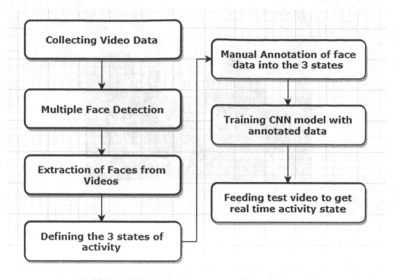

Fig. 2 Workflow of proposed methodology

3.3 Emotion State Detection

In this step; the CNN described above is carried out in Keras framework of Python along with OpenCV. As depicted in Fig. 2, the first step comprises of collecting the classroom video data of the students listening to the instructor, followed by detection of multiple faces by Haar Cascades classifier. The model gets trained on the detected faces of the training video which are manually annotated by five people who classify the images in three activity states which are active, passive, and inactive. After getting trained, the model gets tested on a real-time video and provides a group emotional state score for each of the states per frame.

4 Results and Discussions

Due to a lack of data relating to classroom systems and also the fact that we defined our own labels for the detection system we had to collect data on our own. The data was collected using a standard handheld video camera which was placed in various different classrooms. For the purpose of training the emotion classifier we allocated a training video and extracted all the faces detected by our Haar Cascade face detector. We then had five volunteers annotate all these images after watching the video. This was done to overcome biases that may vary from person to person. Finally for each face an average score is generated for the different classes based on volunteer classifications.

Fig. 3 Snapshot of class emotion detection

70% of this newly generated data was then used to train the CNN as defined in the previous section to perform our emotion detection. This system was then tested on the remaining 30% which showed that it had 83.45% accuracy.

Figure 3 is the output of our final system that gives the emotion scores of the entire classroom for every frame. The system first performs the face detection and then feeds each face image into the trained CNN. This generates a score for each class. The scores generated for all the faces are aggregated using their mean to generate class-wide scores for each emotional state.

5 Conclusion and Future Work

In our proposed method we have incorporated the facial expressions of individuals in a classroom system to detect the activity of the classroom as a whole. We have used a CNN to classify the individual faces into the three activity states defined before. With the current system we have achieved around 83.45% accuracy. Now, to extend our proposed method we could possibly experiment with more standard CNN architectures like VGG or AlexNet to find the optimal network architecture for the classification step. We also plan to incorporate the method in a classroom system with suggestions for teaching staff when class state tends towards inactive.

Declaration Authors have obtained all ethical approvals from appropriate ethical committee and approval from subjects involved in this study.

References

1. Jeon, J., Park, J.-C., Jo, Y.J., Nam, C.M., Bae, K.-H., Hwang, Y., Kim, D.-S.: A Real-Time Facial Expression Recognizer Using Deep Neural Network
2. Zhang, S., Zhang, S., Huang, T., Gao, W.: Multimodal Deep Convolutional Neural Network for Audio-Visual Emotion Recognition
3. Mou, W., Gunes, H., Patras, I.: Automatic Recognition of Emotions and Membership in Group Videos
4. Sun, Y., Sebe, N., Lew, M.S., Gevers, T.: Authentic Emotion Detection in Real-Time Video
5. Soleymani, M., Pantic, M., Pun, T.: Multimodal Emotion Recognition in Response to Videos
6. KalaiSelvi, R., Kavitha, P., Shunmuganathan, K.L.: Automatic Emotion Recognition In Video

A Study on Sclera as a Biometric Authentication System

V. Sandhya and Naga Rathna P. Hegde

Abstract This paper presents the study on sclera as a biometric authentication system. Many of biometric systems from the traditional fingerprint to the iris-based recognition system are used for identifying individuals. Sclera as a biometric recognition system has gained much importance and popularity nowadays. In this study, an overview of sclera found in the literature, different techniques used in extracting, enhancing and matching are discussed. As an identification system that requires rigorous testing databases are used, the databases related to system are discussed briefly. Various biometric parameters that are used in the process of recognition are also presented. This study is concluded with the future scope of the system, limitations, and further research directions.

Keywords Sclera recognition · Biometric · Image

1 Introduction

Security plays a vital role in a real-time system. The different methods of identifying an individual can vary from the traditional methods such as finger print to modern automated systems. Biometrics is a system in which a human being is identified by physical, biological, or behavioral characteristics. The physical characteristics includes fingerprint, palm print, iris, eye, sclera, hand geometry, face. The behavioral characteristics are giant, traits, and habits. Biometric system is of unimodal or multimodal. In a unimodal system only one biometric data is used to identify a person, where as in a multimodal system [1] more than one biometric input is used to authenticate an individual. Among the available biometric systems face recognition

V. Sandhya (✉)
GITAM School of Technology, Rudraram, Medak, Hyderabad, India
e-mail: sandhyav@gitam.in

N. R. P. Hegde
Vasavi College of Engineering, Ibrahimbagh, Hyderabad, India
e-mail: nagaratnaph@gmail.com

© Springer Nature Singapore Pte Ltd. 2019
P. K. Sa et al. (eds.), *Recent Findings in Intelligent Computing Techniques*,
Advances in Intelligent Systems and Computing 707,
https://doi.org/10.1007/978-981-10-8639-7_26

253

Fig. 1 Eye anatomy

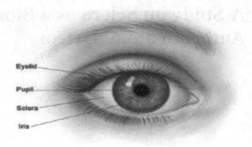

systems are more reliable than other systems. In face recognition the various organs of human being that are used for biometrics are eye, iris, periocular, and sclera. Sclera is a component of eye that is used to identify a person using sclera vein patterns. Sclera as a biometrics recognition system has gained more importance in the modern authentic system [2]. In this study Sect. 2 explains the sclera recognition system, Sect. 3 sclera segmentation, Sect. 4 Sclera vein pattern enhancement, Sect. 5 sclera Extraction and Sect. 6 sclera matching.

2 Sclera Recognition

2.1 Sclera Background

The human eye comprises of various sub parts such as pupil, iris, cornea, upper eyelid, lower eyelid, aqueous humor, sclera, and sclera veins as shown in Fig. 1. Among these, Sclera is nothing but a white part of eye which forms as a supportive wall for the eye. Sclera is covered with a mucus membrane needed for the smooth movement of eye. It is surrounded by optic nerve and is also the thickest one. The various interior parts of sclera are *episclera* present below conjunctiva, *sclera proper* is a thick white color tissue that forms the white color of sclera, *lamina fusca* is an elastic fiber.

Sclera contains blood vessel patterns (Fig. 2) that are unique for all human being. Even twins have different vein patterns. The pattern of blood vessel can be used for recognizing humans in the real-time scenarios. These patterns are visible and will never change [2] over the life time of a person.

2.2 Sclera Recognition System

The first phase in a sclera recognition system is to segment sclera from image of the eye. To extract sclera features Iris localization is applied [2]. After identifying iris

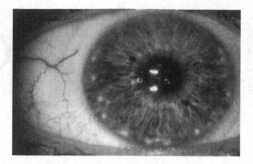

Fig. 2 Sclera vein

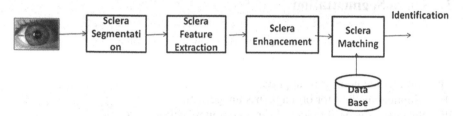

Fig. 3 Sclera recognition system

boundary the remaining left and right part of the eye is used for feature extraction. In a unimodal biometric system only the extracted sclera can be used for identification purpose, whereas in a multimodal biometric system different biometrics can be combined with sclera features to recognize individuals. One such combination is fusing iris and sclera [3]. Integro-differential operator is used to identify the iris boundaries and CLAHE algorithm for sclera region. The resultant image contains iris and sclera information. In [4] iris is detected by unsupervised k-means clustering algorithm and the sclera is detected by r-g, r-b values based on the saturation component. Sclera part of the eye may contain noise as they are captured from distance or on the move. In [4] efficient eye corner detection method has been proposed even for eye gaze direction to avoid eye misclassification.

The second phase is to extract sclera from the segmented sclera eye image. The resultant sclera is enhanced to identify the vein patterns and remove noise present during extraction. The final phase is to match the image that is present in the database with image extracted from the user for identification purpose. If the two images match then the individual is recognized correctly otherwise a mismatch results in authentication failure. Figure 3 the sequence of stages of sclera recognition system is represented.

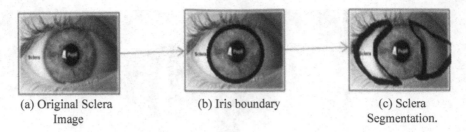

(a) Original Sclera
Image

(b) Iris boundary

(c) Sclera
Segmentation.

Fig. 4 a Original sclera image. b Iris boundary. c Sclera segmentation

3   Sclera Segmentation

To extract the sclera part of the eye first iris boundaries are detected.
Various methods are found in the literature.

I. Circular method for front gaze.
II. Elliptical method for off angle iris images [2].
III. Sober filter is used to identify speculation reflection area.
IV. Center detection method [5].

To fit the iris circular boundary (Fig. 4a, b) least square circle fitting algorithm is used. In [1] fuzzy c means clustering-based segmentation is used to divide sclera area into three clusters one for iris, one for sclera and other one for part of the area outer to eye. Once the iris boundary is identified the 'white' (Fig. 4c) part of the eye sclera can be extracted easily. Sclera image can be gray scale or color. To identify gray-scale images Otsu's method is used where as for color images HSV model is used [4].

4   Sclera Vein Pattern Enhancement

The segmented sclera consists of vascular patterns that may not be clearly visible due to illumination constraints (Fig. 5a, b). The different approaches to enhance the vascular patterns are using gabor filters [6, 7] and adaptive histogram equalization [7]. Figure 5c is the resultant image after applying the enhancement techniques to the sclera vein pattern.

5   Sclera Feature Extraction

The sclera vascular patterns (Fig. 6a) captured from an individual is of different thickness that depends on the mental condition at that moment. To obtain features

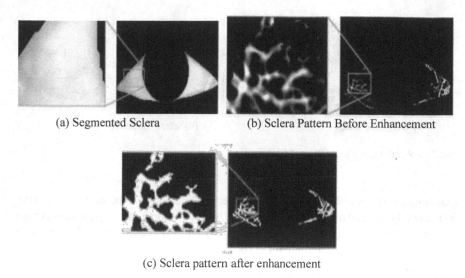

(a) Segmented Sclera   (b) Sclera Pattern Before Enhancement

(c) Sclera pattern after enhancement

Fig. 5 **a** Segmented sclera. **b** Sclera pattern before enhancement. **c** Sclera pattern after enhancement

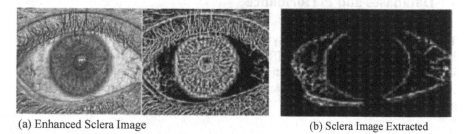

(a) Enhanced Sclera Image   (b) Sclera Image Extracted

Fig. 6 **a** Enhanced sclera image. **b** Sclera image extracted

invariant to movements of sclera line descriptor method is applied. The veins in the sclera (Fig. 6b) are extracted as a line segment [1, 2]. In [1] local descriptor method (LDP) and dense LDP methods are used to identify edge values and convert it into code based image texture.

6 Sclera Matching

Sclera matching is done in two steps. First sclera template registration is done and then template matching. Sclera registration is required because segmented image consists of invariance's due eye and eye lids movement, contraction. Sclera ROI was used initially for registration of templates. In [1, 2] a type-based algorithm RANSAC or random sample consensus is used to estimate sclera matching score. In template

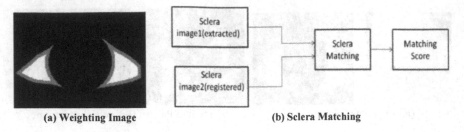

(a) Weighting Image   (b) Sclera Matching

Fig. 7 a Weighting image. b Sclera Matching

matching phase a weighting image of the sclera is created by using masks and setting inside pixels to 1 and out side pixels to 0. In [8] a two-stage matching process coarse to fine was proposed using Y-shaped descriptor and WPL descriptors. Figure 7a, b shows the steps involved in matching process.

7 Databases and Performance

The two databases that are publicly available for sclera recognition are UBIRIS and IUPUI multi wavelength. The UBIRIS consists of 1877 images taken from 241 users in two sessions. It consists of both low- and high-resolution images [2]. In IUPUI database the eye images are of five different colors blue, dark brown, light brown, green, and hazel [1, 2].

7.1 Biometric System Performance

Performance of any biometric identification system is characterized by the following parameters [8].

False Acceptance Rate (FAR)

The probability of accepting an incorrect input to the data that is present in the template database.

False Rejection Rate (FRR)

The probability of rejecting a correct input data to the template data that is present in the database.

Equal Error Rate (ERR)

The rate at which acceptance and rejection is equal.

Failure to Enroll (FTE)

The rate at which an unsuccessful template is obtained from the given input.

8 Conclusion and Future Work

This study expressed the overview of sclera as a biometric recognition system. There are many challenges involved in sclera as an authentication system and its working and practical implementation in a real-time environment. Sclera as a biometric system is a unimodal system, there are many chances to combine this with other biometric methods. In a multimodal system sclera can be combined with iris, palmprint, face, and periocular biometrics.

Declaration Authors have obtained all images of the subjects involved in this study from openly available databases: UBIRIS V2.

References

1. Zhou, Z., Du, E.Y., Thomas, N.L., Delp, E.J.: A Comprehensive Multimodal Eye Recognition. Springer (2013)
2. Zhou, Z., Du, E.Y., Thomas, N.L., Delp, E.J.: A new human identification method: sclera recognition. IEEE Trans. Syst. Man Cybern.-Part A:Syst. Hum. **2**(3) (2012)
3. Gottemukkula, V., Saripalle, S.K., Pasula, R., Ross, A.: Fusing Iris and Conjunctival Vasculature: Ocular Biometrics in the Visible Spectrum. IEEE (2012)
4. Van Huan, N., Binh, N.T.H., Kim, H.: Eye Feature Extraction Using K-means Clustering for Low Illumination and Iris Color Variety. IEEE (2010)
5. Alkassar, S., Woo, W.L., Dlay, S.S., Chambers, J.A.: Efficient Eye Corner and Gaze Detection for Sclera Recognition Under Relaxed Imaging Constarints. IEEE (2016)
6. Oh, K., Toh, K.-A.: Extracting sclera for Canceable Identity Verification. IEEE (2012)
7. Lin, Y., Du, E.Y., Zhou, Z., Thomas, N.L.: An Efficient Parallel Approach for Sclera Vein Recogntition. IEEE (2013)
8. Dahiya, N., Kant, C.: Biometrics security concerns. In: Second International Conference on Advanced Computing and Communication Technologies. IEEE (2012)
9. Das, A., Pal, U., Ballester, M.A.F., Blumenstein, M.: A New Efficient and Adaptive Sclera Recognition System. IEEE (2014)
10. Mahesh, T.Y., Shunmuganathan, K.L.: Detection of diseases based on vessel structure and color changes as viewed on the sclera region of the eye. In: International Conference on Circuit, Power and Computing Technologies. IEEE (2014)

Modulating Properties of Hyperpolarization-Activated Cation Current in Urinary Bladder Smooth Muscle Excitability: A Simulation Study

Chitaranjan Mahapatra and Rohit Manchanda

Abstract Smooth muscle cells from the urinary bladder display enhanced spontaneous electrical activities during the overactive bladder state. It is very well known that the electrical properties of all excitable cells are regulated by the intrinsic active ion channels. In excitable cells like the neurons, cardiac and other smooth muscle cells, the hyperpolarization-activated cation channel has been considered as a potential target to regulate cell's excitability. The primary purpose of this research work is to develop a computational model of the hyperpolarization-activated cation channel in urinary bladder smooth muscle (UBSM) cell to analysis its modulating role in cell's excitability. All required biophysical parameters are adapted from the published experimental studies in UBSM cell. We have successfully simulated and validated the channel model by comparing it with experimental findings in UBSM cells. We have investigated the potential role of this hyperpolarization-activated cation channel in regulating the shape of the evoked action potentials. From this quantitative analysis, we conclude that future pharmacological studies on hyperpolarization-activated cation channel can provide more insights into the underlying bladder overactivity.

Keywords Overactive bladder · Hyperpolarization-activated cation channel
Computational model

1 Introduction

The common physiological function of a urinary bladder is to store and expel urine as a result of voluntary micturition. The activation of parasympathetic innervation pathway is coordinated by the brain and spinal cord to accomplish this micturition process. Overactive bladder (OAB) is a type of urinary incontinence, which is identified by the unintentional contraction of the UBSM cells. From various experimental

C. Mahapatra (✉) · R. Manchanda
Computational Neurophysiology Lab, Department of Bio Sciences & Bio Engineering,
Indian Institute of Technology Bombay, Mumbai 400076, India
e-mail: chitaranjan@iitb.ac.in

© Springer Nature Singapore Pte Ltd. 2019   261
P. K. Sa et al. (eds.), *Recent Findings in Intelligent Computing Techniques*,
Advances in Intelligent Systems and Computing 707,
https://doi.org/10.1007/978-981-10-8639-7_27

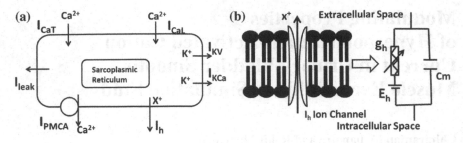

Fig. 1 a Schematic overview of ion channel mechanisms in UBSM cells. **b** Schematic overview of parallel conductance model for I_h current

studies, it is reported that the detrusor smooth muscles in bladder show sponta-neous phasic contraction activities via generation of spontaneous action potentials (sAPs) [1]. Changes in the resting membrane potential (RMP) due to the interplay of intrinsic ion channels in cell membrane play the crucial role in eliciting these sAPs. The T-type and L-type Ca^{2+} channels are essential for the depolarization period, whereas a diverse range of potassium (K^+) channels cause the repolarization and after-hyperpolarization (AHP) period of the AP [2]. Of the large sub-types of molec-ularly and functionally distinct K^+ channels, different types of Ca^{2+} activated K^+ channels (I_{KCa}) and voltage-activated K^+ potassium channels (I_{Kv}) are associated with the L-type and T- type Ca^{2+} channels (I_{CaL}) to generate AP [3] in UBSM cell. Figure 1a shows the overview of all ion channels in UBSM cell, where I_{CaT}, I_{CaL}, I_{KCa}, I_{Kv}, I_{Leak} and I_h represent T-type Ca^{2+} channel, L-type Ca^{2+} channel, Ca^{2+} dependent K^+ channel, voltage-dependent K^+ channel, leakage channel and hyperpolarization-activated cation current respectively. PMCA is known as the Ca^{2+} pump and sarcoplasmic reticulum is the storage house of Ca^{2+} inside the cell.

The non-specific, hyperpolarization-activated cation current (I_h) is involved in modulating rhythmic firing rates in neuron and cardiac cells. By using whole cell voltage clamp techniques, I_h currents had been reported in various spontaneously active smooth muscle cells such as the jejunum, airway, uterus, esophageal and portal vein [4–8]. The presence of I_h channel in rat UBSM cells is documented in Green et al. [9]. The I_h current is also known as inward-rectifying current, funny current and Q current, which is more selective to K^+ than other ions [9]. Therefore, a quantitative analysis of the I_h channel in modulating UBSM cell's excitability can provide key insights towards bladder overactivity.

Biological plausible computational modeling is very crucial for quantitative inves-tigation of complex dynamic systems like brain and muscle cells. Due to complex-ity in the instrumentation system, currently available experimental techniques for investigating real electrophysiological activities in smooth muscle cells do not allow simultaneous studies with multiple parameters. Electrophysiological mimicking of excitable cells makes possible the studies of dynamic systems as a single platform and the applications of physiological functions into artificial systems. Over the past decades, the computational modeling was widely developed in nerve, heart, and smooth muscle cells based on Hodgkin and Huxley formulation, Thermodynamic models and Markov models. In last decades, a few mathematical models concerning the ionic channels and calcium dynamics has been developed for the different types of smooth muscle cells, such as the intestine cell, uterine cell, human jejunal cell, gastric cell, mesenteric cell, and artery cell. However, electrophysiological simulation studies of UBSM cells are at a comparatively primary stage. The aim of this current study is to incorporate I_h channel in a UBSM cell model to investigate its modulating effects in shaping evoked APs and membrane excitability.

2 Methods

The mathematical explanation of a biological dynamic system is the early step in computational modeling. In this study, the mathematical explanation of UBSM cell membrane is established on classical Hodgkin–Huxley-type formalism [10]. The UBSM cell membrane is interpreted as a conductance-based model consisting of multiple variable ion channel conductances and a membrane capacitance C_m. Figure 1b represents the parallel conductance model for I_h current, where C_m is membrane capacitance, X^+ is the cation, g_h is the maximum conductance and E_h is the reverse potential for cation X^+.

The single UBSM cell model is based on single cylindrical compartment. The length and diameter magnitudes of the single cylindrical compartment are 200 μm and 6 μm respectively. In addition to morphological values, the membrane capacitance (C_m), membrane resistance (R_m), and axial resistance values are taken as 1 μF/cm², 138 MΩ-cm², and 181 Ω-cm [11] to simulate the passive electrical properties. The software platform adapted for all kind of simulations and analyses is NEURON [12], because it offers to simulate realistic electrophysiological outputs in biological excitable cells.

The time dependence characteristics of membrane potential (V_m) is represented in equation one.

$$\frac{dV_{m(t)}}{dt} = -\left[\frac{I_{ion(t)} + I_{stim(t)}}{C_m}\right],$$ (1)

where I_{ion} is the ionic current, and I_{stim} is the injected stimulation current.

The I_h current is integrated into a single UBSM cell model [13, 14] consisting the voltage-dependent Ca^{2+} channels, voltage-dependent K^+ channel, Ca^{2+} dependent K^+ channel to elicit APs. The I_h current is modeled as follows:

$$I_h = \overline{g_h} p^2 (V - E_h)$$ (2)

$$p_\infty = \frac{1}{1 + \exp\left(\frac{(V-74)}{21}\right)}$$ (3)

$$\tau_p = 5 * \left(\frac{1}{\left(0.00610779 * \exp\left(\frac{(-V)}{67.0828}\right)\right) + \left(0.0817741 * \exp\left(\frac{(V)}{67.08285}\right)\right)} \right), \quad (4)$$

where $\bar{g}$ is the maximum conductance, E_h is the reverse potential, p is the voltage-dependent activation parameter, τ_p is the time constant parameter and V is the membrane potential.

In Eqs. (2, 3 and 4), I_h ionic current is computed on steady-state activation parameter "p_∞" mentioned in Green et al. [9]. The voltage-dependent value of p_∞ is fitted in Boltzmann distribution with half-maximal activation at -74 mV, a slope factor of 21 and a reversal potential of -40 mV. All these parameters are adapted from experimental recordings documented in literature Green et al. [9]. The maximum conductance was varied to get the modulating effects in AP generation.

3 Results

The simulated activation parameter p_∞ and time constant τ_p were shown in Fig. 2 A and B. The activation parameter p_∞ follows a sigmoidal function in negative voltage range. The voltage-dependent time constant τ_p follows Gaussian function with 74.8 ms at -90.4 mV of membrane potential.

In this model, all ionic conductances are tuned to set the RMP at -50 mV as the experimental RMP values fall in -55 to -45 mV range [15].

The maximum conductance value of I_h channel is 0.0002 mho/cm^2 to maintain this RMP value. The maximum conductance value of I_h channel was varied from the controlled value to investigate the modulating effects in AP shape. A brief square pulse of 0.06 nA amplitude was injected for 10 ms duration to evoke the AP in our single cell model. The evoked AP (solid line in Fig. 3) closely matches with APs recorded in experiments [15] in terms of peak value, the rate of rising (dV/dt) and duration. The peak voltage, duration and AHP values of the AP are 8.3 mV,

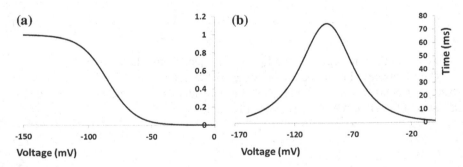

Fig. 2 Simulation of activation parameter p_∞ (a) and time constant τ_p (b) by the voltage clamp protocol in UBSM cell

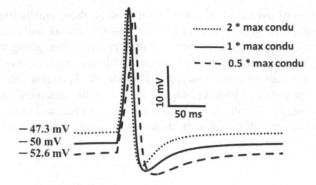

Fig. 3 Modulating effects of maximum conductance (max condu) on AP shapes in UBSM model

Table 1 Comparison of parameters among evoked APs in model cell

Conductance (mho/cm^2)	RMP (mV)	Peak (mV)	AHP (mV)	Duration (ms)
0.0002	−50	8.3	−58	32.5
0.0004	−47.3	8.3	−56.1	32.5
0.0001	−52.6	6.7	−60.2	36.3

32.5 ms and −58 mV respectively. In Fig. 3, the dotted and dashed lines represent the APs for maximum conductance (max condu) of 0.0004 and 0.0001 mho/cm^2. With the maximum conductance value of 0.0004 mho/cm^2, the AP (dotted line) was generated early due to elevated RMP (−47.3 mV) and early crossing of the threshold value. The AHP value of AP was altered to −56.1 mV because of the RMP elevation in the positive (−47.3 mV) direction. So the higher channel density makes the UBSM cell more excitable. By setting the maximum conductance at a lower value of 0.0001 mho/cm^2, the evoked AP (dashed line in Fig. 3) shows reduced RMP, increased duration, reduced peak amplitude and more negative AHP value. All values are mentioned in Table 1. It demonstrated that the lower channel density makes the UBSM cell less excitable.

4 Discussion

This study presents a computational model of a hyperpolarization-activated cation current in UBSM cell. The I_h channel model is integrated into a whole cell model for investigating the modulating role of I_h current in cell's excitability. The cellular excitability is reduced by reducing I_h channel conductance, which causes membrane hyperpolarization. Therefore, I_h channels are the key regulators in eliciting the spontaneous action potential in UBSM cells. In present days, the anticholinergic drugs, such as tolterodine, oxybutynin and trospium chloride are mostly prescribed for the treatment of overactive bladder. Unfortunately, these anticholinergic drugs cause dry mouth, somnolence, complication in visual functions, and constipation as side effects. Scientists are focusing on novel drug compositions acting via different signaling pathways for the treatment of overactive bladder without any chronic side effects. From this computational study, we conclude that I_h channel blockers might be treated as the potential pharmacological target for the treatment of overactive bladder condition. Finally, this I_h model can also be modified and incorporated into the other types of smooth muscle cell models to investigate underlying membrane excitability.

References

1. Brading, A.F.: Spontaneous activity of lower urinary tract smooth muscles: correlation between ion channels and tissue function. J. Physiol. **570**(1), 13–22 (2006)
2. Brading, A.F., Brain, K.L.: Ion channel modulators and urinary tract function. In: Urinary Tract, pp. 375–393. Springer, Berlin, Heidelberg (2011)
3. Petkov, G.V.: Role of potassium ion channels in detrusor smooth muscle function and dysfunction. Nat. Rev. Urol. **9**(1), 30–40 (2011)
4. Benham, C.D., Bolton, T.B., Denbigh, J.S., Lang, R.J.: Inward rectification in freshly isolated single smooth muscle cells of the rabbit jejunum. J. Physiol. **383**, 461 (1987)
5. McGovern, A.E., Robusto, J., Rakoczy, J., Simmons, D.G., Phipps, S., Mazzone, S.B.: The effect of hyperpolarization-activated cyclic nucleotide-gated ion channel inhibitors on the vagal control of guinea pig airway smooth muscle tone. Br. J. Pharmacol. **171**(15), 3633–3650 (2014)
6. Okabe, K., Inoue, Y., Kawarabayashi, T., Kajiya, H., Okamoto, F., Soeda, H.: Physiological significance of hyperpolarization-activated inward currents (I h) in smooth muscle cells from the circular layers of pregnant rat myometrium. Pflügers Archiv Eur. J. Physiol. **439**(1), 76–85 (1999)
7. Ji, J., Salapatek, A.M.F., Diamant, N.E.: Inwardly rectifying K$^+$ channels in esophageal smooth muscle. Am. J. Physiol.-Gastrointest. Liver Physiol. **279**(5), G951–G960 (2000)
8. Greenwood, I.A., Prestwich, S.A.: Characteristics of hyperpolarization-activated cation currents in portal vein smooth muscle cells. Am. J. Physiol.-Cell Physiol. **282**(4), C744–C753 (2002)
9. Green, M.E., Edwards, G., Kirkup, A.J., Miller, M., Weston, A.H.: Pharmacological characterization of the inwardly-rectifying current in the smooth muscle cells of the rat bladder. Br. J. Pharmacol. **119**(8), 1509–1518 (1996)
10. Hodgkin, A.L., Huxley, A.F.: A quantitative description of membrane current and its application to conduction and excitation in nerve. J. Physiol. **117**, 500–544 (1952)
11. Fry, C.H., Cooklin, M., Birns, J., Mundy, A.R.: Measurement of intercellular electrical coupling in guinea-pig detrusor smooth muscle. J. Urol. **161**(2), 660–664 (1999)
12. Hines, M.L., Carnevale, N.T.: The NEURON simulation environment. Neural Comput. **9**(6), 1179–1209 (1997)
13. Mahapatra, C., Brain, K.L., Manchanda, R.: Computational studies on urinary bladder smooth muscle: Modeling ion channels and their role in generating electrical activity. In: 2015 7th International IEEE/EMBS Conference on Neural Engineering (NER), pp. 832–835 (2015)
14. Mahapatra, C., Manchanda, R.: Computational studies on bladder smooth muscle: modeling ion channels and their role in generating electrical activity. Biophys. J. **108**(2), 588a (2015)
15. Meng, E., Young, J.S., Brading, A.F.: Spontaneous activity of mouse detrusor smooth muscle and the effects of the urothelium. Neurourol. Urodyn. **27**(1), 79–87 (2008)

An Experimental Study of Continuous Automatic Speech Recognition System Using MFCC with Reference to Punjabi Language

Nancy Bassan and Virender Kadyan

Abstract Punjabi language has almost 105 million native speakers and faced the challenge of less resource. The Punjabi ASR system has little research as compared to other Indian languages. This paper examines the continuous vocabulary of Punjabi language using Sphinx toolkit. The proposed work has been implemented on speaker-independent and speaker-dependent speakers in different environmental conditions. The Punjabi ASR system has been trained on 442 phonetically rich sentences using 15 speakers (6 Male and 9 female). The system adopts MFCC at the front end and HMM at the modelling phase to extract and classify feature vectors. The simulation result demonstrates the performance improvement of 93.85% on speaker-dependent dataset and 89.96% on speaker-independent dataset.

Keywords Punjabi automatic speech recognition · Mel frequency cepstral coefficient · Hidden Markov model

1 Introduction

The aim of large vocabulary automatic speech recognition (ASR) system is to convert the spoken utterances into corresponding text. Nowadays the requirement of building an ASR system in regional language increases with the increase in demand of the system. Punjabi [1] is one of the Indian languages that do not have any small or medium ASR system. Various vendors developed the commercial ASR system in many different languages like English, Arabic. For successful implementation of speech recognition system in any language depend upon its training and testing phase. Almost from last four decades researcher adopts various feature extraction techniques like MFCC, LPCC, PLP at the front end and acoustic modeling techniques like HMM, GMM, DBN and DNN at the back end. Parallelly language modeling and

N. Bassan (✉) · V. Kadyan
Department of Applied Science, Department of Computer Science & Engineering, Chitkara University Institute of Engineering and Technology Chitkara University, Punjab, India
e-mail: nancy.bassan@chitkara.edu.in

© Springer Nature Singapore Pte Ltd. 2019
P. K. Sa et al. (eds.), *Recent Findings in Intelligent Computing Techniques*,
Advances in Intelligent Systems and Computing 707,
https://doi.org/10.1007/978-981-10-8639-7_28

decoding technique also play key role in success of the system. For implementation of the system, various tools like htk, sphinx and MATLAB has been used for processing of the spoken utterance.

The paper presents the comparative analysis of different speakers involved in training and testing phase of the system using Sphinx toolkit. To overcome the barrier of environment, system performance has been analyzed in different environmental conditions. The rest part of the paper is organised as: Sect. 2 describes the overview of Indian language ASR engines. The corpus design criteria are explained in Sect. 3. Section 4 defines the proposed system architecture and Sect. 5 provides the experimental analysis. The conclusion and discussion of the paper are presented in Sect. 6.

2 Review of Indian ASR Engine

In literature, it has been analyzed that work done in Indian language has been focused upon academic research. Punjabi being spoken by huge mass, it is considered as less resource language and has only limited work done on small isolated, connected or continuous vocabulary Punjabi ASR system. The accuracy of these systems is very poor as compared to other Indian language ASR systems due to unavailability of less or no standard speech and text corpus. Due to the variability of speakers from one region to another region, tonal characteristics and dialectal variation have a major impact on the performance of Punjabi ASR engine. This problem creates an urge for building of large speech corpus to tackle the issue of out of vocabulary word. Table 1 represents the scenario of Indian language in the aspect of the technique, corpus type and accuracy achieved by the proposed system.

3 Corpus Design Criterions

Corpus plays a major role in the development of an efficient Punjabi ASR system. The developed corpus contains 442 phonetically rich sentences that are used in training and testing of the system. The dataset is recorded with the help of sound forge software as shown in Table 2. The utterances are recorded at 44100 Hz and then degraded to 16000 Hz using PRAAT software.

4 Proposed Punjabi Speech to Text (PSTT) Engine

The proposed Punjabi continuous speech recognition engine is implemented using CMU-SPHINX toolkit that consists of sphinx3-0.6.3, cmuclmtk-0.7, sphinxtrain-5prealpha and sphinxbase-5perealpha. The generic architecture for development of

Table 1 Overview of existing Indian ASR system

Language	Author's	Corpus type	Technique		Tool type	Accuracy
			Frontend	Backend		
Hindi	Kadyan et al. (2016) [1]	Continuous	MFCC	HMM	Sphinx 3	75–85%
	Kumar et al. (2012) [2]	Isolated words	MFCC	HMM	HTK	87.01%
	Aggarwal et al. (2011, 2010) [3–5]	Isolated words	MFCC, TRAP and Hybrid, HDLA, RASTA-PLP	GA, HMM	MATLAB, HTK	83.5–99.8% 79.7–97.5%, 78%, 72%, 65%, 56%
Gujarati	Tailor et al. 2016 [6]	Isolated words	MFCC	HMM	HTK	95.9–95.1%
Dravidian	Sangeetha et al. (2016) [7]	Isolated words	MFCC, LPC and SDC	AANN	No	No
Telugu	Mannepalli et al. (2016) [8, 9]	Continuous	MFCC	GMM, NNC	MATLAB	91%, 72%
Punjabi	Ghai et al. (2013) [10]	Continuous	MFCC	HMM	HTK	94.32%
	Dua et al. (2012) [11, 12]	Isolated words, connected words	MFCC	HMM	HTK	94.08–95.63%
	R. Kumar et al. (2011) [13]	Isolated Words	MFCC	DTW	MATLAB	94–96%

Table 2 Speech dataset distribution

Speakers' age category	Male	Female	Total
<30	4	6	10
30>=	2	3	5
Total speakers	6	9	15

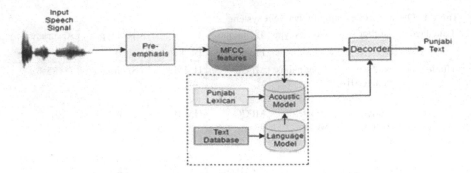

Fig. 1 Generic architecture of PSTT engine

proposed PSTT engine is mentioned in Fig. 1. The framework of PSTT system consists of two main modules: training phase and testing phase.

4.1 Training Phase

The uttered signals are processed by feature extraction phase that is required for both training and testing of the system.

4.1.1 Feature Extraction

Initially, at the front end, the captured speech signals are processed by statistical framework using MFCC. It is [11, 12] used to extract the feature vectors of the spoken utterances. During this phase, the recorded speech signal will undergo framing followed by pre-processing with a framing rate of 100 frames per second and window length of 0.0256 s. Windowing is applied to escape the discontinuity between the two consecutive frames. The windowed signals are derived from DFT and then compute its magnitude. Now, the fetched data is further passed through mel-filterbank (i.e., triangular bandpass filter) and subjected to logarithmic transformation to achieve log melspetral coefficient. Further, these spectral coefficients are carried through DFT to convert it into cepstral coefficient using Eq. (1)

$$C(i) = \sum_{i=0}^{N-1} \log\left(\left|\sum_{i=0}^{N-1} x(n) \exp^{\frac{-2i\pi kn}{N}}\right|\right) \tag{1}$$

The speech feature vectors are extracted feature using 'mfcgen2.sh' script to generate unique representation of each recorded utterance, which act as an input to the classification component.

4.1.2 Acoustic Model Generation

It is used to generate the HMMs [13] (hidden Markov models) which provide an effective dominating frame to estimate the probability of existence of sequence of observations as shown in Eq. (2)

$$W^* = \underset{W}{\text{argmax}} P(W|A) \tag{2}$$

where A is given acoustic observation, W^* is word sequence and P(W|A) is posteriori probability. Byes rule is used to decompose posteriori probability into linguistic and acoustic term by Eq. (3)

$$P(W|A) = \frac{P(A|W)P(W)}{P(A)} \tag{3}$$

For the development of the enhanced acoustic model, CMU-Sphinx 3 uses tri-phone-based Continuous Hidden Markov Models (CHMM) method in order to predict the probability distribution of the emission states generated by the state sequence.

4.1.3 Language Model Generation

The language model offers the grammar used in the classification. It's major role is [14] is to convert the training transcription file to DMP file that is a binary format language model in the standard Advanced Research Projects Agency (ARPA) format. Language modal compute using Eq. (4)

$$P(w_n|w_1, \ldots, w_{n-1}), \tag{4}$$

where w_n is the sequence of words. Joint Probability of words in sentence is given by

$$P(w_1, w_2, \ldots, w_n) = \underset{i}{\pi} P(w_n|w_1, w_2, \ldots, w_{i-1}) \tag{5}$$

To build the language model, uni-gram is computed that is converted into vocabulary with word frequencies, generating tri-grams and bi-grams. Uni-gram model is computed by Eq. (6)

$$P(w_1, w_2, \ldots, w_n) \approx \underset{i}{\pi} P(w_i) \tag{6}$$

<s> ਸਾਨੂੰ ਹਮੇਸ਼ਾ ਰੱਬ ਦੀ ਉਸਤਤ ਕਰਨੀ ਚਾਹੀਦੀ ਹੈ ਸਾਨੂੰ ਆਪਣੇ ਉਸਤਾਦ ਦੀ ਇੱਜ਼ਤ ਕਰਨੀ ਚਾਹੀਦੀ ਹੈ ਨੇਤਾ ਭੜਕਾਊ ਭਾਸ਼ਣ ਦੇ ਕੇ ਲੋਕਾਂ ਨੂੰ ਉਕਸਾਉਣਾ ਚਾਹੁੰਦੇ ਹਨ </s> (filter 1)
<s> ਕਈ ਨੇਤਾ ਦੰਗੇ ਭੜਕਾਉਣ ਲਈ ਉਕਸਾਉ ਭਾਸ਼ਣ ਦਿੰਦੇ ਹਨ ਪੰਜਾਬ ਦੀ ਧਰਤੀ ਬਹੁਤ ਉਪਜਾਊ ਹੈ ਉਗਰਵਾਦ ਨੇ ਕਸ਼ਮੀਰ ਨੂੰ ਤਬਾਹ ਕਰ ਦਿੱਤਾ ਹੈ </s> (filter 2)
<s> ਮੁਹੱਲੇ ਵਿਚ ਸਭ ਤੋਂ ਉੱਚਾ ਘਰ ਸਡਾ ਹੈ ਇਮਾਰਤ ਦੀ ਉਚਿਆਈ ਨੱਬੇ ਫੁੱਟ ਹੈ ਆਪਣੇ ਅਧਿਕਾਰਾਂ ਦੀ ਉਚਿਤ ਚੰਗਾ ਨਾਲ ਵਰਤੋਂ ਕਰੋ </s> (filter 3)

Fig. 2 Punjabi transcription

4.1.4 Punjabi Text Database

- Transcription File

The complete recorded data is accurately transcribed in gedit tool on Linux platform and saved as '.transcription' extension. Any mistake in the transcription file will deceive the training or testing process later. This work is done manually even the noise and silence should be represented in the transcription as shown in Fig. 2.

- Pronunciation Dictionary Creation

The phones-based pronunciation dictionary serves as an intermediate link between the language model and acoustic model. Automatic pronunciation dictionary tool is used and it needs transcription file in order to create the '.dic' file. Dictionary file has all acoustic events and words in the transcription mapped onto the acoustic units that are used to train. Redundancy of words is permitted with marked serial number starts from '2' for the second pronunciation. Filler file contains filler phones: <s>: Beginning silence, </s>: end silence and <sil>: within silence.

4.2 Testing Phase

The trained system is tested for sample dataset taken from Table 2. Initially, the MFCC is computed for all the files that come under the testing phase. The **sphinx_decode** is invoked that will take test transcription file, dmp file, hmm and trees from the acoustic model to produce the required configuration. At the end, the output file is created that contain decoded speech segment, it may be correctly recognised, partially correctly recognised, totally incorrectly recognised or completely not recognised.

5 Experimental Analysis

In this study, the performance of the system is computed by

Performance Correctness (PC) = ((N − D − S)/N) * 100
Percentage Accuracy (PA) = ((N − D − S − I)/N) * 100
Word Error Rate (WER) = 100%—Percentage Accuracy

Here, N = number of spoken words, D = number of deleted words, S = number of substituted words and I = number of inserted words, all in the testing phase.

Table 3 Performance analysis of Punjabi sentences in various environments

Number of spoken sentences	Closed environment			Open environment		
	No. of recognised words	Error rate (in %)	Accuracy (in %)	No. of recognised words	Error rate (in %)	Accuracy (in %)
1000	707	29.25	70.75	658	34.13	65.87
2000	1471	26.45	73.55	1375	31.23	68.77
3000	2271	24.27	75.73	2113	29.55	70.45
4000	3226	19.34	80.66	2874	28.15	71.85
5000	4137	17.25	82.75	3671	26.57	73.43
6000	5218	13.02	86.98	4537	24.38	75.62

5.1 Experiment Analysis with Varying Environment Conditions

It is marked that on increasing the number of spoken utterances in the training set the accuracy of the system increases. Table 3 shows that the performance of the system is higher in the close environment as compared to open environment. As the corpus of the system increases the confusion between feature vectors also increases that affects the accuracy of the system. The huge training data set that includes large utterance of the different speaker using different environmental conditions contribute toward increased inaccuracy of the system.

5.2 Experiments with Variation in Speakers

Simulation results were performed by different speakers (Speaker Independent and Speaker Dependent) those are involved or not involved in training and testing phase of the system. Table 4 shows the experimental results by analyzing the following:

- Comparison of speaker-independent system using same and different sentences
- Comparison of speaker-independent system using same and different sentences

The overall performance of the system shows that the system performed well with the same utterance present in training and testing set of speaker dependent data set. The accuracy of the system is varied from 2 to 3% with speaker independent system as compare to speaker dependent system.

Table 4 Performance analysis with speakers' variability

Testing sentences	Speaker independent						Speaker dependent					
	Same sentences			Different sentences			Same sentences			Different sentences		
	PC	PA	WER	PC	PA	WER	PC	PA	WER	PC	PA	WER
10	100	93.2	6.8	84.5	83.4	16.6	95.8	93.4	6.5	93.5	92.0	8.0
15	92.5	90.5	9.5	92.5	90.4	9.6	94.6	92.8	7.2	94.8	93.2	6.8
20	92.3	92.3	7.7	91.4	89.9	10.1	100	92.1	7.9	94.5	92.7	7.3
25	95.6	91.5	8.5	90.1	88.1	11.9	94.5	92.6	7.4	92.6	91.8	8.2
30	92.7	92.2	7.8	88.7	87.5	12.5	95.8	93.8	6.2	94.5	93.5	6.5

6 Conclusion and Discussion

In this paper, the medium vocabulary Punjabi speech recognition system has been presented by performing different statistics to achieve better results. The experimental result shows a significant improvement in performance of the Punjabi speech to text engine by using MFCC at the front end and hmm at the back end with sphinx toolkit. This approach has been applied to phonetically rich Punjabi continuous sentences using read speech corpus. The simulation shows an average performance of 93.85% on speaker dependent and 89.96% on speaker independent dataset. The work can be further extended on speaker adaptive and spontaneous utterances of Punjabi language.

References

1. Kadyan, V., Singh, A., Wadhwa, P.: Hindi dialect (Bangro) spoken language recognition (HD-SLR) system using Sphinx3. In: Proceeding of International Conference on Intelligent Communication, Control and Devices, pp. 991–998 (2017)
2. Kumar, K., Aggarwal, R.K., Jain, A.: A Hindi speech recognition system for connected words using HTK. Int. J. Comput. Syst. Eng. 1(1), 25–32 (2012)
3. Aggarwal, R.K., Dave, M.: Application of genetically optimized neural networks for hindi speech recognition system. In: World Congress on Information and Communication Technologies, pp. 512–517 (2011)
4. Aggarwal, R.K., Dave, M.: Using Gaussian mixtures for Hindi speech recognition system. Int. J. Signal Process. Image Process. Pattern Recogn. 4(4), 157–170 (2011)
5. Aggarwal, R.K., Dave, M.: Fitness evaluation of Gaussian mixtures in Hindi speech recognition system. In: First International Conference on Integrated Intelligent Computing, pp. 177–183 (2010)
6. Tailor, J.H., Shah, D.B.: Speech recognition system architecture for Gujarati language. Int. J. Comput. Appl. 138(12), 28–31 (2016)
7. Sangeetha, J., Jothilakshmi, S.: Automatic continuous speech recogniser for Dravidian languages using the auto associative neural network. Int. J. Comput. Vis. Robot. 6(1–2), 113–126 (2016)
8. Mannepalli, K., Sastry, P.N., Suman, M.: MFCC-GMM based accent recognition system for Telugu speech signals. Int. J. Speech Technol. 19(1), 87–93 (2016)
9. Mannepalli, K., Sastry, P.N., Rajesh, V.: Accent detection of Telugu speech using prosodic and formant features. In: International Conference on Signal Processing and Communication Engineering Systems (SPACES), pp. 318–322 (2015)
10. Ghai, W., Singh, N.: Continuous speech recognition for Punjabi language. Int. J. Comput. Appl. 72(14) (2013)
11. Dua, M., Aggarwal, R.K., Kadyan, V., Dua, S.: Punjabi automatic speech recognition using HTK. IJCSI Int. J. Comput. Sci. Iss. 9(4), 1694–0814 (2012)
12. Dua, M., Aggarwal, R.K., Kadyan, V., Dua, S.: Punjabi speech to text system for connected words, pp. 206–209 (2012)
13. Kumar, R., Singh, M.: Spoken isolated word recognition of Punjabi language using dynamic time warp technique. In: Information Systems for Indian Languages, pp. 301–301 (2011)
14. Bharali, SrutiSruba, Kalita, Sanjib Kr: A comparative study of different features for isolated spoken word recognition using HMM with reference to Assamese language. Int. J. Speech Technol. 18(4), 673–684 (2015)

Part III
Internet, Web Technology, Web Security, and IoT

Web Browser Analysis for Detecting User Activities

Dinesh N. Patil and Bandu B. Meshram

Abstract The Linux and its distribution are being widely used in the industry and other organizations. It has become essential to perform the log file analysis of the web browser to identify the user activities on the Internet. The log file analysis helps in identifying the malicious insider within an organization. After considering the existing work, this paper suggests an evidence collection and analysis methodology for the Linux web browser forensics. A framework for detecting the suspicious user activities on the Internet is proposed.

Keywords Digital forensic · Web browser forensic · Integrated analysis
Digital evidence · Framework

1 Introduction

With the rise in the use of the Internet worldwide, the misuse of it also has increased. The recent survey has indicated that 37% of the Internet users are using the Linux operating system [1]. The web browsers are used by the user for many purposes like searching information, email, e-commerce, news, e-banking, social media, and blog writing [2].

The digital forensic investigation of the web browser is performed to identify, collect, and analyze the evidence of suspicious activities of the user. The suspect while browsing leaves traces of their activities on the Computer System in various log files maintained by the web browser. This information is useful to the forensic investigator in gathering the evidence against the suspect. The information on the computer system such as history, cookies, download list, and cache can be used by

D. N. Patil (✉) · B. B. Meshram
Department of Computer Engineering, Veermata Jijabai Technological Institute,
Mumbai, India
e-mail: dinesh9371@gmail.com

B. B. Meshram
e-mail: bbmeshram@vjti.org.in

© Springer Nature Singapore Pte Ltd. 2019
P. K. Sa et al. (eds.), *Recent Findings in Intelligent Computing Techniques*,
Advances in Intelligent Systems and Computing 707,
https://doi.org/10.1007/978-981-10-8639-7_29

the forensic investigator to establish the websites frequently visited by the user of the system along with the timeline.

A Linux user uses many kinds of the web browser to access Internet services. The existing web browser forensic studies and tools have focused on the specific web browser. As a user makes use of the multiple web browsers to access the Internet services, it becomes imperative to perform the investigation of multiple web browsers at the same time. The investigation of individual web browser each time causes wasting of the time and the effort. It is essential to extract significant information from the web browser for the digital investigation purpose. As the evidence of the Internet activities by the user are spread over the various log files associated with each browser; it becomes necessary to perform the analysis of all these files together. This paper focuses on the most popular web browsers such as Mozilla Firefox, Google Chrome, Opera, and Vivaldi in Linux as per the statistics and are given in [3]. Figure 1 shows the market share of the various web browsers.

The existing research and tools have focused mainly on doing the individual Linux web browser forensics. The existing tools have explored few log files in extracting the evidence related to the Internet activity. In order to overcome the deficiencies of the existing tool, an improved methodology is proposed. The proposed methodology comprises of an integrated analysis of multiple web browsers used in the Linux, timeline analysis, extraction of user activity related information. A framework based on the extracted information from the various log files associated with the web browsers to detect the traces of suspicious activities of the user on the Internet is proposed. As the existing browser does not support the protection mechanism from the intentional deletion of the browsing history, this paper suggests a mechanism to prevent the unwanted access of the web browser log files from the users. The paper is structured as follows: The existing research on the Linux web browser forensic is covered in Sect. 2. Section 3 discusses the structures of the various web browsers historical log files. The proposed methodology is discussed in Sect. 4. A framework to detect the suspicious Internet activities of the user is proposed in Sect. 5. Section 6 covers the proposed security mechanism for the web browser historical log files. The conclusion and the future work are drawn in Sect. 7.

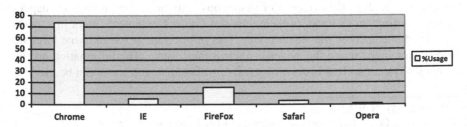

Fig. 1 Market share of web browsers in the Linux

2 Related Work

This section details out the existing research on the web browser forensics and the tool.

2.1 Existing Research

As the Internet is being used widely, the instances of misuse of it also have increased. The web browsers are used to access the various resources on the Internet.

Tracing evidence in the web browser is essential to the digital forensic investigator to convict the criminal. An evidence collection and analysis methodology of the various browsers combined together on the Windows platform has been proposed in [5]. The proposed methodology is implemented in a tool.

In [6], the effectiveness of the privacy mode feature in three different browsers has been investigated. It focused on web browsing history, cached files, and keywords used in queries.

A real crime investigation is carried out in [7]. The artifacts were discovered from the web browsers using the tools specific to the incident. Based on the evidence obtained the criminal was convicted.

2.2 Existing Tool

Autopsy. An autopsy is a digital forensic platform for Windows and Linux. It provides the facility for data carving, timeline analysis, and web artifact analysis. The autopsy extracts the web history, cookies, bookmarks from Firefox, Chrome, and IE.

DEFT. The Digital Evidence and Forensic Toolkit (DEFT) provide tools for Computer Forensic and Incident response on the Linux platform. It also supports extracting the forensic evidence from IE and Firefox web browser.

Pasco and Galleta. The forensic tool Galleta examines the cookies files of internet explorer. Pasco examines the contents of internet explorer's cache file. Pasco and Galleta run on Windows, Linux, Mac OS, and BSD platforms.

FTK. Forensic toolkit (FTK) is digital investigation platform. It provides the facility for advanced volatile memory analysis, the internet and chat analysis, and report generation. It supports the forensic analysis of Internet Explorer, Firefox, Chrome, Safari, and opera web browsers on the Windows platform.

Browser History Examiner. The browser history examiner analyzes web history for Chrome, Firefox, internet Explorer web browsers on the Windows platform.

Encase. Encase is the computer investigation solution for the digital forensic investigator. It performs the acquisition and analysis of data, recovery of data, and automatic report generation. It supports the forensic investigation of the web browser

Table 1 Web browser log file location in the directory structure of the Linux File System

Web browser	URL history file	Cookie file	Cache directory	Location
Firefox	Places.sqlite	Cookies.sqlite	Cache2	/root/.mozilla/firefox/fnf253mz.default
Google Chrome	History.sqlite	Cookies.sqlite	Cache	/home/username/.config/google-chrome/Default
Opera	Global_history.dat	Cookies4.dat	Cache	/root/.opera
Vivaldi	History.sqlite	Cookies.sqlite	Cache	/home/username/.config/Vivaldi/Default

history, cookies and cache files for Internet Explorer, Firefox, Chrome, Opera, and Safari web browsers on the Windows platform.

WEFA. It is the web forensic tool for collecting and analyzing data from the web browsers. It supports web browser such as Internet Explorer, Mozilla Firefox, Safari, Opera, Chrome, Swing, Comodo Dragon. It runs on the Windows platform.

3 Structural Representation of Web Browser Historical Files

This section discusses the historical file structures of the most widely used web browser on the Linux platform. The file formats used to store the historical information about the web browser activities by the user varies from browser to browser. The Firefox, Google Chrome, and Vivaldi use a .sqlite file to store web browser history whereas Opera uses .dat files.

Table 1 gives the summary of the files used to maintain the web browser history, cookies, cache and their location in the Linux directory structure.

3.1 Firefox

The Firefox maintains the historical information about the user activities for accessing the websites in a database named places.sqlite. This database consists of following tables with forensic importance.

- moz_historyvisits: An entry is created in this table each time a page is visited.
- moz_keywords: A unique list of keywords is maintained in this table.
- moz_hosts: The entry about the hosts visited is maintained in this table.
- moz_places: The details about a particular uniform resource locator (URL) visited by the browser are maintained in this table. This table is managed by the history service [4]. It also maintains the long-term download history.
- moz_inputhistory: A history of URL typed is maintained in this table.

As the moz_places table maintains the information about the URL visited by the user, the details of the web site accessed by the user can be known from this table. The table moz_places is having following structure.

Struct moz_places (*id* integer primary key, *url* longvarchar, *title* longvarchar, *rev_host* longvarchar, *visit_count* integer default 0, *hidden* integer default 0 not null, *typed* integer default 0 not null, *favicon_id* integer, *frecency* integer default −1 not null, *last_visit_date* integer, *guid* text, *foreign_count* integer default 0 not null);

The cache details of the Firefox browser are catalog by index.sqlite file.

3.2 Google Chrome

The history.sqlite database maintains the historical information of the websites accessed by the user. The history.sqlite database consists of following important tables for the forensic investigator.

- Downloads: This table keeps a track of all the files that are downloaded.
- URLs: This table maintains the information about the URLs used to access web pages.
- Keyword_search_terms: The keyword searched using the browser is maintained in this table.
- Visits: The information about the websites visited by the user is maintained in this table.

The details of the websites accessed along with the timestamp can be known from the URLs table which is having following structure.

Struct urls(*id* integer primary key, *url* longvarchar, *title* longvarchar, *visit_count* integer default 0 not null, *typed_count* integer default 0 not null, *last_visit_time* integer not null, *hidden* integer default 0 not null, *favicon_id* integer default 0 not null);

The cache details of the Google Chrome browser are catalog by index.bin file.

3.3 Opera

The user data about the web browser activity using Opera is stored in the following files.

- global_history.dat: It is a text file. The information represented in this file is stored in 3 fields, viz., Title, URL, date and time.
- search_field_history.dat: It maintains the history of queries typed in the search bar by the user.
- typed_history.xml: It also is an XML file that contains the history of URL's typed in the address bar by the user.

- Download.dat: It maintains the history of the file whose download is needed to be started again. It maintains the URL from where to do the download.

The vlink4.dat file is used to catalog the cache details of the URL visited by the user.

3.4 Vivaldi

The Vivaldi browser like Google Chrome maintains the information about the web browser history in History.sqlite database. This database consists of the same number of tables as that of History.sqlite database in the Google Chrome. The table URLs contain the information about the websites accessed along with the timestamp. The structure of the URLs table is similar to that of the URLs table in the History.sqlite database of the Google Chrome.

The cache details of the Vivaldi browser are catalog by index.bin file.

4 Evidence Collection Methodology for Web Browser

The web browser is being used by the user to access the websites, downloading and uploading the information on the web server by means of websites. The developers of the websites plant the cookies at the user's Computer while the user is browsing. The cookie maintains the users browsing activity. The caches stores copies of web documents passing through it. If the subsequent request arises for the same web document by the user then it is served through the caches instead of sending the request to the web server. This reduces the load on the web server and also consumption of the bandwidth reduces.

The people with the criminal mindset makes use of the Internet facility in an organization to do industrial espionage by uploading certain confidential documents, access restricted sites and wasting time by browsing during the work hours for a longer duration. The traces of the user activities can be extracted from the log files for the website access history, cookies, downloads, and caches. The proposed methodology for the forensic investigation of the Linux web browser involves the integrated analysis, user activity analysis, and timeline analysis.

4.1 Integrated Analysis

The users select a particular web browser depending on the requirement. The requirement can be high speed, easy to use, web site compatibility, customizability. As there can be more than one browser used by the user on the system, therefore, it becomes

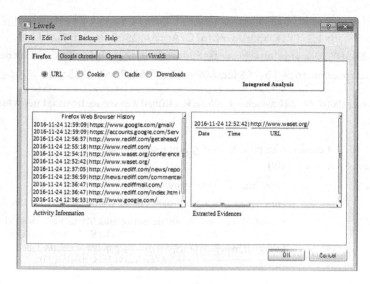

Fig. 2 A snapshot of the proposed tool showing integrated analysis

essential to do the forensic investigation of all the web browsers related log files. The proposed web browser forensic tool for the Linux as shown in Fig. 2 performs the forensic investigation of the log files of Firefox, Google Chrome, Opera and Vivaldi web browser.

4.2   User Activity Analysis

The user performs various activities while using the web browser. These web related activities can be categorized as websites browsed, sending and receiving mail, uploading and downloading of the documents.

The websites browsed. The sites visited by a user of the web browser are determined based on the web browser history log files as in Table 1. Also, the cookies downloaded whenever a user visits a web page provide the evidence of the users browsing history. The cache also provides the name of the URL being visited by the user.

Sending and receiving mail. The details of a user sending a mail using web-based email service are determined based on the contents of the URL in the historical log files. If URL content for the web-based email service of a web browser historical log file consists of the word 'compose' as shown in Fig. 3, it means that the user has used an email service to write an email to a recipient.

The mail account of a web-based email service opened by the user is determined based on the URLs in the historical log files associated with the web browser.

https://f4check.rediff.com/ajaxprism/attach?txtaction=close&login=dinesh_2767
54&session_id=4L24PK1KJKp2rVpxnNfqZQTAryyUcQ90&att_list=%20d.txt%5B162
.7%20KB%5D%20&cancelattachments=66dcc58090a22b3f660c70a002255cd8%3
Ad.txt.attach&compose_key=66dcc58090a22b3f660c70a002255cd8   Attached file

Fig. 3 A snapshot of the URL associated with the Rediffmail web service from the Firefox historical
log file

Table 2 URL's of email service provider during composing of mail

Email service provider	URL
Zoho mail	https://mail.zoho.com/zm/#compose
Gmail	https://mail.google.com/mail/u/0/#inbox?compose=159b15fcc7144f16
Rediffmail	https://f4check.rediff.com/ajaxprism/attach?txtaction=close&login= dinesh_276754&session_id=4L24PK1KJKp2rVpxnNfqZQTAryyUc Q90&att_list=%20d.txt%5B162.7%20KB%5D%20&cancelattach ments=66dcc58090a22b3f660c70a002255cd8%3Ad.txt.attach&comp ose_key=66dcc58090a22b3f660c70a002255cd8
Yandex	https://mail.yandex.com/?ncrnd=65044&uid=456873984&login= dinesh276754#compose/161003686678495236

Uploading and Downloading of the Documents. The documents uploaded by a
user are determined based on the contents of the URLs. In the case of certain web-
based email service, the URL's field of the historical log file consists of the word
'attach' which specifies a file being attached to the mail being composed. Figure 3
shows d.txt file being attached using Rediffmail email service. Table 2 shows the
URL's for various email service providers at the time of mail composing recorded in
the web browser historical log file.

The details of the documents being downloaded by a user are maintained in a
download database as discussed in Sect. 3. In addition, the download directory for
each browser also maintains the files downloaded by the user.

4.3 Timeline Analysis

The timeline analysis helps in sequencing the Web browser related activities that had
occurred on the system. The history log files for each web browser maintains the
URL accessed by the user along with the access date and the time. Figure 2 shows
the date and time of an URL accessed by the user using Firefox browser.

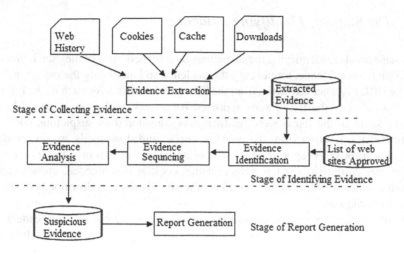

Fig. 4 The proposed framework for detecting suspicious activities

5 Framework for Detecting Suspicious Activities on the Internet

The proposed framework for detecting the suspicious activities carried out by the user accessing internet on the Linux system consists of 3 stages: the stage of collecting evidence, stage of identifying evidence, stage of analyzing evidence. The framework as shown in Fig. 4 can be applied for detecting the user activities on the Windows-based web browser, viz., Firefox, Google Chrome, Opera, and Vivaldi.

5.1 The Stage of Collecting Evidence

In this stage, the evidence of the user activities on the web browser is extracted from the various log files associated with the web browser. These files include web history related files, cookies related information, contents of cache stored in the caches and downloads performed by the user. This evidence is extracted from the log files associated with the Firefox, Google Chrome, Opera, and Vivaldi web browsers as discussed in Sect. 3. These extracted evidence is then stored in a database. The database consists of 4 tables for each of the browser viz., History, Cookies, Cache, and Download.

The structural definition of each of the table is as follows:

Struct History{*url* String, *date* Date, *time* Time};
Struct Cookies{*cookie_name* String, *last_access_date* Date };
Struct Cache{*url* String, *last_access_time* Time, *fetch_count* integer};
Struct Download{*file_name* String, *url_download* String, *date* Date, *time* Time};

5.2   The Stage of Identifying Evidence

This stage involves identifying the suspicious Internet activities of the user. It involves identifying the evidence, sequencing the evidence and analyzing the evidence.

The URL evidence extracted from the history related files for each of the browser is compared with the list of sites approved for access. If the site accessed by the user is not from the listed websites then it is considered as suspicious URL and the related information from the cookies, cache, and downloads are retrieved for analyzing purpose. A report is generated regarding the suspicious websites accessed by the user comprising of the website name, cookies downloaded, caches entries created for the websites, number of times the site visited and downloads operation if any from the site.

If a user had composed a mail and attached some sensitive file, this can be identified by analyzing the URLs of the historical log file associated with the web browsers and searching for the words such as 'compose' and 'attach' in the URL string for the email service provider as in Fig. 3. The name of the file attached then can be used to determine the sensitivity of the document sent by the user. The frequent composing of the mail by the user can be considered as suspicious activity.

5.3   The Stage of Report Generation

In this stage based on the suspicious evidence identified, a report is generated compromising of URL and related information about the cookies, caches, and downloads occurred from the website. A separate report is prepared for each of the web browsers.

6   Protection Mechanisms for Browser Log Files

The suspect on accessing the internet might try to remove his website access history from the database. Each web browser maintains their browsing history in a database which is easily accessible to the user of the system. The access to the web browser remains unprotected. None of the browsers prevents the access of the web browsing history from its users. Therefore, it becomes essential to protect the access to the web browser history and other associated log files.

In order to protect the web browser history database, the authentication mechanism is needed to be set up at the time of the installation of the web browser. Only the system administrator should be given access to the web browser history database. This kind of mechanism is more suitable in the industry where the chances of industrial espionage are often occurrence. Therefore, the insiders in the organization will have a second thought in accessing the sensitive sites or sending the sensitive information via email. The proposed protection mechanism is applicable to the Windows and Linux

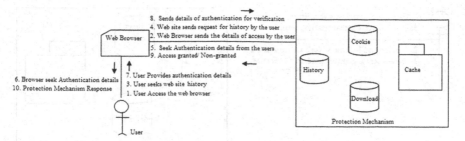

Fig. 5 A protection mechanism for preventing access to the web browser history

based web browser. The proposed protection mechanism suggests two improvements in the existing web browser functionalities.

6.1 Authentication for Web Browser History

Whenever the user opens the web browser to access websites, the details of web access are stored in the database specific to each web browser. Each web browser supports the features for reading the history of its user in accessing the websites. Therefore, any user can access the feature and might remove his websites access trace from the database. In this situation, access to the web browser should be restricted to the normal user. This can be achieved by providing the authentication to the access to the web browser history.

At the time of the installation of the web browser, the password protection should be provided to the access of the history database by the normal user of the system; only the access should be granted to the administrator. The proposed protection mechanism from the access of the web page visit history by the user is given in Fig. 5. As the user access the web pages, the details of the pages visited by the user is stored in the history log files associated with the specific browsers by the protection mechanism. Whenever the user tries to access this history information from the log files through the web browser, the protection mechanism seeks user authentication details. On verifying these authentication details, the protection mechanism grants or deny the access to the historical log files of the web browser to the user.

6.2 Authentication for the Modification/Changes/Deletion of the Web Browser Related Files on the Hard Disk

The malicious insider after accessing the websites might try to remove the traces of the Internet history by accessing the related files on the hard disk or by deleting such files. In order to avoid the potential loss of the web site historical information that

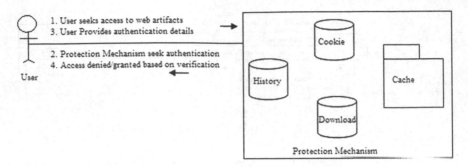

Fig. 6 A protection mechanism for preventing access to the web artifact

can be inflicted by the insider, a protection mechanism is needed to be set up for accessing such web browser related file.

The proposed protection mechanism as shown in Fig. 6 restricts access to the log files of the web browsers on the hard disk. So, whenever the user tries to access these log files on the hard disk, the protection mechanism will seek authentication details from the user. After verification of the user credential, the protection mechanism determines whether to grant or not to grant the access to the user.

7 Conclusion

The paper suggested a methodology to do the collective forensic investigation of various web browsers being used on the Linux platform. The tool such as Autopsy, DEFT, Pasco, and Galleta performs the forensic investigation of limited log files associated with the web browsers on the Linux platform. However, the proposed methodology includes the URL history, Cookies, Cache, and Download log files for various web browsers to detect suspicious activity. A framework to detect the suspicious activities of the user based on the evidence collected from the log files of the web browser is proposed. A protection mechanism for the log files associated with the web browser is suggested to improve the security of the web browser log files.

The future work will focus on exploring more web browser on the Linux platform to gather the evidence of the user activities on the Internet.

References

1. Usage Statistics and Market Share of Linux for Websites (2017). https://w3techs.com/technolo gies/details/os-linux/all/all
2. Keith, J.J., Rohyt, B.: Web Browser Forensics, Part 1 (2005). http://www.securityfoucus.com/i nfocus/1827
3. Browser Statistics (2017). http://www.w3schools.com/browsers
4. The Places Databases (2016). https://developer.mozilla.org/en-US/docs/Mozilla/Tech/Places/D atabases
5. Junghoon, O., Seungbong, L., Sangjin, L.: Advanced evidence collection and analysis of web browser activity. Digit. Investig. (2011) S62–S70
6. Said, H., Mutawa, N.A., Awadhi, I. A., Guimaraes, M.: Forensic analysis of private browsing artifacts. In: Proceedings of IEEE International Conference on Innovations in Information Technology, AbuDhabi, pp. 198–202 (2011)
7. Zsolt, N.: Using forensic techniques for internet activity reconstruction. In: Proceedings of the 16th WSEAS International Conference on Computers, pp. 248–253 (2012)

Phishing Website Detection Using Neural Network and Deep Belief Network

Maneesh Kumar Verma, Shankar Yadav, Bhoopesh Kumar Goyal, Bakshi Rohit Prasad and Sonali Agarawal

Abstract In the internet age, a large number of transactions are performed everyday, so website phishing is a crucial security problem in this age. Website phishing can be explained as stealing some information from the user without let them know that it is going to a non-genuine person. Our focus is on some techniques that make end users to be secured from phishing attack. We are developing neural network-based approach that will be prepared against previous dataset and latest acquired dataset so that it can have every flavor of phishing website data. The output layer of neural network gives the result and that result will determine whether the website is phishing or not.

Keywords Phishing site · URL · Detection · Neural network · Deep belief network1 introduction

1 Introduction

Phishing is attempt by group of people or individual hacker to get the precious information such as account passwords, bank account details or credit card details, etc. [1]. The phishing website are knowingly created to look alike some popular

M. K. Verma (✉) · S. Yadav · B. K. Goyal · B. R. Prasad · S. Agarawal
Indian Institute of Information Technology Allahabad, Allahabad, India
e-mail: iit2013093@iiita.ac.in

S. Yadav
e-mail: iit2013094@iiita.ac.in

B. K. Goyal
e-mail: iit2013092@iiita.ac.in

B. R. Prasad
e-mail: rs151@iiita.ac.in

S. Agarawal
e-mail: sonali@iiita.ac.in

© Springer Nature Singapore Pte Ltd. 2019   293
P. K. Sa et al. (eds.), *Recent Findings in Intelligent Computing Techniques*,
Advances in Intelligent Systems and Computing 707,
https://doi.org/10.1007/978-981-10-8639-7_30

genuine website. At the present time everyday, a large amount of legitimate and phishing websites come on board. So there is need of a latest approaches that could cope with new type of phishing website, which is quite difficult with many previously proposed approaches. Generally, phishing website comes into contact to user by some email or some social media post, and user may find it genuine just because it looks exactly like some popular website that user is already familiar with.

2 Related Work

Work carried out in [1], has developed a phishing detection system. This system works like as human behavior works. This system detects the true form of a website and tells that this website doing something wrong and find all information about the sites and our system. This system uses this information for classification of phishing. In this, they used three feature; WHOIS, URL, and Content. Based on these features, authors compare the real identity as well as claimed identity and on the basis of it, they find that a website is phishing or not. They use Blacklist for a better result. Authors of research work [2] added new features; Google Page Rank, Google Position, and Alexa rank, to improve the accuracy and efficiency of machine learning algorithm in detection of phishing site. In research work [3], authors used a multi-modal approach so that they can reduce the computational and financial cost. In this approach, they used visual and textual classification. This model finds and stores the features like color and histogram as visual features and HTML code as textual feature. Using MapReduce framework [4], they reduce the overall runtime of the process. In another work [5], authors proposed a IPDS system called anti-phishing preventer. Components of IPDS are: Junk Email Analyzer, node of phishing website analysis, and phishing network control center. In addition, they provided some best method for finding anti-phishing. Another research work [6] combined the confidence weight value classifier model with content-based model to create a well diverse and broad area system for catching current and new emerging category of phishing website domains. Researchers of study [7] specified that the phishing URL is detected using ensemble of various classifier, and then can be combined with hierarchical Cluster Training module for phishing classification. A study done by KingSoft Internet Security Lab shows that approach is working in real–time, which fits in the stream data analysis as discussed in various literatures [8, 9]. Their model detects emerging thread as they come to picture, not a long time is needed for a link to be proven as phishing website as is done in blacklist approach. Preprocessing and classification algorithms are used in the approach presented in research paper [10] for feature selection and for classification, respectively. Also, they employed four preprocessing techniques and five different classification algorithm to find the phishing URL as well as compared to their performances. An efficient approach for phishing detection proposed in the work [11], used single-layer neural network technique to calculate the value of heuristics objectively. Also, they used six heuristics function; Alexa rank, primary domain, PageRank, sub-domain, Alexa reputation, and path domain, for the required purpose. Approach given in research paper [12], uses two algorithm; multiclass classification based on association rule and adaptive boosting based on vote of weak classifiers.

3 Proposed Approach

First of all, data will be acquired from pishing tank websites and different types of attributes are gathered as mentioned. Some attributes are collected by URL property. Further, data is converted into categorical values, made in proper format and visualized. Artificial Neural Network and Deep Belief network based algorithm is implemented for classification of phishing URL. Finally, the effect of variation in controlling parameter is assessed on overall performance of algorithm.

3.1 Dataset Description

List of features gathered from the dataset is mentioned in Table 1. We have processed the data and after data preprocessing, a rank is given to different features based on their importance as mentioned in Fig. 1.

3.2 Algorithm Techniques Used

3.2.1 Backpropagation Technique

Backpropagation is a method of training artificial neural networks used in conjunction with an optimization method such as gradient descent as depicted in Fig. 2. In this algorithm, loss function's gradient is calculated with the help of all network weights. There are two phases in backpropagation neural network technique. From training dataset, a single tuple is sent to input layer of model. After this, input is processed on different layers (hidden layer) and forwarded to the output layer which generates final output. When one forwarded step is completed, then backward step starts wherein, at each step, we calculate error by comparing the actual output with gated output and modify weights according to error correction formula [2]. When error is very less, then we stop the process.

Initially, it finds the net weighted input $X = \sum_{i=1}^{n} x_i w_i - \theta$, where n is the number of data in single tuple which would be given to input layer and θ is the maximum threshold value in neural network by which error can be found. After this, data in input layer goes through activation (say sigmoid) function and then unlike a perceptron, neural network in backpropagation follows sigmoid function as $Y^{sigmoid} = \frac{1}{1+e^{-x}}$

Table 1 Feature description

Type of feature	Name of feature
Address bar based feature	Using the IP address Length of URL to hide the suspicious part Using URL shortening (small length) services "TinyURL" URL's having symbol "@" Redirecting using symbol "//" Adding prefix or suffix separated by (–) to the domain Sub-domain and multi sub domains HTTPS protocol (Hypertext transfer protocol with secure sockets layer) Domain registration length Favicon Using nonstandard port The existence of "HTTPS" tokens in the domain part of the URL
Abnormal based features	Request URL URL of anchor Links in <Meta>, <Script> and <Link> tags Server form handler (SFH) Submitting information to email Abnormal URL
HTML and JavaScript based features	Website forwarding Status bar customization Disabling right click Using pop-up window IFrame redirection
Domain based features	Age of domain PageRank Google index

3.2.2 Deep Belief Network

It is a generative graphical model composed of multiple hidden layers with connection between layers. However, there is no connection between units within each layer. It involves two steps; training in unsupervised way and training in supervised way. In unsupervised way, DBN learns to probabilistically reconstruct to input, this layer is called feature detectors on input [3]. In supervised way, DBN works as classifier and does classification. We know that DBN extracts the deep hierarchical representation (knowledge) and learns from this knowledge to make best model. There is joint distribution made between hidden layer 1 and observed vector x. h^k follows as $P\left(x, h^1, \ldots, h^l\right) = \left(\prod_{k=0}^{l-2} P\left(h^k \vee h^{k+1}\right)\right) P\left(h^{l-1}, h^l\right)$, where, $x = h^0$ and $P(h^{k-1}|h^k)$ is a conditional distribution on hidden units of RBM (Restricted Boltzmann machines), shown in Fig. 3, for visible units conditioned at level k and $P(h^{l-1}|h^l)$ is the visible-hidden joint distribution in top-level RBM.

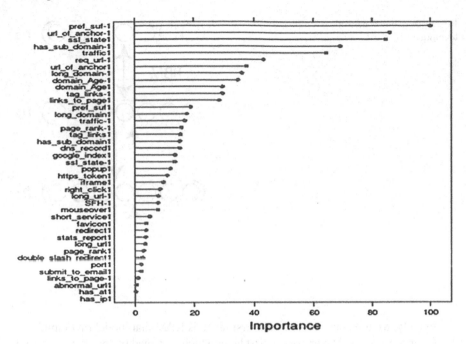

Fig. 1 Feature importance plot

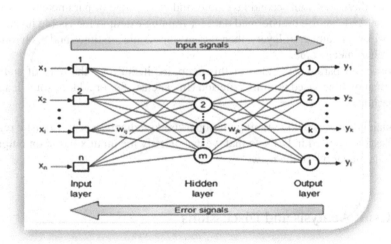

Fig. 2 Backpropagation technique

This technique is also called greedy layer-wise unsupervised training because training of model is done in unsupervised way and this technique can be used to DBN with RBM as building block for each layer. It has the following steps:

Fig. 3 Deep belief network
technique

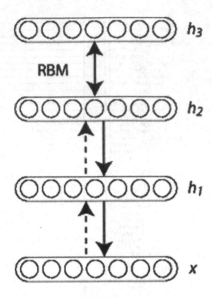

1. Initially, we use some data to train first layer as RBM that model raw input.
2. For input of second layer, we use first layer which is trained in first step. There are two solutions exist and represent like this $P(h^{(1)}=1|h^{(0)})$ or sample of $P(h^{(1)}|h^{(0)})$
3. After these, we train second layer and find exact value of parameters of second layer. And, we use transformed data for training example for visible layer.
4. If we use n number of layers, then iterate steps 2 and 3 n times and each time we reach to mean value.
5. We set all parameters value of this deep architecture by the help of standard format of supervised training criterion (linear classifier or adding some learning machinery).

It mainly focuses on fine-tuning via supervised gradient descent. Logistic regression classifier is used in DBN so that we can classify the input x based on output of last hidden layer $h^{(l)}$ of the network.

4   Result Analysis and Discussions

Figure 4 specifies a sample result snapshot of execution of each of the neural network and deep belief network approaches. Similarly, multiple executions are performed with different values of number of nodes in a layer and different number of layers in order to assess the performance of both the techniques. It is observed that accuracy increases when we increase the number of nodes in hidden layer and increase number of hidden layers. Performance of techniques improves when number of nodes in the

Fig. 4 Sample snapshot for neural network and deep belief network approach

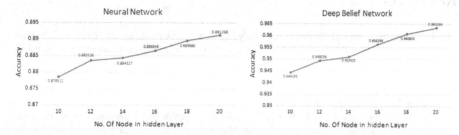

Fig. 5 Performance variation plots with increase in number of nodes in a layer

Table 2 Performance comparison

Algorithmic technique	Prediction accuracy
Neural network	0.8995461
Deep belief network	0.9442695
J48 [10]	0.8854126
Random forest	0.9165258

layers increases as shown in graph specified in Fig. 5. But we increase the nodes for a limit because after this limit the accuracy are linear but the execution time is high.

Artificial neural network approach achieves 89.95% accuracy at number of nodes in hidden layer and number of hidden layers both as five, whereas deep belief network approach gets 96.32% accuracy with similar settings.

Comparison of our approaches with previously used techniques is shown in Table 2, which evidently specifies that deep belief network approach gives better performance over other techniques.

5 Conclusion and Future Scope

Our phishing site recognition strategy recognizes phishing sites with high exactness while the false alert rate is low. In our analyzes, the deferral from WHOIS query was irrelevant contrasted with the stacking time of a site. In any case, still there are false positives and negatives. We plan to utilize another profound learning calculation with the goal that we can get greatest exactness. Another heading of future work is to utilize the site character check for different applications than phishing identification. Because of the tremendous size and unstructured nature of the web, site order is a basic but then difficult errand for web catalogs and centered slithering. Utilizing proposed method, any phishing distinguishing or site checker application can be made.

References

1. Jo, I., Jung, E., Yeom, H.Y.: You're not who you claim to be: website identity check for phishing detection. In: 2010 Proceedings of 19th International Conference on Computer Communications and Networks, Aug 2010
2. Abunadi, A., Akanbi, O., Zainal, A.: Feature extraction process: a phishing detection approach. In: Intelligent Systems Design and Applications (ISDA), 2013 13th International Conference on, pp. 331–335. IEEE (2013)
3. Shrestha, N., Kharel, R.K., Britt, J., Hasan, R.: High-performance classification of phishing URLs using a multi-modal approach with MapReduce. In: 2015 IEEE World Congress on Services (SERVICES), pp. 206–212. IEEE (2015)
4. Prasad, B.R., Agarwal, S.: Comparative study of big data computing and storage tools: a review. Int. J. Database Theory Appl. 9(1), 45–66 (2016)
5. Li, B., Sun, R., Fang, X., Luo, X., Chang, W.: Emergent challenges and IPDS for anti-phishing attack. In: 2014 International Conference on IT Convergence and Security (ICITCS), pp. 1–4. IEEE (2014)
6. Blum, A., Wardman, B., Solorio, T., Warner, G.: Lexical feature based phishing URL detection using online learning. In: Proceedings of the 3rd ACM Workshop on Artificial Intelligence and Security, pp. 54–60. ACM (2010)
7. Zhuang, W., Jiang, Q., Xiong, T.: An intelligent anti-phishing strategy model for phishing website detection. In: 2012 32nd International Conference on Distributed Computing Systems Workshops (ICDCSW), pp. 51–56. IEEE (2012)
8. Prasad, B.R., Agarwal, S.: Stream data mining: platforms, algorithms, performance evaluators and research trends. Int. J. Database Theory Appl. 9(9), 201–218 (2016)
9. Agarwal, S., Prasad, B.R.: High speed streaming data analysis of web generated log streams. In: 2015 IEEE 10th International Conference on Industrial and Information Systems (ICIIS), Peradeniya, pp. 413–418. IEEE (2015)
10. Singh, P., Jain, N., Maini, A.: Investigating the effect of feature selection and dimensionality reduction on phishing website classification problem. In: 1st International Conference on Next Generation Computing Technologies, pp. 388–393. IEEE (2015)
11. Nguyen, L.A.T., To, B.L., Nguyen, H.K., Nguyen, M.H.: An efficient approach for phishing detection using single-layer neural network. In: 2014 International Conference on Advanced Technologies for Communications (ATC), pp. 435–440. IEEE (2014)
12. Kadam, A.S., Pawar, S.S.: Comparison of association rule mining with pruning and adaptive technique for classification of phishing dataset. In: 3rd International Conference on 2013 Computational Intelligence and Information Technology, pp. 61–67 (2013)

Modeling and Mitigation of XPath Injection Attacks for Web Services Using Modular Neural Networks

Gajendra Deshpande and Shrirang Kulkarni

Abstract Injection attacks are considered to impact the most widespread vulnerabilities in web applications by Open Web Application Security Project (OWASP). XML is used as an alternative technology to database systems to store data in XML format, which can be queried to produce the desired results. XPath is a query language for XML which has injection issues similar to SQL. XPath can be used by the attacker to exploit the vulnerabilities in web applications by injecting malicious XPath query. If the web service is injected with malicious XML code, then it affects all the applications which integrate the infected web service. In this paper, we propose a solution, which uses count-based validation technique and Long Short-Term Memory (LSTM) modular neural networks to identify and classify atypical behavior in user input. Once the atypical user input is identified, the attacker is redirected to sham resources to protect the critical data. Our experiment results in over 90% accuracy in classification of input vectors. Our results also show that use of modular neural network results in improved response time of the web application compared to single neural network.

Keywords XPath injection · Web services modular neural networks
Long Short-term memory neural networks

1 Introduction

The Data Security Council of India (DSCI) a not-for-profit and self-regulatory organization considers cyberspace as national asset [1]. According to Open Web Appli-

G. Deshpande (✉)
Department of Computer Science and Engineering, KLS Gogte Institute of Technology,
Belagavi 590008, Karnataka, India
e-mail: gcdeshpande@git.edu

S. Kulkarni
Vellore Institute of Technology, Vellore 632014, Tamil Nadu, India
e-mail: shri1_kulkarni@yahoo.com

© Springer Nature Singapore Pte Ltd. 2019
P. K. Sa et al. (eds.), *Recent Findings in Intelligent Computing Techniques*,
Advances in Intelligent Systems and Computing 707,
https://doi.org/10.1007/978-981-10-8639-7_31

301

cation Security Project (OWASP), injection attacks are considered to be one of the most widely spread web application vulnerabilities [2]. Injection attacks can be classified into several categories such as SQL Injection, Command Injection, and LDAP Injection, etc. Web Services vulnerabilities can be present in Operating System, Network, Database, Web Server, Application Server, Application code, XML parsers, and XML appliances. An XML forms the basis for many prominent technologies such as Web 2.0, Web Services, Service-Oriented Architecture, Cloud Computing, and Semantic Web. XPath injection attacks are considered to be more dangerous than SQL injection attacks, since they lack role-based security. XPath is a standard language and attack can be carried out for any XPath implementation. XPath query can be used by the attacker to exploit the vulnerabilities in web applications by injecting malicious XPath query. Interoperability and reusability, the important characteristics of Web Services may cause security threat to web applications, since web services can be integrated with other web applications. If the web service is injected with malicious XML code, then it affects all the applications which integrate the infected web service. In such scenarios, it is very important to detect the XPath injection attacks and mitigate the risk in web services to protect critical services.

XML can be used as an alternative technology to database management systems to store data. Unlike database systems, XML lacks the feature of access control mechanism, i.e., it is not possible to specify a set of permissions to selected users. In XML, if the attacker gets access to even a single XML element, then the entire XML document structure can be accessed by an attacker. Figure 1 shows the relationship between enterprise networks and its users. In Fig. 1, hacker or the external user tries to access the database through an application and web services. The external user or the hacker supplies malicious input, which bypasses the firewall and application may provide illegal access to the hacker. The malicious input is not barred by the firewall since it is a legitimate HTTP traffic to the firewall. Generally, web services are designed to tunnel through firewalls. Many web servers like Apache, Zend, and Microsoft Internet Information Server (IIS) have a built-in input validation module, which rejects the malicious input. But the input validation module of web servers may reject the valid inputs like O'brian, etc.

In this paper, we propose a new whitelist-based validation technique for validating the user input, i.e., count-based validation. We also propose the architecture, which uses modular neural networks to classify user input as valid or malicious. Using this approach now, it is not possible to reject valid input strings which may comprise a single quote ('), dot (.), etc. The precise working of the system is described in subsequent sections.

Our work has two major contributions:

1. Using a neural network to identify and classify atypical behavior in user input. Once the atypical user input is identified, the attacker is redirected to sham resources to protect the critical data.
2. Use of Modular Neural Networks (MNN) to achieve higher rate of detection of true positives with improved response time of the web application over the existing Single Neural Network (SNN).

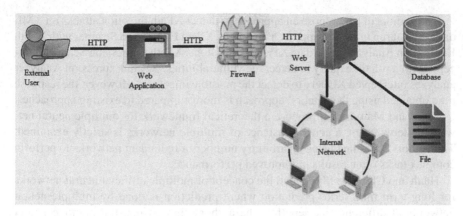

Fig. 1 Enterprise networks and users

2 Related Work

This section shows that there are several techniques used to detect XPath injection attacks but less work is done to analyze the atypical behavior in user input, and protect by misleading the attacker to sham resources using neural networks. It also indicates that the techniques abort the harmful requests and some genuine requests to protect web services.

In [3], the authors applied ontology to build strategy-based knowledge to protect web services from XML injection attack and to mitigate from zero-day attack problem. Strategy-based knowledge technique is a hybrid detection engine, which combines the advantages of both signature and knowledge-based approaches. In strategy-based design, new attack input will be automatically added to the ontology database. As the number of attacks in the ontology database increase, the technique will result in increased response time.

In [4], authors use the representative workload to exercise the web service and large set of SQL/XPath injection commands to disclose the vulnerabilities. The approach is based on XPath and SQL commands learning and posterior detection of vulnerabilities by comparing the structure of the commands issued in the presence of attacks to the ones previously learned. In this approach, results were not promising since the workload generation took few seconds of time, but learning phase took a few minutes of time per operation. The overall time taken by the detection process is approximately 15 min per operation.

In [5], authors discuss an approach consisting of two phases: (i) learning phase to learn valid request patterns and (ii) protection phase to detect and abort potentially harmful requests. The authors used a set of heuristics to accept or reject doubtful instances in the absence of complete learning phase. The authors achieved 76% accuracy in detecting the SQL/XPath injection attacks.

The authors of [6] propose an approach to detect XPath injection attacks on XML databases at run time. The approach integrates XPath Expression Scanner with XPath Expression Analyzer to validate XPath Expressions. XPath Expression Scanner intercepts with a run time query to detect the vulnerabilities. XPath Expression Analyzer analyzes intercepted XQuery to detect the possible injections. However, the response time obtained using the authors' approach is more compared to existing approaches.

Mike and Matthew [7] propose a theoretical framework for multiple neural network systems where a general instance of multiple networks is strictly examined. The authors claim that using an arbitrary number of redundant networks to perform complex tasks often results in improved performance.

Hanh and Christine [8] applied the concept of multiple artificial neural networks for long-term time series prediction where prediction is done by multiple neural networks at different time lengths. The authors showed that the multiple neural network system performed better compared to single artificial neural network for long-term forecast.

Anand et al. [9], used the modular neural network to reduce k-class problems to a set of two k-class problems, where each problem was dealt with separately trained network to achieve better performance compared to non-modular networks.

3 Proposed Approach

Figure 2 depicts the three-tier architecture of the proposed system design. The user or the attacker interacts with the web browser in the presentation tier and supply input to the web application. The input is passed to the web server as an HTTP request. The web server at the business tier consists of a web service (AuthenticationWS) and recurrent neural network as Long Short-Term Memory (LSTM) network. LSTM network [10] is used to classify the user input into three categories, i.e., valid, invalid, and malicious. If the user input is classified as valid, then the web server provides access to the legitimate XML document. If the user input is classified as invalid, then the web server responds with custom error messages. If the user input is classified as malicious, then the server redirects the user to counterfeit XML document, thereby protecting the critical resources.

The proposed algorithm is described below.

Algorithm 1

1. Scan the user input.
2. Determine the length of user input.
3. Count the frequency of every character in the user input [a–z, A–Z, 0–9, ' ". @ # % += ? :].
4. If the frequency of character is below the threshold value set for that particular character in Table 1 then set the error code to 40.
5. Else if the frequency of characters [. @ # % += ' "] is above the threshold value set for that particular character in Table 1 then set the error code to 4000.
6. Else set the error code to 400.

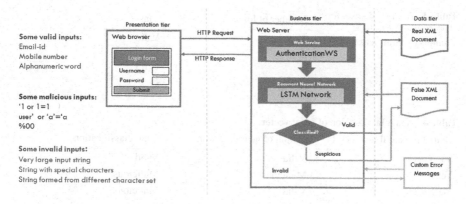

Fig. 2 Three-tier architecture of the proposed system

Table 1 Characters with threshold value

Special character	Threshold	Error Code
Single quotes (')	1	40
Double quote (")	0	4000
Dot (.)	2	40
Alphabets ([a-zA-Z])	Any	40
Digits ([0-9])	Any	40
At the rate (@)	1	40
Equal to (=)	0	400
Square Brackets ([,])	0	400
Round Brackets ((,))	0	400
Curly Brackets ({,})	0	400
Slashes (\,/)	0	400
Asterisk (*)	0	400
Pipe (l)	0	400
Any other character	0	400

Table 2 Training dataset for classification of error codes (Neural network 2)

Error code	Class
40	Valid
400	Invalid
4000	Malicious

7. Build a recurrent neural network 1 consisting of 50 neurons with hidden layer as LSTM network and output layer as SoftMax.
8. Use Rprop—trainer to train the network using the training dataset created using error codes in Table 2.
9. Use the test dataset created in real time to validate against the training dataset.

Table 3 Training dataset for classification of login attempts (Neural network 1)

Number of login attempts	Class
1	Valid
2	Valid
3	Valid
4 or more	Malicious

Table 4 Final classification of input vector

Output of neural network 1	Output of neural network 2	Final classification
Valid	Valid	Valid
Valid	Malicious	Malicious
Malicious	Valid	Malicious
Invalid	Valid	Invalid
Valid	Invalid	Invalid
Invalid	Malicious	Malicious
Malicious	Invalid	Malicious
Malicious	Malicious	Malicious

10. Build a recurrent neural network 2 consisting of 50 neurons with hidden layer as LSTM network and output layer as SoftMax.
11. Use Rprop—trainer to train the network using the training dataset created using number of login attempts in Table 3.
12. Use the test dataset created in real time to validate against the training dataset.
13. If train error and test error of both the networks are 0.0% then

    1. Finally, classify the input vector based on the outputs of both the neural networks in Table 4.
    2. If the user input is successfully classified as "valid" and found in the real XML file then Return the message "login successful".
    3. Else if the user input is classified as "malicious" then Return the contents of the fake XML file.
    4. Else if the user input is classified as "invalid" then Return the "error" message.

14. Else repeat the steps 8 through 13.

4   Results and Discussion

For our experimentation, we have used modular machine learning library PyBRAIN which is an acronym for Python-Based Reinforcement Learning, Artificial Intelligence and Neural Network Library [11]. For web services we used BottlePy micro web framework for Python [12]. BottlePy framework comes with its own web

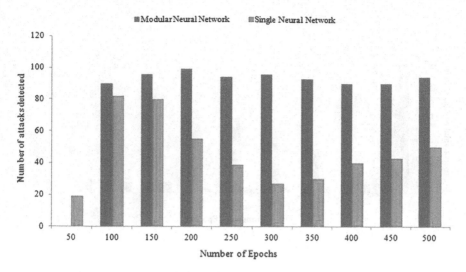

Fig. 3 Comparison of true positives

server known as WSGIRefServer. We used two web servers one for handling normal requests, i.e., and the other for handling malicious requests, i.e., Apache web server.

For our simulation we considered the synthetic dataset where the input sample size is 100. We have performed two experiments to classify atypical behavior in user input: one with single neural network and second with modular neural network. The results are recorded for ten iterations. The graphs are drawn with matplotlib [13] package of Python.

Figure 3 shows a comparison of true positives, i.e., number of attacks detected when malicious input is supplied to the application in the case of modular neural network and single neural network. The results show that modular neural network performs better in detection of malicious inputs compared to single neural network. The average number of attacks detected in the case of modular neural network is 84.2% and that of single neural network is 46.5%.

Figure 4 shows a comparison of false negatives, i.e., number of attacks not detected when malicious input is supplied to the application in the case of modular neural network and single neural network. The results show that modular neural network results in reduced rate of non-detection of malicious inputs compared to single neural network. The average number of attacks not detected in the case of modular neural network is 15.8% and that of single neural network is 53.5%.

Figure 5 shows a comparison of true negatives, i.e., number of valid inputs detected when valid input is supplied to the application in the case of modular neural network and single neural network. The results show that modular neural network performs better in detection of valid inputs compared to single neural network. The average number of attacks detected in the case of modular neural network is 83.8% and that of single neural network is 47.9%.

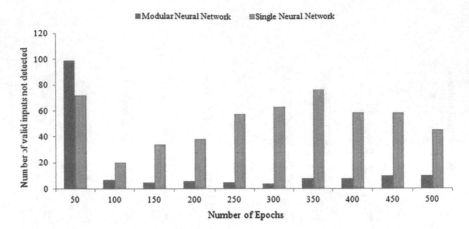

Fig. 4 Comparison of false negatives

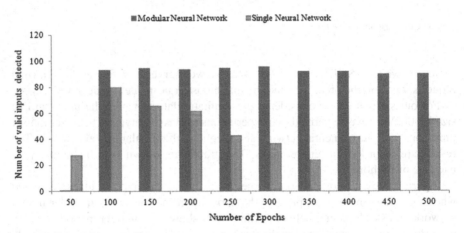

Fig. 5 Comparison of true negatives

Figure 6 shows a comparison of false positives, i.e., number of valid inputs not detected when valid input is supplied to the application in the case of modular neural network and single neural network. The results show that modular neural network results in reduced rate of non-detection of valid inputs compared to single neural network. The average number of valid inputs not detected in the case of modular neural network is 16.2% and that of single neural network is 52.1%.

Figure 7 shows a comparison of response time for modular neural network and single neural network. The response time is recorded for input samples from 10 to 100 for both modular and single neural networks. It is clearly indicated that the response time, i.e., time to react to a given input is less in the case of modular neural network compared to single neural network. The response time for a single request in case of modular neural network is approximately 1 s, and response time for a single request

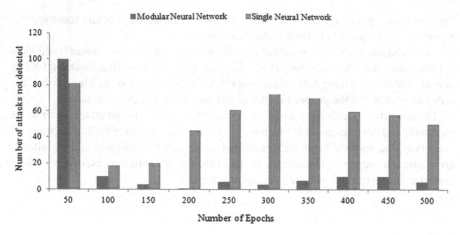

Fig. 6 Comparison of false positives

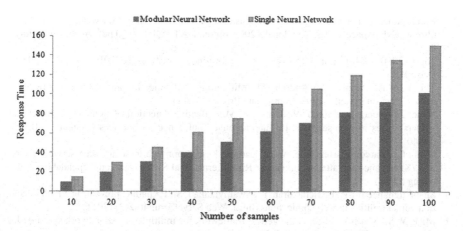

Fig. 7 Comparison of response time

in case of single neural network is approximately 1.5 s. The results also indicate that the response time is 1.5 times more in the case of single neural network compared to modular neural network for 100 epochs. This may not have much effect if the web server is serving tens of requests, but definitely, it will have an influence if the web server is serving hundreds or thousands of requests at a time.

5 Conclusion

The proposed solution offers improved security over existing methods by misleading the attackers to sham resources and custom error pages. Our results also show that the system accepts legitimate input although the user input may contain some special characters and rejects only truly malicious inputs.

Our solution combines modular neural networks and count-based validation approach to filter the malicious input. The results also show that modular LSTM neural networks perform better than single LSTM neural network. The results also serve us to pick out the proper number of network units for computation.

The security systems have to be successful every time. But an attacker has to be successful only in one case. Although we have achieved more than 90% of accuracy in classifying the atypical behavior in user input, still better techniques and algorithms are needed to increase the accuracy in the detection of malicious user inputs while considering the important factor of response time.

References

1. Kamlesh, B.: Cyberspace—global commons or a national asset. http://www.dsci.in/sites/defa ult/files/Cyberspace%20as%20Global%20Common_DATAQUEST_0.pdf. Accessed 08 July 2016
2. Top 10 2010-A1-Injection, https://www.owasp.org/index.php/Top_10_2010-A1. Accessed 02 Dec 2015
3. Thiago, M.R., Altair, O.S., Andreia, M.: Mitigating XML injection attack through strategy-based detection system. IEEE Security and Privacy (2011)
4. Nuno, A., Nuno, L., Marco. V., Henrique, M.: Effective detection of SQL/XPath injection vulnerabilities in web services. In: International Conference on Services Computing. IEEE (2009)
5. Nuno, L., Marco, V., Henrique, M.: A learning based approach to secure web services from SQL/XPath injection attacks. In: Pacific Rim International Symposium on Dependable Computing (2010)
6. Shanmughaneethi, V., Ravichandran, R., Swamynathan, S.: PXpathV: preventing XPath injection vulnerabilities in web applications. Int. J. Web Serv. Comput. 2(3) (2011)
7. Mike, W.S., Matthew, C.C.: A theoretical framework for multiple neural network systems. J. Neurocomput. 71(7–9), 1462–1476 (2008)
8. Hanh, H.N., Christine, W.C.: Multiple neural networks for a long term time series forecast. Springer, Neural Comput. Appl. 13, 90–98 (2004)
9. Anand, R., Mehrotra, K., Mohan C.K., Ranka S.: Efficient classification for multiclass problems using modular neural networks. IEEE Trans. Neural Netw. 6(1) (1995)
10. Hochreiter, S., Schmidhuber, J.: Long short-term memory. Neural Comput. 9(8), 1735–1780 (1997)
11. Tom, S., Justin B., Daan, W., Sun, Y., Martin, F., Frank, S., Thomas, R., Jürgen, S.: PyBrain. J. Mach. Learn. Res. (2010)
12. Bottle: python web framework. http://bottlepy.org/docs/dev/. Accessed 05 Apr 2016
13. matplotlib, http://matplotlib.org/contents.html. Accessed 06 July 2016

Acquisition and Analysis of Forensic Artifacts from Raspberry Pi an Internet of Things Prototype Platform

Nitesh K. Bharadwaj and Upasna Singh

Abstract The emergence of novel devices with admissible computational and communication capabilities has resulted in fabrication of portable and customizable Internet of Things (IoT) products like Raspberry Pi (RPi). The rapid proliferation of such devices has created new opportunities for offensive users and offers considerable challenges to digital investigators. It is necessary for investigators to diligently understand the implicit data formats and the types of evidence present in such devices. In this chapter, we present trade-off triangle which highlights the importance of digital forensics process towards the scenario of IoT enabled services. This chapter also proposes a common methodology for investigating forensic artifacts on IoT prototyping hardware platform. Based on the proposed methodology, a proof-of-concept tool called as RIFT has been developed as an outcome of research that aims the acquisition and preservation of forensically relevant static as well as volatile artifacts from RPi-IoT platform. Several experiments were carried out on the considered platform to reveal the vital evidence that can noticeably assist in forensic investigation. Finally, the preserved evidences are evaluated and presented to illustrate their usefulness and significances in digital forensic.

Keywords Digital forensics · Internet of Things · Raspberry Pi · Evidence acquisition · Evidence preservation

N. K. Bharadwaj (✉) · U. Singh
Department of Computer Science & Engineering, Defence Institute
of Advanced Technology (DU), Pune 411025, India
e-mail: niteshb2k14@gmail.com

U. Singh
e-mail: upasna.diat@gmail.com

© Springer Nature Singapore Pte Ltd. 2019   311
P. K. Sa et al. (eds.), *Recent Findings in Intelligent Computing Techniques*,
Advances in Intelligent Systems and Computing 707,
https://doi.org/10.1007/978-981-10-8639-7_32

1 Introduction

The technological growth in last few decades has resulted in economic and easy accessibility of digital devices. The reduction in technology cost along with an increase in internet proliferation, with negligible financial overheads has led to Internet of Things (IoT) becoming ever pervasive [1, 2]. IoT is defined as a computational object which is more than sensor and actuator that is connected to the internet in order to gain easy monitoring, analytic data acquisition, remote communication, etc., without direct intervention of human. The application range of IoT includes health care oversee, home, and industrial automation, etc.

Although IoT devices can provide a wide range of applications, its risk assessment cannot be ignored [3]. Raspberry Pi[1] (RPi) has been widely used in literature as an IoT device to provide custom applications and services (wireless sensor network gateway [4], communicating power supply [5], SmartLink [6], mobile agents [7], IoT prototype hardware [2], smart home [8], and sensor node [9]). This device can be exploited either as a target or tool to perform or assist the activities, such as denial-of-service attack, e-mail flooding, vulnerability scanning, etc. On the other hand, the examination of illegal activities on the reported devices is carried out by the investigator. An investigator proves or disapproves the existence of reported activities based on the forensically collected, examined, and admissible digital evidences. Digital evidences are the forensically significant information found in almost all type of digital devices. Moreover, IoT platform has created new criminal arena for example, remote burglary of user's personal and sensitive data, remote destruction of user facilities, etc. It is essential for an investigator to understand how IoT devices are actively involved in criminal activities including the type, format, and kind of evidence these devices contain. In this chapter, we propose a novel methodology to collect, preserve, and analyze forensically relevant artifacts from RPi-IoT platform.

This chapter focuses on exploring RPi-IoT prototype platform from the perspective of Digital Forensics (DFs) investigation especially for evidence acquisition and preservation. The contribution of this chapter includes DF challenge trade-off including a new forensics methodology for systematic investigation of customizable IoT platforms. This chapter provides a forensic insight into vital evidences that can assist investigation where RPi-IoT platform is of investigator's interest. As a result, a new forensic toolkit has been developed and presented that collects the static and volatile artifacts collectively from live RPi-IoT platform.

The rest of the chapter is organized as follows. Section 2 discusses the contribution in the field of IoT and system forensics, it also provides insight into the proposed DF trade-off triangle. The proposed investigation methodology for RPi-IoT platform analysis including the discussion on static and volatile artifacts is presented in Sect. 3. Section 4 demonstrates the experimental setup considered for the analysis, where it also provides the evaluation of collected artifacts. Finally, the conclusion and future work is presented in Sect. 5.

---

[1]www.raspberrypi.org.

2 Background and Related Work

Digital forensics in context of IoT becomes a more difficult proposition due to availability of diverse devices that are operated and connected using limited resources, power, and memory requirements of the devices [10]. Such unique capabilities of these customizable IoT devices have also motivated researchers and investigators to develop embedded investigative tool in [11]. Once IoT device is connected to the network it is also prone to cyber compromises, where attackers try to gain unauthorized and illegal access to such devices. IoT is currently being deployed in every sphere of humans life, hence IoT potentially collects and manages large amount of personally sensitive data (e.g., from home, offices, connected cars, or even human bodies). Security and privacy of such large amount of data is at stake and hence, due to less processing power, security issues, lack of encryption, insecure web services management, and other related problems can exist in most of the IoT devices [12]. The proposed trade-off triangle represents the today's technically advanced DF challenges which will persist in future due to proliferation of novel digital devices as shown in Fig. 1.

In the era of novel prototyping hardware platforms, undesirable activities and applications can be easily performed and developed by using such customizable platforms [13, 14]. The reported criminal activities always require investigators response for examining, analyzing and comparing the reported activities with the actual incidents. The scope of DF is enclosed within the bounds of criminal activities because, digital forensic techniques can be used whenever functional and usable environment is affected by undesirable or illegal activities as shown in Fig. 1. Moreover, the trade-off triangle also highlights that, with proliferation of new IoT devices

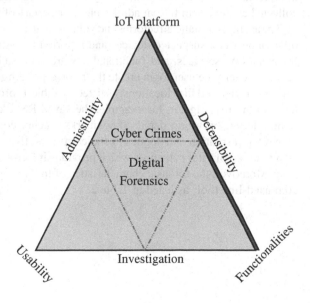

Fig. 1 Proposed trade-off triangle to illustrate DF challenges in IoT application era

more contribution from researchers is required towards DF community including the acceptance of novel DF tools and techniques. Also, the admissibility and justification of new tools and techniques in court-of-law is another factor which mutually substantiate the modern investigation challenges.

3 Proposed Methodology for RPi Investigation

This section discusses a new investigation methodology for IoT platform which encompasses seven stages as shown in Fig. 2. Each phase is equally significant and necessary to investigator, where the initial review phase presents the requirement of identifying IoT device type that exist in the network of heterogeneous devices by OS fingerprinting. The second phase requires investigation plan based on the identified devices. The proposed methodology successfully preserves all the relevant static artifacts, i.e., available at the time of investigation. The limited memory of IoT platform creates state where the relevant files can be overwritten at frequent intervals. Due to the volatile nature of IoT platform it is not always possible to either transfer the important files to another memory or cloud storage, before the evidence are overwritten or compressed. It is good practice to preserves all the volatile artifacts during live state of the suspected system. The following sub-sections highlights the static and volatile artifacts allied to RPi platform. In the similar direction, the authors in [15] proposed a MetroExtractor tool that gathers static and volatile forensic artifacts produced by Windows 8.x store apps. The proposed methodology assesses the integrity of identified artifacts by computing MD5 hash value of each file identified for acquisition. In the analysis and examination phase, the collected evidence are interpreted using available tools. The last phase highlights the necessity to present collected evidences in format admissible in the court-of-law.

The artifacts in static directories and volatile state are the only evidences, whenever relation between suspected devices and reported activities need to be observed. In Raspbian OS, services, and functionalities are accessed from *root* as well as user account, where the user login create their corresponding files at preferred locations. The directories and files locations, as listed in Table 1, provides essential information leads to the investigator, for example, the saved RSA keys for secure shell (SSH) remote login, browsing details (bookmarks, cache, cookies, databases, etc.), and network information. Analogous to static artifacts, live state of suspected devices is also a source of vital evidences which provides information about systems dynamic state. Moreover, the forensic characteristics of the live system is extracted using the command-line tools as listed in Table 2.

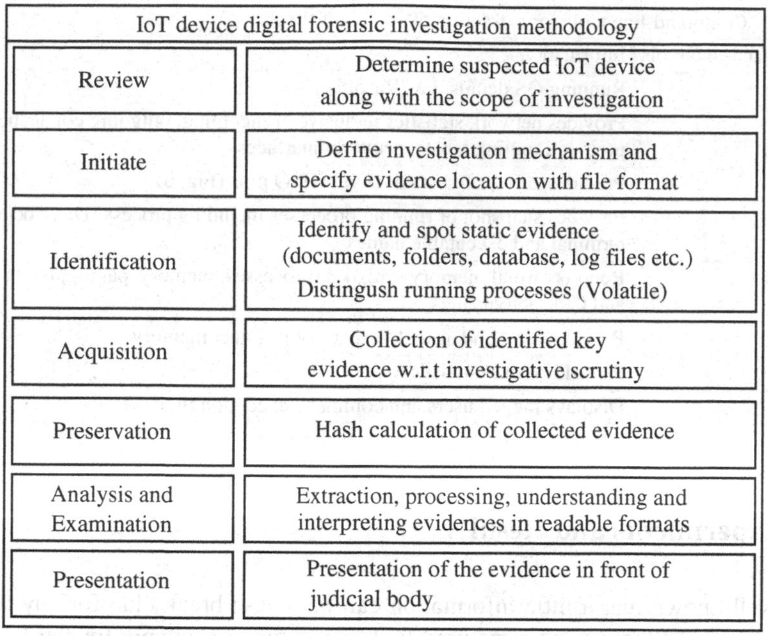

IoT device digital forensic investigation methodology	
Review	Determine suspected IoT device along with the scope of investigation
Initiate	Define investigation mechanism and specify evidence location with file format
Identification	Identify and spot static evidence (documents, folders, database, log files etc.) Distingush running processes (Volatile)
Acquisition	Collection of identified key evidence w.r.t investigative scrutiny
Preservation	Hash calculation of collected evidence
Analysis and Examination	Extraction, processing, understanding and interpreting evidences in readable formats
Presentation	Presentation of the evidence in front of judicial body

Fig. 2 Proposed methodology for forensics investigation of IoT devices

Table 1 Location of relevant static artifacts

File category	Type	Data/Filename/file extension
Command history	Filename	/home/pi/.bash_history
Environment path	Filename	/home/pi/.bashrc
Recently used programs and files	Filename	/home/pi/.idlerc/recent-files.lst
		/home/pi/.local/share/ recently-used.xbel
Browser	Filename and database	/home/pi/.config/epiphany/bookmarks.rdf ephy-bookmarks.xml, cookies.sqlite
	Database	/home/pi/.local/share/epiphany-browser/WebpageIcons.db
	Database	/home/pi/.cache/webkit/icondatabase/Web-pageIcons.db
	Folder	/home/pi/.cache/epiphany-browser/
SSH backdoor-RSA KEYs	Filename	/home/pi/.ssh/authorized_keys.save, /home/pi/.ssh/authorized_keys
SSID and password	File	/etc/wpa_supplicant/wpa_supplicant.conf

Table 2 Command-line tools for extraction of volatile artifacts

Command-line tools	Information
uname	Running OS details, i.e., Raspbian
netstat	Provides network statistics inclusive of incoming, outgoing connections, protocol, routing tables, network interfaces
gpio	Provides current I/O status of all GPIO pins (Fig. 6)
ps	Provides snapshot of running processes including process ID, associated terminal and executable names
vmstat	Reports virtual memory statistics (processes, memory, paging, block IO and CPU activity)
free	Provides used and available space of physical memory
ss	Provides socket statistics
w	Displays logged users and command execution time

4 Experiments and Results

It is well known that a little information can be a case-break clue for any digital investigator. As there is no structured tool or technique available for the forensic analysis of RPi-IoT prototyping platform and in order to bridge the gap between technology advancement and current forensic strength, we have designed a proof-of-concept command-line tool named as RIFT[2] (Raspberry Pi Forensic Toolkit) using Python programming environment. The developed tool acquires and preserves evidence from both static and volatile sources in a time-stamped folder named with present system date and time to ease interpretation and presentation for investigators and law enforcement bodies. To evaluate the efficacy of the proposed tool, the device (RPi B+ model) is configured using official Raspbian OS. The description of the considered OS is *Linux raspberrypi 4.4.11+ # 888 Mon May 23 20:02:58 BST 2016 armv6l GNU/Linux* where, the sequence consists of kernel name, network node name, kernel release, kernel version, machine hardware name, and operating System. Additionally, the corresponding distributor ID, description, release, and codename information includes Raspbian, GNU/Linux, 8.0, and Jessie, respectively. The Raspbian OS is configured to enable RPi with the capabilities of VGA display and Wi-Fi connectivity. A Tenda wireless adapter (Model: W 311MI) has been used for providing wireless access to RPi and connected to internet through the wireless network. Since the python environment is implicitly provided in Raspbian, no extra effort is required for execution of the developed RIFT tool.

The developed RIFT tool can be invoked from external storage device to automatically archive the evidence in the same storage device. The archive file consists of a metadata file named as *audit.csv*, which stores hash value (MD5 Hash) along with the

---

[2]http://github.com/niteshdiat2014/RIFT.

File name and path	Hash Value	Evidence Capture Time	Last Access Time	Last Modified Time	Last Time	File Size (Bytes)
/root/.bash_history	5b80d6e9755e01fce5b1949788a77462	Tue Jul 26 18:56:22 2016	2016-07-24 11:22:33.28	2016-07-26 18:03:17	2016-07-26 18:03:17.61	5501
/root/.profile	54328f6b27a45c51966ed436f3f609bf	Tue Jul 26 18:56:22 2016	2016-05-27 17:14:22.45	2007-11-19 23:27:23	2016-05-27 17:14:22.45	140
/root/.bashrc	e125739f81b08c470f20890304bf53e	Tue Jul 26 18:56:22 2016	2016-05-27 17:14:22.45	2010-01-31 17:22:26	2016-05-27 17:14:22.45	570
/home/pi/.cache/epiphany-browser/3939495368	b7768df63273fd0451e756b913eee29a	Tue Jul 26 18:56:39 2016	2016-07-24 13:26:39.93	2016-07-24 13:26:39	2016-07-24 13:26:39.95	210
/home/pi/.cache/epiphany-browser/1991597057	0c8981d17d4415722c912f0f1ae44c25	Tue Jul 26 18:56:39 2016	2016-07-24 10:54:02.93	2016-07-24 10:54:03	2016-07-24 10:54:03.22	3283
/home/pi/2016-07-26-165513_1024x768_scrot.png	69e7d5c2cdb52637f65a520a30ebed04	Tue Jul 26 18:56:28 2016	2016-07-26 16:55:13.19	2016-07-26 16:55:14	2016-07-26 16:55:14.86	102859
/home/pi/.local/share/webkit/databases/https_m.facebook.com_0.localstorage	bb3728aa4e63298e068271af599b3b62f	Tue Jul 26 18:56:30 2016	2016-07-24 12:13:20.71	2016-07-24 12:14:14	2016-07-24 12:14:14.73	5120
/home/pi/.thumbnails/normal/f5d14e53f1ff42b922e028d48d805b9.png	b5b0713d1e6710cf13fb89cc286016b8	Tue Jul 26 18:56:27 2016	2016-07-26 17:05:11.46	2016-07-26 17:05:11	2016-07-26 17:05:11.60	13381
/etc/default/ssh	500e3cf069fe9a7b9936108eb9d9c035	Tue Jul 26 18:56:23 2016	2016-05-27 17:13:19.08	2016-04-14 23:23:01	2016-05-27 17:13:19.08	133
/etc/default/networking	35cd4a2713961c9239ce4532c1b6c1c7	Tue Jul 26 18:56:23 2016	2016-05-27 17:13:19.08	2012-06-22 03:22:38	2016-05-27 17:13:19.08	306
/etc/default/bluetooth	1e25cf96f29147187311b94f8e0ea89	Tue Jul 26 18:56:24 2016	2016-05-27 17:13:18.81	2014-07-28 20:06:31	2016-05-27 17:13:18.81	845
/etc/network/interfaces	0200b0a5dd541dc426a0d9c7116ad08d	Tue Jul 26 18:56:24 2016	2016-05-27 17:13:20.17	2016-05-27 16:45:58	2016-05-27 17:13:20.17	523
/etc/network/if-pre-up.d/wpasupplicant	4c82dbf7e1d8c5ddd70e40b9665cfeee	Tue Jul 26 18:56:25 2016	2016-05-27 17:13:20.20	2015-11-11 04:11:53	2016-05-27 17:13:20.20	4696
/etc/network/if-pre-up.d/01-wpa-config-copy	24f586d173a334a20b2b22ed5e5d6438	Tue Jul 26 18:56:25 2016	2016-05-27 17:13:20.17	2016-03-31 18:53:44	2016-05-27 17:13:20.17	200
/etc/network/if-pre-up.d/wireless-tools	7bfee52166b1073629eeac6cd1fb7e390	Tue Jul 26 18:56:25 2016	2016-05-27 17:13:20.17	2011-11-14 15:38:34	2016-05-27 17:13:20.17	3839
/etc/network/if-up.d/wpasupplicant	4c82dbf7e1d8c5ddd70e40b9665cfeee	Tue Jul 26 18:56:25 2016	2016-05-27 17:13:20.20	2015-11-11 04:11:53	2016-05-27 17:13:20.20	4696
/etc/network/if-up.d/openssh-server	615af9ea3307b85023a36f4ed3c8f96	Tue Jul 26 18:56:25 2016	2016-05-27 17:13:20.17	2016-04-14 23:23:01	2016-05-27 17:13:20.17	945
/etc/network/if-up.d/mountnfs	528b2efd88285d2f443fc54234b5156	Tue Jul 26 18:56:25 2016	2016-05-27 17:13:20.17	2015-04-06 21:20:43	2016-05-27 17:13:18.81	4958
/etc/network/if-up.d/upstart	dfbcde4fd4a3a2563930605e03e160ab	Tue Jul 26 18:56:25 2016	2016-05-27 17:13:20.17	2013-01-06 23:56:18	2016-05-27 17:13:20.17	1483
/etc/network/if-down.d/wpasupplicant	4c82dbf7e1d8c5ddd70e40b9665cfeee	Tue Jul 26 18:56:25 2016	2016-05-27 17:13:20.20	2015-11-11 04:11:53	2016-05-27 17:13:20.20	4696
/etc/network/if-post-down.d/wpasupplicant	4c82dbf7e1d8c5ddd70e40b9665cfeee	Tue Jul 26 18:56:25 2016	2016-05-27 17:13:20.17	2009-12-18 20:46:25	2016-05-27 17:13:20.17	4696
/etc/network/if-post-down.d/wireless-tools	1f6530d0aee88247fe5001fe2f5f50d0	Tue Jul 26 18:56:25 2016	2016-05-27 17:13:20.17	2009-12-18 20:46:25	2016-05-27 17:13:20.17	1070
/home/pi/.bash_history	02db6206b8c79af91d09cc58a29477ae	Tue Jul 26 18:56:26 2016	2016-07-24 11:07:56.48	2016-07-26 18:03:17	2016-07-26 18:03:17.60	475
/home/pi/.local/share/epiphany-browser/Webpageicons.db	046e101ea1f53a07b76078b65f13e9ef	Tue Jul 26 18:56:29 2016	2016-05-27 17:23:18.12	2016-07-26 17:02:21	2016-07-26 17:02:21.37	200704

Fig. 3 RIFT tool output file, i.e., *audit.csv*

last accessed time, last modified time, last changed time, and file size of all preserved files. The output snippet of created *audit.csv* is provided in Fig. 3. The generated output file provide tremendous amount of vital information that can help examiner during assessment of criminal activity, event timeline creation, timeline validation and finding relationship between two or more activities. For example, the network SSID information is available in */etc/wpa_supplicant/wpa_supplicant.conf* file, if any modification is done in *wpa_ supplicant.conf* file, the investigator can easily find the modification time, access time, and changed time presented in output file. The hash digest also helps investigator to maintain the integrity, identifying known, or identical evidence, for example, the */etc/network/*-(if-pre-up.d/wpasupplicant, if-up.d/wpasupplicant, if-post-down.d/wpasupplicant and if-down.d/wpasupplicant) have identical hash value, time attributes and file size as well, as shown in Fig. 3.

4.1 Analysis of Static Artifacts

Although few evidences are available in formats which require investigators manual interpretation for their understanding but most of the collected evidence can be easily interpreted using existing tools. It may also be possible that the contents of static evidences differ before and after device turn-off/restart. Therefore, it is necessary to collect all important files before turning-off the RPi device. On the similar lines, the developed RIFT tool collects all the vital evidences.

Every browser maintains logging and monitoring mechanism to help robust transaction login and crash recovery support. The RPi-IoT platform ships with *Epiphany* web browser for accessing the information from World Wide Web (WWW). The Epiphany browser has many features such as HTML5 support, hardware-accelerated video decoding, ARMv6-optimized blitting function (several bitmaps are combined

```
<rdf:RDF xmlns="http://purl.org/rss/1.0/" xmlns:dc="http://purl.org/dc/elements/1.1/"
xmlns:ephy="http://gnome.org/ns/epiphany#" xmlns:rdf="http://www.w3.org/1999/02/22-rdf-syntax-ns#">
  <channel rdf:about="file:///home/pi/.config/epiphany/bookmarks.rdf">
    <title>Epiphany bookmarks</title>
    <link>http://www.gnome.org/projects/epiphany/</link>
    <item rdf:about="https://www.google.co.in/search?q=Digital+Forensics&amp;tbm=isch&amp;tbo=
u&amp;source=univ&amp;sa=X&amp;ved=0ahUKEwjnk6ffq4vOAhXEOo8KHYo1BTIQsAQISQ&amp;
biw=1024&amp;bih=631"><title>Digital Forensics - Google Search</title>
    <link>https://www.google.co.in/search?q=Digital+Forensics&amp;tbm=isch&amp;tbo=u&amp;source=univ&amp;sa=X&amp
;ved=0ahUKEwjnk6ffq4vOAhXEOo8KHYo1BTIQsAQISQ&amp;
biw=1024&amp;bih=631</link></item>
    <item rdf:about="https://m.facebook.com/">  <title>Facebook बर आपले स्वागत आहे</title>
    <link>https://m.facebook.com/</link></item>
    <item rdf:about="http://192.168.102.50:3582/">  <title>AJAYKUMAR-PC Web-based Configuration</title>
      <link>http://192.168.102.50:3582/</link></item>  <item rdf:about="http://fe80::b911:7f7e:6389:3f13:3582/">
        <title>IRFAN Web-based Configuration</title>  <link>http://fe80::b911:7f7e:6389:3f13:3582/</link></item>
    <item rdf:about="http://192.168.101.90:80/config/authentication_page.htm">
        <title>switch5ce639-1</title>  <link>http://192.168.101.90:80/config/authentication_page.htm</link></item>
    <item rdf:about="http://192.168.100.221:80/config/authentication_page.htm">
```

Fig. 4 Browsing details and IP addresses information residing in bookmark.rdf file

into one using a Boolean function), etc. The browser artifacts provide remarkable contribution in reconstructing the suspicious events, which help the investigator to prove the reported activities. The developed RIFT tool collects and preserves browser artifacts from forensically relevant static sources, i.e., *.cache* and *.config*, whereas the Epiphany browser also stores bookmarks, cache files, cookies, webpage-icon databases, and filter list. For example, traces of nearby hosts and network devices were found in RPi browser artifacts as shown in Fig. 4. In addition, visited web pages, terminal-device IP address, and port number were also reported in this file (Fig. 4). The IP addresses and device name can help investigators in identifying suspicious events, constructing, and validating activity time line. The browsing artifacts regarding the history id, visited urls, titles, visit counts, last visit time (Unix), and thumbnail update time (Unix) are demonstrated using Table 3. The time-stamp information can be successfully decoded using available database interpreter like DB browser for sqlite[3] and Sqlite manager.[4] Thus, the investigator can reveal the most frequently used web pages as well as most recently used web visits as desirable. It is possible that SSH-RSA public key can be remotely copied within the target devices using the network connection. As a result the attacker can use this RSA keys to gain unauthorized access into RPi device without any authentication. The developed RIFT tool successfully preserves the RSA keys artifacts as shown in Fig. 5 along with the corresponding hash digest, access time, modification time, and last changed time of *authorized_keys* file.

---

[3] http://sqlitebrowser.org.

[4] http://sqlabs.com/sqlitemanager.php.

Table 3 Most frequently used host details from ephy-history.db database

Id	Url	Title	Visit count	Last visit time (Unix)	Thumbnail update time (Unix)
1	http://10.10.20.1/	10.10.20.1	3	1469342782	0
2	http://www.google.co.in/	www.google.co.in	13	1464350087	0
3	http://www.youtube.com	www.youtube.com	11	1469346444	1469337863
4	http://www.digitalforensicsmagazine.com/	www.digitalforensicsmagazine.com	4	1469345975	1469341609
5	http://m.facebook.com/	m.facebook.com	1	1469342584	0
6	http://192.168.101.248/	192.168.101.248	0	1469529220	1469347260
7	http://192.168.250.200/	192.168.250.200	9	1469346999	0
8	http://www.diat.ac.in/	DIAT, Pune	6	1469529220	1469347260

```
ssh-rsa AAAAB3NzaC1yc2EAAAADAQABAAAABAQDru/4MpIZQmFmhCPZgUnZcUxN6xEjvJ1QPPzeugkiwTqKVapDS/2DeCP8R8KUk
bRl11d2tB8egMB4w8zVWr4KNBE5eSo2zsiqpfGGp+CYczFJIcRq9KwC993/c+EoKqKilSYEupyoGWXZuPGWV0XpTPoQ47sLDfthT
vhgQKnlg+8UhTY9pBKEzH8bFObyNT/mlSkKtW5KG0pJfO8KLgCH7cAuheDwfo98q12qKg6zFp5y80sESOLz2RWaP5+BQ5ICEvxN7
6B2sNZ+d+nhYKao3YwCeinZ6VIukRTglF5xTtqeCOSSwN99cSOUrJ9qd/S29t8nFfznxr5MBIA+UGGWp root@kali
ssh-rsa AAAAB3NzaC1yc2EAAAADAQABAAAABAQDOUEWU7Brpx3flTU+uQ/WNPJA+irYgcxSjH7ehXZqbex7mpOGNN80kK8ukAl0H
mB82jgKXQYUD8e1DT4UvTbzpHYStoEcAKxH24JJxDxkAiT/VAVVa/NuRvGxuUI8jYl7n6doDR+QMeU95TjTGfDAvBzSFuTot4qiu
D68buV31BKoPj7/enUxDv/0J+7E/njEb97o7q07UKkm8GaPlu/qVhx7J8FRgceZKUSLc/8PXvbh4cIKnCNnn970hCPBAHBkZzEGE
HgH4LI+ijIxipgd0rOmJWqr7/pXD6Ub8UvMGiS8TNCztulo6001H2qFeCAsPdRKvg3Ulp1RMGqu8xupd nitesh@kali
```

Fig. 5 Saved RSA keys for remote SSH login

4.2 Analysis of Volatile Artifacts

Along with static evidences it is equally important for investigator to consider volatile evidences of the system. The RPi's external world communicates is enabled via GPIO pins (40 pins) with the help of sensors, actuators, and other integrated IoT platforms. Raspbian is provided with GPIO library for accessing and programming each GPIO pin according to programmer's need. The users can integrate almost any type of digital sensors or actuators, for example temperature sensors, motion sensors, etc. Therefore, the GPIO pin status can help in correlating running process and configured pins. The change in default state of GPIO may be helpful in categorizing undesirable activities. In order to analyze the usability of particular pin, the RIFT tool also captures the running configuration status of GPIO. An example of default and modified state of pins are demonstrated and highlighted using Fig. 6. Moreover, the statistics of process, memory, network connectivity and remote login can provide goldmine artifacts about suspected RPi device for example; network service set identifier, IP tables, running processes, creation or deletion of files, etc.

(a) Default GPIO pin status

BCM	wPi	Name	Mode	V	Physical		V	Mode	Name	wPi	BCM
		3.3v			1	2			5v		
2	8	SDA.1	IN	1	3	4			5V		
3	9	SCL.1	IN	1	5	6			0v		
4	7	GPIO. 7	IN	1	7	8	1	ALT0	TxD	15	14
		0v			9	10	1	ALT0	RxD	16	15
17	0	GPIO. 0	IN	0	11	12	0	IN	GPIO. 1	1	18
27	2	GPIO. 2	IN	0	13	14			0v		
22	3	GPIO. 3	IN	0	15	16	0	IN	GPIO. 4	4	23
		3.3v			17	18	0	IN	GPIO. 5	5	24
10	12	MOSI	IN	0	19	20			0v		
9	13	MISO	IN	0	21	22	0	IN	GPIO. 6	6	25
11	14	SCLK	IN	0	23	24	1	IN	CE0	10	8
		0v			25	26	1	IN	CE1	11	7
0	30	SDA.0	IN	1	27	28	1	IN	SCL.0	31	1
5	21	GPIO.21	IN	1	29	30			0v		
6	22	GPIO.22	IN	1	31	32	0	IN	GPIO.26	26	12
13	23	GPIO.23	IN	0	33	34			0v		
19	24	GPIO.24	IN	0	35	36	1	OUT	GPIO.27	27	16
26	25	GPIO.25	IN	0	37	38	0	IN	GPIO.28	28	20
		0v			39	40	0	IN	GPIO.29	29	21

(b) Preserved modified GPIO pin status

BCM	wPi	Name	Mode	V	Physical		V	Mode	Name	wPi	BCM
		3.3v			1	2			5v		
2	8	SDA.1	IN	1	3	4			5V		
3	9	SCL.1	IN	1	5	6			0v		
4	7	GPIO. 7	IN	0	7	8	1	ALT0	TxD	15	14
		0v			9	10	1	ALT0	RxD	16	15
17	0	GPIO. 0	IN	0	11	12	0	IN	GPIO. 1	1	18
27	2	GPIO. 2	IN	0	13	14			0v		
22	3	GPIO. 3	IN	0	15	16	0	IN	GPIO. 4	4	23
		3.3v			17	18	0	IN	GPIO. 5	5	24
10	12	MOSI	IN	0	19	20			0v		
9	13	MISO	IN	0	21	22	0	IN	GPIO. 6	6	25
11	14	SCLK	IN	0	23	24	0	OUT	CE0	10	8
		0v			25	26	1	IN	CE1	11	7
0	30	SDA.0	IN	1	27	28	1	IN	SCL.0	31	1
5	21	GPIO.21	IN	1	29	30			0v		
6	22	GPIO.22	IN	1	31	32	0	IN	GPIO.26	26	12
13	23	GPIO.23	IN	0	33	34			0v		
19	24	GPIO.24	OUT	0	35	36	1	OUT	GPIO.27	27	16
26	25	GPIO.25	OUT	0	37	38	0	IN	GPIO.28	28	20
		0v			39	40	0	OUT	GPIO.29	29	21

Fig. 6 The preserved status of the GPIO pins

5 Conclusions and Future Scope

The advancement in capabilities and easy availability of small, low cost, and high-performance digital devices has opened new opportunities for both purposeful and offensive users. RPi-IoT platform is selected for the study and analysis since RPi is one of the most frequently used customizable devices due to its reliability, performance, and portability features. The possibilities that RPi can be easily customized to be used as a tool or target for performing undesirable activities have motivated for experimenting with RPi to explore what kind of forensic artifacts are necessary to preserve and what information is important for forensics purpose. In this chapter, a new trade-off triangle and IoT forensics methodology is proposed to explain the technology shifts from DF perspective. Since no standard tools are available for investigation of RPi-IoT platform; a systematic approach for evidence preservation from RPi-IoT platform has been presented. During preservation of relevant evidence, the developed RIFT tool collects both static and volatile evidences including the hash value which is used to maintain the integrity and authenticity of collected evidences. The time attribute extracted by the proposed tool can be a goldmine artifact in creating activity timeline, i.e., recreating crime scene to study the relationship between reported events. The future exploration of IoT platforms other than RPi device can provide new forensic insights to the digital investigator.

References

1. Stankovic, J.A.: Research directions for the Internet of Things. IEEE Internet of Things J. **1**(1), 3–9 (2014)
2. Maksimović, M., Vujović, V., Davidović, N., Milošević, V., Perišić, B.: Raspberry Pi as Internet of Things hardware: performances and constraints. Des. Iss. **3**, 8 (2014)
3. Williams, M.G.: A risk assessment on Raspberry Pi using NIST standards. Int. J. Comput. Sci. Netw. Secur. (IJCSNS) **15**(6), 22 (2015)
4. Kruger, C., Hancke, G.P.: Benchmarking Internet of Things devices. In: 2014 12th IEEE International Conference on Industrial Informatics (INDIN), pp. 611–616 (2014)
5. Lanzisera, S., Weber, A.R., Liao, A., Pajak, D., Meier, A.K.: Communicating power supplies: bringing the Internet to the ubiquitous energy gateways of electronic devices. IEEE Internet of Things J. **1**(2), 153–160 (2014)
6. Perera, C., Jayaraman, P.P., Zaslavsky, A., Georgakopoulos, D., Christen, P.: Sensor discovery and configuration framework for the Internet of Things paradigm. In: 2014 IEEE World Forum on Internet of Things (WF-IoT), pp. 94–99 (2014)
7. Järvenpää, L., Lintinen, M., Mattila, A.-L., Mikkonen, T., Systä, K., Voutilainen, J.-P.: Mobile agents for the Internet of Things, system theory, control and computing (ICSTCC). In: 2013 17th International Conference, pp. 763–767. IEEE (2013)
8. Casey, E.: Smart home forensics. Digit. Investig., Elsevier **13**(C), A1–A2 (2015)
9. Vujović, V., Maksimović, M.: Raspberry Pi as a sensor web node for home automation. Comput. Electr. Eng. Elsevier **44**, 153–171 (2015)
10. Hegarty, R., Lamb, D., Attwood, A.: Digital evidence challenges in the Internet of Things. In: Proceedings of the Tenth International Network Conference (INC), pp. 163–172 (2014)

11. Souvignet, T., Prüfer, T., Frinken, J., Kricsanowits, R.: Case study: from embedded system analysis to embedded system based investigator tools. Digit. Investig. Elsevier **11**(3), 154–159 (2014)
12. Oriwoh, E., Jazani, D., Epiphaniou, G., Sant, P.: Internet of Things forensics: challenges and approaches. In: 9th International Conference Conference on Collaborative Computing: Networking, Applications and Worksharing (Collaboratecom), pp. 608–615. IEEE (2013)
13. Arquilla, J., Adams, J.C.: Controlling cyber arms, and creating new LEGOs. Commun. ACM **59**(1), 18–19 (2015)
14. Dudas, R., VandenBussche, C., Baras, A., Ali, S.Z., Olson, M.T.: Inexpensive telecytology solutions that use the Raspberry Pi and the Iphone. J. Am. Soc. Cytopathol. **3**(1), 49–55 (2014)
15. Murtuza, S., Verma, R., Govindaraj, J., Gupta, G.: A tool for extracting static and volatile forensic artifacts of windows 8.x apps. In: IFIP International Conference on Digital Forensics, pp. 305–320. Springer (2015)

IoT-Based Automated Solution to Irrigation: An Approach to Control Electric Motors Through Android Phones

Vaishnavi Bheemarao Joshi and R. H. Goudar

Abstract In recent years, agriculture is getting automated with the use of application of information technology and the Internet of Things (IoT). In IoT, physical devices are interconnected to form a network, and these are enabled by the internet and the web that interact with these devices. One major problem in a developing country like India is the availability of electricity at the farming areas, which has resulted in problems of irrigation. This paper proposes a system, which notifies the availability of electricity at the farming areas by sending a SMS to the users mobile, thus helping the user to know about the availability of electricity and thus, allowing to switch ON/OFF the electric motor at the agricultural area from any remote location using IoT. This paper also involves an enhanced system for one user having farming areas located at different locations and control all these areas using only one device.

Keywords IoT · Temperature sensors · GSM · Arduino microcontroller

1 Introduction

The IoT is a "Network of Internet enabled objects, together with web services that interact with these object". The principle of IoT is that when a physical object that is virtual gets connected to the internet it becomes an IoT. Here, these things are actually not computers, but their working principle is like a computer. This idea is not a new one but it has only become more closely connected to the practical world due to progress in hardware development, and also due to decline in the size, cost, and energy consumption which has resulted in the production of smaller and low-cost computers. This IoT helps various devices and sensors to send data over the internet.

V. B. Joshi (✉) · R. H. Goudar
Department of Computer Networks Engineering, Centre for PG Studies, Visvesveraya
Technological University, Belagavi 590018, Karnataka, India
e-mail: vaishi.joshi@gmail.com

R. H. Goudar
e-mail: rhgoudar.vtu@gmail.com

© Springer Nature Singapore Pte Ltd. 2019
P. K. Sa et al. (eds.), *Recent Findings in Intelligent Computing Techniques*,
Advances in Intelligent Systems and Computing 707,
https://doi.org/10.1007/978-981-10-8639-7_33

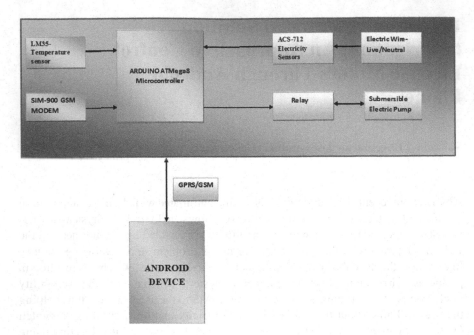

Fig. 1 System block diagram

Agriculture also plays a major role in a developing country like India. About 70% of India's population is dependent on agriculture. This agriculture uses 85% of fresh water, and this percentage will increase with the increase of population and food demand. Electricity is one of the major concerns for agriculture productivity. Due to nonavailability of electricity in rural and farming areas, this has resulted in irrigation problems due to which results in crop failure and decrease in the crop yields. In developing country like India, the availability of electricity at the farming areas is very low and is not timely hence, this paper proposes a system which uses IoT to provide solution to the above problem. The system designed provides an automated facility of notifying the user of the availability of the electricity and thus, allowing the user to make irrigation to be done easily from any remote location.

2 Proposed System

The proposed system consists of the temperature sensors, electric current sensors, a microcontroller, an Android device, a GSM Modem, a relay circuit, and an submersible electric motor. The difference in this system is that of the presence of electricity sensor to know the presence of the electricity. Figure 1 represents the block diagram of the proposed system.

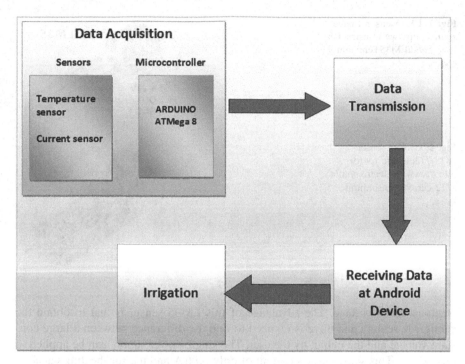

Fig. 2 System design showing the different modules involved

The system as a whole can be divided into different modules like the sensor modules which has two types of sensors, the GSM Modem, i.e., the SIM900 being used, the microcontroller which will be set up at the agricultural area and the Android device, which will be used by the user wherein this device communicates using GPRS/GSM and the user can operate from any location.

The two sensors LM35 (temperature sensor) and ACS 712 (current sensor) are used. These sensors monitor the data and send the data to the microcontroller. The system is designed in such a way that as soon as there is presence of electricity, the electricity sensor senses the current, and this will be sent to microcontroller which in turn will notify the farmer's Android device wherein he can turn ON the motor from any location provided this device is connected to the microcontrollers GSM Modem. The system as a whole can be divided into modules like the data acquisition module, transmission module, and the receiving and the irrigation module. This system design representation is shown in Fig. 2.

A. Data Acquisition

There are sensors used wherein data from these sensors need to be sent to controller. There are two sensors used, namely the temperature sensor and the current sensor. LM35 is the temperature sensor used here. The LM35 is a precision integrated-circuit temperature devices whose temperature measured in centigrade will be linearly pro-

Fig. 3 LM35sensor. Image *source* http://www.instructab les.com/id/LM35Temperatur eSensor

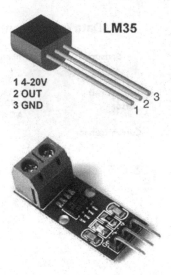

Fig. 4 Current sensor ACS-712. Image *source* http://www.alselectro.com/ac s712-current-sensor.html

portional to its voltage. The advantage of this LM35 sensor is that to obtain the centigrade scaling and there is no need to find the difference between a large constant voltage and the output by the user. The single power supply can be applied to this device. This sensor draws current of only 60 μA and the self-heating capacity of this temperature sensor is 0.1 °C in still air. Figure 3 depicts the LM35 sensor.

The second sensor used is the current sensor, i.e., ACS-712. Both AC and DC current sensing is done by this ACS712. The device is made up of linear Hall circuit with a conduction path and is made up of copper which has a low-offset. Of copper, which will be sensed by the IC and should be converted to a proportional voltage. Figure 4 shows the ACS-712. A magnetic field is generated when the electric current is applied, and is flowing through the conduction path made up of copper.

B. Data Transmission

The data received from the sensors need to be sent to the Android device that is being used by the farmer through a wireless transmission. This data transmission can be done by the GPRS/GSM Module. The transmitter and the receiver are connected to the microcontroller. The SIM900 GSM Module is used here for transmitting the data. This GSM SIM900 has a very low power consumption of about 1.5 mA and operates in a 900 MHz band, which is same for all the network operators in India, and can operate between −40 and +85 °C temperatures and can be controlled via ATcommands.

C. Data Receiving and Irrigation

The transmitted data is sent to the Android device of the user and the user can operate from any remote location. Here, the transmitted data contains the availability of power supply at the field and the sensor values of temperature that will be recorded by the sensor. As soon as a notification or a SMS is sent to the users device, the user can turn ON and OFF the electric motor using his Android device provided the user must have a prior knowledge of the time it takes to irrigate the field, depending on which he can turn OFF the motor.

2.1 Interfacing the Devices

The Arduino ATMega8 is a 28-pin microcontroller with 6 analog pins and 14 digital pins, it has a 8 kb of memory that can be reprogrammed and also erase which is called flash memory is present, 512 bytes of pulsed voltage erasable memory also known as EEPROM and also has a 1 kb of high-speed internal RAM and 3 internal timers are available. Pin 14 and 15 are digital pins connected to receiver and the transmitter of the SIM900 GSM Modem which in turn is connected to a +5 V power supply. Pin number 8 is ground connected to the ground of relay and is same with the pin number 7, i.e., vcc connected to the vcc of relay. The Normally Open (NO) pin is connected to the electric motor with one connection from the power supply. Pin number 23 is connected to the temperature sensor pin 2 output and pin 28 to the current sensor, i.e., ACS712 which in turn is connected to the power supply. The interfacing of sensors to the microcontrollers is shown in Fig. 5.

The above interfacing circuit gets connected to the user's Android device through the SIM900 GSM Modem that has been interfaced. This SIM900 operates in 900 MHz band which is same as that of all the network operated bands in India. A SIM card needs to be inserted into this modem for enabling the connection to the users' Android device. The only requirement is that the user needs to be connected to this modem to enable the working of motor from any remote location.

2.2 Enhanced System for Multiple Farming Locations for a Single User

This proposed system can be further enhanced to one particular users fields which are located at different locations far away but still can be operated using his Android device. Figure 6 depicts the enhanced design for multiple areas which are located at different places.

This enhanced system can be used for people whose agricultural areas are located at different places and want to have control among all of them. There are four farm locations shown in Fig. 6 which can be communicated using only one Android device, i.e., only single user. This system can be applicable to "n" number of field locations. Every location is configured with the whole system, i.e., shown in Fig. 1 except for Android device which is only one with the user and hence, all the locations can be controlled using only one Android device.

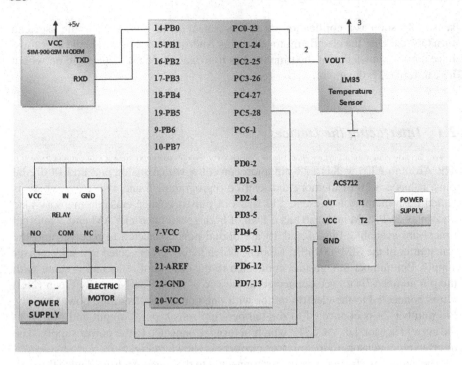

Fig. 5 Interfacing the microcontroller and the sensors

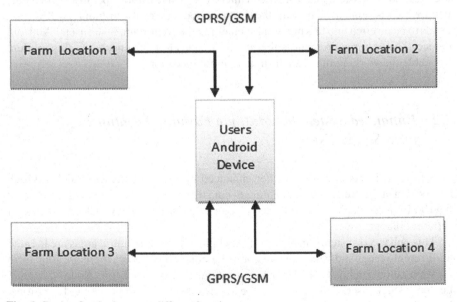

Fig. 6 Design for single user at different locations

3 Related Work

Shenoy and Pingle [1] describe the solution to reduce the transport cost, the prices, and market conditions for the agriculture produces, and also helps in reducing the number of middle hops between farmer and end consumer using IoT-based solution.

Baranwal and Pateriya [2] developed a solution using IoT in agriculture for the security and protection of resources and also agricultural products. An IoT-based device is used to analyze this information and control the problems from a remote location.

Zhao et al. [3] have described a solution for automation of agriculture using IOT. This solution consists of a "greenhouse production environment measurement and control system" which consists of different sensors like temperature, humidity, and are controlled using IoT.

Rajlakshmi and Devi Mahalakshmi [4] proposed an IoT-based automatic irrigation system and also a crop field monitoring system, which consists of processing of data from all the sensors that are used like temperature, humidity, soil moisture, light, and decision-making is done which makes the system to be fully automated.

Lee et al. [5] have proposed a prediction system for the growth of crops. This was done by gathering the information from the environment which will be recorded by the sensors. This system also quantitatively predicts the consumption of the agricultural products, changes in the weather, diseases and insect damages.

Yan-e [6] proposed an intelligent system based on IoT for the management of information system for the intelligent agriculture. This system provides a solution for the crops production, analysis, and storage of the products.

Ma et al. [7] developed and deployed a model for connecting agriculture through the sensor networks using the IoT. This system is a low cost, reliable, interoperable, and commercial system to be used using IoT and this is done by the sensor networks.

The above works describe different techniques that technology can be used in agriculture using IoT. They also provide solutions to certain problems in agriculture, but they lack the solution to notify the availability of electricity in the farming area. Hence, this paper provides a solution by notifying the user for the availability of electricity and can automate irrigation from any remote location.

4 Conclusion

This paper provides a solution for agriculture problem in India specially for the availability of the electricity. A notification or a SMS will be sent to users to notify the availability of electricity at the farming area. This system becomes very useful in areas where electricity supply is not available round the clock and the person needs to be notified for the availability of electricity. Further, it eliminates the manual work of the person going to the farming area and manually do the irrigation. Also, an enhanced system for a single-user multiple farming locations was also described wherein all the motors can be controlled using a single device. This system is a solution, provided the user has an Android device, which has become cheaper nowadays and limits for the availability of electricity 24/7.

References

1. Shenoy, J., Pingle, Y.: IoT in Agriculture. In: 3rd International Conference on Computing for Sustainable Global Development (INDIACom), pp. 1456–1458 (2016)
2. Baranwal, T.N., Pateriya, P.K.: Development of IoT based smart security and monitoring devices for agriculture. In: 6th International Conference on Cloud System and Big Data Engineering (confluence), pp. 597–602 (2016)
3. Zhao, J., Zhang, J., Feng, Y., Guo, J.: The study and application of the IoT Technology in agriculture. In: 3rd International Conference on Computer Science and Information Technology, vol. 2, pp. 462–465 (2010)
4. Rajalakshmi, P., Devi Mahalakshmi, S.: IoT based crop-field monitoring and irrigation automation. In: 10th International Conference on Intelligent Systems and Control (ISCO), pp. 1–6 (2016)
5. Lee, M., Hwang, J., Yoe, H.: Agricultural production system based on IoT. In: 16th International Conference on Computational Science and Engineering, pp. 833–837 (2013)
6. Yan-e, D.: Design of intelligent agriculture management information system based on IoT. In: 4th International Conference on Intelligent Computation Technology and Automation, pp. 1045–1049 (2011)
7. Ma, J., Zhou, X., Li, S., Li, Z.: Connecting agriculture to the Internet of Things through sensor networks. In: IEEE International Conference on IoT and Cyber, Physical and Social Computing, pp. 184–187 (2011)

Utilization-Based Hybrid Overlay for Live Video Streaming in P2P Network

Kunwar Pal, M. C. Govil and Mushtaq Ahmed

Abstract Over the last few years, there has been a surge of popularity in live video streaming using Peer-to-Peer network. Today, a significant amount of traffic uses the P2P approach for media transmission. It is a decentralized media communication method, which is more cost-effective and scalable. However, the traditional Peer-to-Peer network suffers various limitations and overlay construction is one of them. Different unstructured overlay approaches like tree and mesh have already been discussed. In this paper, we will provide the utilization-based hybrid overlay for the P2P network. There have been basically two prime considerations for creating the overlay-the upload bandwidth of the peer and utilization of that peer. We have used them as a crucial part in our approach.

Keywords Peer to peer network · Resource utilization · Live video streaming
Overlay

1 Introduction

The exponential growth of number of users over the internet in the last few years has contributed to a rapid increase in the demand for online video streaming. A study conducted by Cisco, has revealed that video transmission will comprise about 90% of the total data, over the internet till 2019 [1]. The increase in demand of

---

K. Pal (✉) · M. C. Govil · M. Ahmed
Department of Computer Science and Engineering, Malaviya National Institute
of Technology Jaipur, Jaipur, India
e-mail: 2013rcp9568@mnit.ac.in

M. C. Govil
e-mail: govilmc@gmail.com

M. Ahmed
e-mail: mahmed.cse@mnit.ac.in

© Springer Nature Singapore Pte Ltd. 2019
P. K. Sa et al. (eds.), *Recent Findings in Intelligent Computing Techniques*,
Advances in Intelligent Systems and Computing 707,
https://doi.org/10.1007/978-981-10-8639-7_34

331

video streaming over the network also increases the cost and complexity at the server side in the traditional client–server architecture [2]. To solve the issues of traditional client–server architecture, a new Peer-to-Peer (P2P) computing approach has been gaining popularity from the last few years. CoolStreaming [3] represents the milestone work that has been with respect to P2P media streaming. The basic idea of CoolStreaming, its implementation and the design issues are discussed by Xie [4]. An improvement design, theory, and practice of CoolStreaming are discussed in [5]. Live video streaming using Peer-to-Peer network introduces new challenges for its implementation like overlay construction, data scheduling, selfish peer/free riders, etc. In this paper, we will concentrate on overlay construction, and discuss different overlay construction techniques. We also propose a utilization-based overlay approach to solve the problem of overlay construction in P2P network.

2 Related Work

An overlay network is a network of computers which is built on top of the already existing network. Peers in the overlay network are connected by a logical or virtual link. Traditionally, unstructured overlay construction in P2P network further can be classified as tree and mesh overlay.

2.1 Tree Overlay

Tree overlay is easy to maintain and less complex as compared to the other overlay approaches available in the P2P network. Source or video server is available at the zero level, and a hierarchy of peers is created that comprises of multiple levels. The lower level peer works as a parent to the peers which are immediate above its level. Parent peer uploads the media content to all its child peers that are available in the tree overlay. Due to the fixed structure between the peers like the parent–child structure in the tree overlay, start-up delay is less compared to that in the mesh overlay. Upload capacity of each peer is limited and that also affects the number of children a peer can have. ESM and NICE are some of the examples of tree-based overlay [6].

2.2 Mesh Overlay

Another solution to the overlay construction problem is mesh overlay. The peers connected to each other follows the property of complete mesh. Some basic problems in live video streaming using P2P network are bandwidth and content bottle-neck. PRIME describes the mesh-based P2P live video streaming in detail [7]. The swarming-based content delivery mechanism is used for data transmission between the peers, and it tries to provide the solution to both the problems discussed above. To solve the content bottleneck problem, author uses the efficient pattern delivery scheme. The bandwidth bottleneck problem is solved using the bandwidth degree condition scheme [8]. Some examples of mesh-based overlay are Bullet, Anysee, CoolStreaming, and Chainsaw [9].

2.3 *Hybrid Overlay*

Both tree and mesh overlays have some pros and cons. Some authors have tried to combine both the overlays and have proposed a hybrid overlay to address the limitations of tree and mesh overlay. A novel hybrid overlay approach is defined by Huang et al. [10], which is a combination of mesh and tree overlay. Tree overlay is used for control packets and mesh overlay is used for data transfer. For tree overlay, geographical location of peer is used and for mesh overlay, layered peer selection mechanism is used. Problem with the approach is it only uses the geographical location of peer, and other important properties like bandwidth of peer is not taken into account. Hybrid Live P2P Streaming Protocol (HLPSP) is described by the author to solve the problem of hybrid overlay network [11]. Upload bandwidth of peers are used for creating the overlay in HLPSP. Using the upload bandwidth of peer groups is formed and the peers that have maximum upload bandwidth are nearest to source, and work as a parent to other peers available in the network. Simulation analysis shows that HLPSP provides better data flexibility, less control overhead, and less start-up delay. Different factors which are also crucial for the overlay construction are not considered in HLPSP. A new hybrid overlay approach is discussed in our previous work which takes into consideration all crucial factors such as the dynamic nature of peer, as well as its geographical location and upload bandwidth [12]. It is a hybrid overlay which combines both mesh and tree overlay property. Simulation result of the approaches shows that start-up delay, playback delay, and end-to-end delay is minimum for the approach as compared to Denacast [13].

3 Utilization-Based Overlay

Above discussion shows that the upload bandwidth of the peers is a prime factor in P2P overlay network. The upload bandwidth of peer plays a major role but if a peer does not have good connectivity to the other peers or peer does not upload sufficient amount of data to other peers, then its upload bandwidth does not remain as efficient. So, the utilization of peer in the overlay also plays a significant role instead of only its upload bandwidth. In this paper we will provide a new utilization-based approach for creating an overlay in the P2P network. The peer that uses its resources more or utilizes its resource more are given priority, instead of the peers which have more upload bandwidth but do not utilize their resources.

The procedure to follow new peer is described in Algorithm 1.

Algorithm 1[A1]:

1. $REQ_i^T \langle B_{TU}^i, C_{id}, G_i \rangle$
2. If Request is new goto 3 else 4.
3. Position find for new peer $P_i(B_{TU}^i, G_i)$[A2]
4. Neighbor/Parent List Creation L_i. [A3]
5. $RES_T^i(L_i)$
6. Check update Periodically
7. $C\_req_i^j$ where $(j \subseteq L_i)$
8. $C\_res_j^i$ Positiv/Negative
9. if $(C\_res_j^i)$ goto 10 else 7 for new j from L_i
10. According to RTT, time finds best response
11. If peer continue goto step (12) else peer leave network.
12. Periodically update information to Tracer and goto step (2).

When a new peer enters in the P2P network, it will follow the procedure as shown in Algorithm 1. The new peer i sends the request to tracker peer T. This request is a tuple which consists of three variables $< B_{TU}^i, C_{id}, G_i >$ where B_{TU}^i is the total upload bandwidth of peer i, C_{id} is the content id of media which peer i wants to view, G_i is the geographical location of peer i. Tracker receives the request tuple and calculates the level for peer i in the overlay, which is shown in Algorithm 2.

Algorithm 2[A2]: *Find Level* (B_{TU}^i, G_i)

1. Find B_{TU}^S
2. Find Range R of network
3. Search location according to (C_{id}, G_i)
4. $L\_vel_i = Ceil\big[\{B_{TU}^S - B_{TU}^i\}/R\big];$
5. Return $L\_vel_i;$

Algorithm 2 finds the level of peers in P2P overlay using the geographical location and upload bandwidth. After finding the level for peer i, tracker finds the list of best available peers for peer i. From this list, peer i can choose an appropriate neighbor/parent. Algorithm 3 is used for list creation by tracker.

Algorithm 3: Neighbor/Parents Creation [A3]:

1. if $(L_S < Max\ (S)\ )$
2. L_i.insert(S)
3. $L\_vel_c = 1$
4. While($L\_vel_c <= L\_vel_i$ && $L_i < Th_1$)
5. {   For(x=0; x<n; x++)
6.   {   If($P_U^i > Th_2$ && $P_o^i! = P\_active_{max}^i$)
7. L_i.insert(P_U^i)   }
8.   If($L_i < Th_1$)
9.   {   For(x=0; x<n; x++)
10.   {   If(If($P_U^i > Th_2$ and at lower level there is one peer than L_vel_i and pi has space in L_i)
11. L_i.insert(P_U^i)   }
12.   level = level+1;
13.   } }
14. Start L_vel_c from 1
15. While($L\_vel_c < L\_vel_i$ and $L_i < Th_3$)
16. {   For(x=0; x<n; x++)
17.   {   If(P_U^i has at least one place remain left)
18. L_i.insert(P_U^i)   }
19.   If($L_i < Th_3$)
20.   {   For(x=0; x<n; x++)
21.   {   If(P_U^i has at least one place less than L_vel_i and P_U^i has one place remain left)
22. L_i.insert(P_U^i)   }
23.   level = level+1;
24. } }

For creating the list of possible parents, tracker uses the level of each peer and utilization of each peer. Server is considered at the highest level in the overlay. So, first priority is given to the server (line 1–2, Algorithm 3) but if peer is lower in the overlay, then updating also takes place (line 11–20, Algorithm 3). The list contains the peers only for lower levels (server is at the 0 level and as the height of the tree increases the level also increases). And for each level, only those peers are added in the list that are having utilization value greater than the threshold (Th_2). Tracker adds the new peer to the list, if there is sufficient space in the list (Th_1) and if the level of old peer is less than the level of new peer (line 4–13, Algorithm 4). For calculating the utilization of each peer, only the data of its last three transactions is considered. And for a new peer that comes for the first time in the network, the average utilization value of the network is assigned to that peer. Overall overlay is created in such a way that the peers which have sufficient upload bandwidth are placed above in the hierarchy and for parent selection, only those peers are considered which have a higher utilization ratio. So that new peer can receive the media content as fast as possible. After creating the list of possible parents/neighbor, tracker sends that list to requester peer i. Peer i sends the connection request to neighbor peer from the list. Peer i can send connection request to more than one peer and according to RTT (Round-Trip Time) of the responding peer, it chooses the parent peer (line 5–10,

Algorithm 1). Tracker provides the list according to the utilization, upload bandwidth, and geographical location; however, RTT time gives the idea of real-time congestion in the network.

4 Theoretical Analysis if Utilization-Based Overlay Scheme

Some of the major characteristics of Utilization-based overlay are:

Bandwidth Utilization and Scalability: In a distributed network, the bandwidth utilization of each peer plays a significant role. Utilization-based overlay recognizes the importance of this factor. Every peer that has higher bandwidth and shares its bandwidth in the network must get priority. Utilization-based overlay is a hybrid overlay, i.e., result of combination of tree and mesh overlays. The bandwidth utilization of leaf peer in case of tree overlay can also be used in hybrid overlay. And thus, an increase in the overall bandwidth utilization of the network also affects the scalability of the P2P network.

Start-up Delay: Start-up delay is a crucial factor for live video streaming in P2P network, it directly depends on overlay construction and buffer management in the network. Utilization-based overlay approach is built around a fragile structure in which high bandwidth and more utilized peers are placed above in the hierarchy. In the parent selection process, the tracker also gives priority to the highly utilized peer. So, a new peer can easily receive the chunk from the highly utilized peers and the start-up delay at new peer is also minimized, the receiver enjoys a smooth play out. Thus, the average start-up delay of overall network is improved.

Network Congestion: Maintenance of mesh overlay is more as the control packets are very large in number. So, the congestion in the case of mesh overlay is very high. However, utilization-based overlay is a hybrid overlay and is not a complete mesh overlay. So, the control packets are very less as compared to the complete mesh overlay. Packets that miss their deadline are not usable in live streaming and only increase the congestion over the network. Thus, if the packets are received before the deadline, then the congestion over the network also decreases. Utilization-based overlay provides the chunks to the peer as soon as possible, giving priority to the highest utilized peer and considering it as a parent node. In this manner, the overall congestion in the utilization-based overlay decreases.

Reliability: Reliability of the network depends on the selection of parent peers and mobility of peers in case of P2P live streaming. Reliability of tree overlay is less as compared to that of the mesh overlay due to the dynamic nature of the parent peers. If a parent peer is more dynamic, then not only its immediate child is affected but also the dependent of its child is affected. But in the case of hybrid overlay, child peer connects to more than one parent peer and if a parent peer leaves the network, then the child peer can still connect to other parent peers. Thus, there is no need to repeat the parent selection process. This contributes to an increase in the reliability of the overall network.

Transmission Delay: Transmission delay between the peers in P2P network is a combination of start-up delay, end-to-end delay, and playback delay. In the case of utilization-based overlay as the congestion in the network decreases, the start-up delay also decreases so; the overall transmission time of the network also decreases. Utilization-based overlay gives priority to the highly utilized, high bandwidth peers while parent selection. And thus, the overall transmission delay in case of utilization-based overlay is also affected.

5 Conclusion and Future Work

Peer-to-peer network is getting popularity in live video streaming transmission due to its scalability, resource utilization, less complexity, and cost-effective attitude. However, there are various fundamental limitations available in the traditional P2P network approaches and one of them is overlay creation. In this paper, we have defined a new utilization-based overlay constriction in P2P overlay. Here, upload bandwidth and utilization of each peer is used for creating the overlay. Both of these parameters are very useful for overlay creation. If the highly use and uploaded bandwidth peer is near the source, then it can provide better media quality to other peers as well. Different levels are created in the overlay and highly use peers are created for mesh overlay. Our future work is to provide the implementation of utilization-based hybrid overlay with different existing hybrid approaches.

References

1. Summary, E.: Cisco visual networking index: forecast and methodology, 2014–2019 White Paper—white_paper_c11-481360.pdf (2015)
2. Venkataraman, V., Francis, P., Calandrino, J.: Chunkyspread: multi-tree unstructured peer-to-peer multicast. In: Proceedings of the 14th IEEE International Conference on Netwok Protocols—ICNP'06, pp. 2–11 (2006)
3. Zhang, X., Liu, J., Li, B., Yum, T.S.P.: CoolStreaming/DONet: a data-driven overlay network for efficient live media streaming. In: Proceedings of the IEEE Infocom, vol. 3, no. C, pp. 13–17 (2005)
4. Xie, S., Li, B., Member, S., Keung, G.Y., Zhang, X., Member, S.: Coolstreaming: design, theory, and practice. IEEE Trans. Multimed. **9**(8), 1661–1671 (2007)
5. Li, B., et al.: An empirical study of the coolstreaming plus system. IEEE J. Sel. Areas Commun. **25**(9), 1627–1639 (2007)
6. Awiphan, S., Su, Z., Katto, J.: ToMo: a two-layer mesh/tree structure for live streaming in P2P overlay network. In: 2010 7th IEEE Consumer Communications and Networking Conference, CCNC 2010, pp. 1–5 (2010)
7. Magharei, N., Rejaie, R.: Understanding mesh-based peer-to-peer streaming. In: Proceedings of the 2006 International Workshop on Network and Operating Systems Support for Digital Audio and Video, p. 10 (2006)
8. Magharei, N., Rejaie, R.: PRIME: Peer-to-Peer Receiver-Driven Mesh-Based Streaming, pp. 1415–1423 (2007)

9. Byun, H.B.H., Lee, M.L.M.: HyPO: a peer-to-peer based hybrid overlay structure. In: 2009 11th International Conference on Advanced Communication Technology, vol. 1, pp. 840–844 (2009)
10. Huang, Q., Jin, H., Liao, X.: P2P live streaming with tree-mesh based hybrid overlay. In: Proceedings of the International Conference on Parallel Processing Workshops, no. 60433040 (2007)
11. Hammami, C., Jemili, I., Gazdar, A., Belghith, A., Mosbah, M.: Hybrid live P2P streaming protocol. Procedia Comput. Sci. **32**, 158–165 (2014)
12. Pal, K., Govil, M.C., Ahmed, M.: A new hybrid approach for overlay construction in P2P live streaming. In: ICACCI, pp. 431–437 (2015)
13. Pal, K., Govil, M.C., Ahmed, M.: Comparative analysis of new hybrid approach for overlay construction in P2P live streaming. In: ERCICA (2016)

Web Service Discovery Approach Among Available WSDL/WADL Web Component

Vikalp Sharma and Pranay Yadav

Abstract Today, web service plays a huge part where multiple applications used by people in their daily lives such as weather information or healthcare for accessing data. Therefore, our study takes an important study on why the web service discovery algorithm is a better option over available web services on the internet. Web service and various web platforms in tracking and updating information; for instance, in recent research, a high-efficiency interface for web service was discovered to make use of semantics, abbreviation, and fragment combination of user input which is OPD (Operation Discovery Algorithm), rather than the traditional algorithm. However, a problem identification where the end-user still needed a refined solution of using high trusted value. On the other hand, a re-ranking approach among the output result in between the recommended web services discovery need to perform by the organization. Further, OPD can be enhanced using efficient web service discovery, along with the highest public recommendation.

Keywords Web service · OPD · Web service discovery · Web mining · Ranking and web discovery

1 Introduction

Web services and its discovery is the part of data mining, where several web services and its application are available to solve the daily usage and problems that affects end-users.

In past years, lots of data were collected and stored on servers across the globe, which was mainly from information-related companies and social networking sites.

V. Sharma (✉)
Spring S Technologies, B-69, Kasturba Nagar, Chetak Bridge, Bhopal, India
e-mail: er.vikalp10@gmail.com

P. Yadav
Ulta Light Technology, B-69, Kasturba Nagar, Chetak Bridge, Bhopal, India
e-mail: pranaymedc@gmail.com

© Springer Nature Singapore Pte Ltd. 2019
P. K. Sa et al. (eds.), *Recent Findings in Intelligent Computing Techniques*,
Advances in Intelligent Systems and Computing 707,
https://doi.org/10.1007/978-981-10-8639-7_35

However, there was a need to extract and classify useful ones. Moreover, from this process of finding the hidden information or patterns from the repositories, they can be used on fields that use data mining techniques; this includes medical research, marketing, telecommunication, and stock markets, health care, etc. Meanwhile, these techniques consist of the various technical approaches, which include machine learning, statistics, database system, etc. However, the target is to discover information from these huge databases and transform them into a human understandable format. Both data mining and knowledge discovery are essential components to any organization due to its decision-making strategy. Besides, classification, regression, and clustering are three approaches of data mining in which they are grouped into identified classes. First, classification is a popular task in data mining, especially in knowledge discovery and future planning. However, it provides the intelligent decision-making. It not only examines the existing sample data, but also predicts the future behavior of that sample data. In this case, it maps them into the predefined class and groups. In addition, it can be used to predict group membership for data instances.

Second, web service discovery systems in a typical e-commerce situation, involves millions of users, as well as, of products. The incredible increase in information has put users to face situations to choose, from a variety of products. For this reason, recommendation and discovery system becomes a hot spot for researchers soon after its emergence. However, there are several issues related to discovery and recommendations, which can be understood in form of basic problems suffered by various recommended techniques.

2   Related Work

In 2016, Baldonado et al. [1] presented his web service discovery model, then an OPD algorithm for the website discovery and output of the user with high precision. In addition, an architectural model which had been taken from the input from various online resources and processed the discovery approach. Therefore, the paper discusses in detail the approach on the other previous technique such as, cosine-based approach, semantic-based and Annotation approach that lead in finding a better solution. Moreover, he worked on fragment and abbreviation set for the input generation and for the output generation. However, an algorithm defined by him was efficient in terms of data processing, precision, and recall. Therefore, a further enhancement can be made to this algorithm in order to find the best approach in getting the result given by OPD approach.

In 2014, Cheng et al. [2] had presented and integrated the service for web service extraction and its technique that was used in a newspaper scenario. In that case, they did a content level search and discovery approach which process HTTP and DOM data model, with feeding and execution phase as a part of running approach. A resource and data level discovery was being performed by their system which demonstrates the proper outcome by its model.

Fern´andez-Villamor et al. [3] proposed an approach for web service discovery which used probabilistic machine-learning techniques to extract latent factors from semantically enriched service descriptions for obtaining the output results. They had specified results with a large dataset and its effectiveness with processing its components. In addition, they tested their algorithm with logic-based and keyword-based service search and discovery solutions. On computing that precision, entropy was over the dataset and comparison with other algorithm was attained and could later be used in real-time system directory.

In this paper [4] 2012, the author presented his work through semantic analysis and semantic-based service, categorization, and semantic enhancement for service discovery approach. Besides, they achieved a functional level of service categorization. Therefore, a matching was performed by Latent Semantic Indexing (LSI) technique. Meanwhile, a synonym-based approach was established to make use of data dictionary, which could take advantage of rich text and search. Thus, an improved solution was provided by the proposed algorithm over the existing one in terms of F-measure, precision, and recall computed by them.

In this paper [5] 2012, they had worked on WSDL and web service discovery where the two model single and composite discovery was done by the authors. They had designed the graph-based algorithm for both discovery web service methods. They have worked on predicting the similarity in between the input and further over the discovery outcomes such that the efficient outputs should be provided to the end-users. In addition, they computed the service aggregation graph model. Furthermore, their result showed the highest precision as compared with other approach such as schema matching, woggle, service pool, etc.

In this paper [6] 2012, the authors have presented the web service discovery approach using global social service network. They had used various social links and services published over the internet for investigation and finding the usable service. Finally, a result is shown by them that their technique could reduce the threshold and provide a better result over the platform.

In this article [7] 2012, the author have proposed a work which is cosine-based algorithm for the web service discovery and it produced efficient results. The theory is based on the computation which takes advantage of different vector and variables. Besides, the algorithm was given in existing base paper that exhibited the competitive result with the OpD; hence, a part from this model also can be taken for further approach and work.

Here in Table 1, below is the complete comparison of available solution for the web.

In existing work uses OpD algorithm in such a way that the service discovery produces high parameter values.

The existing approach exhibits the following points:

1. Worked on conceptual description model.
2. Introduced the interaction input framework alongside the traditional textual search generally provided by other platform.

Table 1 Different mechanism performed by the previous authors

Author	Approach performed	Work description
Bo Cheng, Chang bao Li, Junliang Chen [1]	OpD approach	Highest precision, recall, and low time
Jos´e Ignacio Fern´andez -Villamor, Carlos A. Iglesias, and Mercedes Garijo [2]	HTTP and DOM model	Contributing in structure analysis
Cassar, G.; Barnaghi, P. Moessner, K. Probabilistic Paliwal [3]	Matchmaking method	A syntax-based approach with exactness term
A.V., Shafiq, B., Vaidya, J. Hui Xiong, Adam, N	Semantic-based service discovery	A data dictionary-based approach for large number of results
Vaidya, J., Hui Xiong; Adam, N	Semantic-based service discovery approach	A data dictionary-based approach for large number of results
Xuanzhe Liu, Gang Huang, Hong Mei, W. Chain and I. Paik	Graph-based algorithm, global social service network	A large number of results with multiple dimension
E. Garcia	Cosine-based approach	A work on social platform, thus a real platform is used. A computation-based approach

3. Used synonyms, abbreviation, as well as, fragment combination parameter in evaluation of web service discovery.
4. OpD algorithm gave results in single operation and combine approach operation outputs, such that a variant result could be an option as per requirement.
5. Extracted the web service dataset from well-known registry sites such as *who.is* and *whois.net*.
6. Downloaded 1084 WSDL/WADL files for the experiment evaluation and discovery. Further, upon the duplicated service, they worked on 1026 web services.
7. Provided a work comparison of OpD with other algorithm using time cost, precision, and recall as parameter.

3 System Challenges

They [1] were claims that their work with different web mining algorithm such as OpD, OpD and single, service pool posed a different web service discovery generator and other optimized technique which was present in this area. Although the technique would consume less time and produce accurate discovery using the available techniques; still while dealing with them, there were few limitation and challenges that occurred while dealing with these technique. Therefore, to move with automated discovery generation technique, the following points should be kept in mind to settle down on accuracy and result. There are IOT devices and services which needs to follow web services to deal with future innovation which required effective web service discovery over available system [8].

1. There are available tools which are costly and all the program and company cannot use this technique.
2. Web service discovery and language can be difficult to understand by the team, and passing of argument and testing would not be easy as compared to manual testing; since it is carried out by the team member only.
3. In order to understand the web stuff of specific tool, a proper training for that product should be done so that all the annotation and other related detail it generate automatically can be seen.
4. Stack holders keep secrecy of the awareness, sometimes code mismatch can occur with the tool, thus a complete understanding of stack holder is required with automated technique.
5. Developer or web engineer would expect to generate web service requirement up to some process of execution.

4   Further Enhancements

Here, the presentation for optimizing our technique is far more straight and user-friendly, and also an effective ready solution which gives best among the approach given in field of web service discovery over Internet world.

In order to increase parameter efficiency with time cost, precision, and recall. An improvement in the algorithm can be done in the following ways:

1. An approach for the proposed given implementation of proposed scheme with implementation of any format such as WSDL/WADL [9–11], XML, GET, POST API format released by different vendor. To Presentation of optimization in result approach to draw a re-ranking approach on performing computation with discovery outcomes.
2. The algorithm first going to extract the web service discovery using OpD algorithm and then further extraction ranking approach over the discovered item [12] is going defined by us.

The ranking optimization process is going to perform by us, which is going to perform using sample inputs and output, thus it also can determine which service is efficient while working with real-time scenario.

5 Conclusion

The web service plays an important role while walking through the number of options today in different categories. There are systems that take advantage and use them efficiently to discover proper result to the user. Currently, there is a technique such as OpD, semantic-based approach, pattern-based approach which utilizes the keywords and produce results. In our dissertation work, algorithms were used to verify and recommend the best service among the availability of choice and those to opt out. Data fetching and usage in our working model was very important decision to provide best customer service. A study in web service discovery and exactness to the system is going to derive with its improved efficiency, precision, recall, and detection rate.

A further extension can be done in order to find better prediction over search results obtained over the web service discovery approach technique. Our aim is going to perform the similar requirement with a real-time requirement usage.

References

1. Baldonado, M., Chang, C.-C.K., Gravano, L., Paepcke, A.: The Stanford digital library metadata architecture. Int. J. Digit. Libr. **1**, 108–121 (1997)
2. Cheng, B., Li, C., Chen, J.: A web service discovery approach based on mining underlying interface semantics. IEEE (2016)
3. Fern´andez-Villamor, J.I., Iglesias, C.A., Garijo, M.: A framework for goal-oriented discovery of resources in the RESTful architecture. IEEE Trans. Syst. Man Cybern. **44**(6) (2014)
4. Cassar, G., Barnaghi, P., Moessner, K.: Probabilistic matchmaking methods for automated service discovery. IEEE Trans. Serv. Comput. **7**(4), 654–666 (2014)
5. Paliwal, A.V., Shafiq, B., Vaidya, J., Xiong, H., Adam, N.: Semantics-based automated service discovery. IEEE Trans. Serv. Comput. **5**(2), 260–275 (2012)
6. Liu, X., Huang, G., Mei, H.: Discovering homogeneous web service community in the user-centric web environment. IEEE Trans. Serv. Comput. **5**(2), 167–181 (2012)
7. Chen, W., Paik, I.: Improving efficiency of service discovery using linked data-based service publication. Inf. Syst. Front. 1–13 (2012)
8. IOT devices web report. http://springstrategies.in/uploads/88794523051788483152.pdf
9. Garcia, E.: Cosine similarity and term weight tutorial. http://www.miislita.com/information-retrievaltutorial/cosine-similarity-tutorial.html
10. WeatherForecast. http://www.webservicex.net/WeatherForecast.asmx
11. WebServiceX. http://www.webservicex.net
12. Web Service Search Engine. http://www.servicexchange.cn

Designing a Virtual Environment Monitoring System to Prevent Intrusions in Future Internet of Things

Rajendra Patil and Chirag Modi

Abstract Security is the major concern in the current advancements of virtualization-based computing technologies. To offer secure computing services over the internet, the underlying virtual environment should be secured from system level and network level attacks. In this paper, we design a virtual environment monitoring system to prevent intrusions. It scans newly joined virtual machine for potential vulnerabilities, threats, and attacks. It detects the possibility of attacks and their root vulnerabilities at system and network level. In addition, it predicts the future attacks with the help of the scanned vulnerabilities.

Keywords Virtualization · Virtual network · Vulnerability · Security
Intrusion prevention · Virtual live patching

1 Introduction

Virtualization is an abstraction of hardware or software resources that allows heterogeneous architectures to run on the same hardware. In the design of recent computing technologies, virtualization has played a vital role [1]. However, it has various vulnerabilities which brings additional security challenges. These vulnerabilities allow to disturb availability, integrity and confidentiality of the underlying resources and services [2]. A VM is the potential target to gain access to the hypervisor or other installed VMs. In virtual environment, VM state can be inactive, moving (migrating), and running state. It is vulnerable to different types of attacks in all the states. The well-known attacks on VMs are the VM side channel attack, VM covert channel attack, VM-to-VM denial of service attack, VM rollback attack, VM sprawl attack,

R. Patil (✉) · C. Modi
Department of Computer Science and Engineering, National Institute of Technology Goa,
Farmagudi, Ponda 403401, India
e-mail: rajendrapatil@nitgoa.ac.in

C. Modi
e-mail: cnmodi@nitgoa.ac.in

© Springer Nature Singapore Pte Ltd. 2019
P. K. Sa et al. (eds.), *Recent Findings in Intelligent Computing Techniques*,
Advances in Intelligent Systems and Computing 707,
https://doi.org/10.1007/978-981-10-8639-7_36

etc. [3, 4]. In addition, different types of network attacks port scanning, IP spoofing, network sniffing, DoS/DDoS, etc., are possible on VMs [5]. An infected VM allows to monitor the activities and data of other users. Another entry point is its interface with internal and external network.

To address above security problems, the common security measures viz.; firewall, access control, security groups, and intrusion detection systems have been investigated [6–8]. These solutions face additional challenges due to the dynamic and shared nature of virtualization [8, 16]. Thus, there is a need of an automated recovery system against the attacks. In addition, it is required to predict the possible attacks.

In this paper, we design a virtual environment monitoring system to prevent both system level and network level attacks. It performs predictive analytics of the virtual environment using three methods viz.; early detection, future prediction, and early prevention of the attacks. The proposed system looks for potential vulnerabilities, threats, and attacks to and from a newly joined, newly created, or newly started VM and attempts to detect and prevent both system level and network level attacks. In addition, it predicts the future attacks and attempts to recover possible vulnerabilities.

The rest of this paper is organized as follows: Sect. 2 presents the existing proposals of securing virtual environment. A detailed discussion of the proposed system is given in Sect. 3, while Sect. 4 analyzes the features of the proposed system. Finally, Sect. 5 concludes our research work with references at the end.

2 Existing Proposals to Virtualization Security

There have been various proposals to date for securing virtual environment.

Payne et al. [9] have presented a XenAccess for the monitoring of guest OSes running on the Xen hypervisor. It accesses the state of the guest OS. It is developed based on the manually retrieved data structure knowledge of the monitored kernel. Jiang et al. [10] have presented the VMwatcher to get semantic information about the VMs. It exploits the rootkits and achieves stronger tamper resistance. Dastjerdi et al. [11] have proposed scalable, flexible and cost-effective DIDS for cloud applications regardless of their locations using Mobile Agent (MA). VMs are attached to MA which collects evidences of an attack from all the attacked VMs for further analysis and auditing. Lo et al. [12] have developed a cooperative IDS framework. Here, an IDS is deployed in each cloud region. If any region suffers from severe attack defined in its block table, an alert message is sent to other IDS to identify distributed attack. Here, communication effort is high. Modi and Patel [13] have proposed Hybrid NIDS (H-NIDS) which uses the combination of three classifiers. First, it detects known attacks using snort based on rules derived by signature apriori algorithm. For anomaly detection, the preprocessed packets are applied to three classifiers. The final decision about intrusions is taken based on weighted averaging method.

Li et al. [14] have proposed a distributed intrusion detection model based on several Intrusion Detection Agent (ID Agent) subsystems and a data aggregation subsystem. ID agent subsystem detects intrusions based on cloud intrusion detection algorithm and sends it to the data aggregation subsystem for the global decision. Gupta et al. [15] have proposed a profile-based distributed network intrusion detection system. For each VM, a profile Database (DB) is created to describe the attack patterns. Here, VM is monitored for a specific amount of time for detecting packet flooding attacks such as DoS attacks.

As per our observation, there is a need of detecting and preventing the attacks to and from VM at initial stage of its lifecycle. In addition, there is a need of fulfilling Cloud IDS requirements viz.; handling large-scale computing systems, detecting variety of attacks, fast detection and prevention of attacks, automated self-adaptive capability, scalability, synchronization of autonomous IDS sensors and resistance to compromise and performance and efficiency metrics, as suggested in [16].

3 Proposed Virtual Environment Monitoring System

The objective of the proposed monitoring system is to secure virtual environment by identifying and preventing attacks to and from newly joined VM, while fulfilling the security needs of virtual environment. It should predict the future attack vectors in virtual environment. It should fulfill the security requirements at both running state and migrating state of VM. It should protect the security system itself. It should be able to achieve complete introspection of VM. As shown in Fig. 1, the proposed system named as "Virtual Environment Monitoring Tool" is deployed on hypervisor. It has three modules viz.; early scanning, future attack prediction, and live patching.

3.1 Early Scanning

When a new VM joins the virtual environment, this module performs Virtual Machine Introspection (VMI) to inspect its runtime state [17]. It works in both directions viz.; detecting threats from the VM and detecting threats to the VM. To detect threats from VM, it performs system level scan for entire VM in order to find the malwares, malicious process running inside the VM or malicious traffic going out to the VM. It consists of two sub-modules viz.; profile generation and attack detection. The profile generation has four VMI read-only interfaces viz.; CPU interface, memory interface, disk interface. and network interface. The CPU interface examines the CPU registers of the vCPU. The memory interface obtains raw memory contents, examines them, and extracts OS data. The disk interface obtains raw disk data, examines it, and interprets the actual file system of the VM. The network interface captures the packets from the VM, analyzes them, and identifies the attack patterns. Then, it generates the profile for each virtual component (vCPU, vRAM, vDisk, and vNIC). The attack detection module compares each profile with corresponding intrusion profile and looks for any infection. If it is infected, it lists the corresponding system level vulnerabilities leveraged by an attacker.

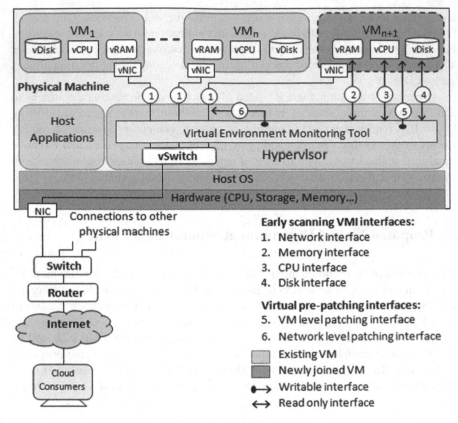

Fig. 1 VMI-based virtual network monitoring system

To detect threats to VM, malicious traffic coming to newly joined VM is inspected here. It captures the traffic from virtual network, internal network, and external network, and performs signature-based detection to detect the network attacks on the newly joined VM. It lists the root vulnerabilities. In addition, it finds the additional vulnerabilities that can be exploited in future.

3.2 Future Attack Prediction

The early scanning module lists the vulnerabilities in four components (vCPU, vRAM, vDisk, and vNIC) of the VM. Based on the reported vulnerabilities, it creates the test data set for all four virtual components of the newly joined VM. Let, $VM_T = \{W_T, X_T, Y_T, Z_T\}$ is the test datasets of vulnerabilities for corresponding virtual components vCPU, vRAM, vDisk, and vNIC, respectively.

$W_T = \{w_1, w_2, w_3 \ldots w_n\}$, $w_1 \ldots w_n$ represent the CPU data
$X_T = \{x_1, x_2, x_3 \ldots x_n\}$, $x_1 \ldots x_n$ represent the RAM data
$Y_T = \{y_1, y_2, y_3 \ldots y_n\}$, $y_1 \ldots y_n$ represent the Disk data
$Z_T = \{z_1, z_2, z_3 \ldots z_n\}$, $z_1 \ldots z_n$ represent the NIC data.

These test datasets become the reference for the further analysis. It is observed that either of four components of test dataset may be vulnerable or all components may not be vulnerable. Thus, system maintains $2^4 - 1 = 15$ sets of the vulnerable VMs and a normal VM as training datasets. The dataset component with subscript 1 represents the corresponding vulnerable components of that VM and components with subscript 0 represents the non-vulnerable component of that VM.

$VM_0 = \{W_0, X_0, Y_0, Z_0\}$ $VM_5 = \{W_1, X_1, Y_0, Z_1\}$ $VM_{10} = \{W_1, X_0, Y_1, Z_0\}$
$VM_1 = \{W_0, X_0, Y_0, Z_1\}$ $VM_6 = \{W_0, X_1, Y_1, Z_0\}$ $VM_{11} = \{W_1, X_0, Y_1, Z_1\}$
$VM_2 = \{W_0, X_0, Y, Z_0\}$ $VM_7 = \{W_0, X_1, Y_1, Z_1\}$ $VM_{12} = \{W_1 X_1, Y_0, Z_0\}$
$VM_3 = \{W_0, X_0, Y_1, Z_1\}$ $VM_8 = \{W_1, X_0, Y_0, Z_0\}$ $VM_{13} = \{W_1, X_1, Y_0, Z_1\}$
$VM_4 = \{W_0, Y_1, Y_0, Z_0\}$ $VM_9 = \{W_1, X_0, Y_0, Z_1\}$ $VM_{14} = \{W_1, X_1, Y_1, Z_0\}$
$$VM_{15} = \{W_1, X_1, Y_1, Z_1\}$$

The proposed system learns from test datasets and performs analysis with training datasets based on feasible techniques, as mentioned in [16]. If any dataset component (W_T, X_T, Y_T, Z_T) crosses the vulnerability boundary/limit, it is classified to corresponding vulnerable VM and predicts the possibility of corresponding attacks.

3.3 Virtual Live Patching

It applies prevention mechanism before the attacks infect the virtual environment. It uses two interfaces viz.; VM interface and network interface.

VM level patching interface processes the inputs given by CPU, memory, and disk interfaces of early scanning to identify the root cause of an attack. It applies virtual patching for all known attacks possible from the same VM. It provides the guarantee that even infected VM joins the virtual environment; it will be patched for known vulnerabilities. Network level patching interface takes the inputs from network interface of the early scanning and applies patches for known network attacks. It includes closing specific ports, blacklisting particular IP addresses, disabling the connection, etc. It enforces Network Access Control (NAC) to isolate the newly joined VM from the rest of virtual environment till scanning and patching is performed.

4 Analysis of the Proposed System

The proposed system is deployed at the hypervisor level. It helps the monitoring system to scan the VM at entry level and to detect and possibly prevent the intrusions before their infection in virtual environment. It combines both signature- and anomaly-based techniques as like in [16], and thus it is capable of detecting a variety of system level and network level attacks. In addition, it scans the external as well as internal network traffic and applies prevention at network level.

It performs the vulnerability scan for newly joined VMs at early stage of their lifecycle and generates the report of system level and network level vulnerabilities. This report helps in predicting future possible attacks to and from target VM.

The proposed system can handle the newly migrated VM, newly created VM, and newly started VM. It works at hypervisor level with four VMI interfaces to get complete internal state of the target VM. Thus, it is capable to bridge the semantic gap which has been a big challenge for VMI-based security approaches.

The proposed system works at hypervisor level, outside of the VMs. Thus, even if an attacker is able to compromise the VMs, it is very difficult to affect the functionality of the proposed system through malwares running in the VMs. It decides specific traffic to be inspected rather than handling whole network traffic. It first looks for the network traffic to and from newly joined VM, and then it moves towards the traffic from external and internal network skipping network communication among the running VMs. It uses lightweight techniques as in [16] for detection and prevention. Here, only newly joined VM is under consideration. The detection and prevention of attacks to/from the VM needs only minimum number of hosted VMs and target VM to be patched. In addition, system uses predefined signature database for known attacks and it applies anomaly-based detection for unknown attacks. Thus, the detection and prevention rate is faster.

The new research from Kaspersky Lab declares that the cost of recovering from a security incident doubles when the attack affects virtual infrastructure [4]. The proposed system has minimum or no recovery cost.

5   Conclusions and Future Work

To address the security problem in the virtual environment, we have designed a virtual environment monitoring system, which performs the detection and prevention of system level and network level attacks at early stage of the newly joined VM. It attempts to fulfill security needs of the virtual environment at both system and network point of view. The analysis of the proposed system is very encouraging and it seems to be fit very well for securing virtual environment.

References

1. Xing, Y., Zhan, Y.: Virtualization and cloud computing. In: Future Wireless Networks and Information Systems, pp. 305–312 (2012)
2. Li, S.H., Yen, D.C., Chen, S.C., Chen, P.S., Lu, W.H., Cho, C.C.: Effects of virtualization on information security. Comput. Stand. Interfaces **42**, 1–8 (2015)

3. Dawoud, W., Takouna, I., Meinel, C.: Infrastructure as a service security: challenges and solutions. In: 7th International Conference on Informatics and Systems, pp. 1–8 (2010)

4. Kaspersky-Lab: Security of virtual infrastructure (2015). http://media.kaspersky.com/en/busi ness-security/enterprise/IT_Risks_Survey_Report_Virtualization.pdf

5. Bays, L.R., Oliveira, R.R., Barcellos, M.P., Gaspary, L.P., Madeira, E.R.M.: Virtual network security: threats, countermeasures, and challenges. J. Internet Serv. Appl. 6(1), 1–19 (2015)

6. Liao, H.J., Lin, C.H.R., Lin, Y.C., Tung, K.Y.: Intrusion detection system: a comprehensive review. J. Netw. Comput. Appl. 36(1), 16–24 (2013)

7. Modi, C., Patel, D., Borisaniya, B., Patel, H., Patel, A., Rajarajan, M.: A survey of intrusion detection techniques in cloud. J. Netw. Comput. Appl. 36(1), 42–57 (2013)

8. Modi, C.N., Acha, K.: Virtualization layer security challenges and intrusion detection/prevention systems in cloud computing: a comprehensive review. J. Supercomput. 1–43 (2016)

9. Payne, B.D., Martim, D.D.A., Lee, W.: Secure and flexible monitoring of virtual machines. In: 23rd Annual Computer Security Applications Conference, pp. 385–397 (2007)

10. Jiang, X., Wang, X., Xu, D.: Stealthy malware detection through vmm-based out-of-the-box semantic view reconstruction. In: 14th ACM Conference on Computer and Communications Security, pp. 128–138 (2007)

11. Dastjerdi, A.V., Bakar, K.A., Tabatabaei, S.G.H.: Distributed intrusion detection in clouds using mobile agents. In: 3rd International Conference on Advanced Engineering Computing and Applications in Sciences, pp. 175–180 (2009)

12. Lo, C.C., Huang, C.C., Ku, J.: A cooperative intrusion detection system framework for cloud computing networks. In: 39th International Conference on Parallel Processing Workshops, pp. 280–284 (2010)

13. Modi, C.N., Patel, D.: A novel hybrid-network intrusion detection system (H-NIDS) in cloud computing. In: IEEE Symposium on Computational Intelligence in Cyber Security (CICS), pp. 23–30 (2013)

14. Li, H., Wu, Q.: A distributed intrusion detection model based on cloud theory. In: 2nd International Conference on Cloud Computing and Intelligence Systems, pp. 435–439 (2012)

15. Gupta, S., Kumar, P., Abraham, A.: A profile based network intrusion detection and prevention system for securing cloud environment. Int. J. Distrib. Sens. Netw. 1–12 (2013)

16. Acha, K., Modi, C.: An efficient security framework to detect intrusions at virtual network layer of cloud computing. In: 19th International ICIN Conference Innovations in Clouds, Internet and Networks, pp. 133–140 (2016)

17. Nance, K., Bishop, M., Hay, B.: Virtual machine introspection: observation or interference? IEEE Secur. Priv. 32–37 (2008)

Survey of Security Mechanisms in Internet of Things

L. Vidyashree, Mokhtar A. Alworafi, Sheren A. El-Booz and Suresha

Abstract Internet of Things (IoT) has made considerable changes in the real world and penetrates all aspects of human life, controls on many things such as car, mobile, etc., it would be treated as future of the Internet. It affects a new world acquired by small, smart objects interacting with the environment and controlled over the Internet. Securing an IoT is a major thing. At the security system, there are two major mechanisms in IoT, i.e., Authentication and Authorization. Authentication is an important issue in IoT, where identifying who can access the system is the most effective issue. This paper aims to compare various existing authentication mechanisms for securing IoT. We deduced form this study that IoT accomplished an efficient communication overhead security rate.

Keywords Internet of things (IoT) · Security · Authentication · Authorization

1 Introduction

The Internet of Things (IoT) is a concept that interprets totally interconnected world. Configuration and size of devices are manufactured with "smart" applications that potentiate them to interact and communicate with other devices, exchange data, produce its own decisions and perform useful tasks based on preset conditions. It

L. Vidyashree (✉) · M. A. Alworafi · Suresha
DoS in Computer Science, University of Mysore, Mysore, India
e-mail: Vid14.1987@gmail.com

M. A. Alworafi
e-mail: mokhtar119@gmail.com

Suresha
e-mail: sureshabm@yahoo.co.in

S. A. El-Booz
Faculty of Electronic Engineering, CSE Department,
Menoufia University, Shibin El Kom, Egypt
e-mail: eng.sheren1975@gmail.com

© Springer Nature Singapore Pte Ltd. 2019
P. K. Sa et al. (eds.), *Recent Findings in Intelligent Computing Techniques*,
Advances in Intelligent Systems and Computing 707,
https://doi.org/10.1007/978-981-10-8639-7_37

too refers to the universe where technology will make life safer, more comfortable, more productive, and more comfy. The IoT has begun to shape our modern world, where smart devices communicate not only with humans, but also with other objects, smart devices, infrastructure, and environments [1]. IoT security is the most area of endeavor concerned with protecting all the connected devices and networks [2]. IoT security is the effort to secure connected devices communicating over one or several networks. As IoT is broadly used in many areas, securing an IoT is becoming especially important and will take great effects on it. Device identity and device authentication mechanism is one of the fundamental elements in securing an IoT infrastructure. The process of checking individual identity and promises the reliability of communication is known as Authentication and it acts as a gateway in front of a secure system to prevent malfunctions, it is one of the principal goal of security [3]. The three interrelated pillars that form a security system are 'Identification', 'Authentication' and 'Authorization'. Identification is the communication of identity to an Information System (IS). Enrollment is an important concern and should also be carefully handled [4].

1.1 Basic Steps for Authentication

The common basic steps for authentication are: (1) **Initial step**: the heir is not authenticated; (2) **Acceptance step**: to get acceptance, the heir requires to IS and use the function that requires an authentication. (3) **Authenticated step**: the heir is authenticated. The IS provides the user required functions; (4) **Rejection step**: the users rejected from the monitor and the state returns to the initial step. This step can be initiated on a timeout or by an action of the user [4].

The process of individual authentication should be based on a combination of more than one of authentication terms There are mainly four extensively recognized factors for user authentication as: (1) Something the user knows: a password, a passphrase, a PIN code, the mother's maiden name; (2) Something the user owns: a USB token, a phone, a smart card, software token, a navigator cookie etc.; (3) Something qualifies the user: a fingerprint, DNA fragment, voice pattern, hand geometry, etc.; (4) Something the user can do: a signature and a gesture [4].

1.2 IoT Architecture

As shown in the below Fig. 1. the IoT ecosystem composed of three layers named (perception layer, network layer, and application layer): (1) **Perception Layer**: It act as a sensing component and the main important working period of IoT, i.e., all the information is collected from perception layer with the help of different devices like RFID tag, smart card reader, sensor networks, GPS and camera, etc. [5]; (2) **Network Layer**: It is the brain of IoT with a function of transmitting and processing data. First,

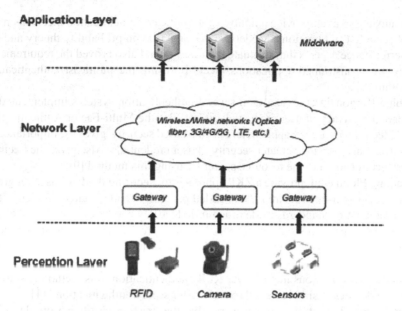

Fig. 1 IoT architecture [22]

the data gathered by sensors be sent to internet through network layer with the aid of wireless/wired network, computer, or other components [5]; (3) **Application Layer**: The assigned functions to that layer are analyzing the received information and then give the decision control to reach their intelligent processing feature [5].

2   Related Work

Moosavi et al. proposed an efficient contribution of authentication and authorization architecture over IoT-based healthcare systems, when using distributed smart e-health gateways. The proposed contribution architecture had been verified by developing a prototype IoT-based healthcare system [5].

Farash et al. proposed an improved user authentication and key agreement scheme for heterogeneous WSN and it is matched for the IoT world and also specifies first authentication model for four-step sensor node. By using this authentication model without involving with a gateway node, each registered user from the IoT can directly contact with a sensor node from the WSN [6].

Parikshit N et al. proposed an identity Authentication and Capability-based Access Control (IACAC) for the IoT focused on access control and authentication. It provides a ECC-based secured integrated approach for access control and authentication. In the form of computational time proved that the method is efficient [7].

Xuanxia Yao et al. provide a lightweight multicast authentication mechanism for small-scale IoT applications. Model evaluation based on probability theory and for the performance aspects they evaluate their design and also proved the requirements of resource constrained applications meets by using the multicast authentication algorithm [8].

Tuhin Borgohain et al. analyzed several authentication systems implemented to preserve the privacy of user credentials in IoT. But the Multi-Factor Authentication (MFA) techniques are not applicable. For IoT-based security, gives the importance of OAuth and also provides greater security to user credentials. Also gives the Secured experience for login to the resource server by using this method [9].

Padraig Flood et al. provide a ZKP approach for securing the IoT based on graph theory. Security infrastructure for embedded processors and resource-efficient alternative for existing standards is identified in IoT [10].

Swathi Kumari proposed a new security layer in real-time authentication system for RFID applications. In order to granting access to the system, the application captures location information then matches it with predefined authorized location. Compared with previous methods for RFID authentication, this method is suitable for RFID devices. Using back end servers, offer secure authentication [11].

Jae-Kyung Park et al. claimed to resolve the existing certificate problems for authentication service using certification device. It is based on public key cryptography. To verify personal identity and transaction service, an alternative method with total authentication service is introduced [12].

Rehiman KA et al. proposed a unique trust management framework for mobile devices by considering security-related issues. For smart mobile devices, a new lightweight method for trust management specific is proposed. The security goals —confidentiality, privacy, and access control proposed improves the trustworthiness of mobile devices in IoT [13].

Yi-Pin Liao proposed an ID-verifier transfer protocol integrated with a secure ECC-based RFID authentication scheme. The proposed scheme satisfies the basic requirements of RFID system, as well as achieving mutual authentication. Function comparison and performance evolution demonstrate that the proposed scheme is well suited for RFID tags with the scarceness of resources [14].

Sheikh Iqbal Ahamed proposed an ECC-based authentication protocol but the major drawback is that it is not Denial of Service (DoS) attack resistant. The billions of devices in IoT, resistance to DoS attack is of vital importance [15].

Guanglei Zhao et al. proposed an ECC-based mutual authentication protocol for IoT using hash functions. Using secret key cryptosystem, storage, mutual authentication is achieved between platform and terminal node by introducing the problem of key management [16].

C. Jiang et al. presented a self-certified keys cryptosystem based on distributed user authentication scheme for WSN, where only user nodes are authenticated [17].

Shafagh et al. provide the evaluation results depict that the certificate-based authentication causes three times higher transmission overhead and 28 times higher processing overhead than a preshared-key-based authentication [18].

Chatzigiannakis et al.provide a generic implementation of ECC that runs on different host operating systems, such as Contiki, TinyOS, iSenseOS, ScatterWeb, and Arduino. The method runs on smartphone platforms such as Android and iPhone and also any linux-based systems (e.g., raspberryPi). But in the proposed implementation does not contain any platform-specific specializations. It allows a single implementation to run natively on heterogeneous networks [19].

Mahmood et al. proposed an authentication scheme which depends on Diffie–Hellman key establishment protocol. While reducing overall communication and computation overheads the scheme provides mutual authentication, thwarting replay and man-in-the-middle attacks and achieves message integrity [20].

Amin et al. proposed an architecture which is applicable for distributed cloud environment and based on it, an authentication protocol using smartcard. AVISPA tool and BAN logic model is used to proof security strength of protocol. The performance analysis and comparison confirm that the proposed protocol is superior to its counterparts [21].

3 Discussion and Comparison

In the previous section, different Security Mechanisms proposed by various researchers have been discussed. Security is the most important issue in IoT, where everyone must be authenticated efficiently. Table 1 gives a comparative analysis of different security Mechanisms.

Security Mechanisms: (1) **Authentication**: It is a process or technology in which both entities in a medium authenticate with each other; (2) **Privacy**: protect the information of individuals and maintaining the privacy information in communication and network (3) **Integrity**: the assurance that information can only be accessed or modified by those authorized users; (4) **Confidentiality**: It gives the limit access to information and prevent sensitive information from reaching the wrong people; (5) **Access Control**: It is the selective restriction of access to a place or other resource.

Obviously, in Table 1 we summarize all the security mechanisms used in all the related work, while all the existing security solutions do not fulfill all the secrecy requirements for IoT. We can illustrate all the table parameters in Fig. 2.

We must illustrate how these security mechanisms in Table 1 influence the IoT performance, where there are many factors must be measured to evaluate the performance such as (Communication overhead, Processing overhead, Computational time, Storage consumption, and Cost) as shown in Table 2. The definition of all the factors measured in Table 2 are as follows: (1) **Communication overhead**: It is the proportion of time user spend communicating with the team instead of getting productive work done; (2) **Processing overhead**: It is the processing time required by a device prior to the execution of a particular application; (3) **Computational time**: The overall time taken in processing a complete application; (4) **Storage Consumption**: The amount of memory consumed in a particular application; (5) **Cost**: It is the total number of amount for any application in IoT.

Table 1 Comparison of existing security mechanisms

Papers	Authentication	Privacy	Integrity	Confidentiality	Access control
Moosavi [5]	✓	✓	✗	✗	✗
Farash [6]	✓	✗	✗	✗	✗
Mahalle [7]	✓	✗	✗	✗	✓
Yao [8]	✓	✗	✗	✓	✗
Borgohain [9]	✓	✓	✗	✗	✗
Flood [10]	✓	✗	✗	✗	✗
Kumari [11]	✓	✓	✗	✓	✗
Park [12]	✓	✗	✗	✗	✗
Rehiman [13]	✓	✓	✗	✓	✓
Liao [14]	✓	✗	✓	✓	✓
Ahamed [15]	✓	✓	✗	✗	✗
Zhao [16]	✓	✗	✗	✗	✗
Jiang [17]	✓	✗	✗	✗	✗
Shafagh [18]	✓	✗	✗	✗	✗
Chatzigiannakis [19]	✓	✓	✗	✓	✗
Mahmood [20]	✓	✗	✓	✗	✗
Amin [21]	✓	✓	✗	✓	✓

Table 2 Performance analysis of security mechanism

Papers	Communication overhead	Processing overhead	Computational time	Storage consumption	Cost
Moosavi [5]	✓	✗	✗	✗	✗
Farash [6]	✓	✗	✗	✓	✓
Mahalle [7]	✗	✗	✓	✗	✗
Yao [8]	✓	✗	✓	✗	✓
Kumari [11]	✓	✗	✓	✗	✗
Rehiman [13]	✓	✗	✗	✗	✗
Liao [14]	✓	✓	✓	✓	✓
Ahamed [15]	✗	✓	✓	✓	✗
Zao [16]	✓	✗	✗	✓	✓
Jiang [17]	✓	✗	✓	✓	✗
Shafagh [18]	✓	✓	✓	✓	✗
Chatzigiannakis [19]	✗	✓	✓	✓	✗
Mahmood [20]	✓	✗	✗	✗	✓
Amin [21]	✓	✗	✗	✓	✓

Comparison of Existing Security Mechanisms

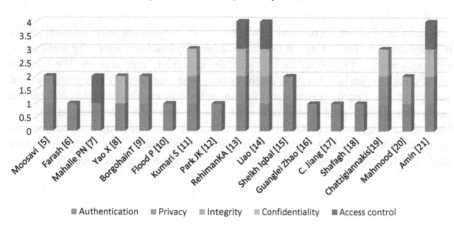

Fig. 2 Comparison of existing security mechanism

Performance Analysis of Various Security Mechanisms

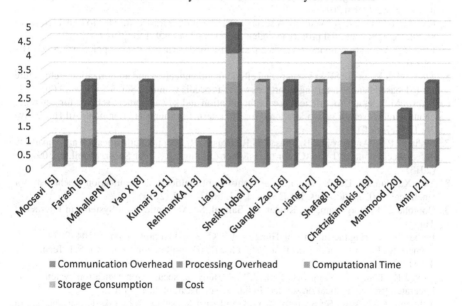

Fig. 3 Performance analysis of various security mechanisms

Briefly, we can summarize the results for the performance analysis of the security mechanism used in earlier papers as shown in Fig. 3. where some papers achieved an efficient level of security without influence the performance parameters.

4 Conclusion

The main goal of the current study was to survey various security mechanisms in IoT environment. Increasing the information security had been done with some security mechanisms. This paper tackled with these security mechanisms. We focused on efficient comparison between all the papers from the security mechanisms used and how these security mechanisms influence the System Performance. Where the performance of the system analyzed based on some factors as Communication overhead, Processing overhead, Computational time, Storage consumption and the how these mechanisms influence the Cost of the hardware used. We deduced form this study that IoT accomplished an efficient communication overhead security rate.

References

1. http://www.futureworldsymposium.org. Accessed 22 Dec 2015
2. http://internetofthingsagenda.techtarget.com/definition/IoT-security-Internet-of-Things-security. Accessed 20 Jan 2017
3. Rafidha Rehiman, K.A., Veni, S.: A secure authentication infrastructure for IoT enabled smart mobile devices–an initial prototype. Indian J. Sci. Technol **9.9**, 1–6 (2016)
4. Idrus, S.Z.S., et al.: A review on authentication methods. Aust. J. Basic Appl. Sci. **7.5**, 95–107 (2013)
5. Moosavi, S.R., et al.: SEA: a secure and efficient authentication and authorization architecture for IoT-based healthcare using smart gateways. Procedia Comput. Sci. **52**, 452–459 (2015)
6. Farash, M.S., et al.: An efficient user authentication and key agreement scheme for heterogeneous wireless sensor network tailored for the internet of things environment. Ad Hoc Netw. **36**, 152–176 (2016)
7. Mahalle, P.N., Anggorojati, B., Prasad, N.R., Prasad, R.: Identity authentication and capability based access control (IA-CAC) for the internet of things. J. Cyber Secur. Mobil. **1**, 309–348 (2013)
8. Yao, X., Han, X., Du, X., Zhou, X.: A lightweight multicast au-thentication mechanism for small scale IoT applications. IEEE Sens. J. **13**, 3693–3701 (2013)
9. Borgohain, T., Borgohain, A., Kumar, U., Sanyal, S.: Authentica-tion Systems in Internet of Things (2015)
10. Flood, P.: Securing the Internet of Things—A ZKP Based Approach, pp. 1–104 (2014)
11. Kumari, S.: Real time authentication system for RFID ap-plications. Indian J. Sci. Technol. **7**, 47–51 (2014)
12. Park, J.K., Lee, H.S., Kim, S.J., Park, J.P.: A study on secure authen-tication system using integrated user authentication service. Indian J. Sci. Technol. **8**, 1–6 (2015)
13. Rehiman, K.A., Veni, S.: Security, privacy and trust for smart mobile devices in internet of things—A literature study. IJARCET **4**, 1775–1779 (2015)
14. Liao, Y.-P., Hsiao, C.-M.: A secure ECC-based RFID authentication scheme integrated with ID-verifier transfer protocol. Ad Hoc Netw. **18**, 133–146 (2014)
15. Ahamed, S.I., Rahman, F., Hoque, E.: ERAP: ECC based RFID authentication protocol. In: 12th IEEE International Workshop on Future Trends of Distributed Computing Systems, pp. 219–225 (2008)
16. Zhao, G., Si, X., Wang, J., Long, X., Hu, T.: A novel mutual authentication scheme for Internet of Things. In: Proceedings of 2011 IEEE International Conference on Modelling, Identification and Control (ICMIC), pp. 563–566 (2011)

17. Jiang, C., Li, B., Xu, H.: An efficient scheme for user authentication in wireless sensornetworks. In: 21st International Conference on Advanced Information Networking and Applications Workshops, pp. 438–442 (2007)
18. Shafagh, H., et al.: Leveraging Public-key-based Authentication for the Internet of Things, pp. 1–98 (2013)
19. Chatzigiannakis, I., Vitaletti, A., Pyrgelis, A.: A privacy-preserving smart parking system using an IoT elliptic curve based security platform. Comput. Commun. **89**, 165–177 (2016)
20. Mahmood, Khalid, et al.: A lightweight message authentication scheme for Smart Grid communications in power sector. Comput. Electr. Eng. **52**, 114–124 (2016)
21. Amin, R., et al.: A light weight authentication protocol for IoT-enabled devices in distributed Cloud Computing environment. Future Gener. Comput. Syst. 1–27 (2016)
22. Abdmeziem, M.R., Tandjaoui, D., Romdhani, I.: Architecting the internet of things: state of the art. Robots and Sensor Clouds. Springer International Publishing, pp. 55–75 (2016)

Framework to Build Wireless Communication in IoT

Jayalaxmi Naragund, Madhu Asundi and R. M. Bankar

Abstract Internet of Things (IoT) is emerging technology and it plays a key role in developing smart cities. This new technology includes embedded devices, which are connected to the internet. IoT applications are spread over various fields like agriculture, healthcare, automotive industry, home automation, and security, etc. Nowadays, Wireless Communication (WC) is becoming an essential tool to link the devices, organizations, and people in the world. In IoT, WC is budding technology and more exploration is needed. This issue motivates the authors to propose generalized framework for WC in IoT. Framework is divided into five phases according to the architecture of IoT. First two phases focus on the configuration of sensors and gateways; third phase explains the cloud integration, fourth phase is about configuration of the server, and fifth phase considers the user interface. Hence, proposed framework does ease the work of IoT application developer. Authors illustrated the framework using the case study of electricity utilization in domestic. The application uses sensors and Raspberry pi for computing and providing electricity usage consumed by home or building using wireless communication between the devices. This provides the awareness to the user about power consumption pattern and better utilization of the electricity.

Keywords IoT · Wireless network · Cloud

J. Naragund (✉) · M. Asundi · R. M. Bankar
K L E Technological University, Hubballi 580031, India
e-mail: jaya_gn@bvb.edu
URL: http://www.kletech.ac.in/

M. Asundi
e-mail: madhuasundi7792@gmail.com

R. M. Bankar
e-mail: bankar@bvb.edu

© Springer Nature Singapore Pte Ltd. 2019   363
P. K. Sa et al. (eds.), *Recent Findings in Intelligent Computing Techniques*,
Advances in Intelligent Systems and Computing 707,
https://doi.org/10.1007/978-981-10-8639-7_38

1 Introduction

IoT is a network of things like sensors, smartphones, or any device that is able to communicate with other devices using the internet [1]. Things collect and share data over the internet, so that processes are automated with minimal energy and less cost. The main salient feature of IoT is automation, monitoring, and on-the-fly communication between devices [2]. Wider application are coming under the umbrella of IoT such as healthcare, production industry, home automation, agriculture, defense, and automobiles, etc.

Wireless communication transfers information between two or more devices using radio signals, microwaves, and other wireless media. Bluetooth, Zigbee, 6LoWPAN, Wi-Fi, Cellular and BLE [3], etc., are different wireless technologies used by IoT applications. These technologies are the means of connection and communication between devices. However, the choice of these technologies depends on the application requirement for data transmission and application constraints. For instance, for huge data transmission and higher transmission range one may choose Wi-Fi over Zigbee.

The author developed five phases generalized framework for WC in IoT, so that IoT application developers can concentrate more on their business logic using this framework. The first phase tells about sensor specifications like energy level and transmission range to deploy them on communicating devices. The second phase configures the communicating devices. These devices collect the data from sensors and process it by measuring the required parameter. The third phase explores the cloud integration with IoT application. It provides an efficient and scalable platform for continuous streaming of data over TCP connection using HTTP protocol. In the fourth phase, authors use Flask framework [4] to configure the communicating device like Raspberry pi as server. The last phase shows the development of interactive user interface for IoT applications using programming languages like Python, PHP, Perl, etc. To demonstrate the proposed framework, authors consider electricity utilization in domestic as a case study.

The article is organized as follows: Sect. 2 explains related work regarding building of IoT applications, Sect. 3 describes proposed work, Sect. 4 explains case study energy monitoring system using IoT. The paper is concluded by Sect. 5.

2 Literature Survey

In this section, authors explain the survey on framework of IoT applications and development process.

Alkhamisi et al. [1] discuss on a distributed Cross-Layer Commit Protocol (CLCP) for data aggregations. The first stage of protocol concerned with stage Aggregating Node (AN) identification. Authors show that due to the usage of energy in each round, residual energy drops by about 0.05 J.

Jover and Murynets [3] comparse the performance of MQTT (Message Queuing Telemetry Transport), CoAP (Constrained Application Protocol), DDS (Data Distribution Service), and a custom UDP-based protocol for IoT applications using a network emulator. DDS requires higher bandwidth that of MQTT but has better performance with respect to latency and reliability. MQTT has better performance under poor network conditions, whereas DDS outperforms MQTT in terms of telemetry latency in poor network condition. In case of low bandwidth and low latency application, UDP-based CoAP and custom UDP are feasible but not practical due to unpredictable packet loss.

Jaiswal et al. [5] proposes a framework for the advancement of IoT applications, which is executed in Python cloud simulator. Framework uses APIs (Application Program Interface) to provide services and allows clients to select a specific programming model. Their study shows that the speedup factor increases upto seventh worker and then this improvement is constant, similarly execution time also improves drastically in the beginning then the improvement is stable.

Chauhan et al. [6] demonstrates IoT suite for developing an application using a set of tools. It supports automatic code generation and integration with a set of modeling languages. Domain, architecture, user interaction, and deployment specifications can be specified by the developer.

Litunpatra et al. [7] discuss on building blocks of IoT. It includes specifications of RFID, WSN, addressing scheme, visualization, storage for data, and analytics. Authors propose five-layer architecture for IoT, which are device layer, network Layer, middleware layer, application layer, and business layer.

Nowadays, usage of IoT technology is extending in many different fields. WC is demanding mode of communication and it is necessary for IoT applications. This factor motivates the authors to present the framework for WC in IoT, and next section narrates the framework.

3 Proposed Work

This section gives a brief idea about the real implementation of hardware and software in IoT for WC along with cloud computing. The Sect. 3.1 describes the architecture of IoT applications and Sect. 2 explains the proposed framework for WC in IoT.

3.1 Architecture of IoT Application

IoT consists of things formed as a network, which is capable of collecting and exchanging data. The things are devices, building, home appliance, vehicle, etc. To make these devices operational, sensors are being deployed in each interested things. Wireless Sensor Network (WSN) is a key enablers of IoT system [8] and performs sensing and actuating of sensors. Another complexity arises when sensor begins to generate huge amount of real-time data, and cloud technology is used as alternative storage for this large set of data.

Cloud computing and IoT are tightly coupled, it offers on demand access to data from anywhere, at any time [1]. IoT with wireless connectivity and cloud computing is hot research area, and it requires interdisciplinary skills like circuit design, micro-controller/microprocessor, cloud computing, big data, wireless network, operating system, and programming.

Figure 1 depicts the architecture of IoT applications, and it follows three-layer architecture. Further part of this section explains each layer components and their functionalities.

Layer 1. This consists of sensor nodes deployed on the field for collecting environmental data such as temperature and pressure. This forms sensing station, where all the sensors regularly transmit data to the gateway nodes.

Layer 2. In this layer, gateway nodes are connected to the internet and cloud for storing the data collected from previous layer. Cloud can be integrated with gateway nodes using cloud storage APIs. This layer also consist of servers, which provide services to the user present at layer 3. Whenever the user request for an action or service, the server runs user commands and pass it on the cloud platform to retrieve the data, in turn the retrieved data is sent to the layer 3 of the architecture for representation. Arduino and Raspberry pi are used as gateway nodes and those are discussed in further part of this subsection.

(i). Arduino and wireless network connection: It is based on open source platform designing, building of hardware and software components. It is built with programmable board and also piece of software that can be used to upload and design

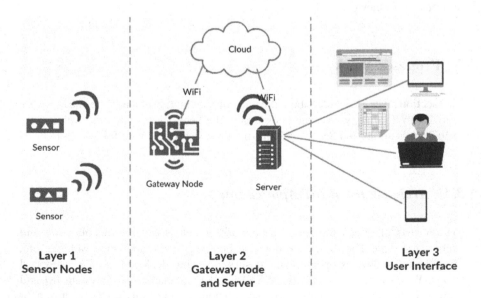

Fig. 1 Three-layer architecture of IoT

application. Ethernet shield can be used by the Arduino to connect it to the modem or local network, which in turn again connected to internet. To achieve wireless communication from Arduino to any other device, Wi-Fi shield or Wi-Fi Module (ESP8266) can be used. The ESP8266 Wi-Fi Module is integrated with TCP/IP protocol stack which gives access to Wi-Fi network. Wired or wireless media can be used to connect sensors.

(ii). Raspberry pi: Raspberry pi is small computer, which can be used for various operations like web application implementation and hosting, providing services, and data processing. The Raspberry pi board contains General-Purpose Input-Output (GPIO) pins to create programs for input or output processing. C++ or Python programming can be used to create programs in Raspberry pi.

(iii). Flask framework: Flask is a web application framework written in Python, based on the Werkzeug WSGI toolkit and Jinja2, which is popular templating engine for Python [4]. Web Server Gateway Interface (WSGI) is standard for Python web application development. A web templating system combines a template with a some data source to display dynamic web pages. Figure 3a, b represents web application, which was designed using Flask framework.

(iv). Cloud: Once the IoT application becomes operational, sensors begins to generate lots of data. There must be organized way of storing and processing this large set of data (Bigdata). Many cloud platform can be used for storing data which can be further used for analysis and visualization. For our application, we have used open source cloud Thingspeak for storing sensor values. Raspberry pi can be used for accessing cloud platform. Using Python programming, HTTP connection can be established to cloud server for storing and retrieving data.

Layer 3. The data that is being generated continuously is stored in cloud platform and can be accessed for continuous monitoring, controlling, and analyzing the data. The representation of this data must be convenient for the user to analyze it. This layer consists of User Interface (UI) for accessing and representing the data. This data can be viewed in mobile, computer, or any device with the help of GUI.

3.2 Framework for WC in IoT

Authors propose the novel framework for designing IoT applications with WC, which includes five phases according to the three-layered architecture specified in subsection 3.1. Next part of this subsection gives the details of these phases.

Phase 1: Deploying Sensor. Sensors are the devices that sense the environment and sends those values as signals to the application. Installation and maintenance of these sensors is a complex task. Wireless connectivity and energy consumption requires vital importance. The quality of sensor depends on the climate and range of transmission, so additional care must be taken while choosing the sensor. Sensors can be of digital or analog, digital sensors can be directly interfaced with microcontroller and analog sensors may need analog to digital converters. Sensors can contain set of pins for input, output, and power supply. Sensing station continuously transmits data, depending on the available resources and environmental factors. The system can use GPRS, Wi-Fi, or Ethernet gateways to transfer data. Voltage and current requirement also vary depending on the sensors. Various types of sensors are pressure sensor, temperature sensor, humidity sensor, and current sensor.

Phase 2: Configuring Gateway Nodes. Gateway nodes are data collectors and are connected to the internet directly or through other devices. Gateway nodes collect data from a sensor deployed on the field and use that data for the further processing.

(i). *Configuring Wi-Fi*: To get wireless connection, additional hardware may be required. For example, for Arduino board, we can use Wi-Fi shield or low-cost Wi-Fi module. This additional hardware enables Arduino to a establish connection to Internet or any network.

To establish Wi-Fi Connection, Arduino IDE must be installed on the computer and to write ESP8266 programs, we need to have ESP8266 Library installed and Board manager set as Generic ESP8266 Module.

For Raspberry pi, we can use command line or GUI for Wi-Fi settings. The following command establishes Wi-Fi Communication to specified SSID of the network.

```
WiFi.begin(ssid, password)
```

The following command will scan and list all available Wi-Fi networks available.

```
Iwlist
wlan0 scan
```

The following Wi-Fi configuration must be added in wpa_supplicant configuration file. This configuration will enable Raspberry pi to get Wi-Fi connection with specified SSID. This configuration will be noticed immediately after saving the file.

```
network = { ssid="SSID" psk="password" }
```

(ii). *Uploading Sketch*: These nodes can be programmed directly burning codes in their memory using IDEs, for Arduino, we can use open source Arduino IDE. This IDE can be installed on the computer, and once the program is written in IDE's editor, the code can be uploaded selecting Upload option in Sketch menu.

Phase 3: Cloud Integration. Data collected from sensor must be stored in such a way that it should be easy to retrieve and process, another requirement is scalability, to meet this requirement cloud is integrated in IoT application. Gateways and server access cloud to store and retrieve data. Cloud provides an efficient and scalable platform for continuous streaming data. Introducing cloud also reduces the cost in setting up the infrastructure. For instance, to upload data to cloud using Raspberry Pi, we need to have a channel for storing the values. This can be done after creating an account in the cloud platform. Any number of channels can be created depending on the number of application. Each channel consists of a number of fields for storing the sensor data. A channel will have API key through which we can fetch and store data on the cloud using Raspberry pi.

Phase 4: Configuring and Running Server. Raspberry pi is a small computer which can be used for various operations. It can be used as client or server. Server is the one which provides service to the client's request. We can set up Raspberry pi as a server using Flask framework. Flask is a web application framework written in Python.

Phase 5: Accessing User Interface. Interactive applications can be designed for IoT application. The application can be run as web-based application or Mobile android applications. To develop User Interface in Raspberry pi we can use many programming languages like Python, PHP, Perl, etc. Choice of this language is left to the developer. The user interface should be ease of use and must be convenient to for representing real-time data values.

4 Case Study: Energy Monitoring Using IoT

This system is IOT-based energy monitoring system, It continuously reads electricity usage and uploads to the cloud, The usage details can be viewed by the consumer. Consumers can benefit from continuous monitoring of energy usage, because monitoring is the key to identifying areas where consumption can be reduced, and significant cost savings can be handled by controlling consumption. The system starts with sensing the current consumed by the load using ACS712 sensor, we calculate power and energy of connected load, this result is sent to the cloud using ESP8266 Wi-Fi module. Users can access web application to view the usage details. Raspberry pi is used as a server for displaying the usage details, and the data is fetched from the cloud.

4.1 Workflow of System

The architectural diagram depicted in Fig. 2 represents the working procedure of the system (Fig. 3).

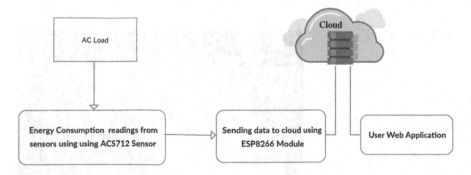

Fig. 2 Workflow of the system

(a) Home Page   (b) Monitoring Page

Fig. 3 Web application

Continuously sensor values are read for processing and calculating energy consumption. Wi-Fi communication is established between ESP8266 and cloud to transmit the energy readings over the internet. Results that are calculated are sent to cloud for storing the sensor data.

Web application displays energy readings fetched from cloud to the user which contains current, voltage, power, energy. For each 10 s, page refreshes itself to load dynamic data.

4.2 Hardware Setup

This section explains hardware and software setup for designing IoT application. Testbed is shown in Fig. 4. Table 1 lists hardware components required to build energy monitoring system. Table 2 consists of steps for connecting system hardware.

Ardino Programming IDE should be installed on machine to load arduino with the code. To develop web page and display arduino data on the web page Python, Flask framework is installed. To program ESP8266 Module, Generic ESP8266 must be installed in the Arduino IDE.

Fig. 4 Testbed setup

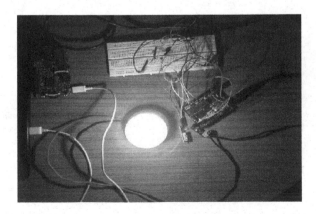

Table 1 Hardware components

Components	Remarks
Raspberry pi	Running raspbian OS/Server
ESP8266	Wi-Fi module
ACS712	30 A current sensor model
60 W Bulb	Connected to ACS712 sensor as prototype for current load

Table 2 Connection setup

Step	Connection
1	Connect the Arduino's 3.3 V pin output to ESP8266 VCC. The ESP8266 works with 3.3 V and not 5 V
2	Connect GND (ground) to Arduino's GND
3	Connect the RES or RESET pin to the Arduino's GND
4	Connect the RXD pin of the Arduino to the RX of the ESP8266
5	Connect the TXD pin of the Arduino to the TX pin of the ESP
6	Connect the GND pin of the ESP to the GND and the VCC pin to the VCC
7	Finally CH_PD goes to the Arduino's 3.3 V
8	Connect load to ACS712 sensor
9	Connect ACS712 OUT to AO pin of Arduino
10	ACS712 VCC to 5 V
11	ACS712 GND to Arduino's GND

4.3 Software Setup

ACS712 Sensor sense current flow in the load, power, and energy is calculated by fetching this sensor values. ESP8266 is used to establish Wi-Fi connection to upload data on the cloud by making HTTP connection. Raspberry pi acts as a server and retrieves the cloud data making HTTP connection to open source cloud and displays values in GUI as shown in Fig. 3a, b.

4.4 Results and Analysis

Observing the usage of energy consumption regularly consumers can come to know where most of the energy is wasted, and can think of a solution like using less energy

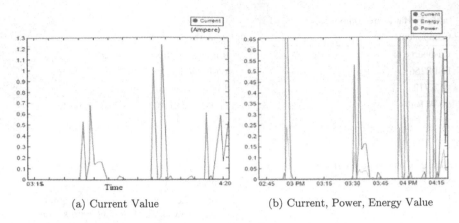

(a) Current Value   (b) Current, Power, Energy Value

Fig. 5 Web application

consuming appliances, switching to CFL bulbs, or checking for energy leaks. This can significantly reduce the energy consumption.

The deployed system is equipped with sensors and is used to collect information from the field. The main goal of the energy monitoring system is to help the consumer to analyze the data on daily basis and identify where they can save energy and cost. Figure 5a, b represents the graph of the data read over a period of time, and it can be used for identifying the usage patterns, key influencers, and outliers.

5 Conclusion

The proposed work gives guidelines for designing IoT applications using five phases and also states necessary prerequisites for designing IoT Application. Wireless communication has become the major platform for building IoT application. And also plays a crucial part in establishing communication and data transmission.

The application energy monitoring using IoT provides continuous electricity energy usage details, consumers can benefit from continuous monitoring of energy usage, and user can access electricity usage details as and when required from the cloud.

The results show that continuous monitoring can benefit in reducing energy consumption. Because detailed interval energy consumption data makes it possible to see patterns of energy waste that it would be impossible to identify otherwise. The future scope of application is to use it in agricultural field for analyzing varying electricity usage pattern and also providing predictive analysis of energy consumption using Bigdata Analytics in distributed environment.

References

1. Alkhamisi, A., Nazmudeen, M.S.H., Buhari, S.M.: A cross-layer framework for sensor data aggregation for IoT applications in smart cities. In: 2016 IEEE International Smart Cities Conference (ISC2), pp. 1–6, Sept 2016
2. Chen, Y., Kunz, T.: Performance evaluation of IoT protocols under a constrained wireless access network. In: International Conference on Selected Topics in Mobile Wireless Networking (MoWNeT), pp. 1–7, Apr 2016
3. Jover, R.P., Murynets, I.: Connection-less communication of IoT devices over LTE mobile networks. In: 12th Annual IEEE International Conference on Sensing, Communication, and Networking (SECON), pp. 247–255, June 2015
4. About flask. http://flask.pooco.org/docs/0.11/
5. Jaiswal, A., Domanal, S., Reddy, G.R.M.: Enhanced framework for IoT applications on python based cloud simulator (pcs). In: IEEE International Conference on Cloud Computing in Emerging Markets (CCEM), pp. 104–108, Nov 2015
6. Chauhan, S., Patel, P., Sureka, A., Delicato, F.C., Chaudhary, S.: Demonstration abstract: Iotsuite—a framework to design, implement, and deploy iot applications. In: 15th ACM/IEEE International Conference on Information Processing in Sensor Networks (IPSN), pp. 1–2, Apr 2016
7. Patra, L., Rao, U.P.: Internet of things 2014 architecture, applications, security and other major challenges. In: 2016 3rd International Conference on Computing for Sustainable Global Development (INDIACom), pp. 1201–1206, Mar 2016
8. Han, D.M., Lim, J.H.: Smart home energy management system using IEEE 802.15.4 and zigbee. In: IEEE Transactions on Consumer Electronics, vol. 56, pp. 1403–1410, Aug 2010

LD-IoT: Long-Distance Outdoor Networking for 802.11ah-Based IoT

Rakesh Kumar Ambhati, Shashikanth Y. Chaudhari and Manoj Jain

Abstract With the advent of IoT revolution, installation and deployment of various outdoor sensors in multiple technology verticals is wide rampant. Typical outdoor LANs cases include sensor backhaul, urban–rural broadband connectivity, and emergency management networks. To facilitate these scenarios, the Medium Access Layer (MAC) needs to enable wireless long range, improved power management, enhanced scalability of associated Stations (STAs), and interference management schemes. The IEEE 802.11ah (Wi-Fi HaLow) standard aims at long distance, scalable and low data rate network. In this paper, the design of 802.11 channel access and media layer for long-distance outdoor networks is discussed.

1 Introduction

Multitude of wireless (short-, medium-, and long range) technologies are widely adopted and interwoven into human life, to the extent their presence is disappeared from our general purview. The era of heterogeneous wireless connectivity is at cusp of pragmatic realization, leading to ubiquitous processing and networking. In last decade, both academia and industry synergize their efforts in various aspects of IoT like connectivity, digital and analog communication, and networking, which are indeed very important for IoT revolution. IEEE 802.11ah [1, 2] technology is aimed at long distance, outdoor, and scalable IoT sensor deployment. In this paper, sensor backhaul networks (high throughput) supporting afore-mentioned networks are taken into consideration. Wireless channels are time varying in nature,

R. K. Ambhati (✉) · S. Y. Chaudhari
Member Research Staff, Central Research Lab, Bengaluru, India
e-mail: arakeshkumar@bel.co.in

S. Y. Chaudhari
e-mail: cshashikantyeshwant@bel.co.in

M. Jain
Principal Scientist, Central Research Lab, Bengaluru, India
e-mail: manojjain@bel.co.in

© Springer Nature Singapore Pte Ltd. 2019   375
P. K. Sa et al. (eds.), *Recent Findings in Intelligent Computing Techniques*,
Advances in Intelligent Systems and Computing 707,
https://doi.org/10.1007/978-981-10-8639-7_39

therefore rudimentary understanding of channel, especially in outdoor conditions needs to be assessed for reliable communication. Electromagnetic waves traveling in air meet obstacles and also needs to bear with affects of medium, like reflection, refraction, absorption, and scattering. Such influence splits waves and each independent wave travel in different paths to reach destination. These multiple waves can get added constructively or destructively at receiver affecting characteristics of electromagnetic waves like amplitude and phase. Methods like interleaving, scrambling, forward error correction, etc., exists in digital domain to tackle multipath affects countered in analog medium, however, all such techniques do aid to a certain extent only. However, indoor path loss models are not valid in outdoor [3–6] scenario. New baseband processing algorithms, novel PHY layer headers designed for long-ranged are of paramount importance. While physical layer is about processing electromagnetic wave received via antenna, Medium Access Layer (MAC) deals with channel access mechanisms (Lower MAC, mostly implemented in hardware), Frame assembling and disassembling, MAC address management cum networking. Typically, MAC layer does access channel condition parameters like received power levels, receiver sensitivity, etc. from physical layer via control and status registers and can change medium access mechanism variables accordingly. MAC layer maintain several timers, packet count metrics, jitter values, etc., to maintain flow control of communication. MAC layer for long-distance communication also needs to be redesigned for outdoor coverage. Apart from PHY and MAC layers issues, antenna deployment in outdoor communication is of significant importance. Antenna with suitable characteristics like directivity, polarization, and gain needs to selected for consistent communication. Some critical germane to outdoor networks in general are discussed below.

1.1 Organization of This Paper

The purpose of this paper is to throw some limelight on various PHY and MAC layer design issues encountered in Long distance and High Throughput 802.11 networks. Section 2 informs about MAC Layer issues under consideration for aforementioned purpose. Section 3 does simulate outdoor propagation scenarios in NS3 and rest follows conclusion and references.

2 MAC Layer Design Issues

2.1 802.11 Physical Layer MAC Variables

Rudimentary Physical Layer Lower MAC variables for Throughput and data rate are T_{slot} and T_{SIFS}. T_{slot} is basic time duration used by random back_off timer in CSMA/CA operation. However, T_{SIFS} (as indicated in Fig. 2) is duration of time transmitter and needs to wait to receive acknowledgment from receiver, conforming the successful reception. Both variables are calculated as below.

$$T_{SIFS} = Rx_{Rf} \text{ delay (Receiver Radio Front End Delay)}$$
$$+ Rx_{PLCP} \text{ delay (Receiver PHY layer delay)}$$
$$+ Rx_{MAC} \text{ processing delay} + \text{RxTx turnaround time.}$$

$$T_{slot} = \text{Clear channel assessment time}$$
$$+ \text{RxTx Turnaround Time (Receiver to transmitter switching delay)}$$
$$+ \text{air Propagation delay} + \text{MAC processing delay}$$

All other inter-frame spacing timings used in DCF channel access are calculated as follows.

$$T_{DIFS} = T_{SIFS} + 2 * T_{slot}$$
$$T_{PIFS} = T_{SIFS} + T_{slot}$$

Above-mentioned variables and corresponding timers are typically implemented and maintained in hardware using hardware description languages (Verilog, VHDL, RTL, etc.) as they are time critical. For long-distance wireless networks, propagation delays also play a vital role in calculating IFS timing values. Medium propagation delay is typically neglected in indoor networks.

Let us consider a typical 802.11n standard parameters for general indoor networking at 2.4 GHz is Table 1 below. However, for outdoor scenario, all mentioned parameters need to be recalculated.

Assuming coverage of 1200 m, propagation delay would be 4 us (assuming electromagnetic wave speed in equivalent to light speed). Clear channel assessment mechanism in outdoor needs to also detect non-Wi-Fi signal, hence energy detection-based CCA needs to be employed in addition to currently functional preamble-based carrier sense. Therefore, CCA time also inflates, however, it could be assumed as more than double of previous value, i.e., 30 us. RxTx turnaround time would remain same.

Table 1 802.11 PHY MAC variables

$T_{slotnew}$	39 us (long) + PA_{delay}
$T_{SIFSnew}$	10 us (long) + PA_{delay}
$T_{DIFSnew}$	50 us + PA_{delay}
CCA time	Less than 25 us
RxTx turnaround time	Less than 5 us
CW_{min}	16 (min. contention window)
CW_{max}	1023 (max. contention window)
PA_{delay}	Unknown

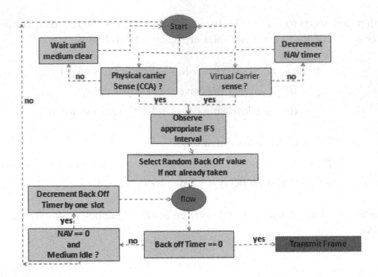

Fig. 1 Distributed CSMA/CA

An additional delay in long-distance wireless would be transmit power amplifier delay, which varies across hardware provided by different vendors. Recalculated variables for outdoor are shown in below Table 1.

$$T_{SIFS\,new} = T_{SIFS} + PA_{delay}$$
$$T_{slot\,new} = T_{slot\,new} + PA_{delay}$$

Distributed Coordinated Function (DCF): DCF [7–10] is the fundamental building block of contention-based carrier sense multiple access by collision avoidance state machine present in Fig. 1. Mandatory physical carrier sense (at physical layer using clear channel assessment function) and optional virtual carrier sense (at MAC layer using RTS/CTS mechanism) aids in channel access in 802.11 networks.

2.2 Suitability of DCF-Based CSMA/CA Outdoor Links

DCF ACK timing requirements are bit relaxed and can be adjusted as per need to deployment (for example, Atheros-based Wi-Fi chips can set with ACK timeout as long as 746 us using madwifi driver). This lenience makes CSMA/CA [11–14] usable in outdoor networking. As mentioned in previous section, both SIFS and time slot values significantly increased in outdoor networks. Channel access delays increased as the duration of overall backoff time is increased drastically, which have a noticeable effect on data rate, henceforth on overall throughput as well. In spite of shortcomings, DCF can be used if the following conditions are met.

Reduced DIFS Interval DIFS interval includes two-time slots, to facilitate special IFS interval, namely PIFS used in PCF mode. However, such constraint can be relaxed to one-time slot and DIFS interval can be made 30 us, as this duration is decreased channel access delay could be improved to some extent as shown in CSMA/CA state machine.

Reduced Backoff Interval As in outdoor networking medium propagation delay and CCA delay increases, this can be compensated by reducing backoff interval by modifying contention window parameters, for 802.11n currently contention window interval cmin, cmax is 32–1024. However such change also mandate to reduce number of nodes supported per cell coverage, as more nodes leads to more contention.

Small Payload Size Small packets are to be transmitted as compared to long packets with improved Forward Error Correction Techniques (FEC). As re-transmission of long packets leads to increased transmission delay with very less scope of error correction.

Bulk Acknowledgments Typical MAC flow control mechanism follows sliding window protocol for ACK recovery, however, in long-distance protocols like selective repeat needs to be implemented. Acknowledging multiple MAC frames at once would avoid spectrum wastage and also leads to increased throughput.

Fine-Tuning TCP parameters Transport layer needs to be informed about nature of underlying link, it is very much of use as TCP protocol will try to avoid unnecessary re-transmissions by expanding its congestion window size. TCP timeouts also need to be readjusted in accordance with base band processing, transmission, and medium propagation delays.

3 NS-3 Simulation Results of Outdoor Networks

Above-mentioned model is simulated in open source and state-of-art Networks Simulator third version (NS-3). NS-3 [15–17] identifies 802.11 nodes as "WIFINetDevice", it does support 802.11 a/b/g/n/ac/Ah MAC and PHY models [18–21]. Notable modules of NS-3 for Wi-Fi-based IoT simulation are shown below.

WifiChannel: A near approximate characterization of wireless medium with support for various propagation models.
WifiPhy: Base Band processing model for various Modulation and Coding Sequence (MCS) combinations, supporting multiple data rates. Preamble and header for 802.11ah, considered in this simulation is implemented here.
MacLow: Modeling channel access mechanisms, namely DCF and EDCA.
MacHigh: Implementation of MAC layer management entity protocols (like authentication, association, roaming, etc.)

Fig. 2 Simulation topology

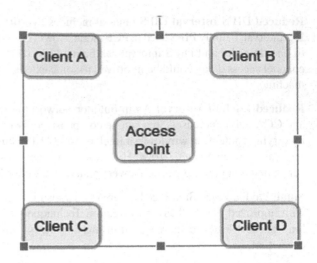

NS3 provides data analysis and visualization extensions to it simulation code. "FlowMonitor" class is used in simulation to collect throughput and packet loss ratio statistics. Three different scenarios for outdoor long distance wireless networks are simulated for the below-mentioned topology of Access Point (AP) and four client nodes positioned at 1200 m from AP with CSMA/CA as channel access mechanism. In all simulation (with three different simulation settings) runs, maximum throughput and minimum packet loss ratio are observed. Variation pattern of mentioned variables is also noted (Fig. 2).

3.1 Scenario 1

802.11ah-based IoT sensor network, with ordinary nodes (low power and low data rate) and Access Point node (high-speed processing capabilities) are created in NS-3. All nodes in network are configured with PHY and MAC parameters as shown in Tables 2 and 3. Simulation is run for 25 s for 15, 25, and 50 number of stations. From simulation results (Fig. 3a, b), maximum throughput of 680 kbs and minimum packet loss ratio of 18.75% is achieved and can be observed. However, throughput and PLR values over multiple simulation runs are not predictable, as these values did not stabilize at end of simulation time.

3.2 Scenario 2

In this scenario, the following changes in Table 4 are made in simulation configuration as compared to Scenario 1. With the introduction of rate adaption, both maximum throughput (695 kbs) and minimum PLR (24.75%) improved to a moderate extent

Table 2 NS3 features

Propagation model	Log distance
Propagation delay	5 us
Rate control	Constant rate
Mobility model	Constant position
Mac high model	Access point (at AP), station (at Client)
Mac low layer	DCF
PHY model	802.11ah standard
Packet length	No restriction
SIFS	40 us
DIFS	50 us
Contention window	32 to 512 slots
Bandwidth	20 MHz
Gaurd interval	800 ns

Table 3 Physical layer parameters

Propagation model	Long distance
Propagation delay	5 us
Transmit power	0 dbm
Transmission gain	0 dBm
Receiver gain	3 dB
Noise figure	3 dB
Coding method	BCC
Propagation model	Outdoor (Macro)
Error detection model	Yans error rate

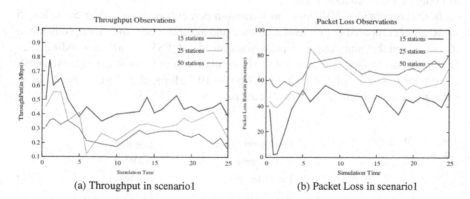

(a) Throughput in scenario1   (b) Packet Loss in scenario1

Fig. 3 Scenario 1

Table 4 Changes in scenario 2

Propagation model	Long distance
Rate control	AARF
Packet length	512 bytes

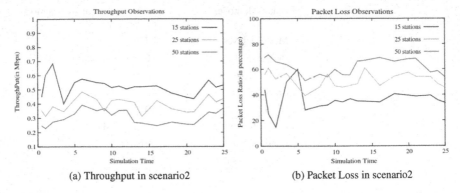

(a) Throughput in scenario2   (b) Packet Loss in scenario2

Fig. 4 scenario 2

(Fig. 4a, b for 15 stations). Significant observation here is both throughput and PDR stabilized to the end of simulation run.

3.3 Scenario 3

In this scenario, the following changes in Table 5 are made in simulation configuration as compared to Scenarios 1 and 2.

Both throughput (750 Kbps) and minimum percent) improved (Fig. 5a, b for 15 stations) as compared to previous scenarios and also have become more predictable. It is observed that simulator in all scenarios opted to QPSK 1/2 and 3/4 coding rates (it is well-known fact, that default QAM modulation is not robust in long distances). Maximum data rate of QPSK modulation is 19.5 Mbps, therefore at MAC layer throughput is approximately 60% of data rate, i.e., 11.7 Mbps.

Table 5 Changes in scenario 3

Propagation model	Log distance
Rate control	AARF
Packet length	256 bytes
Bulk acknowledgment	Enabled
Mac model	EDCA

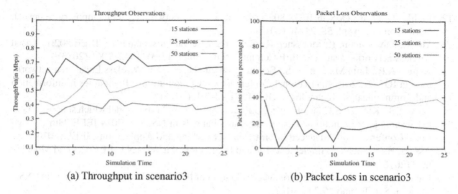

(a) Throughput in scenario3   (b) Packet Loss in scenario3

Fig. 5 Scenario 3

4 Conclusion

PHY and MAC layer issues, for outdoor 802.11 wireless networks are discussed. Appropriate values for PHY and MAC variables are analyzed for long-distance networking. Simulation in NS-3 simulator for various scenarios is done to for throughput and PLR analysis to verify proposed throughput modes.

References

1. Stefan, A., Prasad, R.V., Niemegeers, I.G.M.M.: Outdoor long-range WLANs: a lesson for IEEE 802.11ah. IEEE Commun. Surv. Tutor. **17**(3), 1761–1775 (2015)
2. Jain, R.: Low Power WAN Protocols for IoT: IEEE 802.11ah, LoRaWAN (2016)
3. Aust, S., Prasad, R.V., Niemegeers, I.G.M.M.: Outdoor long-range WLANs: a lesson for IEEE 802.11ah. IEEE Commun. Surv. Tutor. **17**(3), 1761–1775 (2015)
4. Rani, P., et al.: A review on wireless propagation models. Int. J. Eng. Innov. Technol. (IJEIT) **3.11** (2014)
5. Biswas, S., et al.: Large-scale measurements of wireless network behavior. In: ACM SIG-COMM Computer Communication Review, vol. 45, No. 4. ACM (2015)
6. Abdelgader, A.M., Wu, L.: The physical layer of the IEEE 802.11 p WAVE communication standard: the specifications and challenges. In: Proceedings of the World Congress on Engineering and Computer Science, vol. 2 (2014)
7. Bhoyar, R., Ghonge, M., Gupta, S.: Comparative study on IEEE standard of wireless LAN/Wi-Fi 802.11 a/b/g/n. Int J. Adv. Res. Electron. Commun. Eng. (IJARECE) **2.7** (2013)
8. Deng, D.-J., Chen, K.-C., Cheng, R.-S.: IEEE 802.11 ax: next generation wireless local area networks. In: 2014 10th International Conference on Heterogeneous Networking for Quality, Reliability, Security and Robustness (QShine). IEEE (2014)
9. Abichar, Z., Chang, J.M.: Group-based medium access control for IEEE 802.11 n wireless LANs. IEEE Trans. Mob. Comput. **12.2**, 304–317 (2013)
10. Banerji, J.S., Chowdhury, R.S.: On IEEE 802.11: Wireless LAN Technology (2013). arXiv:1307.2661
11. Prasetya, S., Rahmat, B., Susanto, E.: Quality of service improvement with 802.11 e EDCA scheme using enhanced adaptive contention window algorithm. In: 2015 IEEE International Conference on Communication, Networks and Satellite (COMNESTAT). IEEE (2015)

12. Malik, A., et al.: QoS in IEEE 802.11-based wireless networks: a contemporary review. J. Netw. Comput. Appl. **55**, 24–46 (2015)
13. Yu, X., Navaratnam, P., Moessner, K.: Resource reservation schemes for IEEE 802.11-based wireless networks: a survey. IEEE Commun. Surv. Tutor. **15**(3), 1042–1061 (2013)
14. Kumar, A.R., Jain, M., Ungati, S.: Cognitive channel access for Wireless Local area networks used in IOT. In: 2016 International Conference on Computational Techniques in Information and Communication Technologies (ICCTICT). IEEE (2016)
15. Ernst, J.B., Kremer, S.C., Rodrigues, J.J.P.C.: A Wi-Fi simulation model which supports channel scanning across multiple non-overlapping channels in NS3. In: 2014 IEEE 28th International Conference on Advanced Information Networking and Applications. IEEE (2014)
16. Assasa, H., Widmer, J.: Implementation and Evaluation of a WLAN IEEE 802.11 ad Model in ns-3 (2016)
17. Ravindranath, N.S., et al.: Performance evaluation of IEEE 802.11 ac and 802.11 n using NS3. Indian J. Sci. Technol. **9**.26 (2016)
18. Gupta, P., et al.: Link-level measurements of outdoor 802.11 g links. In: 2009 6th IEEE Annual Communications Society Conference on Sensor, Mesh and Ad Hoc Communications and Networks Workshops. IEEE (2009)
19. Huang, K.D., Malone, D., Duffy, K.R.: The 802.11 g 11 Mb/s rate is more robust than 6 Mb/s. IEEE Trans. Wirel. Commun. **10**(4), 1015–1020 (2011)
20. Paul, U., et al.: Characterizing WiFi link performance in open outdoor networks. In: 2011 8th Annual IEEE Communications Society Conference on Sensor, Mesh and Ad Hoc Communications and Networks (SECON). IEEE (2011)
21. Hazmi, A., Jukka, R., Mikko, V.: Feasibility study of IΐΤΐΤΐΤ 802.11ah radio technology for IoT and M2M use cases. In: 2012 IEEE Globecom Workshops. IEEE (2012)

Trends of Publications and Work Done in Different Areas in IoT: A Survey

Nagma, Jagpreet Sidhu and Jaiteg Singh

Abstract IoT is transforming the world into smart world. Current works in IoT are directing towards creation of smart things. However, to the best of our knowledge there is a lack of systematic literature review of work done in various domains in IoT. To bridge this gap, literature survey is provided which classifies the works done in IoT in four different domains according to the research contribution. The proposed classification furthermore provides applications of work done by different authors, techniques, datasets and tools used by peer group to analyze objectivity/subjectivity of problem domain. This paper also examines the status of numerous publications according to bibliographical citations on the basis of diverse global regions involved in research in IoT, various journals citing the work, year of publishing, research community involved in exploration, fund provisioning or self-promotion factors. Trends are illustrated in form of graphs. The outcomes will offer great assistance to peer research community.

Keywords Internet of things · Security · Message passing · Scheduling
Workload balancing · Energy saving · Smart applications · Trends
Bibliographical citations

1 Introduction

Internet of Things (IoT) is the next revolution in the field of technology. It facilitates the interconnection of various smart devices that are equipped with sensors, software, actuators to collect and transmit data. IoT has applications in assisted living,

Nagma (✉) · J. Sidhu · J. Singh
Department of CSE, CUIET, Chitkara University, Chandigarh, India
e-mail: nagma@chitkara.edu.in

J. Sidhu
e-mail: jagpreet.sidhu@chitkara.edu.in

J. Singh
e-mail: jaiteg.singh@chitkara.edu.in

© Springer Nature Singapore Pte Ltd. 2019
P. K. Sa et al. (eds.), *Recent Findings in Intelligent Computing Techniques*,
Advances in Intelligent Systems and Computing 707,
https://doi.org/10.1007/978-981-10-8639-7_40

385

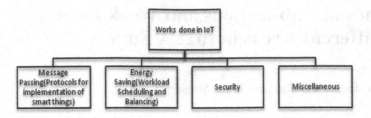

Fig. 1 Classification of work done in IoT

e-health, enhanced learning, automation, industrial manufacturing, transportation, business/process management, and logistics [1]. IoT has given a direction towards creation of smart world by providing models of smart parking systems, smart vending machines, and smart surveillance systems [2–4]. Current research in IoT includes work on different protocols in IoT using Radio Frequency Identification (RFID) [2]. Semantic oriented approach is also used in IoT [5]. Many works have been found in which combination of two or more techniques is used. So there are many visions of perception of IoT [1]. IoT is playing a major role in today's world by conquering nearly every field. This survey provides a clear picture of work done in IoT by classifying the work done in IoT in four different areas as shown in Fig. 1 and trends according to bibliographical citations have been shown. Analysis of large number of papers mainly published from 2014 to 2017 has been done. Although we had to include some papers from 2011 to 2013 in order to get 10 papers of each domain in IoT. Because of being at a very initial stage of research, this analysis has been done to formulate a problem definition in cloud computing paradigm so only papers of highly reputed and relevant journals are taken just to know the current scenario and to check the feasibility of pursuing work in future in IoT. This work may be extended later on by taking papers from various other journals and conferences too.

2 Literature Review

The analysis has been done on the subset of 10 papers that have been selected in each domain in IoT to find out the techniques, datasets and simulators used by researchers. Tables 1, 2, 3 and 4 describes the techniques, datasets, tools used and applications of the work done in the areas of message passing, energy saving, security and miscellaneous fields respectively.

Table 1 Techniques/algorithm, dataset, simulation tools used and applications of work done in message passing in IoT

R. no.[a]	Dataset	Simulation tools	Applications/use	Technique/algorithm used
[2]	Synthetic dataset has been used	Had set up a mobile device; RFID readers; iBeacon; ESP 8266 have been used on parking slots	Can be implemented in any park	Google API; RFID; GPS; ESP8266
[3]	Any message in base 64 encoding	Arduino components; OpenStack	To reduce total cost of ownership	Near field communication; open source technologies
[4]	Sensors to produce packets	Two different network topologies, network scenarios	Proposed protocol provided superior results	HEVC encoding; adaptive packet frame grouping; adaptive quantization
[6]	Seismic data of ecuador	Smartphones	Earthquake early warning system	Message queue telemetry transport
[7]	14 end users	Android postgres database	Developed a system which enables wheel chair users to interact with items placed beyond their arm's length	AR and RFID
[8]	Not mentioned	Game model to evaluate performance	Proposes game based model for behavior analysis of IoT	Tree analysis method; linear time algorithm
[9]	Heterogeneous data provided by different data sources such as sensing devices	Test environment developed by authors	Data decomposition storage method is developed	Ontology based data fusion
[10]	Real sensor dataset	Real world environment; built an application for crowd sensed air quality monitoring	Developed an application for air quality monitoring	CUPUS

(continued)

Table 1 (continued)

R. no.[a]	Dataset	Simulation tools	Applications/use	Technique/algorithm used
[11]	Analyzed the data flows of simple capability wEASEL services converted from the OWLS-TC4 collection	Developed a test bed of simple and composite services	Presented a framework which supports dynamic reasoning	wEASEL
[12]	Used sensor nodes	Cooja simulator	Proposed the service provisioning architecture for smart objects with semantic annotation to enable the integration of IoT applications into the Web	RESTful service provisioning

[a]R.No. stands for reference number

Table 2 Techniques/algorithm, dataset, simulation tools used and applications of work done in energy saving/scheduling in IoT

R. no.[a]	Dataset	Simulation tools	Applications/use	Technique/algorithm used
[13]	Collected from COEX Mall (Korea) Environment data; Weather data has been collected from year 2010–2012 Randomly selected 1500 energy datasets	Implemented using WISE MP; WISE A, WISE GW; SOAP	Proposed web of object based on architecture which contains various objects for providing web based IoT applications	WISE MP and WISE GW
[14]	Project open fridge Pilots with real users have been run and eventually the data has been collected in a collaborative crowd sourcing manner by surveying active users	Active users were given plug wise tool kits Testing on refrigerators and a portal	Combines IoT, web and semantic web technology to create a platform engaging users in collecting, processing, linking and opening the data contributed by users and adding value services on top of this data	Advanced Encryption Security REST interface
[15]	Collected from sensor nodes	NS2 Xilinx V5 platform	Designed TESN Proposed WLC for parallel computing and load balancing	Weighted least connection

(continued)

Table 2 (continued)

R. no.[a]	Dataset	Simulation tools	Applications/use	Technique/algorithm used
[16]	Devices used transmitted data simultaneously	Network was made with gateways	MU-MIMO capability is implemented in IoT gateway device	Multiple input and multiple output scheduling scheme
[17]	Data for the Fyffe's redundancy allocation problems [18]	Taguchi method	Solves redundancy allocation problem in smart sensors	Detailed computational results from solving a series-parallel redundancy allocation problem with a mix of components is presented
[19]	Real time data by medical sensors	Panda board; smart module RF06 board; UT gateway CC2538	End to end security scheme using concept of fog computing, decreased overhead and thus energy consumption	Session resumption technique
[20]	Sensors are employed in unattended environment	MATLAB for mathematical analysis	To balance load in a network	Deep learning algorithms
[21]	An instance of 100 jobs is randomly generated	C++	Showed pareto optimization scheduling problem in polynomial time	Multi agent scheduling
[22]	Got data from smart city, smart grid [23], smart building and smart home [24]	Developed a detailed event-driven simulator	Reduced the overhead of data analytics	Object based storage
[25]	Data generated from sensors	Panda board; Smart RF06 board integrated with CC2538 module and MOD-ENC28J60	Presented fog-based mobility support to enable seamless connectivity for mobile sensors	Presented a prototype of a smart e-health gateway called UT-GATE

Table 3 Techniques/algorithm, dataset, simulation tools used and applications of work done in security in IoT

R. no.[a]	Dataset	Simulation Tools	Applications/use	Technique/algorithm used
[26]	Experimental files with small sizes are selected	Machine with 2.40 GHz, INTEL Pentium 4 CPU, 489.0 Mb memory and 2.628–17 linux kernel	To meet communication and storage capabilities for data mining and analytic applications	Elliptic curve cryptography
[27]	Not mentioned	Not required	Comparison is made	Shamir's system
[28]	Smart Project in which grid readings were continuously collected from households in every second within 2 months	Wrote two software modules using C language to represent client and cloud Experiment data was stored on cloud	To calculate bill Preserved user privacy of smart meta data in smart grid by cryptography	Homomorphic asymmetric key cryptosystem
[29]	Used pictures taken with canon IP camera inside lab manually and sent by camera	Computer vision modules with models prepared	Showed the use of Computer Vision in IoT to improve security	Computer Vision
[30]	Real time data by medical sensors	Panda board; smart module RF06 board; UT gateway CC2538	Decreased overhead and thus energy consumption	Session resumption technique
[31]	Not mentioned	Cryptographic tools; Equations	ABS scheme was described	Attribute based encryption
[32]	Random numbers	Implemented a real RFID system using omni key (5421), Built in elliptic curve calculator tool, IDE Eclipse (JAVA)	New RFID authentication protocol to remove vulnerabilities	RFID; elliptic curve cryptography
[33]	Nothing mentioned	Compared with block chain of bit coin bit torrent for implementing	Tells that system is up to date or not	Block chain technology
[34]	Data is sensed from network of users	Monticore; Xtest	Enables developers to integrate privacy in cloud services	PDL model
[35]	Not mentioned	Public cloud by Amazon; private smart home platform by Guangdong University and Canbo Co. Ltd.	A multi-layer cloud architectural model is developed	Ontology method

Table 4 Techniques/algorithm, dataset, simulation tools used and applications of work done in miscellaneous areas in IoT

R.no.[a]	Dataset	Simulation tool	Applications/use	Technique/algorithm used
[26]	Real time through medical sensors	Environment containing various sensors	To provide health care services to patients with medical sensors	Analytic hierarchy process
[27]	5–50 devices are employed and they generate data periodically	MATLAB and made FSIIoT Sim	Can be used in real ISN environment	Scheduling algorithm in MAC
[28]	Used Midgar and Things Speak to collect data	Prototype made	Temperature monitoring	Fuzzy logic; IoT
[29]	Subject systems taken from DaCapo benchmark	JAVA	To find similar code fragments	Liveliness analysis; reaching application
[30]	Data collected through surveys	Ziegler corporate interlocks	Detected community using graphs	graphs
[31]	Sensors were used	Realistic environment	Proposed accurate localization scheme	Blind node localization process
[32]	Data collected from previous research	Not mentioned	Defined requirements for smart homes	Not required
[33]	Data collection from cloud scalable databases	OpenStack system and FIWARE GEs	Presented a service for data collection from Internet of Things	RESTful API, JSON
[34]	Collected requests over pre-determined time interval	MATLAB	Proposes a coalition formation game	Coalition game theory; greedy approach
[35]	Randomly generated applications	Eclipse MARS; TOSSIM; COOJA	Created a development environment	Domain specific language

3 Findings

Analysis have been done on the subset chosen to evaluate the status of numerous publications according to bibliographical citations on the basis of diverse global regions involved in research in IoT, various journals citing the work, year of publishing, research community involved in exploration, fund provisioning or self-promotion factors. The interpretations explained below are our way of seeing the things according to the sub set chosen. Peer researchers may interpret the trends differently.

Section (a) of Fig. 2 illustrates that in Asia maximum funds are granted in the domain of energy saving and scheduling in IoT while in Europe, maximum funds are in domain of message passing in IoT. Section (b) of Fig. 2 illustrates that Asia is publishing maximum number of papers in energy saving/scheduling in IoT while Europe is publishing maximum papers in message passing and there is a minor contribution of rest of continents in research in IoT. Section (a) of Fig. 3 shows the combined comparison of published and funded work according to continents making it clear that maximum works done in domains of energy saving/scheduling and security are funded. Section (b) of Fig. 3 indicates that more publications are from Asia and Europe so more funding is provided in these continents only. Thus, it could be interpreted that Asia and Europe are maximum contributors in the field of research in IoT in different domains and hence they receive maximum number of grants which can be validated from Figs. 2 and 3.

Section (a) of Fig. 4 illustrates that highest number of papers are published in year 2016 in IoT and Future Generation Computer Systems is publishing maximum papers as compared to other journals. Section (b) of Fig. 4 indicates that Asia and Europe have prominent citations as compared to other continents, thus producing qualitative research articles. So it could be interpreted that 2016 is the year in which IoT has gained prominent importance and Asia and Europe are pioneer in research which could be validated from Fig. 4. Figure 5 illustrates that there is no research in IoT by industry, maximum research is done by academicians and there is a minor contribution by putting collaborative effort.

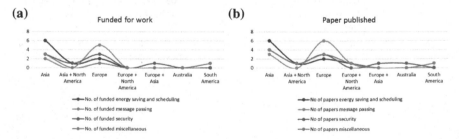

Fig. 2 Trends in publishing and funding of work done in IoT in different domains **a** Funding of work in different domains in Iot according to continents **b** Continent wise publication in different domains in IoT

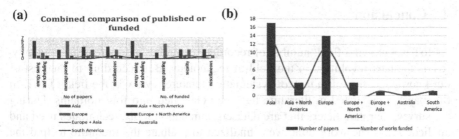

Fig. 3 Trends in publications in all domains in IoT and comparison of published and funded work **a** Number of papers published in different domains in IoT in different continents and number of funds granted **b** Combined continent wise publications of all domains in IoT

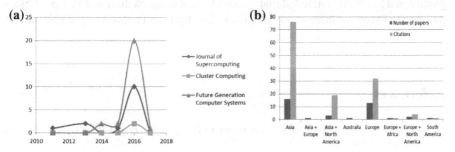

Fig. 4 Trends in citations and year wise publications of papers by different journals **a** Articles published year wise in different journals **b** Continent wise citations of articles

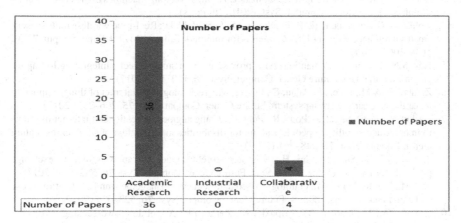

Fig. 5 Trends in publications according to involved research group

4 Conclusion

As IoT is a growing field, number of recorded researches are increasing gradually. IoT is the next revolution in the field of technology. But according to the best of our knowledge, there is a lack of systematic literature review in the field of IoT. In this paper, work listed in literature has been classified into four domains. During the survey, four parameters that are datasets, simulation tools, techniques used and applications of the work done were finalized to evaluate the literature to find the objectivity of problem domain. To validate the facts objectively, different interpretations have been made which indicate the trends in publications in IoT according to bibliographical citations and research contribution by different groups across the globe and it has been concluded that there is lack of research done in IoT from objective point of view so there is a need of standard tools and datasets which should be used to turn it into objective work.

References

1. Atzori, L., Iera, A., Morabito, G.: The internet of things: A survey. Comput. Netw. **54**(15), 2787–2805 (2010)
2. Tsai, M.F., Kiong, Y.C., Sinn, A.: Smart service relying on Internet of Things technology in parking systems. J. Supercomput. 1–24 (2016)
3. Solano, A., Duro, N., Dormido, R., González, P.: Smart vending machines in the era of internet of things. Future Gener. Comput. Syst. (2016) (In Press)
4. Kokkonis, G., Psannis, K.E., Roumeliotis, M., Schonfeld, D.: Real-time wireless multisensory smart surveillance with 3D-HEVC streams for internet-of-things (IoT). J. Supercomput. **73**(3), 1044–1062 (2017)
5. Han, S.N., Crespi, N.: Semantic service provisioning for smart objects: Integrating IoT applications into the web. Future Gener. Comput. Syst. **76**, 180–197 (2017)
6. Zambrano, A.M., Perez, I., Palau, C., Esteve, M.: Technologies of internet of things applied to an earthquake early warning system. Future Gener. Comput. Syst. **75**, 206–215 (2017)
7. Rashid, Z., Melià-Seguí, J., Pous, R., Peig, E.: Using augmented reality and internet of things to improve accessibility of people with motor disabilities in the context of smart cities. Future Gener. Comput. Syst. **76**, 248–261 (2017)
8. Tao, X., Li, G., Sun, D., Cai, H.: A game-theoretic model and analysis of data exchange protocols for internet of things in clouds. Future Gener. Comput. Syst. **76**, 582–589 (2017)
9. Tao, M., Ota, K., Dong, M.: Ontology-based data semantic management and application in IoT and cloud-enabled smart homes. Future Gener. Comput. Syst. **76**, 528–539 (2017)
10. Antonic, A., Marjanovic, M., Pripuzic, K., Zarko, I.P.: A mobile crowd sensing ecosystem enabled by CUPUS: Cloud-based publish/subscribe middleware for the internet of things. Future Gener. Comput. Syst. **56**, 607–622 (2016)
11. Urbieta, A., Gonzalez-Beltran, A., Mokhtar, S.B., Hossain, M.A., Capra, L.: Adaptive and context-aware service composition for IoT-based smart cities. Future Gener. Comput. Syst. (2017) (In Press)
12. Han, S.N., Crespi, N.: Semantic service provisioning for smart objects: integrating IoT applications into the web. Future Gener. Comput. Syst. (2017) (In Press)
13. Yu, J., Lee, N., Pyo, CS., Lee, Y.S.: WISE: web of object architecture on IoT environment for smart home and building energy management. J. Supercomput. 1–16 (2016)

14. Fensel, A., Tomic, D.K., Koller, A.: Contributing to appliances' energy efficiency with Internet of Things, smart data and user engagement. Future Gener. Comput. Syst. (2016) (In Press)
15. Qiu, T., Zhao, A., Ma, R., Chang, V., Liu, F., Fu, Z.: A task-efficient sink node based on embedded multi-core soC for internet of things. Future Gener. Comput. Syst. (2016) (In Press)
16. Kim, T.Y., Kim, E.J.: Uplink scheduling of MU-MIMO gateway for massive data acquisition in internet of things. J. Supercomput.. 1–15 (2016)
17. Yeh, W.C., Lin, J.S.: New parallel swarm algorithm for smart sensor systems redundancy allocation problems in the Internet of Things. J. Supercomput. 1–27 (2016)
18. Fyffe, D.E., Hines, W.W., Lee, N.K.: System reliability allocation and a computation algorithm. IEEE Trans. Reliab. 64–69 (1968)
19. Moosavi, S.R., Gia, T.N., Nigussie, E., Rahmani, A.M., Virtanen, S., Tenhunen, H., Isoaho, J.: End-to-end security scheme for mobility enabled healthcare internet of things. Future Gener. Comput. Syst. **64**, 108–124 (2016)
20. Kim, H.Y., Kim, J.M.: A load balancing scheme based on deep learning in IoT. Clust. Comput. **20**(1), 873–878 (2017)
21. Wan, L., Wei, L., Xiong, N., Yuan, J., Xiong, J.: Pareto optimization for the two-agent scheduling problems with linear non-increasing deterioration based on internet of things. Future Gener. Comput. Syst. (2016) (In Press)
22. Xu, Q., Aung, K.M.M., Zhu, Y., Yong, K.L.: Building a large-scale object-based active storage platform for data analytics in the internet of things. J. Supercomput. **72**(7), 2796–2814 (2016)
23. Boumkheld, N., Ghogho, M., Koutbi, M.E.: Energy consumption scheduling in a smart grid including renewable energy. J. Inf. Process. Syst. **11**(1), 116–124 (2015)
24. Vanus, J., Smolon, M., Martinek, R., Koziorek, J., Zidek, J., Bilik, P.: Testing of the voice communication in smart home care. Hum. Center Comput. Inf. Sci. **5**(15), 1–22 (2015)
25. Rahmani, A.M., Gia, T.N., Negash, B., Anzanpour, A., Azimi, I., Jiang, M., Liljeberg, P.: Exploiting smart e-Health gateways at the edge of healthcare internet-of-things: a fog computing approach. Future Gener. Comput. Syst. (2017) (In Press)
26. Yao, X., Chen, Z., Tian, Y.: A lightweight attribute-based encryption scheme for the internet of things. Future Gener. Comput. Syst. **49**, 104–112 (2015)
27. Jiang, H., Shen, F., Chen, S., Li, K.C., Jeong, Y.S.: A secure and scalable storage system for aggregate data in IoT. Future Gener. Comput. Syst. **49**, 133–141 (2015)
28. Mai, V., Khalil, I.: Design and implementation of a secure cloud-based billing model for smart meters as an Internet of things using homomorphic cryptography. Future Gener. Comput. Syst. (2016) (In Press)
29. García, C.G., Meana-Llorian, D., G-Bustelo, B.C.P., Lovelle, J.M.C., Garcia-Fernandez, N.: Midgar: detection of people through computer vision in the internet of things scenarios to improve the security in smart cities, smart towns, and smart homes. Future Gener. Comput. Syst. (2017) (In Press)
30. Moosavi, S.R., Gia, T.N., Nigussie, E., Rahmani, A.M., Virtanen, S., Tenhunen, H., Isoaho, J.: End-to-end security scheme for mobility enabled healthcare internet of things. Future Gener. Comput. Systs. **64**, 108–124 (2016)
31. Su, J., Cao, D., Zhao, B., Wang, X., You, I.: ePASS: an expressive attribute-based signature scheme with privacy and an unforgeability guarantee for the Internet of Things. Future Gener. Comput. Syst. **33**, 11–18 (2014)
32. Alamr, A.A., Kausar, F., Kim, J., Seo, C.: A secure ECC-based RFID mutual authentication protocol for internet of things. J. Supercomput. 1–14 (2016)
33. Lee, B., Lee, J.H.: Blockchain-based secure firmware update for embedded devices in an internet of things environment. J. Supercomput. 1–16 (2016)
34. Henze, M., Hermerschmidt, L., Kerpen, D., Haubling, R., Rumpe, B., Wehrle, K.: A comprehensive approach to privacy in the cloud-based internet of things. Future Gener. Comput. Syst. **56**, 701–718 (2016)
35. Tao, M., Zuo, J., Liu, Z., Castiglione, A., Palmieri, F.: Multi-layer cloud architectural model and ontology-based security service framework for IoT-based smart homes. Future Gener. Comput. Syst. (2016) (In Press)

36. Jeong, Y.S., Shin, S.S.: An IoT healthcare service model of a vehicle using implantable devices. Clust. Comput. 1–10 (2016)
37. Lee, H.H., Kwon, J.H., Kim, E.J.: FS-IIoTSim: a flexible and scalable simulation framework for performance evaluation of industrial Internet of things systems. J. Supercomput. 1–18 (2016)
38. Meana-Llorian, D., Garcia, C.G., G-Bustelo, B.C.P., Lovelle, J.M.C., Garcia-Fernandez, N.: IoFClime: the fuzzy logic and the internet of things to control indoor temperature regarding the outdoor ambient conditions. Future Gener. Comput. Syst. (2016)
39. Tekchandani, R., Bhatia, R., Singh, M.: Semantic code clone detection for internet of things applications using reaching definition and liveness analysis. J. Supercomput. 1–28 (2016)
40. Barthwal, R., Misra, S., Obaidat, M.S.: Finding overlapping communities in a complex network of social linkages and Internet of things. J. Supercomput. 66(3), 1749–1772 (2013)
41. Chen, Z., Xia, F., Huang, T., Bu, F., Wang, H.: A localization method for the internet of things. J. Supercomput. 63(3), 657–674 (2013)
42. Hui, T.K., Sherratt, R.S., Sanchez, D.D.: Major requirements for building smart homes in smart cities based on internet of things technologies. Future Gener. Comput. Syst. (2016) (In Press)
43. Douzis, K., Sotiriadis, S., Petrakis, E.G., Amza, C.: Modular and generic IoT management on the cloud. Future Gener. Comput. Syst. (2016)
44. Farris, I., Militano, L., Nitti, M., Atzori, L., Iera, A.: MIFaaS: a mobile-IoT-Federation-as-a-Service model for dynamic cooperation of IoT cloud providers. Future Gener. Comput. Syst. 70, 126–137 (2017)
45. de Farias, C.M., Brito, I.C., Pirmez, L., Delicato, F.C., Pires, P.F., Rodrigues, T.C., Batista, T.: COMFIT: a development environment for the internet of things. Future Gener. Comput. Syst. (2016) (In Press)

Part IV
Applications of Informatics and Data Privacy

A Novel Technique of Feature Selection with ReliefF and CFS for Protein Sequence Classification

Kiranpreet Kaur and Nagamma Patil

Abstract Bioinformatics has gained wide importance in research area for the last few decades. The main aim is to store the biological data and analyze it for better understanding. To predict the functions of newly added protein sequences, the classification of existing protein sequence is of great use. The rate at which protein sequence data is getting accumulated is increasing exponentially. So, it emerges as a very challenging task for the researcher, to deal with large number of features obtained by the use of various encoding techniques. Here, a two-stage algorithm is proposed for feature selection that combines ReliefF and CFS technique that takes extracted features as input and provides us with the discriminative set of features. The n-gram sequence encoding technique has been used to extract the feature vector from the protein sequences. In the first stage, ReliefF approach is used to rank the features and obtain candidate feature set. In the second stage, CFS is applied on this candidate feature set to obtain features that have high correlation with the class but less correlation with other features. The classification methods like Naive-Bayes, decision tree, and k-nearest neighbor can be used to analyze the performance of proposed approach. It is observed that this approach has increased accuracy of classification methods in comparison to existing methods.

Keywords Bioinformatics · Gene data · Feature selection · Protein sequence data · Filter · ReliefF · CFS · Classification

K. Kaur (✉) · N. Patil
Department of Information Technology, National Institute of Technology Karnataka Surathkal,
Mangalore, India
e-mail: kaur.kiranpreet05@gmail.com

N. Patil
e-mail: nagammapatil@nitk.ac.in

© Springer Nature Singapore Pte Ltd. 2019
P. K. Sa et al. (eds.), *Recent Findings in Intelligent Computing Techniques*,
Advances in Intelligent Systems and Computing 707,
https://doi.org/10.1007/978-981-10-8639-7_41

1 Introduction

Bioinformatics is emerging as an active research area. The main aim of bioinformatics is to obtain the data, store it in databases, and analyze it to obtain important information about structure and functions of proteins. Sequence database [1], is a biological database that stores various protein sequences, nucleic acid sequences, or other polymer sequences. With the advancement in learning, the size of protein database is increasing tremendously. This huge increase in the size of protein data makes it a challenging task to classify this data. Proteins is the most fundamental element of an organism that consists of 20 amino acid molecules. There are many encoding techniques as given in [2] to extract features from a protein sequence. These techniques provides huge data. Feature selection is a dimensionality reduction technique that selects important features from the original feature set. There exist three broad categories for feature selection methods, namely filter, wrapper, and embedded algorithms [3]. In this work, we used the n-gram method for feature construction from the protein sequences. A two-stage feature selection algorithm, consisting of ReliefF and CFS is proposed to select important features from the extracted feature set. The classification models such as Naive-Bayes, decision tree and k-nearest neighbor are used to evaluate this proposed approach. The paper comprises of five sections. In Sect. 2, some previous work related to feature selection and protein sequence classification is discussed. Section 3, explains the proposed approach. In Sect. 4, the experiments carried out and the results obtained are explained. The conclusion and future work is discussed in Sect. 5.

2 Related Work

There exist a number of approaches as described in Saidi et al. [2] to extract feature vector from the given protein sequence data. Iqbal et al. [4], used n-gram feature extraction technique and measured the statistical significance of features using mean and variance of each feature of a superfamily with all other superfamilies. A new approach based on the concept of hashing is introduced in [5, 6]. In these works, a hash function is used that maps high-dimensional vector into low-dimensional vector. Feature selection techniques have been proposed to reduce the dimensionality. Bolon-Canedo [7], covers different feature selection techniques for different microarray datasets. Patil et al. [8], proposed a genetic algorithm-based wrapper method and SVM for feature selection and prediction of sequence data. Dash et al. [9], used four different feature selection techniques in a pipeline fashion with all 24 possibilities. Feature selection using information theory measures has also gained a wide importance. Song et al. [10], proposed a clustering-based approach that uses a metric based on information-gain. Bennasar et al. [11], proposed a joint mutual information (JMI)-based technique. It follows "maximum of the minimum" criterion and selects feature that is most beneficial for the already selected subset.

It can be observed that there exist different methods for feature selection, but these failed in case of sequence data. We need to devise a technique for protein sequence data that can handle the variations and complexity of sequence data.

3 Proposed Method

In this work, a two-stage algorithm is proposed for feature selection. This algorithm employs ReliefF algorithm at first stage and generate a candidate feature set. This candidate feature set is then passed to second stage at which CFS approach is used to select final set of features. In this paper, four datasets, namely DS1, DS2, DS3, and DS4 have been collected from SWISS-PROT [12, 13]. It comprises of total 1327 protein sequences. These datasets vary in size, number of classes, and complexity. The datasets DS1 and DS3 are multi-class dataset. DS2 and DS4 are binary-class dataset. The other details corresponding to these datasets is shown in Table 1. The methodology used for proposed approach is shown in flow diagram in Fig. 1. The details of each module is given in next subsections.

Table 1 Protein sequence dataset

Dataset	Classes	#Sequences
DS1	Hydrogenase nickel incorporation protein HypA, High-potential Iron–Sulfur protein, Glycine dehydrogenase	60
DS2	Melanocortin, Chemokine	510
DS3	Homodimer, Monomer, Homotrimer, Homopentamer, Homooctamer, Homotetramer, Homohexamer	717
DS4	Nonhuman TLR, human TLR	40

Fig. 1 Flow diagram

```
>1MYL:A|PDBID|CHAIN|SEQUENCE
MKGMSKMPQFNLRWPREVLDLVRKVAEENGMSVNSYIYQLVMESFKKEGRIGA
```

Fig. 2 Protein sequence

3.1 Feature Construction

Given a protein sequence data consisting of 20 amino acid molecules, we need to extract feature vector from it. **N-Gram** method parses the sequence data and extracts subsequences consisting of n-characters from that data. It slides a window on the given sequence and the window size is kept equal to n-characters. For 1-Grams, there can be 20 possibilities A, C, D, ..., Y. For 2-Grams, it will be AA, AC, AD, ..., YY, and so on. One example of the sequence is shown in Fig. 2.

3.2 Feature Selection Using ReliefF and CFS

The feature vector obtained here contains 420 features. Among these, there exist irrelevant as well as redundant features that are not going to provide any important or additional information to the class. These features also degrade the performance of the algorithms. Feature selection is a technique that can be used to select only the important features and remove all the unnecessary features from the dataset. The feature selection algorithm proposed here consists of two stages for selection of features involving ReliefF and CFS approach. These approaches are explained as follows.

3.2.1 ReliefF

It is an efficient approach that select features based on how well they can differentiate among its neighboring instances. It uses the two parameters, one is nearest hit and the other is nearest miss. It picks up a random instance i, with class c and find k-nearest neighbors (where k is predefined) from same class called as nearest hits h_i, and similarly neighbors from different classes are considered as nearest misses m_i. A quality parameter (q) is considered for each feature (f). If instance i and h_i instances have different values for feature f, value of q in decreased, and if instance i and those in m_i has different values for feature f, value of q increased. This whole process is repeated multiple times and in the end, features with good quality parameter values are selected.

3.2.2 Correlation-Based Feature Selection (CFS)

It is a filter method, which provides a discriminative set of features based on correlation measure. It considers different subset of features and selects the subset with features that are uncorrelated with each other but highly correlated with class label. The score for a subset S, can be calculated using Eq. 1.

$$Score_s = \frac{n\overline{corr}_{cf}}{\sqrt{n + n(n-1)\overline{corr}_{ff}}} \tag{1}$$

where $Score_s$, represents the calculated score for subset S having n number of features, $\overline{corr}_{ff}$ is the average value of feature-feature inter-correlation, and $\overline{corr}_{cf}$ represents the average value of feature-class correlation.

3.2.3 Class Imbalance

It refers to the problem where total number of data instances that belong to one class is far less or far more than number of data instances that belong to another class. In this case, classification accuracy obtained is generally biased. There exist many approaches like resampling, subsampling, SMOTE [14], that balances the number of data instances for all the classes. These may increase the minority class data instances or decrease the majority class data instances or a combination of both.

In this work, first feature vectors are constructed using n-gram method. After this, ReliefF and CFS approach is applied to select important features. In the end, three classifiers, Han et al. [15], namely Naive-Bayes, decision tree and k-nearest neighbor are used for performance evaluation of proposed technique.

4 Experiment Results and Discussion

In this work, four datasets as described in Table 1, are considered for experiments purpose. Each sequence is read and feature vector is constructed for each sequence using frequency count of each 1-Gram and 2-Gram in the sequence. For datasets DS2 and DS4, resampling technique is used to solve class imbalance problem. The proposed approach is implemented using Weka tool. The data is passed to the ReliefF algorithm for feature selection. The ranker search method is used and number of neighbors to be considered is set to 10. It provides a ranking of all 420 features, out of which top 200 features are selected for the next stage. These 200 features are passed to CFS algorithm. The greedy search algorithm is used along with backward selection to obtain final set of features. For different datasets, different number of features are obtained as shown in Table 2. For this final set, tenfold cross-validation is used for classification purpose. The results for classifiers Naive-Bayes (NB), 3-NN, and decision tree are shown in Table 2. It can be observed that number of features

Table 2 Experimental results

Dataset	Classifier	Accuracy (%) without feature selection	Accuracy (%) using ReliefF (#features = 200)	Accuracy (%) using ReliefF+CFS (#features)	Accuracy (%) using DDSM in [2]
DS1	NB	96	96.4	**100 (14)**	80
	IB3	88.3	90	**100 (14)**	–
	C4.5	91.6	90	95 (14)	96.7
DS2	NB	98.8	99.2	**100 (32)**	100
	IB3	99.8	100	**100 (32)**	–
	C4.5	97.1	98	98.5 (32)	99.4
DS3	NB	45.1	77.3	**76 (7)**	59.4
	IB3	36.8	68.5	**67.7 (7)**	–
	C4.5	35.7	75	76.4 (7)	79.2
DS4	NB	60	87	**98 (16)**	95
	IB3	55	83	**87 (16)**	–
	C4.5	52.5	74.5	76 (16)	82.5

obtained in final set are very less than in original set and accuracy has improved by a considerable amount. The results are compared with substitution matrices feature extraction approach used in [2]. It can be observed that using basic n-gram method, our approach has performed better in most of the cases. Also, classifier performance has improved by using feature selection method.

5 Conclusion

In this paper, feature selection approach based on ReliefF and CFS is introduced. It combines the benefits of both ranker method (ReliefF) and subset method (CFS). The performance is evaluated by use of different classifiers. It can be observed that using this combined approach, results have improved in comparison to classification without feature selection as well as for classification using only ReliefF method. With the increase in number of features, more subsets will be there and it will raise computational complexity. This work can be optimized by using parallel approach for subsets evaluation and combining results from parallel parts to provide one final set of feature.

References

1. Sequence Database. https://en.wikipedia.org/wiki/Sequence_database
2. Saidi, R., Maddouri, M., Nguifo, E.M.: Protein sequences classification by means of feature extraction with substitution matrices. BMC Bioinform. **11**(1), 1 (2010)

3. Ladha, L., Deepa, T.: Feature selection methods and algorithms. Int. J. Comput. Sci. Eng. (IJCSE) (2011)
4. Iqbal, M.J., et al.: Efficient feature selection and classification of protein sequence data in bioinformatics. Sci. World J. **2004** (2014)
5. Caragea, C., Silvescu, A., Mitra, P.: Protein sequence classification using feature hashing. Proteome Sci. **10**(1), 1 (2012)
6. Forman, G., Kirshenbaum, E.: Extremely fast text feature extraction for classification and indexing. In: Proceedings of the 17th ACM Conference on Information and Knowledge Management. ACM (2008)
7. Boln-Canedo, V., et al.: A review of microarray datasets and applied feature selection methods. Inf. Sci. **282**, 111–135 (2014)
8. Patil, N., Toshniwal, D., Garg, K.: Effective framework for protein structure prediction. Int. J. Funct. Inf. Pers. Med. **4**(1), 69–79 (2012)
9. Dash, R., Misra, B.B.: Pipelining the ranking techniques for microarray data classification: a case study. Appl. Soft Comput. **48**, 298–316 (2016)
10. Song, Q., Ni, J., Wang, G.: A fast clustering-based feature subset selection algorithm for high-dimensional data. IEEE Trans. Knowl. Data Eng. **25**(1), 1–14 (2013)
11. Bennasar, M., Hicks, Y., Setchi, R.: Feature selection using joint mutual information maximisation. Expert Syst. Appl. **42**(22), 8520–8532 (2015)
12. Bairoch, A., Apweiler, R.: The SWISS-PROT protein sequence database and its supplement TrEMBL in 2000. Nucleic Acids Res. **28**(1), 45–48 (2000)
13. National Center for Biotechnology Information. http://www.ncbi.nlm.nih.gov
14. Sun, Y., Wong, A.K.C., Kamel, M.S.: Classification of imbalanced data: a review. Int. J. Pattern Recogn. Artif. Intell. **23**(04), 687–719 (2009)
15. Han, J., Pei, J., Kamber, M.: Data Mining: Concepts and Techniques. Elsevier (2011)

An Analysis of the Role of Artificial Intelligence in Education and Teaching

Garima Malik, Devendra Kumar Tayal and Sonakshi Vij

Abstract The contribution of Artificial Intelligence (AI) in the field of education has always been significant. From robotic teaching to the development of an automated system for answer sheet evaluation, AI has always helped both the teachers and the students. In this paper we have done an in depth analysis of the various research developments that were carried out across the globe corresponding to artificial intelligence techniques applied to education sector so as to summarize and highlight the role of AI in teaching and student's evaluation. Our study shows that AI is the backbone of all the NLP enabled intelligent tutor systems. These systems helps in developing qualities such as self reflection, answering deep questions, resolving conflict statements, generating creative questions, and choice-making skills.

Keywords Artificial intelligence · Education · Teaching · Assessment
Intelligent tutor system

1 Introduction

Artificial Intelligence can be described as the association of intelligence with machines, i.e. machines which can exhibit human intelligence and take decision with human capabilities. Its major focus is on creating super advanced systems which can think strategically. Artificial intelligence complements computer science by creating efficient programs that help to develop virtual machines with capabilities of reasoning, problem solving and learning. Machines based on artificial intelligence possess linguistic, mathematical, logical, inter- and intra-personal intelligence. The domain

G. Malik (✉) · D. K. Tayal · S. Vij
Department of CSE, IGDTUW, New Delhi, India
e-mail: annu.2353@gmail.com

D. K. Tayal
e-mail: dev_tayal2001@yahoo.com

S. Vij
e-mail: sonakshi.vij92@gmail.com

© Springer Nature Singapore Pte Ltd. 2019
P. K. Sa et al. (eds.), *Recent Findings in Intelligent Computing Techniques*,
Advances in Intelligent Systems and Computing 707,
https://doi.org/10.1007/978-981-10-8639-7_42

of artificial intelligence covers a wide range of fields such as Natural Language processing, expert systems, fuzzy logic, neural networks and robotics [1]. Combination of natural language processing and artificial intelligence builds the fundamentals of any expert system such as flight tracking systems or medical oriented systems. NLP helps the system to achieve language translation and language generation through which virtual agents can behave in an intended manner according to the environment. Robotics is one of the most significant domains in artificial intelligence that demonstrates the expert robots acting as artificial agents in a real-time environment. AI programs works as core units of robots which enables them to function as a virtual agent in computer—simulated environment. AI promotes independent generation of codes which can be easily modified and it is also trying to address few twenty-first-century skills such as self reflection, self direction and teamwork. The growing importance of AI can be seen as it is deriving more and more complex systems which can successfully understand human speech compete at higher levels of gamification and interpret the interactive media.

Recent work in the education sector is the new mythologies for teaching advancements through teaching agents, courser like web based engines are offering best courses of respective topics to generate interest among students. AI is instructing and assisting a judgement free environment for education in the world. It is aiming to increase the lifelong and life-wide technology by learning outside the school. Global classrooms implementations are leading to increase in interconnectedness and accessibility of classrooms worldwide. In order to construct the intelligent tutor systems AI plays a vital role in tracking the mental steps of learners such as self regulation, monitoring and explanation. It also decides the most appropriate content for the learner. AI is helping in the transformation of tell and practice culture to deep reasoning and learning systems. It also helps in organizing and synthesizing interactive content such as e-books, video lectures, natural games and individual assessments of teaching agents.

This paper describes the emerging role of Artificial intelligence in education which leads to computational creativity and social intelligence in education. Building super intelligent systems in education which can revolutionize the education system like Virtual mentors in education such as Siri in iPhone, Facebook suggestion of friends, Google self driving cars are significant examples of changing lives through artificial intelligence [2]. An automated online shopping assistant minimizes the human efforts to a large extent. AI is also penetrating the medical field bringing doctor assistants as bots or agents. This was seen in the recent Microsoft project "Hanover", based on cancer, in which the bots can track the medical history of cancer patients so that prescription of medicines and cancer-related drugs can be done intelligently [3]. Automotive industry is also hugely affected by AI in today's reality by creating driverless cars resulting disruptive changes in the industry.

In this paper we have surveyed various intelligent tutor systems and major characteristics can be described as all the systems implement virtual agents which can advice, instruct and assist human in decision making, derives a solution, explanatory powers, predicts the result, justify the conclusion and suggests the alternative solution for the problems. New systems can also evaluate the quality of curriculum's and teaching materials. Recommendation engines are leading to adaptive learning and dynamic feedback generation which is totally different from traditional lecture-based methods. Each and every subject can be digitalized even the practical experimentation of subjects like physics, robotics and statistics can be viewed as a recent development in the web based education. These web-based educational applications rely on two principles Learning-By-Doing (LBD) and Learning-By-Teaching (LBT) [4]. Extensive research has been done in this field to significantly improve web-based education. AI plays a key role in designing these systems through semantic web and interoperability of Intelligent Tutor System (ITS) are considered to be efficient and effective for users.

This paper is organized as follows Sect. 2 describes the role of artificial intelligence in education with brief introduction to the papers surveyed by us in two categories, i.e. teaching with AI and Qualitative Assessments with AI. Section 3 represents an exhaustive survey of the papers pertaining to the concepts and significance of various ITS systems listed in the papers. Section 4 will be conclusion and future scope.

2 Role of Artificial Intelligence in Education

This section discusses the significance and contribution of various artificial intelligence techniques adopted for successful teaching and corresponding assessment.

2.1 Teaching with AI

Artificial intelligence is organizing the education systems in strategic manner to provide a digital learning platform with deep learning systems. Interactive graphical representations, enhanced gaming patterns to solve natural problems, teaching through virtual agents, context specific feedback generation, precision in curriculum all these factors are leading to a new era of education. Teaching with AI means learning with simulating agents and incorporating human-inspired tutoring strategies [5].

BEETLE II, Assessment ecosystem, Reasoning Mind Genie 2 and AutoTutor and Family represents intelligent tutor systems with all the capabilities of human intelligence. All the systems includes multiple 3D representation with dynamic graphical user interfaces, various degrees of problems (lower, intermediate and high), logical reasoning platform, analysis of simulated and real data, teaching with virtual agents and fusion of human and computer intelligence with strong and context related decision making qualities.

BELLA, Teachable agents game for primary school children, Robot laboratory for teaching AI for undergraduate students of computer science are all significant examples of introduction of real-time gaming in ITS [4]. In these systems problems are formed in the form interesting adventure games which engages students deep into the problem and increase the learning capabilities.

2.2 Qualitative Assessment with AI

Artificial intelligence techniques and algorithms such as predictive modelling and artificial neural networks introduce advanced techniques in data science to optimize the process of data analytics [6]. Bayesian networks provide more accurate prediction and classifications in recent intelligent tutor systems. Collaborative filtering of data also uses advanced AI principles such as dimensionality reduction and prediction elicitation [7]. Assessments of real-time data obtained from these systems are very important to give user a real-time environment where user can compare the performances with respect to others.

The papers which are surveyed and underlying in this category are as follows:

- Analysing problem solving time.
- Comparison of interleaved and blocked approaches of multiple graphical representation.
- Importance of semantic web in ITS.

These papers are using various prediction and classification techniques to analyse the simulated and real data of the system such as problem solving time of any problem is predicted through user's interaction with the system and the collaborative filtering of data. Comparison of above-mentioned two approaches is done through analysis of data collected from the ITS systems in which they have shown that learning gains were consistent in interleaved approach. Creation of large-scale systems requires robust support from web-based engines and data management. In the process of making web more understandable by machines AI is playing a major role in all aspects of web-based education.

3 Exhaustive Literature Survey

This section presents an exhaustive literature survey as shown in Table 1 to highlight the role of artificial intelligence in various domains.

4 Conclusion and Future Scope

In this paper we have surveyed various research paper broadly based on the role of AI in education field and making of intelligent tutor system for web-based learning environment. We have concluded that the student's learning capabilities have increased through a large extent after the introduction of web-based education. Teachers are also utilizing these platforms in curriculum designing which can benefit all other students who still not using these systems natural math problems in the form of enhanced gaming techniques attracts most of the primary school students to learn basic mathematics concepts in more interactive way. These platform enables students, teachers

Table 1 Literature Survey

S. no	Title of paper	Concept	Significance
1.	"BEETLE II: Deep Natural Language Understanding and Automatic Feedback Generation for Intelligent Tutoring in Basic Electricity and Electronics" [8]	• Developed an Intelligent Tutor System or a learning environment with adaptive feedback generation • A Natural Language Processing and contextual interpreter was used for the same • It basically provides an integrated platform for those students whose conceptions are highly resistant to traditional methods of teaching • It aims to provide an interactive experimentation engine of physics which can increase deep reasoning in students	• Interactive experimentation—it implies that system also consists of dynamic diagrams of electric circuits for better understanding • Reflective dialogue components- • Natural language processing is used to understand the student explanations of problems and hence provide context specific feedback • Tactic labels for the students such as "keep going" "Try again" to encourage the students when they made mistakes in problem solving • Dynamic feedback generation—A dialogue manager us maintained in the system which keeps track of the problem context provides dynamic feedback • Promote self monitoring and self explanatory learning
2.	"The ASSISTments Ecosystem: Building a Platform that Brings Scientists and Teachers Together for Minimally Invasive Research on Human Learning and Teaching" [9]	• Multipurpose platform for teachers, students and researchers • Main agenda is "put the teacher incharge not the computer" • It incorporates content of science, english and statistics • Initially over 50,000 students used this and doubling since last 8 years • It brings teachers and researchers on a same platform so that both can benefit each other explicitly and implicitly • It tries to improve education through scientific research and artificial intelligence • Student and teachers uses the assessment part while the researches uses data mining tools to analyse the data generated while users interact with the system	• Flexible system that allows teachers to use the tool with classroom routine • It allows teachers to write individual assessment such as Question and answers, hints, solutions, web based videos • Gives immediate feedback to students or users • This platform is also used for prediction of state test scores and to judge the emotional state of students • It also includes: teacher skill builder report which signifies the students understanding of concepts • It also derives automated peer review system • Differential Instruction- teacher can decide the difficulty levels of problem given to respective students • EdRank- Randomized controlled experiments is also done to compare the value of different web pages to teach different topics • Placements- computer adaptive tests are also the highlights of the system that uses prerequisite knowledge or skills of students to decide the questions

(continued)

Table 1 (continued)

S. no	Title of paper	Concept	Significance
3.	"Reasoning Mind Genie 2: An Intelligent Tutoring System as a Vehicle for International Transfer of Instructional Methods in Mathematics" [10]	• Intelligent tutor system based on mathematics • As sharing of mathematical information is restricted and professional knowledge is undocumented so this system implements a platform by interviewing human experts to extract their knowledge in field of mathematics • It clearly focuses on curriculum and tries to implement cross-cultural curriculum and basically implemented and tested with Russia and U.S.A • It defines that using international curriculum adoption the way of teaching mathematics can be changed in various countries by sharing of knowledge • They initially interviewed Russian teachers in order to create the model of approach	• The system has the ability to change the classroom configuration by rendering the existing cultural scripts • System comprises of explicit rules and definitions as it considers the importance of theoretical knowledge in mathematics • It also incorporated mathematical precision as statements made in mathematics should be formulated in mathematical way only • Initially it was designed for only 5th grade students • Automaticity through spaced practice–each component of the system contains some automatic skill part • System also facilitates the students by breaking the difficult concept into constituent component • Interactive way of studying mathematics for 5th grade students • Initially 24,000 students used the system in testing period but now it is using in 8-7 states
4.	"Reinforcing Math Knowledge by Immersing Students in a Simulated Learning-By-Teaching Experience" [4]	• It is the model based on Learning-By-Teaching(LBT) especially made for 6th grade students named as BELLA • It is also based on mathematics concepts but made in a slightly different way as it is not build from scratch • It provides new technology such as virtual tutors • It builds up mental model of the human student by observing them interacted with Elle • It has ontology of 500,000 terms of mathematics and 10 million rules and definitions • It simulates a learning agent and user plays the role of teacher and it also introduces new gamification element	• Interaction with the system take place in the context of adventures games • All the problems of mathematics are designed in the form of games which attracts students of all age groups • The system works on currents user model as it describes the state of student in the game • It provides Teacher and administrator interfaces and utilities as well as student utilities • Simulating tutor and user both solves or handle the problem simultaneously so that student or user can feel the real time statistics of game • It is very effective and efficient way of learning and solving game—emergent mathematics problems

(continued)

Table 1 (continued)

S. no	Title of paper	Concept	Significance
5.	"AutoTutor and Family: A Review of 17 Years of Natural Language Tutoring" [11]	• This paper shows the journey of AutoTutor—a Software designed with help of artificial intelligence • It attempts to combine the capabilities of virtual tutor and human tutor • It incorporates 3 strategies 1. Human inspired tutoring strategies 2. Pedagogical agents 3. Natural Language tutoring • It primarily focuses on teaching through discourse as discourse contributes to wide range of new learning activities • The emergence of Autotutor from a simple intelligent tutor system to interactive simulating system with 3D representations • It also used semantic Analysis, authoring and delivery tools and provides robust semantic matching algorithm	• It uses the strategies of human tutor or ideal strategies derived from fundamental research • AutoTutor is indistinguishable when the conversations is evaluated by third person which proves that it is close to a human responsive system which can simulate all different emotions and feeling of a human tutor • It encourages students for answering deep reasoning questions as well as it provides self reflection which helps students to resolve conflicting statements in problems and tries to inculcate the behaviour of generating interesting questions for the specific problem • The dialogue system of Auto Tutor provides Domain—Relevant Dialogue • As it relies on lexicons, syntactic parsers, part-of-speech • It also provide the facility of comparison between students verbal input to Auto tutor's expectations • AutoTutor Lite (ATL) is recent web based advancement which includes all the interactive GUI, spatial orientation of 3D objects and more deeper understanding of science subjects like physics
6.	"A Teachable Agent Game Engaging Primary School Children to Learn Arithmetic Concepts and Reasoning" [5]	• A platform for the younger school children which is based on Arithmetic concepts and mental reasoning • It is primarily a learning environment which focuses on 2 motives Learning-By-Doing Learning-By-Teaching • It also features Teachable agents which are simply computer programs and can be taught by users to behave in a certain manner • It was tested in 9 schools among 443 students from the grade of 2–6 • It basically consists of 48 games which are 2 players game one is teachable agent and other is user and this system helps in achieving deeper levels of learning • It is the amalgamation of motivational capabilities of games and reflective power of virtual agents (tutor) • It runs in two mode show mode—in this users shows agent how to play and try mode-in which users letting the agent try once and then choose to accept and reject the decision of agent	• It implies the best way to promote strategic thinking among students and inventing new computational strategies • Learning environment can encourage students for advanced mathematical thinking and it also helps to increase the choice—making skill • It gives complete advantage over tell and practice approach as the strategies adopted by students are examined through multiple studies • It also studies the change in mental modes of arithmetic before and after the game play • Teachable agent in the system is capable to implement or simulate the peer-tutoring system and questioning system which increases the interpretability of system • Graphical modes of system are mesmerizing for students or users which enables them to play the game • Sometimes student and teaching agent work together on a same problem just to enhance the experience of gaming • It uses regression models for studying the cumulative growth of concepts of students

(continued)

Table 1 (continued)

S. no	Title of paper	Concept	Significance
7.	"Student Modelling Based on Problem Solving Times" [12]	• It implements the system which focuses on the problem solving time in intelligent tutor systems • The model for this paper describes the timing information associated with each and every problem such as Sudoku—it is a puzzle game where solution is not just important but the time user take sot reach the solution is important • It uses both simulated and real data to predict the problem solving time • It relies on the fact that time taken to solve the problem can also be the good factor to judge the goodness of performance of students instead of correctness of problem • This paper suggests the basic and advance model of mathematics which is based on the function that is the relationship between problem solving skill and a logarithm of time to solve a problem is linear • It includes the problem such as graphs and functions, robot and standard programming and automata design	• It facilitates both models basic and advanced as basic model deals with only fixed problem skills whereas advance model deals with multi dimensional skills • It provides recommendation engine based on collaborative filtering as it uses users data or user behaviour to select the next problem • It also compares the problem solving time of different students using data mining tools which increase the natural competitiveness amongst students • It can also predicts the average difficulties of problem and tries to discriminate the problem according to degree of student learning • It also implements the automatic problem selection procedure with the help of recommendation system in ITS • It proves to be a effective platform which relies or extensive research for learning curves as it is difficult to assume the students ability of solving the problem initially so it provides multidimensional skills so that every strategy applied by students to solve the problem can be effectively captured

8.	"How Should Intelligent Tutoring Systems Sequence Multiple Graphical Representations of Fractions'? A Multi-Methods Study" [13]	• This paper shows the comparative study of two approaches in representing the multiple graphical representations
		  1. Interleaved practice
		  2. Blocked practice
		• It basically conducted a research using Fraction tutor— an intelligent tutor system
		• This paper answered the question as how to sequence the multiple spatial and 3D representations
		• It deals with domain-specific multiple graphical representations especially in mathematics
		• It initially tested 230 students to conduct the research on classroom experiences with both the approaches above mentioned
		• Furthermore, it uses Bayesian knowledge to analyse the data collected from the above experiment
		• It describes the learning sciences research on how to sequence the representations for different tasks such as addition and multiplication in mathematics
			• The primary beneficiaries of this projects are designer who designs the respective curriculum for the subjects and for intelligent tutor systems also
			• Results shown in the paper proves that interleaved approach is way better than blocked approach in representing multiple graphical pictures
			• It emphasized on the importance of multiple graphical representations as it enhances student learning skills.
			• It enables the student to recognize the complex representation by breaking it into individual representation and tries to make connections between single representations
			• It also believes in the potential benefits of interleaved approach in adaptive and automatic sequencing of multiple graphical representations.
			• It concludes that interleaved process of representation lays a vital role in deep cognitive learning and it also promotes robust learning in ITS systems
			• In this paper they are also trying to address several issues regarding the designing of syllabus for ITS systems for this some research questions were also analysed doing the process and hence proves that learning gains were effective and considerable in interleaved approach

(continued)

Table 1 (continued)

S. no	Title of paper	Concept	Significance
9.	"Education and the Semantic Web" [7]	• This paper deals with the challenges in improving web based education • This is basically a survey paper which describes the importance of web based semantics in intelligent tutor systems • It gives brief description about ontologies used in semantic web, services, semantic mark-ups and language for the semantic web (e.g.: HTML,XML,) • Ontological engineering is also discussed to provide relevance to semantic web	• This paper suggests various methods which can be seen as potential advancements in the field of Artificial intelligence in education • Summarized the functions of semantic web or web based applications which can provide more adaptivity and intelligence to future ITS systems • It aims to develop reliable and large-scale system which completely utilizes the capabilities of web based services to better provide better user interaction with the system • Implementation of semantic web simply implies that making the web more understandable by machines and its essential features derives the importance in advance web based educational applications
10.	"A robot Laboratory for teaching artificial intelligence" [14]	• This paper proposed a system which is based on robot-centred approach • This systems offers the core syllabus of artificial intelligence and its significance and computer science curriculum • It primarily solve the problem of building virtual agents that receive some kind of natural language as input and output appropriate action based on inputs • This paper shows a robotics laboratory setup which provides students a robotic kit consist of a operating system, robot controller board and control softwares	• It basically motivates and encourage students to learn robotics with concrete hands on exercises • The most fundamental advantage of these laboratories is that it enables student to make physical agents not the virtual one which adds an additional dimension of complexity • It provides unique opportunity for students to tackle many issues in building of robot such as algorithm designing, memory management and domain specific programming • This lab setup is easy to install and operate. All the equipments used in the lab belongs to standard category and easily available • This system solves the problem of curriculum designing of AI for undergraduate students of computer science

and researchers sharing the same space, knowledge and concepts, it further establish cross-culture curriculum worldwide such as an Indian student can learn various courses from Russian teachers via these platforms. Usage of multiple graphical representations leads to better long-term retention of information and better transfer of fundamental concepts. Comparison of different approaches or strategies adopted by the students to solve the problem increase the natural competitiveness and advanced mathematical thinking in primary education. All these expert systems encourage the students to construct new computational strategies to solve the problem correctly. It promotes multidimensional skills of students which lead to increase in deep learning. These tools encourage the web-based learning over traditional lecture-based learning. Usage of voice and speech recognition techniques and natural language processing in tutoring agents helps the students to realize the power of virtual reality in education. Learning from these systems is based on repeated reactivation, i.e., representation of specific knowledge. Interactive experimentation of subjects like physics, chemistry and statistics makes these systems more understandable for students. Setup of robotics laboratory in institutions is a new way introducing physical robots instead of virtual robots which can exhibit all human capabilities. For future generation AI is becoming the need of the hour it is trying to surpass the human perception or intelligence in a more précised way.

References

1. http://www.tutorialspoint.com/artificial_intelligence/
2. http://fortune.com/2016/03/22/artificial-intelligence-nvidia/
3. http://www.theverge.com/2016/9/20/12986314/microsoft-ai-healthcare-project-hanover-canc er
4. Lenat, D.B., Durlach, P.J.: Reinforcing math knowledge by immersing students in a simulated learning-by-teaching experience. Int. J. Artif. Intell. Educ. **24**(3), 216–250 (2014)
5. Pareto, L.: A teachable agent game engaging primary school children to learn arithmetic concepts and reasoning. Int. J. Artif. Intell. Educ. **24**(3), 251–283 (2014)
6. http://www.artificialintelligencealgorithms.com/
7. Devedzic, V.: Education and the semantic web. Int. J. Artif. Intell. Educ. **14**(2), 165–191 (2004)
8. Dzikovska, M., Steinhauser, N., Farrow, E., Moore, J., Campbell, G.: BEETLE II: deep natural language understanding and automatic feedback generation for intelligent tutoring in basic electricity and electronics. Int. J. Artif. Intell. Educ. **24**(3), 284–332 (2014)
9. Heffernan, N.T., Heffernan, C.L.: The ASSISTments ecosystem: building a platform that brings scientists and teachers together for minimally invasive research on human learning and teaching. Int. J. Artif. Intell. Educ. **24**(4), 470–497 (2014)
10. Khachatryan, G.A., Romashov, A.V., Khachatryan, A.R., Gaudino, S.J., Khachatryan, J.M., Guarian, K.R., Yufa, N.V.: Reasoning Mind Genie 2: an intelligent tutoring system as a vehicle for international transfer of instructional methods in mathematics. Int. J. Artif. Intell. Educ. **24**(3), 333–382 (2014)
11. Nye, B.D., Graesser, A.C., Hu, X.: AutoTutor and family: a review of 17 years of natural language tutoring. Int. J. Artif. Intell. Educ. **24**(4), 427–469 (2014)

12. Pelánek, R., Jarušek, P.: Student modeling based on problem solving times. Int. J. Artif. Intell. Educ. **25**(4), 493–519 (2015)
13. Rau, M.A., Aleven, V., Rummel, N., Pardos, Z.: How should intelligent tutoring systems sequence multiple graphical representations of fractions? A multi-methods study. Int. J. Artif. Intell. Educ. **24**(2), 125–161 (2014)
14. Kumar, D., Meeden, L.: A robot laboratory for teaching artificial intelligence. ACM SIGCSE Bull. **30**(1), 341–344 (1998)

A Novel Approach for Forecasting the Linear and Nonlinear Weather Data Using Support Vector Regression

Gaurav Chavan and Bashirahamad Momin

Abstract Since past decades, tremendous amount of data is being generated and transmitted because of Internet of Things revolution. In agriculture as well as many industrial services, close monitoring of surrounding helps in taking effective decisions at near real time. Focus is much upon forecasting weather in such environments. Many algorithms are being utilized for performing predictions. Linear regressions give satisfactory results. Also time-series models tend to give poor results for nonlinear nature of data. To overcome this drawbacks support vector machines as a regression is being used. This paper gives a brief insight into the way support vector machines can be used for prediction which gives much better prediction results for linear as well as nonlinear nature of data. With proper tuning of model and appropriate kernel selection, around 98% of accurate prediction has been achieved.

Keywords Internet of things (IoT) · Support vector machine (SVM) · Support vector regression (SVR) · Kernels · Root mean square error (RMSE)

1 Introduction

Forecasting of weather data is considered as an important and essential investigation in data analytics. Since past decade, Internet of Things (IoT) has emerged and replaced human labour with automation [1]. Sensors are being utilized which are connected over internet by means of which data is collected and transmitted to a central node or server. This data is continuous and time bound. Data after being received at server end is being processed and prediction algorithms are applied to infer future events depending upon present and past records.

G. Chavan (✉) · B. Momin
Department of Computer Science and Engineering, Walchand College
of Engineering, Sangli, Maharashtra, India
e-mail: gauravgchavan91@gmail.com

B. Momin
e-mail: bfmomin@yahoo.com

© Springer Nature Singapore Pte Ltd. 2019   419
P. K. Sa et al. (eds.), *Recent Findings in Intelligent Computing Techniques*,
Advances in Intelligent Systems and Computing 707,
https://doi.org/10.1007/978-981-10-8639-7_43

The time-series data collected can be utilized by numerous structural and time-series model. Also, the concept of neural network as well as Back Propagation Algorithms can be applied to perform predictions. It was developed for forecasting weather and giving information related to cultivating seasonal crops to yield predictions as well [2]. But time-series model cannot perform proper prediction if data is random and nonlinear in nature. Also, regressions like linear or multiple linear regression give forecasting results but with relatively less accuracy [3]. Semi-supervised algorithms such as Support Vector Regression (SVR) can be used to perform prediction over nonlinear data. SVR can be used to perform predictions of weather data for different periodic trends with fewer amounts of forecasting errors.

Section 2 describes Support Vector Regression. Section 3 explains the experimental results of implementing the best tuned SVR algorithm followed by conclusion.

2 Support Vector Regression

The SVM [4] can be utilized for regression purpose also. So in regression, goal is to define the loss function which does not consider the errors, which are situated at certain distance $\in$ of the actual values of data points. The function is often defined as $\in$-intensive loss function. Figure 1 shows linear regression in one-dimension with $\in$-intensive band. The variables compute the cost of the errors on the training samples and are considered to be zero for all samples which are located inside the band.

As of SVM, in regression we use constraints $y_i - (wx_i) + b \leq \in$ and $(wx_i) + b - y_i \leq \in$ to allow for some deviation $\in$ between the eventual targets y_i and the function $f(x) = wx + b$, modelling the data. $||w||^2$ is minimized to penalize over-complexity.

To account for training errors, slack variables ξ_i, ξ_i^* are also introduced. The slack variables is always assigned zero inside the band and progressively increase for the point outside the band (Fig. 2).

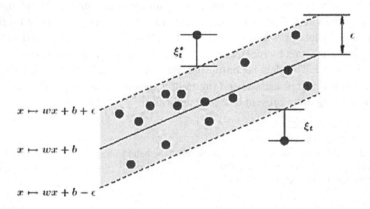

Fig. 1 $\in$-Intensive loss function for support vector regression

Fig. 2 ∈-intensive loss function

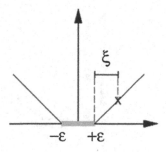

Thus, in linear ∈-intensive loss function, the objective is to minimize

$$min\left[\ ||w||^2 + C\sum_{i=1}^{m}(\xi_i + \xi_i^*)\right]$$ (1)

In support vector regression, the algorithm will output a value that minimizes some error function. However, it considers that for a given minimal distance ϵ, the error is still zero, so the regression line, in terms of error function is more like a tube (∈-intensive tube). Thus, in support vector regression, the input is first mapped onto m-dimensional feature space with the help of some mapping which is nonlinear in nature, and then a linear model is constructed in this feature space.

3 Experiments and Observations

To forecast weather, annual Average temperature of India starting from 1900 till 2014 was used for analysis [5]. Tuning of model is done at training stage in which parameters of SVR like ∈ and Cost C are computed. The standard way of model selection is by doing a grid search which will train a lot of models for the different sets of ∈ and cost, and choose the best one.

While designing model, the γ has to be tuned correctly in order to fit the hyperplane to the data. The hyperplane is going to look like a straight line if γ is small. If γ value is too high, the hyperplane will be curvier which will cause overfitting. Another parameter to be tuned which will improve accuracy is cost C. It decides for the size of the "soft margin" of hyperplane. Points inside this soft margin are not classified. To increase the soft margin, the value for C should be minimal. The ∈-SVR was implemented for kernels of linear, polynomial, sigmoid as well radial basis kernel.

The models were tuned accordingly for different cost; gamma and epsilon values and Root Mean Square Errors (RMSE) were computed for different types of kernels based on training-holdout method. The best kernel was the radial basis kernel with RMSE of 0.21. The model was evaluated using 10-fold cross-validation in which some part of data is removed before training begins. The Fig. 3. shown below gives

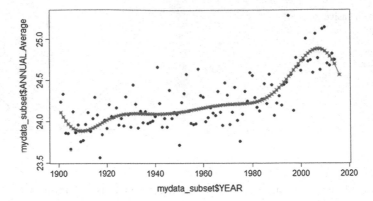

Fig. 3 Tuned SVR plot for weather dataset

tuned SVR plot for weather dataset. As shown, the SVR is much better in modelling nonlinear data points.

4 Conclusion

The present time-series models and regression models show less accuracy when dealing with nonlinear data points. In case of SVR, the nonlinear predictions can be made with ease by mapping data points in higher dimensional feature space. Although various factors like cost, gamma, epsilon should be taken care to find the best SVR model, the prediction errors are significantly reduced. Also, SVR is less complex as compared to unsupervised learning algorithms. But, it takes more computation efforts to find the best tuned model for performing prediction. Also, if data points are constant, then margin cannot be constructed and SVM model cannot be built. For annual weather forecasting in terms of temperature, the best model came out to be the one which uses the radial basis kernel.

As part of future the SVR model will be compared with much evolving Holt-Winter time-series models. Also, the behaviour of SVR will be monitored for forecasting weather depending upon different periodic trends and the results of prediction will be compared with time-series models.

References

1. Gusmeroli, S., Piccione, S., Rotondi, D.: IoT@ Work automation middleware system design and architecture. In: 2012 IEEE 17th Conference on Emerging Technologies & Factory Automation (ETFA). IEEE (2012). Zhang, C., et al.: A multimodal data mining framework for revealing common sources of spam images. J. Multimed. **4**(5), 313–320 (2009)
2. Rani, R.U., Rao, T.K.R.K.: An enhanced support vector regression model for weather forecasting. IOSR J. Comput. Eng. (IOSR-JCE) **12**, 21–24 (2013). Zhang, M.: Application of data mining technology in digital library. JCP **6**(4), 761–768 (2011)
3. Paras, S.M.: A simple weather forecasting model using mathematical regression. Indian Res. J. Ext. Educ. **12**(2), 161–168 (2016)
4. Xue, H., Yang, Q., Chen, S.: SVM: support vector machines. Chapman & Hall/CRC, London, UK (2009)
5. Radhika, Y., Shashi, M.: Atmospheric temperature prediction using support vector machines. Int. J. Comput. Theory Eng. **1**(1), 55 (2009)
6. Kapoor, P., Bedi, S.S.: Weather forecasting using sliding window algorithm. ISRN Signal Process. **2013** (2013)
7. Suykens, J.A.K., et al.: Weighted least squares support vector machines: robustness and sparse approximation. Neurocomputing **48.1**, 85–105 (2002)
8. Smola, A.J., Schölkopf, B.: A tutorial on support vector regression. Stat. Comput. **14**(3), 199–222 (2004)
9. Meyer, D.: Support Vector Machines: The Interface to libsvm in Package e1071 (2004)
10. Karatzoglou, A., Meyer, D., Hornik, K.: Support Vector Machines in R (2005)
11. Jain, R.K., et al.: Forecasting energy consumption of multi-family residential buildings using support vector regression: investigating the impact of temporal and spatial monitoring granularity on performance accuracy. Appl. Energy **123**,168–178 (2014)
12. Yang, H.-T., et al.: A weather-based hybrid method for 1-day ahead hourly forecasting of PV power output. IEEE Trans. Sustain. Energy **5**(3), 917–926 (2014)

An Efficient Approach for Privacy Preserving Distributed K-Means Clustering in Unsecured Environment

Amit Shewale, B. N. Keshavamurthy and Chirag N. Modi

Abstract In this paper, we propose an efficient approach for privacy preserving distributed k-means clustering in unsecured horizontally distributed data. We use an elliptic curve cryptography to offer data privacy and security against involving sites and an external adversary. We analyze the proposed approach in terms of privacy, security, communication, and computational cost.

Keywords Data mining · Distributed k-means clustering · Privacy preservation Elliptic curve cryptography

1 Introduction

Nowadays, data mining has variety of applications in different areas like statistical analysis, marketing analysis, prediction of trends in different businesses, etc. In recent years, privacy preservation for individuals has been one of the biggest concerns in data mining [1]. Data mining techniques are applied on centralized data environment as well as distributed data environment. In centralized data environment, data is stored at the central location and data mining techniques are applied on these stored data to derive meaningful information while in distributed data mining scenario, the data is stored at different sites due to privacy reasons. Here, mining techniques are applied locally on each individual site and local information is shared among different sites to derive the global results. In distributed environment, each site holds different types of data distribution. It can be horizontally distributed data, where

A. Shewale (✉) · B. N. Keshavamurthy · C. N. Modi
Department of Computer Science and Engineering, National Institute of Technology Goa,
Ponda 403401, India
e-mail: Shewaleamit22@gmail.com

B. N. Keshavamurthy
e-mail: bnkeshav.fcse@nitgoa.ac.in

C. N. Modi
e-mail: cnmodi@nitgoa.ac.in

© Springer Nature Singapore Pte Ltd. 2019
P. K. Sa et al. (eds.), *Recent Findings in Intelligent Computing Techniques*,
Advances in Intelligent Systems and Computing 707,
https://doi.org/10.1007/978-981-10-8639-7_44

every site has same number of attributes but different number of transactions. In case of vertically distributed data, each site has different set of attributes but same number of transactions. In addition, there exists arbitrary partitioned data model [2] which generalizes above two models where different attributes of different entities in the same database are owned by various sites.

In distributed environment, each site sends its information to other sites to compute global mining results. The existing approaches assume the secure communication channel among the involving sites. However, the communication channel is not secured in real-world scenario. Thus, an external adversary can read the channel and can affect the mining results. In this paper, we propose an elliptic curve cryptography (ECC) approach for privacy preserving k-means clustering in distributed environment. We have used Elliptic Curve Diffie-Hellman (ECDH) protocol which is used for key sharing and Elliptic Curve Digital Signature Algorithm (ECDSA) for authentication. This approach provides security and privacy against both involving sites and external adversaries.

Rest of this paper is organized as follows: Sect. 2 discusses background work done in privacy preserving k-means clustering. In Sect. 3, the proposed approach is discussed in detail. In Sect. 4, security and performance analysis of proposed approach is given. Finally, Sect. 5 concludes our work with references at the end.

2 Background and Related Work

K-means clustering [3] is one of the most significant mining technique. Many privacy preserving k-means clustering algorithms are proposed on different data distributions such as horizontal and vertical partitioning. Consider two parties named as Alice and Bob. They are having some data samples. Each party wants to compute clusters jointly without disclosing their information to each other. Consider that, there is a third party. Each party sends the sum of data points of each cluster and total number of samples of each cluster to TP which calculates the global mean and sends the mean values to parties. This is an iterative approach so at each iteration, new cluster mean is calculated by communicating with third party. However, in unsecured environment, the involving parties can reveal the information of each other. Thus, there is need of privacy of individual party's data.

Consider data D is horizontally distributed among different sites. Let the sites be $S_1, S_2, \ldots, S_n$. These sites want to compute the global clusters so that only final result can be seen by all sites without compromising any other site's data. It should have following objectives: (1) Any involving site should not be able to reveal the data or information of other sites. (2) External adversaries should not be able to affect the security and privacy of any site's data. (3) It should have an affordable computational and communication cost.

The goal of privacy preserving data mining (PPDM) is to protect the privacy of the individual site's information while deriving global mining results. The first work of PPDM was proposed in [4, 5]. In [4], privacy is maintained by randomization on the dataset. In [5], secure multiparty computation [6, 7] model is used which is very rigorous in nature. Later, many theories were put forth on the issue of PPDM [8, 9] in order to generalize the result for data mining techniques. Clustering is the technique [10] that has not taken actual part in the previous works [10]. The most prominent approach in clustering is k-means algorithm which is studied and analyzed thoroughly in the context of preservation of privacy [11]. Oliveira and Zaiane [12, 13] have proposed several data transformation techniques that enables data owner to share the mined data with other sites. In [14, 15], data holder parties build local model and then third party generates global model from local models. Further, clustering is done using the global model. The privacy issue of clustering is important as compared to other data mining techniques. Vaidya and Clifton [16] have given a solution for vertically partitioned data on multiparty scenario. Here, sites are kept as confidential. Distances between the data points are calculated using homomorphic schemes [17]. Secure distance comparison is achieved by Yao in [18]. Samet et al. [19] gives an algorithm that preserves privacy on vertically distributed data using new comparative primitive based on secure multiparty computation (SMC) and secure sum which is computing sum of the distances. In horizontally distributed data, distance computation between the sites does not affect the privacy since each party holds all attributes of the entity. Jha et al. [20] have proposed protocols for privacy preservation in k-means algorithm. Each party's entities are kept secret. The intermediate cluster centers are revealed to parties and distance is calculated locally by each party. Security primitives such as oblivious polynomial evaluation and homomorphic encryption are used in order to preserve the privacy of each site. Samet et al. [19] have proposed secure division method. This approach can be used for multiparty scenario but intermediate centers are always revealed.

In existing approaches, communication channel among involving sites is assumed as secure. However, in real time this assumption is invalid. An external adversary can affect the security and privacy of involving sites by monitoring communication channel among them.

3 Proposed Approach

The objective of proposed approach is to derive global clustering result in the unsecured horizontally distributed data environment while preserving the data privacy of the involving sites against other sites and external adversaries. In addition, it should have low computational and communication cost.

In the proposed approach, we use an Elliptic curve cryptography (ECC) [21] as it requires much shorter key length than that of RSA and offers same security level as that of RSA. We have used ECDH protocol for sharing the keys and ECDSA for the authentication and verification of the involving sites.

Assume that there exists n sites and a third party (TP) as shown in Fig. 1. TP computes the global clustering result. The data points at each site are clustered using k-means algorithm. Each site produces its own local clusters and final global clusters are generated after applying the proposed algorithm. In this approach, each site shares sum and count as information to other sites. The proposed approach works as follows:

(1) As shown in Fig. 1, Sites S_i ($1 \le i \le n$) share a key K_i ($1 \le i \le n$) respectively with TP by using ECDH key sharing protocol.

(2) Site S_1 encrypts the sum of its data points by adding Key K_1 and sends {Sum_1 $+ K_1$} to site S_2 after digitally signing it by using ECDSA. This is shown as (1) in Fig. 1. Site S_2 verifies the digital signature and adds its own encrypted sum to the received data and sends digitally signed {$Sum_1 + K_1 + Sum_2 + K_2$} to Site S_3. This step is marked as (2) in the Fig. 1.

(3) Similar to Site S_2, Site S_3 verifies the signature and adds its encrypted sum {$Sum_3 + K_3$} and sends digitally signed {$Sum_1 + K_1 + Sum_2 + K_2 + Sum_3 + K_3$} to TP.

(4) TP verifies the signature and calculates global sum by subtracting the shared keys K_1, K_2, K_3. Thus, TP gets the global sum as below:

$$G_{Sum} = \{Sum_1 + K_1 + Sum_2 + K_2 + Sum_3 + K_3\} - \{K_1 + K_2 + K_3\}$$

(5) Now, site S_1 encrypts the count of all data points of cluster 1 of site 1 by adding Key K_1 and sends digitally signed {$Count_1 + K_1$} to site S_2. This is shown as (4) in Fig. 1.

(6) Site S_2 verifies the digital signature and adds its own encrypted count to the received data and sends {$Count_1 + K_1 + Count_2 + K_2$} to Site S_3. This step is marked as (5) in the Fig. 1.

(7) Similar to Site S_2, Site S_3 verifies the signature and adds its encrypted count {$Count_3 + K_3$} and sends digitally signed {$Count_1 + K_1 + Count_2 + K_2 + Count_3 + K_3$} to TP.

(8) TP verifies the signature and calculates global count by subtracting the shared keys K_1, K_2, K_3. Thus, TP gets the global count as below:

$$G_{Count} = \{Count_1 + K_1 + Count_2 + K_2 + Count_3 + K_3\} - \{K_1 + K_2 + K_3\}$$

(9) TP calculates global mean of data points using G_{Sum}/G_{Count} and shares with all the sites.

(10) All the sites find distance between their data points and global mean values. According to the distance, they generate new local clusters using k-means algorithm and follows steps (5)–(10) iteratively until there is no change in mean values.

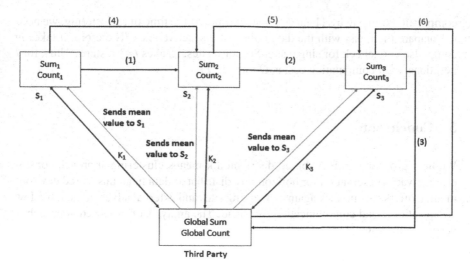

Fig. 1 Proposed approach for privacy preserving distributed k-means clustering

4 Analysis of the Proposed Approach

In the proposed approach, each site sends the encrypted and digitally signed sum and count of the data to the neighbor site instead of original data. Hence, the site which receives this information is unable to know the actual content. Also, each site sends encrypted data to other site by using the keys which are shared with TP, so any other site cannot predict the original value. Even TP cannot know the individual site's data as it gets the encrypted sum of data from the last site. Therefore, this approach offers privacy against involving sites and TP.

If an external adversary tries to read the communication channel between the sites, it cannot alter the original contents as information sent to next site is digitally signed using ECDSA. In addition, an adversary cannot affect the privacy of the information since it is in encrypted form. Thus, it offers the privacy against external adversaries.

Let the number of sites participating be n, number of initial clusters required be k and I be the number of iterations for k-means algorithm. Initially, all sites share keys with TP. This will take n communications. For all sites to send sum to neighboring site is $k * n$ since n communications for each k clusters. Similarly, for sending count to neighboring sites, $k * n$ communications are required. TP sends the calculated means to all sites which takes n communications. Thus, communication overhead for one iteration is $(n + k * n + k * n + n)$. These number of steps continue for I iterations until we get the same cluster assignments. Thus, overall communication overhead is $O(n * k * I)$.

Let the number of sites be n and m be the number of data samples in each site. Let I be the number of iterations and k be the number of clusters required for k-means algorithm. The computation of sum by each party takes constant time. Similarly, the computation of mean by TP takes constant time. At the time of reclustering, we need to compare the mean with the data points of respective sites. Therefore, it takes m comparisons to check for single site. So, for n sites, it takes $m * n$ steps. Thus, for I iterations, total computation cost is $O\ (m * n * I)$.

5 Conclusions

We have proposed an ECC-based distributed k-means clustering approach considering privacy preservation for horizontally distributed data in an unsecured environment. It offers data privacy against involving sites and external adversaries. It has low computational and communication overhead. The analysis of proposed approach is very encouraging.

References

1. Borhade, S.: A survey on privacy preserving data mining techniques. Int. J. Emerg. Technol. Adv. Eng. **5**(2) (2015)
2. Jagannathan, G., Wright, R.N.: Privacy-preserving distributed k-means clustering over arbitrarily partitioned data. In: Proceedings of the Eleventh SIGKDD International Conference on Knowledge Discovery in Data Mining, pp. 593–599 (2005)
3. Lloyd, S.: Least squares quantization in PCM. IEEE Trans. Inf. Theory **28**(2), 129–137 (1982)
4. Agrawal, R., Srikant, R.: Privacy-preserving data mining. SIGMOD Rec. **29**(2), 439–450 (2000)
5. Lindell, Y., Pinkas, B.: Privacy preserving data mining. In: Annual International Cryptology Conference, pp. 36–54 (2000)
6. Goldreich, O.: Draft of a chapter on general protocols. In: Foundation of Cryptography. The Press Syndicate of the University of Cambridge (2003)
7. Goldreich O., Micali S., Wigderson A.: How to play any mental game—a completeness theorem for protocols with honest majority. In: Proceedings of the Nineteenth ACM Symposium of the Theory of Computing, pp. 218–229 (1987)
8. Vaidya, J.: A survey of privacy-preserving methods across vertically partitioned data. In: Privacy-Preserving Data Mining, pp. 337–358 (2008)
9. Verykios, V.S., Bertino, E., Fovino, I.N., Provenza, L.P., Saygin, Y., Theodoridis, Y.: State-of-the-art in privacy preserving data mining. ACM SIGMOD Rec. **33**(1), 50–57 (2004)
10. Jain, A.K., Murty, M.N., Flynn, P.J.: Data clustering: a review. ACM Comput. Surv. (CSUR) **31**(3), 264–323 (1999)
11. MacQueen, J.: Some methods for classification and analysis of multivariate observations. In: Proceedings of the Fifth Berkeley Symposium on Mathematical Statistics and Probability, vol. 1, no. 14, pp. 281–297 (1967)
12. Oliveira, S.R., Zaiane, O.R.: Achieving privacy preservation when sharing data for clustering. In: Workshop on Secure Data Management, pp. 67–82 (2004)
13. Oliveira, S.R., Zaïane, O.R.: Privacy preserving clustering by data transformation. In: Proceedings of the Eighteenth Brazilian Symposium on Databases, pp. 304–318 (2003)
14. Merugu, S., Ghosh, J.: Privacy-preserving distributed clustering using generative models. In: Proceedings of the Third IEEE International Conference on Data Mining, pp. 211–218 (2003)

15. Klusch, M., Lodi, S., Moro, G.: Distributed clustering based on sampling local density estimates. In: Proceedings of the Eighteenth International Joint Conference on Artificial Intelligence, pp. 485–490 (2003)
16. Vaidya, J., Clifton, C.: Privacy-preserving k-means clustering over vertically partitioned data. In: Proceedings of the Ninth ACM SIGKDD International Conference on Knowledge Discovery and Data Mining, pp. 206–215 (2003)
17. Du, W., Atallah, M.J.: Privacy-preserving cooperative statistical analysis. In: Proceedings of the Seventeenth Annual Computer Security Applications Conference, pp. 102–110 (2001)
18. Yao, A.C.C.: How to generate and exchange secrets. In: Proceedings of the Twenty-seventh IEEE Annual Symposium on Foundations of Computer Science, pp. 162–167 (1986)
19. Samet, S., Miri, A., Orozco-Barbosa, L.: Privacy preserving k-means clustering in multi-party environment. In: Proceedings of the International Conference on Security and Cryptography, pp. 523–531 (2007)
20. Jha, S., Kruger, L., McDaniel, P.: Privacy preserving clustering. In: European Symposium on Research in Computer Security, pp. 397–417 (2005)
21. Koblitz, N.: Elliptic curve cryptosystems. Math. Comput. **48**(177), 203–209 (1987)

A Dynamic Trust Model Based on User Interaction for E-commerce

Shashi Shreya and Kakali Chatterjee

Abstract Evaluating trust of customer of e-commerce websites is an important paradigm as per their privacy is concerned. Here the division of the level of privacy is taken which we conform by the trust value. This paper provides three categories of customers based on their trust value. Trust value is calculated based on the customer behavior of interaction to the corresponding e-commerce site. Here interactions focus the weightage based on the factors like utilization factor, unauthorized access, and access time. In this paper, we have proposed a trust evaluation framework based on fuzzy technique. In this framework after identifying the set of customers, their trust degree and weight is calculated. On their basis, final trust value is generated. This generated value divides our customer in three different level for providing privacy preservation.

Keywords Trust value · Fuzzy-based calculation · Interaction-based trust evaluation

1 Introduction

Today the scenario of the world is dealing with advance digitalization which is proliferating from past few years [1]. For this reason, the business-to-consumer (B2C) service model of e-commerce are emerging as a boom. These e-business is doing dramatical development of the living standards and comforts of human being. There are various B2C websites competing each other for their growth in the market. This requires analyzing the market growth requirements in B2C industries. B2C includes online retailing as well as services such as cash on delivery (COD) and online banking. This new way of doing online shopping which is entirely different from traditional shopping, like going out physically and buying things. E-commerce websites first

S. Shreya (✉) · K. Chatterjee
Department of Computer Science and Engineering, National Institute
of Technology Patna, Patna, Bihar, India
e-mail: shashishreya88@gmail.com

© Springer Nature Singapore Pte Ltd. 2019   433
P. K. Sa et al. (eds.), *Recent Findings in Intelligent Computing Techniques*,
Advances in Intelligent Systems and Computing 707,
https://doi.org/10.1007/978-981-10-8639-7_45

need to maintain the participants and try to attract more and more customer. To achieve this, e-commerce websites must provide a good strength privacy, baits, offers, deals, etc., to the customer. But to do so the service provider must conform customer's loyalty and dedication to the website. There are various ways of conformation. In this paper, our concern is the trust value of each customer in this paper. Per 5% customer loyalty growth rate will promote 25–85% average profit gains in the industry [1]. Recently in a study, consumer behavior explains that the growth of e-commerce is primarily depended on the repeat purchase, that somewhere shows the loyalty and satisfaction of the consumer. The complete destruction of trust in e-commerce applications may cause business operators to be prudent and forgo the use of the Internet to the client and revert the traditional methods of doing business [2]. Another drawback found is that, personal data is flowing from one node to another for various purpose. For this, privacy enforcement is mandatory for protection of data for the customer concern and trust. In this direction, we have proposed a framework for calculating the trust of the customer and categories them in different level to provide privacy preservation. The calculation is mainly based on customer interaction for given period from where weightage and trust degree is calculated and then final trust value is generated. This value divides three level of customer and provide the three different privacy enforcement rule.

2 Related Work

This section discusses concepts that are used in our research such trust model, different types of customer, fuzzy-based techniques. Many trust models had been proposed to reduce complexity and uncertainty for transections and relationship with customer in e-commerce. Some of the existing trust are

Reputation-based trust:
These types of trust based on item reputation in the market. It depends on the demands or feedback of the product by the people [3].
Cognitive-based trust:
Trust is based on rational and behavior of the user [4].
Emotional-based trust:
It helps customer to go beyond the available evidence or you can say that based on the reviews just for their feelings and loyalty [4].
Feedback-based trust:
Trust is based on feedback of the customer.
Institutional-based trust:
It is within the organization, that is providing an environment which helps to establish cooperative behavior [5].

If a vendor interacts with its customers, then only demonstrate to the customer that they are benevolent, competent, honest, and predictable in nature. The interaction in between them provides the customer evidence that the vendor has positive attributes, and hence it is strengthening trust beliefs [6]. Lack of trust is the important deterrent. It is one of the fundamental requirements for establishing online exchange relationships [7]. Bella et al. [8] proposed an approach of exhaustive representation for the privacy enforcement technologies. His approach has explained the brief outline about the approaches that deals with main paradigm to conserve the privacy from being public over the network. First case is that, customer's trusting the audience for conservation of their privacy and the second one depends on the customer's anonymity so that they become unknown to the public network. They have shown these two paradigms are totally opposite to each other, but still they balance between them for improving the privacy techniques. Schunter and Waidner et al. [9], design a suite of protocols that helps the customer to maintain their private data throughout the network of trusted nodes. First, they make sure about the nodes that must be trusted and after that user will be conformed for any customer's privacy policy directed by the service provider. Ahmadi [10], proposed an approach where he describes e-customer behavior in B2C relationships and as per their behavior they form a technique for evaluating B2C e-customer importance. They have overcome the problems related to simple net present value (NPV) based CLV models. The problems that they dealt with is big amount of customer data and market risks along with the good value for CLV. Firstly, they illustrate about behavior or activity of customer in e-relationships on behalf of B2C. In B2C model, the customer buys products and termed as consumer. They consume the trade that are less in amount than the trade in B2B models or they have mentioned as relationships, without any predefined contract. But the number of customer is more in B2C as compared to that B2B relationships. So, to make a simplified model, they divided the customer into different types depending on average time interval period between two sequential purchases. They have improved the CLV equation for more desirable results. Folorunso and Mustapha [11] has given a fuzzy framework of trust metric calculation with the help of TBAC. In his approach, Trust Based Access Control (TBAC) based on fuzzy expert system is used for enhancing the computation quality in crowdsourcing environment. Crowdsourcing environment contains two group of users, i.e., requester and workers. Here we are focusing mainly on the trust base access control model which includes the finite set of worker $W_r = \{wr1, wr2, \ldots wrn\}$ and classify them in highly skilled, skilled, semiskilled and unskilled worker with their trust degree (Td) which value is derived from trust metric, i.e., $td \in Td; td\{high, average, low, verylow\}$ [11]. Trust value is given after every visiting of host. The trust metric has an equation that computes trust value to each worker.

3 Proposed Approach

The proposed approach provides the trust value of the consumers. First, our target is to calculate this trust value in bulk, that is weekly, monthly or yearly. Second, we categories the consumer with high level, medium level, and low level. In our

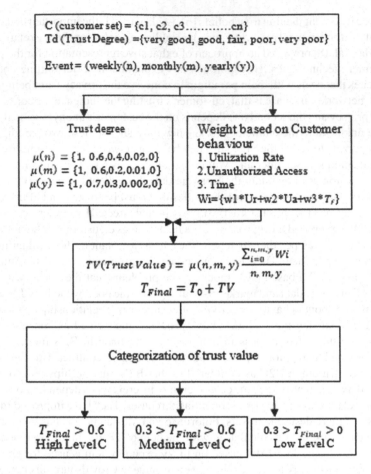

Fig. 1 Block diagram of trust generation

approach, we use fuzzy techniques. Here, a framework is designed to calculate the trust based on interactions (Fig. 1).

Following steps are involved in the framework:

Step 1. **Identify** the set of customers

$$C = \{c1, c2, c3 \ldots cn\}$$

Step 2. Calculate Trust degree
Trust degree (td) of customer is calculated based on the range decided by the line formula, which classify them in linguistic form.

$$Td = \{very\ good,\ good,\ fair,\ poor,\ very\ poor\}$$

Table 1 Trust degree of customer evaluated weekly

Weekly	Days's range	Trust degree	Fuzzy set
	$n > 50$	1	Very good
	$49 \geq n \geq 30$	0.6	Good
	$29 \geq n \geq 20$	0.4	Fair
	$19 \geq n \geq 2$	0.02	Very poor
	$n < 2$	0	Poor

Table 2 Trust degree of customer evaluated monthly

Monthly	Days's range	Trust degree	Fuzzy set
	$n \geq 200$	1	Very good
	$200 > n \geq 130$	0.6	Good
	$130 > n \geq 50$	0.4	Fair
	$50 > n \geq 2$	0.02	Very poor
	$n < 2$	0	Poor

Table 3 Trust degree of customer evaluated yearly

Yearly	Days's range	Trust degree	Fuzzy set
	$n > 600$	1	Very good
	$600 > n \geq 450$	0.6	Good
	$420 > n \geq 200$	0.2	Fair
	$200 > n \geq 2$	0.02	Very poor
	$n < 2$	0	Poor

Approach takes trust degree based number of interaction done and computes the fuzzy value. We are dividing these range with the help of line slop formula.

$$\text{Line Formula:} \ \frac{y2 - y1}{x2 - x1} = \frac{y - y1}{x - x1}$$

Here Y **gives** the value provided, that is μ. Tables 1, 2, and 3, provide the crisp range of the trust degree. Variable **n** is the number of interaction (weekly/monthly/yearly).

Step 3. **Calculate Weight of the interaction of customer**

Weight is also provided, so that proper interaction is counted and ignore the frequent visiting's only. This is because there may be a case that some person join as customer just for malicious activity. Apart from that other cases are also considered like customer interact to the website just for few seconds and logout repeatedly. Therefore, we are considering few factors such as Utilization factor, Unauthorized access, and timing of online shopping. Here we are discussing about these three factors. First of them are Utilization factor which defines that how many hours are utilized by the customer within the fixed number of hours.

$$U_f = \frac{Utilization\ hour}{fixed\ hour}$$

' that refers to the malicious action taken by the consumer. This action either be yes or no and provide fuzzy value.

$$U_a = \begin{cases} 0, if\ access > 3 \\ 1, if\ access \leq 3 \end{cases}$$

And the third factor is Time, this specify, in which time of 24 h of a day customer interacted the website. Time is important because website have their maintenance time, in our approach we have considered 2 h from 3 A.M to 5 A.M in the morning. At the time of maintenance hour, no trust value will be provided.

Now we calculate the final T_f to get the overall weekly or monthly or yearly value of time factor for the customer.

$$T_f = \frac{\sum Time\_Fuzzy\ Value\ of\ particular\ customer}{Fixed\ hour}$$

Finally, the weightage is given as follows:

$$W_i = w1 \times U_r + w2 \times U_a + w3 \times T_f$$

Here w1, w2, w3 are the initial weight provided to each factor.

Step 4. **Trust Value generation**

$$TV\ (Trust\ Value) = \mu(n, m, y)\frac{\sum_{i=0}^{n,m,y} Wi}{n, m, y},$$

$$T_{Final} = T_0 + TV$$

where $T_0 = 0.1$ is, the initial trust provided by the service provider, $\mu(n, m, y)$ is the fuzzy value of number of interaction done weekly, monthly, or yearly, and n, m, y represents the number of days interacted.

Step 5. **Categorization of customer**

Customer are then categories in three different level. Table 4 shows the customer level. According to our approach high level customer are very important to the service provider whereas low level customer is not so important (Table 5).

Table 4 Fuzzy value based on time interval

Time interval	Meridiem	Time fuzzy value
3–5	A.M	0
5–10	A.M	0.25
10–4	P.M	0.5
4–3	P.M	1

Table 5 Level of customer

Levels	Final trust value
High	$T_{final} > 0.6$
Medium	$0.3 > T_{final} > 0.6$
Low	$0.3 > T_{final} > 0$

Table 6 Primary database containing basic details of customer

Customer name	Reg_NO	Registration date	State	Phone number
John	001	1-1-13	Delhi	12345679
Alice	002	1-1-14	Mumbai	12345689
Bob	003	1-1-15	Kolkata	12345916
Micky	004	1-1-16	Goa	12345657
April	005	1-1-17	Kerala	12345679

4 Illustration

This section illustrates a scenario following the steps for calculating the trust value and then customer are kept in their respective categories. We have a primary database or you can say it as parent table as shown in Table 6. This parent table shows the basic details of customers.

We also have another table which is secondary database or you can say it as child table, this table contain the internal details like utilization rate, customer's unauthorized access, time of their interaction converted in fuzzy range as per Table 4.

Our scenario taken here is to calculate the trust value of *John* and place him in the categories as specified by the Table 4. John has registration number 001. Here registration number tuple is the primary key which links both the table. With the help of secondary table, we will now calculate the trust value step by step,

Step 1 Let the customer be john.

Step 2 John's number of interaction in the week as per Table 7 is 25. Now according to Table 1, we provide him fuzzy range as 0.4.

Step 3 Weight calculation

1. *John* utilization hour of the week = 70 h

$$U_f = \frac{70}{154} = 0.45,$$

Table 7 Secondary database containing internal details

Reg_No	Timing	Total hour	Time (fuzzy value)	U_a	Transection status
001	5–10 A.M	2	2.5	Yes	No
001	5–10 A.M	3	2.5	No	No
003	4–3 P.M	3	1	No	No
001	10–4 P.M	1	0.5	Yes	Yes
002	5–10 A.M	4	2.5	No	Yes
004	3–5 A.M	2	0	Yes	No
001	10–4 P.M	2	0.5	No	Yes
005	5–10 A.M	33	2.5	No	No
005	5–10 A.M	33	2.5	No	No
001	10–4 P.M	2	0.5	No	Yes
001	5–10 A.M	3	2.5	No	Yes
001	5–10 A.M	3	0.5	No	Yes
002	5–10 A.M	2	2.5	Yes	No
002	3–5 A.M	1	2.5	No	No
001	10–4 P.M	3	2.5	No	Yes
001	10–4 P.M	5	0.5	No	Yes
001	5–10 A.M	2	1	No	No
001	3–5 A.M	3	2.5	No	No
001	10–4 P.M	1	2.5	No	No
001	10–4 P.M	3	2.5	No	Yes
001	3–5 A.M	2	2.5	No	Yes
001	10–4 P.M	2	2.5	Yes	Yes
003	10–4 P.M	1	0	No	No
003	5–10 A.M	1	0.5	No	No
001	10–4 P.M	5	0.5	No	No
004	10–4 P.M	2	2.5	No	Yes
001	5–10 A.M	5	0.5	No	No
001	10–4 P.M	4	0.5	No	No
001	10–4 P.M	3	0.5	No	No
005	10–4 P.M	1	1	No	Yes
005	4–3 P.M	1	1	No	Yes
001	4–3 P.M	4	1	No	No
002	4–3 P.M	1	1	No	Yes
003	4–3 P.M	1	0	No	No
001	3–5 A.M	1	1	No	Yes
001	4–3 P.M	1	0.5	No	Yes
001	10–4 P.M	4	1	No	Yes
001	4–3 P.M	3	1	No	Yes
001	5–10 A.M	3	2.5	Yes	No

Where 154 is the total fixed hour in a week excluding the maintenance hours.

2. Unauthorized access done by *John* is 3 times, therefore fuzzy value assign to him is 1 as per the range illustrated above.

3. Time fuzzy value for the entire week to *John* as per Table 4 is calculate. From Table 7, we can see that *John* 10 interaction done between 5 A.M to 10 A.M, 10 interactions are done between 10 A.M to 4 P.M and 5 interactions between 4 P.M to 3 A.M.

   Therefore, final calculate time fuzzy value will be

$$T_f = \frac{\sum 2.5 \times 10 + 0.5 \times 10 + 1 \times 5}{154} = 0.23$$

4. Now W_i is calculated by putting all the values with w1, w2, w3 as 0.1:

$$W_i = \frac{0.45 + 1 + 0.23}{25} = 0.0672$$

Step 4 We than calculate the Trust value

$$\text{TV(Trust Value)} = 0.4 \times 0.067 = 0.2688$$
$$T_f = 0.25 + 0.2688 = 0.5188$$
$$T_f = 0.52$$

Step 5 We get to know that final trust value of John is 0.52 which comes in the category of medium level customer as per the Table 5. Therefore, John is the customer of medium valued customer.

5 Analysis

We have analyzed that trust is often mentioned reason in B2C. For promoting the trust in the e-commerce website, different trust models have been built. Some of them are discussed above in the paper. From the survey, we find that most of the trust model are built as items specific or organization specific. Our approach is related one of the most important factor, that is human factor. We can build many trust models related to human, among which activity of human on website is totally unpredictable. Their activity is entirely depending upon their perception. With our approach, we can track the interaction and plot trust to the customer. Here graph is plotted for showing john interaction done weekly (Fig. 2).

Fig. 2 Weekly activity graph of john

6 Conclusion

We have formed a trust model which have effective method that ensures the customer's priority in the eyes of the service provider. Here the evaluation or priority check is done based on the human behavior over the website. This priority check is nothing but the trust value of the customer. Our approach provides the simple and effective methods of trust calculation. This model familiarizes or you can say brings into light about the time factor, unauthorized access and period interval to reflect the activity of user behavior. With the help of these factors, reliability of these user is depicted. The depiction can be form many scenarios that helps to make our model more realistic.

Acknowledgements This research was supported by Information Security Education and Awareness Project (ISEA Project Phase-II) funded by Ministry of Electronics and Information Technology, Government of India.

References

1. Chern, Y., Tzeng, G.-H.: A consumer e-loyalty assessment model: B2C service management by fuzzy MCDM techniques. In: 2012 7th International Conference on Computing and Convergence Technology (ICCCT). IEEE (2012)
2. Marchany, R.C., Tront, J.G.: E-commerce security issues. In: Proceedings of the 35th Annual Hawaii International Conference on System Sciences, 2002 (HICSS). IEEE (2002)
3. Xiong, L., Liu, L.: Peertrust: supporting reputation-based trust for peer-to-peer electronic communities. IEEE Trans. Knowl. Data Eng. **16**(7), 843–857 (2004)
4. Basso, A., et al.: First impressions: emotional and cognitive factors underlying judgments of trust e-commerce. In: Proceedings of the 3rd ACM conference on Electronic Commerce. ACM (2001)
5. Pavlou, P.A., Gefen, D.: Building effective online marketplaces with institution-based trust. Inf. Syst. Res. **15**(1), 37–59 (2004)
6. McKnight, D.H., Chervany, N.L.: Conceptualizing trust: a typology and e-commerce customer relationships model. In: Proceedings of the 34th Annual Hawaii International Conference on System Sciences, 2001. IEEE (2001)
7. Palvia, P.: The role of trust in e-commerce relational exchange: a unified model. Inf. Manag. **46**(4), 213–220 (2009)

8. Bella, G., Giustolisi, R., Riccobene, S.: Enforcing privacy in e-commerce by balancing anonymity and trust. Comput. Secur. **30**(8), 705–718 (2011)
9. Schunter, M., Waidner, M.: Simplified privacy controls for aggregated services—suspend and resume of personal data. In: International Workshop on Privacy Enhancing Technologies. Springer, Berlin, Heidelberg (2007). Patnaik, A.R., Srivastava, A.G., Nayaka, A.R.: Building flexible trust models for e-commerce applications (2006). http://tecsis.ca/services/2.Trust%20 Model.pdf. Accessed 21 Nov 2013
10. Ahmadi, K.: Predicting e-Customer behavior in B2C relationships for CLV model. Int. J. Bus. Res. Manag. **2**(3), 128–138 (2011)
11. Folorunso, O., Mustapha, O.A.: A fuzzy expert system to trust-based access control in crowd-sourcing environments. Appl. Comput. Inform. **11**(2), 116–129 (2015)

Stipulation-Based Anonymization with Sensitivity Flags for Privacy Preserving Data Publishing

K. Ashoka and B. Poornima

Abstract Privacy is a major concern for organizations that release Microdata for informal analysis. Most of the Privacy Preserving Data Publishing (PPDP) techniques anonymize data based on personalized privacy requirements or based on some general utility specification. The consequence is that, either the record owner's privacy requirements or the data miner's (analyst's) data efficacy requirements are considered for data anonymization, which leads to tainted accuracy in several data mining tasks. Motivated by this we propose a novel approach which considers privacy requirements in the form of Sensitivity Flags from the record owners end, as well as Application Specific Requirements from the data miners (analysts) end. Our proposed method is theoretically analyzed and the mathematical analysis outperforms the earlier works with sufficient experiments.

Keywords Privacy preserving data publishing · Data anonymization
Data utility · Information loss

1 Introduction

In recent years Data Mining has been used successfully in different areas like weather forecasting, Market Prediction, Financial Fraud recognition, Medical Data Analysis and also in Counterpunch Terrorism. It is often essential to share/publish data for research/analysis purpose. But there is serious privacy issues if the detailed, person specific data (raw data) published as it is, may consists of sensitive personal information. To overcome these privacy issues, numerous Privacy Preserving Data Publishing (PPDP) techniques have been developed. Few significant works for this research are, Randomization method [1], Data swapping [2], Cryptographic approach

K. Ashoka (✉) · B. Poornima
Bapuji Institute of Engineering and Technology, Davangere 577004, Karnataka, India
e-mail: ashoka_kkd1@yahoo.com

B. Poornima
e-mail: poornimateju@gmail.com

© Springer Nature Singapore Pte Ltd. 2019
P. K. Sa et al. (eds.), *Recent Findings in Intelligent Computing Techniques*,
Advances in Intelligent Systems and Computing 707,
https://doi.org/10.1007/978-981-10-8639-7_46

Table 1 Input microdata table

Name	Gender	Age	Zip code	Disease	Sensitivity flag value
Arun	M	9	90001	Cough	Φ
Basu	M	5	90010	Asthma	Lower respiratory system disorder
Carl	M	6	90036	Pneumonia	Φ
David	M	8	90001	Tuberculosis	Any disease
Edwin	M	10	93305	Cold	Respiratory system disorder
Frank	M	7	93304	Bronchiectasis	Respiratory system disorder
Geeta	F	35	18001	Fever	Φ
Hema	F	39	18023	Tonsillitis	Respiratory system disorder
Isha	F	40	18006	Fever	Any disease
James	M	48	93301	Asthma	Lower respiratory system disorder
Kini	M	44	93302	Tuberculosis	Tuberculosis

Table 2 The 2-anonymous table

Gender	Age	Zip code	Disease
M	1–10	900**	Cough
M	1–10	900**	Asthma
M	1–10	900**	Pneumonia
M	1–10	900**	Tuberculosis
M	1–10	9330*	Cold
M	1–10	9330*	Bronchiectasis
F	31–40	180**	Fever
F	31–40	180**	Tonsillitis
F	31–40	180**	Fever
M	41–50	9330*	Asthma
M	41–50	9330*	Tuberculosis

* and ** are the generalization of an attribute values based on the generalization tree

[3] and Anonymization techniques. Most popular PPDP techniques are Anonymization techniques, because they have lesser information loss and high data utility. For example, Generalization [4, 5] for k-anonymity, Bucketization for ℓ-diversity [6], t-closeness [7], m-invariance [8], Personalized Privacy [9] etc. These techniques preserve the data privacy at the cost of losing some amount of data utility. Hence "data privacy" and "data utility" are the two contradictory goals [10] where we have to lose one for gaining the other.

An improvement of the k-anonymity/ℓ-diversity was done in [11, 12], where the authors suggest a systematic clustering technique for k-anonymization. Slicing [13] and SLOMS [14] are some of the recent improvements of bucketization. For illustration, Table 1 shows the sample input microdata and Table 2 shows the 2-anonymous table.

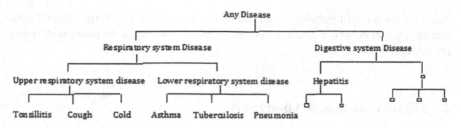

Fig. 1 Taxonomy tree of 'Disease'

1.1 Motivation

In Personalized Privacy [9], record owner's privacy requirements in the form of guarding node are considered for anonymization. Here each sensitive attribute has a taxonomy tree and each record owner selects a guarding node in this tree. For example the taxonomy tree for the sensitive attribute "Disease" is presented in Fig. 1. Although personalized privacy has less information loss than the other methods, it has some limits. The record owner may *play safe* by setting his/her guarding node as "Any disease" which will keep him/her in safer privacy zone. But this play safe leads to imprecise results in many data mining tasks. For example, suppose a medical researcher want to study the patterns of Lower_respiratory_system disorders in some geographical areas like Los Angeles or Bakersfield (as the pollution level is comparatively high in those cities), suppose the record owner sets the guarding node as "Any disease" or "Respiratory_system_disease" (as in fourth and sixth tuple of Table 1) the Data miner/Analyst will not get accurate results. Also the distribution of Lower_respiratory_system disease was seriously affected in personalized anonymization, the same will not be affected in our proposed stipulation-based anonymity technique.

1.2 Contribution and Paper Organization

In this paper, we suggest a novel technique "Stipulation based Anonymization". Along with the person who specify the level of privacy protection for his sensitive attribute values, we will also consider the utility of the data by taking the Application Specific Requirements from the Data miner/Analyst.

We also avoid play safe, by maintaining Sensitive Attribute List (SA_LIST). If the record owners SA is in SA_LIST then only the personalized sensitivity flag is accepted, otherwise it will be published as it is, which will improve the data utility.

The rest of the paper is organized as follows. In Sect. 2 we formalize the Stipulation based Anonymization. In Sect. 3 we present preliminary concepts and notions.

Section 4 gives the Stipulation based Anonymization Algorithm. Experimental evaluation of our approach is given in Sect. 5. Section 6 concludes the paper with future research directions.

2 Stipulation-Based Anonymity

In personalized privacy the privacy necessities of the record owner is considered for anonymization, on the other hand the Data miner/Analyst may also have his own specific obligations that needs to be considered for anonymization. Here we observe the chance that, if the Data miner's/Analyst's requirements are considered for anonymization so that to yield an anonymized data with much higher data utility which is formalized as shown below.

Let T be the input Microdata table to be published. Assume T contains d no. of attributes $A = \{A_1, A_2...A_d\}$ and their attribute domains are $\{D[A_1],$ $D[A_2]...D[A_d]\}$. A record $t \in T$, represented as $t = (t[A_1], t[A_2]...t[A_d])$ where $t[A_i]$ $(1 \le i \le d)$ is the attribute values of t.

Definition-1 Application Specific Requirements (ASR)

Let P be an application taking the table T as input, the application Specific Requirements of P is denoted as $R = [(A_1, v_{1,ij}), (A_2, v_{2,ij})...(A_d, v_{d,ij})]$ $(0 \le j \le h - 1)$ $(1 \le i \le MAX\_VAL_j)$.

Where Ai values are the attributes of T, h is the maximum height of the taxonomy tree of the attribute Ai and Vij $\in$ D[Ai] are the potential values of the attribute Ai and also it is a node/leaf of a taxonomy tree of the attribute Ai. Here if $R = \Phi$ then there is no application specific requirements for the Data miner/Analyst. Any number of such requirements can be specified in R, but the priority will be taken as "first specified first served" basis.

Example: Suppose a Data miner/Medical researcher want to learn the patterns of Lower_Respiratory_System disorders around the locality Los Angeles or the Bakersfield then he can specify his requirements as $R = [(Disease,$ Lower_respiratory_system_disorder), (Zip code, Los Angeles), (Zip code, Bakersfield)]. Then, even if the record owner to "play safe" sets his sensitivity level value as "Any disease", his disease attribute value is set to Lower_respiratory_system_disorder by more generalizing the corresponding QI attributes of this tuple. (as it has to satisfy k-anonymity and ℓ–diversity constraint). The ordering of the requirements should be sensitive attributes first and QID attributes next. This ordering itself gives us the preference of the attributes for generalization.

Some diseases like HIV, Lung cancer, stroke, etc., are more sensitive that the record owner do not want to disclose them, whereas the diseases like fever, cough, tonsillitis, etc., are very common diseases and most of the record owners do not mind to disclose them. But, some people have the tendency to play safe by setting a more generalized (higher privacy protected) sensitivity level value that affects the

Table 3 The sensitive attribute list. (SA_LIST)

Sensitive attribute-1	V11	V12	...	V1x
Sensitive attribute-2	V21	V22	...	V2x
–	...	...	...	...
Sensitive attribute-z	Vz1	Vz2	...	Vzx

Table 4 Example sensitive attribute list (SA_LIST)

Disease	HIV	Lung cancer	Stroke	Tuberculosis

data utility. To prevent this, we created a two dimensional list of sensitive attribute and their sensitive values which really needs privacy; as shown in Table 3.

Before executing anonymization the value of the input sensitive attribute is examined in the SA_LIST, if it is present then only personalized sensitivity level value is taken from the record owner, else it will be published as it is by considering it as less sensitive. In our example we consider for simplicity only Disease as sensitive attribute with values shown in Table 4.

3 Probability of Privacy Breach

It is the probability that an adversary can deduce from the published table T* that any of the correlations $\{I, v_1\}, \ldots \{I, v_x\}$ exists in the input data table T, where $v_1, v_2, \ldots$ v_x are the attribute values of the sensitive attribute that comes under the sensitivity level value specified by the record owner and I is an individual record owner.

Suppose an adversary tries to infer the sensitive data of an individual I from the published table T*, in the worst case scenario he will have all QID values $I \cdot A_1^{qi}$, $I \cdot A_2^{qi}, \ldots I \cdot A_d^{qi}$ of I. Hence he checks only the tuples $t* \epsilon$ T* with QID values $t* \cdot A_i^{qi}$ covers $I \cdot A_i^{qi}$ for all $1 \leq i \leq d$. These tuples form the necessary QID group. An adversary by referring external databases like voter registration list, can get those QID values of I. If I does not appear in voter registration list then I's privacy preserved by default. Otherwise the adversary attempts combinatorial methodology to infer the sensitive attribute SA of I.

Definition-2 Necessary QI-Group

For an individual I, the Necessary QI-Group (I) is the only QI-Group in T* that covers $I.QI_i$ for all $i \epsilon [1, d]$. The set of individuals who have same QID values after generalization are denoted as $S_{real}(I)$. Set of individuals from the external database (like voter registration list) that have same QID values of some i-th necessary QI-Group(I) is denoted as $S_{ext}(I)$. Here

$$S_{real}(I) \subseteq S_{ext}(I)$$

To illustrate the above concepts, suppose an adversary tries to infer the disease value of Basu from the published table T* with his age 5 and zip code 90010. The Necessary QI-Group consists of tuples 1–4, i.e., $S_{real}(Basu) = \{Anil, Basu, Carl, David\}$. In an attempt to derive $S_{real}(Basu)$ the adversary may refer to database like voter registration list where he obtain $S_{ext}(Basu)$ set consists of many individuals with same QID values of Basu.

In deriving the probability breach of an individual I, let us assume m as the size of the Necessary QI-Group(I) and n be the cardinality of $S_{ext}(I)$. For simplicity assume no record is repeated in Necessary QI-Group (I), i.e. every tuple belongs to a distinct person. Let r be the number of tuples t_j* ($1 \leq j \leq m$) in necessary QI-Group (I) that comes under sub tree of the sensitivity flag. Then Breach Probability (BP) is given by

$$BP = \frac{ACTN}{T\_RECON},$$

where *ACTN* is the actual number of combinatorial reconstructions violating the privacy constraints imposed by sensitivity level flag and *T_RECON* is the total number of possible reconstructions. Here *T_RECON* can be calculated by

$$T\_RECON = Permu\,(n, m) \tag{1}$$

And *ACTN* can be calculated by

$$ACTN = r \cdot Permu\,(n - 1, m - 1) \tag{2}$$

If the SA of the target person is in subtree of the sensitivity level flag, then With (1) and (2)

$$BP = r/n \tag{3}$$

To illustrate the above concept, assume the adversary tries to infer the disease value of Basu, i.e., tuple t_2 from the published table. As the sensitivity level flag of Basu is Lower_Respiratory_Sysytem disease, Necessary QI-Group includes first four tuples of Table 1. Here $r = 3$ because the SA values of Basu, Carl and David (tuples 2, 3 and 4) overlaps the subtree under Lower_Respiratory_System Disorder. Suppose $n = 8$, i.e., eight people from the external database has same QID values of t_2 then

$$BP(Basu) = 3/8 = 0.375.$$

4   Stipulation-Based Anonymization Algorithm

Our Stipulation-based Anonymization Algorithm has two parts. In the first part the greedy-Personalized-Generalization Algorithm [9] is modified to take care of "play safe" by the record owner. The sensitive attribute Disease value of each tuple is searched in the SA_LIST. If it is found, then only sensitivity flag is considered for anonymization. Next in the second part, generalization of QIDs and then generalization of SA was done. The ASR Algorithm takes as input the application specific requirements in the form of list as mentioned in Definition 1.

Algorithm: Stipulation_based_Anonymization
Input: Private Microdata with sensitivity level flag values, requirement specification R and the SA_LIST.
Output: The publishable table T*.

1.   If ($n < TH_{min}$) then return with warning message // T should contain minimum TH_{min} records
2.   For each tuple $t_i \in T$ ($1 \leq i \leq n$)
3.   Search t_i.SA in SA_LIST   //To avoid play safe
4.   If found then
5.   If t_i.A_j is not in R ($1 \leq j \leq d$)
6. t_i*= GEN(t_i);
7.   else t_i*.A_j = R_i.V_j for all requirements R_i
8.   If BP (t_i) > P_{breach} then
9. t_i*.SA = The parent of t_i*.SA
10.   else $t_i = t_i$* ;
11.   i= i + 1;
12. Publish T*

The output of this algorithm for the input Table 1 is shown in Table 5.

5   Experiments

In this section we experimentally evaluate the efficacy of our approach as compared to k-anonymity, ℓ-diversity and personalized privacy. We used adult dataset from UCI machine learning repository [15]. This adult data set contains fifteen attributes. After eliminating tuples with missing and null values, there are 45,222 effective tuples in total. In our experiments we have used Age, Education, Marital status, Occupation, Race, Sex and Country as Quasi Identifiers.

The sensitive attribute "Salary" has been converted into disease. Personalized sensitivity level flag values are generated for one class of data set with more than 40% of disease values in SA_LIST and nearly 50% of them are leaf values, 30% of them are their parent nodes and 20% of them are either root or the second level

Table 5 Potential result of stipulation anonymity

Gender	Age	Zip code	Disease
M	9	90001	Cough
M	5–10	900**	Lower respiratory system disorder
M	6	90036	Pneumonia
M	5–10	900**	Lower respiratory system disorder
M	10	93305	Cold
M	5–10	933**	Lower respiratory system disorder
F	35	16001	Fever
F	39	18000	Tonsillitis
F	40	19000	Fever
M	41–50	933**	Lower respiratory system disorder
M	44	93302	Tuberculosis

** indicates the generalization of an attribute values based on the generalization tree

intermediate nodes. We also created SA_LIST consisting of Disease attribute with values Pneumonia, HIV, Tuberculosis, Lung-cancer etc. The other class of dataset does not have any personalized sensitivity level flag values. In such case our algorithm will be merely a generalization algorithm.

The Information Loss (I-Loss) of the anonymized table T* can be calculated by

$$I - \text{Loss}(T^*) = ws \frac{(The\ number\ of\ values\ in\ t^* \cdot SA) - 1}{The\ number\ of\ values\ in\ the\ domain\ of\ SA}$$

$$+ \sum_{i=1}^{d} wq \left(\frac{(The\ number\ of\ values\ in\ t^* \cdot QI(i)) - 1}{The\ number\ of\ values\ in\ the\ domain\ of\ A(i)} \right)$$

For all tuples t* ∈ T*.

Here *ws* and *wq* are the positive penalty factors for loosing precision upon generalization.

In our experiment these values are set to 1. For k-anonymity the value of k is taken as $1/P_{breach} = 4$. Same will be the value of ℓ for ℓ-diversity. We conducted the experiments for three different cases. In the first case the application specific requirements R will be specified on QIDs only, like (Age = 25–30, Country = India). In the second case the application specific requirements R will be specified on SA only, like (Disease = Lower Respiratory System Disease). In the last case the requirements are specified for QIDs as well as SAs. Figure 2a, b and c demonstrates the value of information loss calculated with $P_{breach} = 0.25$ for increasing input load of up to 40k records for each of these cases. The results reveals that Information Loss is very less compared to k-anonymity and ℓ-diversity and it increases with increasing requirements.

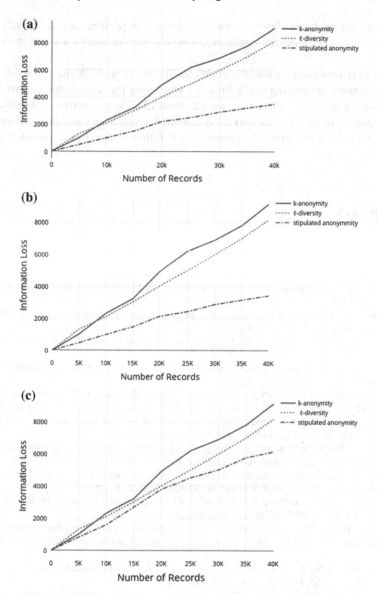

Fig. 2 **a** Information loss for Case-1. **b** Information loss for Case-2. **c** Information loss for Case-3

6 Conclusion and Future Work

In this paper we presented a new concept "Stipulation based Anonymization" for preserving privacy in Microdata publishing. The use of SA_LIST improves the data utility as the entire records do not require privacy, also "play safe" effects of the record owner on data utility are avoided. Along with personalized sensitivity flags, our approach takes input the application specific requirements for anonymizing the data.

Our work motivates various directions for future research. We have considered an aggregate query answering application for specifying the requirements. This can be extended for different data mining applications like cluster analysis, classification, co-relation analysis, etc., we carried the work for single release data, whereas for sequential release and multiple release data still there exists lots of research opportunities.

References

1. Agrawal, D., Aggarwal, C.C.: On the design and quantification of privacy preserving data mining algorithms. In: Proceedings of the 20th ACM Symposium on Principles of Database Systems (PODS), pp. 247–255, Santa Barbara, CA, May 2001
2. Reiss, S.P.: Practical data-swapping: the first steps. ACM Trans. Database Syst. (TODS) 9(1), 20–37 (1984)
3. Pinkas, B.: Cryptographic techniques for privacy-preserving data mining. ACM SIGKDD Explor. Newsl. 4(2), 12–19 (2002)
4. Samarati, P.: Protecting respondent's privacy in microdata release. IEEE Trans. Knowl. Data Eng. 13(6), 1010–1027 (2001)
5. Sweeney, L.: k-anonymity: a model for protecting privacy. Int. J. Uncertain. Fuzziness Knowl.-Based Syst. 10(5), 557–570 (2002)
6. Machanavajjhala, A., Gehrke, J., Kifer, D., Venkitasubramaniam, M.: ℓ-diversity: privacy beyond k-anonymity. In: IEEE International Conference on Data Engineering, p. 24 (2006)
7. Li, N., Li, T.: t-closeness: privacy beyond k-anonymity and ℓ-diversity. In: IEEE International Conference on Data Engineering, pp. 106–115 (2007)
8. Xiao, X., Tao, Y.: m-invariance: towards privacy preserving re-publication of dynamic datasets. In: ACM SIGMOD International Conference on Management of Data, pp. 689–700 (2007)
9. Xiao, X., Tao, Y.: Personalized privacy preservation. In: Proceedings of ACM International Conference on Management of Data (SIGMOD), Chicago, IL (2006)
10. Rastogi, V., Suciu, D., Hong, S.: The boundary between privacy and utility in data publishing. In: VLDB'07, 23–28 Sept 2007, Vienna, Austria
11. Kabir, Md.E., Wang, H., Bertino, E.: Efficient systematic clustering method for k-anonymization. Acta Inf. 48(1), 51–66 (2011). http://dx.doi.org/10.1007/s00236-010-0131-6
12. Bhaladhare, P.R., Jinwala, D.C.: Novel approaches for privacy preserving data mining in k-anonymity model. J. Inf. Sci. Eng. 32(1), 63–78 (2016)
13. Li, T., Li, N., Zhang, J., Slicing, M.: A new approach for privacy preserving data publishing. IEEE Trans. Knowl. Data Eng. 24(3), 561–574 (2012)
14. Han, J., Luo, F., Lu, J., Peng, H.: SLOMS: a privacy preserving data publishing method for multiple sensitive attributes micro data. J. Softw. 8(12), 3096–3104 (2013)
15. The UCI machine learning repository. http://mlearn.ics.uci.edu/MLRepository.html

Context Aware Content Delivery System in Conversational Interfaces

Suraj Manjesh, Tushar Kanakagiri, P. Vaishak and Vivek Chettiar

Abstract With the increased penetration of mobile-based chat applications among the customer base at large, there has been an ever-increasing demand across various industries to leverage the humongous user base for advancing their interests in a targeted fashion to achieve effective conversion. To achieve the above, it is of importance to provide timely and accurate content, for which there is need for identifying and maintaining static as well as dynamic context. This paper discusses a novel approach involving action items and content triggers to achieve the above result. The experimental section demonstrates the implementation of the approach in a real-life travel planning example.

Keywords Dynamic context detection · Content delivery · Chatbots
Artificial intelligence · Ad delivery

1 Introduction

In today's age, application-based group messaging and conversations have become an integral part of our lives. With the advent of accessible touch screen phones and popularity of a few platforms, businesses are identifying this as a new avenue for product placement, advertisements, any other content that businesses deem important enough for customers to see. Currently, the most popular chatting application, WhatsApp, has a monthly active user base of over 1 billion, with a growth of 42%

S. Manjesh (✉) · T. Kanakagiri · P. Vaishak · V. Chettiar
R.V. College of Engineering, Bengaluru, Karnataka, India
e-mail: smshet1995@gmail.com

T. Kanakagiri
e-mail: tusharkanakagiri@gmail.com

P. Vaishak
e-mail: vaishak.prashanth@gmail.com

V. Chettiar
e-mail: vivekrchettiar@gmail.com

© Springer Nature Singapore Pte Ltd. 2019
P. K. Sa et al. (eds.), *Recent Findings in Intelligent Computing Techniques*,
Advances in Intelligent Systems and Computing 707,
https://doi.org/10.1007/978-981-10-8639-7_47

over the last calendar year. The enormous number of users, coupled with a very useful group chatting feature, has made businesses very aware of the potential, and these businesses are investing heavily to make use of the opportunities afforded by this growth. Group chats are used extensively today to plan out various events, like vacations, luncheons, birthday parties, marriages and many more.

When there exists a group of people that need to collaborate, they set up a WhatsApp group with ease and add concerned members into the group so all participants can plan and discuss details such as event locations and activities, instead of communicating with all members individually. That method is obsolete and very inefficient, with the additional danger of accidentally omitting an important member.

A chat setting between several people discussing about a particular scenario tends to bring up several important elements that businesses can leverage to increase brand presence and sales. For example, when a WhatsApp group has been created to plan a trip to a particular location, the participants commonly discuss about points of interest, places to stay, modes of transport, etc. This information, if leveraged effectively, can be used to push targeted, context aware advertisements.

Content delivery is done by continuously calculating the context of the chat. In the event that the chat is still missing a piece of vital information that is required for accurate context detection, the system will try to help in a way suitable to the current scenario.

Further, the content that has to be served has to be served unobtrusively, hence a conversational interface can be used. Conversational workflows is a new, emerging paradigm of human–computer interaction that can greatly aid in solving the above requirement. Participants tend to find it easier, faster and more intuitive to talk to interfaces that mimic human interaction. Thus, content delivery in the form of messages through a bot is ideal in this scenario.

The process of conversational content delivery has three main parts; context identification, identification of action items and lastly, content delivery triggering mechanism.

2 Related Work

Context aware systems have seen a rise in popularity in recent years. The same question may have different answers based on the context in which it is to be considered. As we try to move towards a human–computer interaction system that is as close to natural conversation, one of the pitfalls is to not consider existence of varied meanings, hence it becomes increasingly important to factor in the current context.

Gediminas Adomavicius and Alexander Tuzhilin have built a recommendation system that uses contextual information such as location, time, and other factors to provide better recommendations. This context aware recommender uses contextual prefiltering, post-filtering, and modeling to include contextual information as a factor to consider when recommending [1].

Xinyou Zhao and Toshio Okamoto use a context aware system to build an adaptive multimedia content delivery for u-learning. The adaptive learning system takes into consideration many contextual factors such as user history, user preferences, behavior, location, time, user's device, etc., when considering how best to deliver the content to the user. For example, if the user is on the move, he may prefer an audio lesson versus a graphical one as it is hard to grasp visual input on a small handheld device [2].

Pichon Dominique and Seite Pierrick, Bonnin Jean-Marie have attempted to analyze the environment and figure out the best long-term adaptation strategy, mainly in video streaming services [3].

Schmidt Albrecht and Van Laerhoven Kristof explain a methodology to build context aware devices by utilizing a structured method for sensor selection [4].

Biegel and Cahill describe a framework that helps optimize the integration of information from various sources and representation of the application with reasonable context [5].

Xiong Caiming, Merity Stephen, and Socher Richard draw inspiration from the context aware system to improve a dynamic memory network that is capable of answering questions by maintaining the context of the previous questions [6].

3 Methodology

To be able to achieve good levels of accuracy, we essentially identify the need to determine context, following which we provide responses based on some conditions that trigger the algorithm.

The basic real time conversation analyser has three different parts that work in tandem to produce the desired result. First, the context detection algorithm detects the probable context and returns a list of key-value pairs which define the context and its probability. Second, we identify action items which are defined by the developer, this is done using a variety of methods. Third, we identify action triggers. This is the component that maintains real time state variables that are continuously updated with incoming messages. These variables are keys in ensuring timely display of messages, and are vital to prevent erroneous or out of context messages.

3.1 Context Identification

In [1] the authors propose a method to detect topic by clustering keywords. Clustering of keywords involves using distance measure between words. Distances between these keywords are based on the statistical distribution of words in a corpus of text. This corpus of text is predefined. The approach in [1] requires no predefined set of possible contexts that need to manually sent. However, in this context, the developer

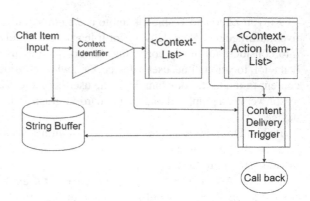

Fig. 1 Basic workflow of the context identification system

using the platform can manually define a set of key contexts which are likely to be encountered in the conversation, which increases accuracy of the model.

Since the contexts of a conversation changes dynamically, we need a method divide the conversations into parts when analyzing the context. This is done by maintaining a string buffer, where recent chat items are appended. This buffer is cleared when Content Delivery Trigger is fired or when context values do not tend to a particular context. However there may occur a situations where the following chat items have the same context as the set that had been cleared. In Fig. 1 we can see that to properly address this condition we maintain a temporary register that stores the value of the last cleared context and compare the value stored in this register with the newly calculated value. If the contexts match the cleared chat item, they are restored. This leads to succinct set of chat items grouped together which have a common context.

3.2 Identification of Action Items

Identifying context is not enough to deliver suitable content. We propose the concept of "action items" which are specific verbs or related words that define what the users are trying to ask. These Actions items in Fig. 2 are to be manually defined by the developer for each context that needs to be served. We store a map of context to action items. This map is a m-n relation. Once the context has been determined, the action items list is drawn from the map.

Once the action item list is available, we can use multiple methods to check which one pertains to conversation. A simple frequency analysis of the words in the chat items after stop word removal yields acceptable results. These action items are returned to the developer in a callback function. Action word analysis leads to determining what the user is trying to imply. In case of multiple action items being identified, they can all be returned to the developer in a list.

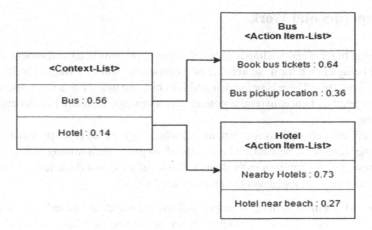

Fig. 2 Context list and action item lists for an example conversation about travel

3.3 Content Delivery Triggers

This component decides when to fire the implemented callback by the developer and return the calculated actions.

Data structures used

1. <Context-List>: A list of Key-Value Pair that stores current contexts sorted in descending order by the value.
2. <Context-Action Item-List>: Stores the action items and associated value along with which context it belongs to.

Process

1. The string buffer is empty initially, as chat items are added the context of each individual chat item is calculated and appended to the <Context-List>.
2. Action items for each context identified and added to the Context-Action Item-List>.
3. Once the normalized value for and Action item crosses a predefined value (Epsilon) the context and Action item is returned to the callback function. the string buffer is cleared and all the variables are reinitialized.
4. Simultaneously a counter is started, and incremented every time a chat items comes in. If the threshold is not crossed before the counter reached a predefined value (Theta), the string buffers is cleared and all variables are reinitialized.

4   Experimental Work

One of the biggest and leading sources of revenue streams is advertisement. Google despite being known for its search and other products, still generates over 90% of its income from Ads. There is rising demand for ensuring that the ads reach specifically to the targeted audience, to reduce potential aversion and maintain healthy consumer relations.

The effectiveness of ads is compounded when they are far more personalized and this is the motivation beyond vast amounts of surges in data mining industries. The chat conversations between individuals is the ideal space for these targeted ads.

For the chat taking place in the given example

1. The string buffer is initially empty and the messages are added one by one. In Fig. 3 the second message leads to the context to be calculated as "Trip" with a probability of 0.6. The third message recognizes the context as "Ladakh".
2. By now the action items defined by the developer has accumulated in <Context-Action Item-List>.
3. As the fourth message comes in, the normalized value for the action item "Trip" crosses the Epsilon value and the context is returned to the developers callback

Fig. 3 Ads: travel plan use case

Fig. 4 Ads: travel
attractions use case

function. Here the developer's callback system returns an advertisement for a
package trip to Ladakh.

4. The string buffer is cleared and the process starts again. In Fig. 4 the context
   "Ladakh" has been moved to the temporary buffer. As the 5th chat message
   comes in and the context is calculated as "Travel" and the temporary buffer is
   cleared.

5 Conclusion

The paper discusses an approach that can be used for context aware content delivery
through a conversational chat interface. The research explores a model that maintains
a static context as well as a dynamic context, which has been proven to be effective
at leveraging group chat flows for attaining business goals through effective, targeted
content delivery.

References

1. Laakko, T.: Context-aware web content adaptation for mobile user agents. In: Studies in Computational Intelligence, vol. 130, pp. 66–69
2. Zhao, X., Okamoto, T.: Adaptive multimedia content delivery for context-aware u-learning. Int. J. Mob. Learn. Organ. 5(1), 46
3. Pichon, D., Seite, P., Bonnin, J.-M.: Context-aware delivery of video content to mobile users. In: Proceedings of the 6th International Conference on Mobile Technology, Application & Systems—Mobility'09, pp. 1–9
4. Schmidt, A., Van Laerhoven, K.: How to build smart appliances? IEEE Pers. Commun. 8(4), 66–71
5. Biegel, G., Cahill, V.: A framework for developing mobile, context-aware applications. Pervasive Comput. Commun. 26031, 361–365 (2004)
6. Xiong, C., Merity, S., Socher, R.: Dynamic Memory Networks for Visual and Textual Question Answering, arXiv, p. 8

Analyzing Online Location-Based Social Networks for Malicious User Detection

Ahsan Hussain and Bettahally N. Keshavamurthy

Abstract An increasing popularity of Location-Based Social Networks (LBSNs) has been recently observed due to the advent of smart-phones. These networks provide particular suggestions to users based on their check-in interests. But this disclosure of information can have adverse effects on the security of LBSN users. In this paper, we analyze the user-patterns of major world cities based on the geographical check-ins by online LBSN users. Later, we put down the framework for malicious social user detection using Machine Learning approaches. We implement our methodology over *Foursquare* social network datasets for *New York* and *Tokyo* cities and evaluate the classification accuracy. It is shown that the proposed framework can provide effective malicious user detection in LBSNs, while preserving the personalized abstract view.

Keywords Foursquare Check-ins · Location-Based Social Networks
Machine Learning classifiers · Malicious user detection · Social network threats

1 Introduction

Web-based social networking is progressively turning out to be a piece of our regular day to day existences, from associating with companions and sharing pictures to exploring geographical areas through Location-Based Social Networks (LBSNs). These new applications provide an alternate view-point from which we can comprehend, investigate, explore, and geographically record the spots we live in. LBSNs, like, *Foursquare* and *Gwalla* permit us to spatially check-in to famous places of a city, updating rich databases that hold computerized engravings of our associations

A. Hussain (✉) · B. N. Keshavamurthy
Department of Computer Science and Engineering, National Institute of Technology Goa,
Ponda 403401, India
e-mail: ahsan.hussain@nitgoa.ac.in

B. N. Keshavamurthy
e-mail: bnkeshav.fcse@nitgoa.ac.in

© Springer Nature Singapore Pte Ltd. 2019   463
P. K. Sa et al. (eds.), *Recent Findings in Intelligent Computing Techniques*,
Advances in Intelligent Systems and Computing 707,
https://doi.org/10.1007/978-981-10-8639-7_48

at the same time. To break down these follows, the LBSNs Application Programming Interfaces (API's) are utilized to get to area based information to figure out where web-based social networking users communicate "where they are". This geographic information analysis can uncover the psycho-topography and financial territory of online networking clients of a geographic location.

Online social networking app clients update what they are doing, and how they are feeling. Clients uncover when and where they are experiencing a passionate emergency, encountering their very own paradise or damnation, having a good time or a prophetically calamitous occasion. Overlaying this information with more conventional government informational indexes, uncover the monetary examples natural in the way these applications are utilized. For instance the larger part of *check-in* information originates from those city regions that have the most noteworthy proportions of business utilize.

Our objective is to analyze social users' behavior over LBSNs. We give a detailed reasoning of the geographically associated social users attachments and their linkage-modes. Further, we apply various Machine Learning approaches to detect malicious users over the LBSNs, who are a major threat to security of social networking sites. The rest of the paper is organized as follows. The related work is presented in Sect. 2 while the analysis of online LBSNs is given in Sect. 3. Proposed framework is given in Sect. 4. Experimental Evaluation and Result Analysis is presented in Sect. 5. Finally the conclusions and future work are given in Sect. 6.

2 Related Work

Security issues in LBSNs have been widely studied over the past decade. Most of the research efforts focus on location privacy and malicious user detection using Machine Learning approaches. Online LBSNs can be severely affected by various notorious people who have the sole intention of causing harm to social users. In the following section, we show the work done related to malicious activity detection in LBSNs.

2.1 Malicious Activities in Online LBSNs

The effect of pernicious exercises has expanded enormously due to the expansion in the quantity of malevolent records made on Social Networks (SNs) every day. As per the *Nexgate* report in 2013, vindictive exercises on social organizations, for example, spam conveyance has ascended to around 355% in the principal half of the year 2013. 15% of all social spam messages contain a URL to spam content, malware, or obscene sites [1]. It revealed that vindictive records have been utilized to take the characters of countless clients crosswise over various SNs. In a current report by Javelin Strategy and Research, the aggregate number of personality extortion casualties has developed to around 13 million every year and around $112 billion has been stolen in the previous 6 years [2]. Social spammers make about $200 million consistently constituting to a misfortune in social trust, efficiency, and benefit. Actually, the development in explicit spam has multiplied on most prevalent SNs, planning to lure clients to click vindictive connections or download malware [3].

It moreover recommends that people are rarely anxious to forego some security for a sufficient level of peril. By using SNs [4], individuals open themselves to various sorts of risks that have consistent effects of breaking their guard. In a practically identical strand, the theory stipulates that exposure is sure to strong instruments that allow clients to control the sum they reveal in light of their goals, learning and mindsets toward security. In the association of online social range communication, such control can be proficient using security settings [5].

2.2 Machine Learning

Machine Learning (ML) has assumed critical parts in distinguishing malicious accounts in SNs. ML fuses an assortment of strategies, for example, supervised, unsupervised, and semi-supervised learning. To predict the class label for new data, Supervised ML algorithm acquires a labeled dataset and learns a model as output [6]. A classification model is used to distinguish malicious and legitimate accounts after completion of training [7]. Some important characteristics of malicious users, such as following large users with less followers, posting duplicate and illogical contents, excessive redirections, etc., were used for manual data-labeling [8, 9].

It is essential to note that few reviews that connected regulated ML techniques are roused by the need to present new features. Once a feature set has been distinguished, ML classifiers are prepared. With the end goal of clarity, Singh et al. [7] prepared five characterization calculations: BayesNet, NB, Sequential minimal optimization (SMO), J48, and Random Forest using content/behavioral features, with Random Forest beating other four classifiers. Almaatouq et al. [10] prepared six distinct classifiers: ZeroR, Bayesian system, NB, strategic relapse, choice trees, and Random Forest utilizing both substance/behavioral and arrange highlights. They additionally surveyed that the execution of choice tree and Random Forest have been very fascinating. Along these lines, we examine distinctive ML classifiers and their different classifications to distinguish malicious users. The emphasis is especially on the classifiers with better execution in light of the measurements used to assess them.

3 Analysis of Online LBSNs

LBSNs can have varying inspirations driving communications. *Foursquare* clients disclose more about the ordinary subtleties of life—where people relax, work, or the place they get their weekend dinner at. While Facebook clients tend to utilize the site

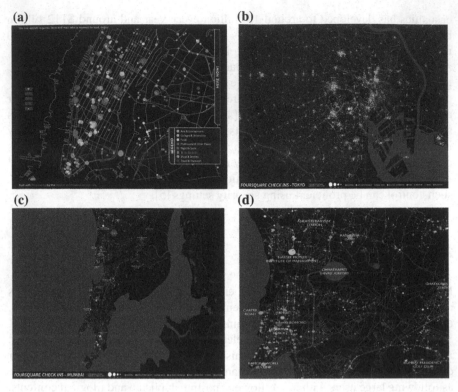

Fig. 1 Foursquare check-in visualizations of: **a** New York City, **b** Tokyo City, **c** Mumbai City and **d** Mumbai City—zoomed

to boast about the relaxing spaces in the respective cities of their check-ins, they have been to. Both destinations reveal to us how online social networking users investigate the city or all the more essentially, how they communicate their endeavors.

Apart from sharing their feelings, *Foursquare* users additionally reveal land utilization and financial information for cities, where this data can be difficult to get. The given maps of cities like Mumbai, Tokyo, etc., in Fig. 1, allow us to look at how *Foursquare* users operate in these cities, more significantly disclose generally concealed urban financial patterns. *Foursquare* is utilized more in Tokyo than Mumbai and it discloses to us somewhat about the city's way of life while uncovering the basic urban financial examples. Tokyo's travel organized check-ins are unmistakably pictured as registration of various kinds bunch around travel hubs. The maps show that *Foursquare* is utilized more by foreign guests than locals in Mumbai. *Foursquare* is all the more a need in huge urban communities like Mumbai, where new places are opening constantly and its difficult to monitor them all. On the contrary in smaller cities, users are more inspired by old top picks instead of investigating new places.

The thought behind *Foursquare* ratings feature is that individuals are voting by measuring signals about spots, for loyalty, expertise, and notion, with their feet in real world, not just leaving a star or a like on a site. LBSNs give another way to inquire about how individuals react and cooperate with urban areas in a path at no other time accomplished. From the majority of this information about where individuals are going, the best places visited can be observed. This stunning sign/signal about what millions of individuals are doing in this present reality at each snapshot of the day in urban areas all around the world can be captured. The real-time pulse of a city can be a valuable resource for understanding cities, giving a larger view of the collective movement patterns of millions of people. Taking a gander at these cities gives a feeling of how individuals go around, where there are, employment and business centers, and occasional inclinations. In Tokyo and Mumbai, the transportation framework is unmistakably noticeable as the nervous system of the cities.

The signs incorporate where you are, the spots you get a kick out of the chance to go, the time, the inclinations of your companions, and what is famous around you. Mapping some of these relationships is straightforward like when its a weekend night, individuals jump at the chance to get to bars, and a few connections are more perplexing like when its warm in Mumbai, individuals will probably go to cool Cafes. Historical data can enable very interesting services—for example, tracing when two users happened to be in the same place or attended the same event in the past, or helping predict future user locations. However, such data can also expose users to new threats; inferences on historical location data, for instance, can lead to loss of anonymity. Association of location with users and content is provided by context-aware services known as LBSNs that offer various services, like photo sharing, friend tracking, and *check-ins*. New privacy threats arise with the revealing of users' locations, which thus call for new protection assurance strategies.

3.1 Threats to Online LBSNs

An ever-increasing number of organizations are utilizing online networking adver-tisements not withstanding Google ads to promote their products. Soon, the News feed will be ruled by advertisements and organizations will spend more cash on them. It's additionally expected that online networking platforms will permit all the more promoting capacities, including computerized advertisements and pop-ups.

As a result of the rising online social networking patterns, more clients have gotten to be casualties of phishing and other digital violations. Such concerns call for better security in unveiling individual data on the web. In the coming years, online social networking platforms will likely have better security features which will counteract data fraud, digital stalking, and phishing. Table 1 lists some major threats along with the possible solutions that have been recognized in online social networks [11].

Table 1 Major online LBSNs threats with possible solutions

Online LBSNs threats	Possible solutions
Stolen identity	Careful status updates and strong passwords
Social profile hacking	Don't go for shortened urls
Increased stalking susceptibility	Limit posts on social networks
Location security	Limit posting outdoor travel plans in advance
Being over-confident using firewall and antivirus	Avoid posting personal information like birthdays, etc.
Cyberbullying	Take threats seriously, file messages as evidences
Physical attacks	Make proper use of privacy settings options

Fig. 2 Proposed framework for malicious social user detection

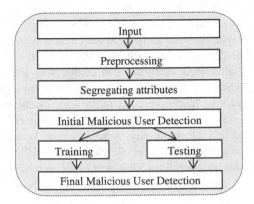

4   Proposed Framework: Malicious Social User Detection

The proposed framework, as shown in Fig. 2, begins with the information gathering from *Foursquare* social network. Information gushing technique considers nonstop information openness, which gives a stage to ongoing information accumulation. The datasets obtained are cleaned during the preprocessing. The motivation behind this part is to extricate important attributes from client metadata and distinctive messages posted on the system. The ML module can be used to learn the user behavioral patterns around the metadata and textual contents extracted.

To detect fake and compromised accounts, the initial Malicious User Detection (MUD) module verifies the accounts of users based on their profiles and respective check-ins and then passes the data to the classifier module. The classifier module processes and generates a classification accuracy based on the user check-ins. Finally the tested output is generated that is fed to the decision-maker at the end. Here again the ML approach is adopted to accurately detect the malicious social users. It is important to note here that this system is non-specific and abstracts main functions of every module. By joining content/behavioral and social user data with real-time analysis of data, an enhanced malicious social user detection system is feasible.

5 Experimental Evaluation and Result Analysis

5.1 Datasets Considered

Foursquare social network dataset [12] has been considered to obtain the user node check-ins of *New York* and *Tokyo* cities. The raw dataset consists of more than 2,150,000 nodes with more than 27,000,000 social-connections or directed edges. The directed edges are converted into undirected ones and only bi-directional edges are considered for evaluation. Thus, the dataset is trimmed to nearly 1,500,000 valid nodes with 8,650,000 undirected edges. The average degree and clustering coefficient of the network is calculated to be 11.5 and 0.3, respectively.

5.2 Implementation Procedure

As far as the implementation of the proposed framework for MUD in LBSNs is concerned, Python 3.6.0, a dynamic object-oriented programming language, is used. We divided our datasets into three ranges of training and testing data (10, 5 and 4-fold). We calculated the results using most of the ML Classifiers on the *Foursquare* datasets for New York and Tokyo cities. After rigorous experimentation, the best accuracy is obtained using the *Decision-Tree Classifier* and *Regressor* followed by *NB_Bernoulli*. On the other hand, *Nearest Centroid* and *K_Neighbors Regressor* give the least accuracy in detection of malicious users. We give the detailed results of the classification accuracy for MUD using our proposed framework in Table 2.

6 Conclusions and Future Directions

The results uncover that creating framework for malicious user detection in LBSNs is very challenging due to the avoidance strategies postured by malignant social users. We thoroughly analyzed the check-in patterns of social users in major world cities. The analysis draws peculiar patterns for the life-styles of the users with respect to the location-based services. We developed a novel procedure for *MUD* and obtained the best accuracy using the *Decision-Trees (Classifier and Regressor)* and *Bernoulli-NB* classifiers. In any case, there are spaces for improvements, for example, identifying features based on content/behavioral and network information to capture a large number of malicious accounts and their behaviors. A model for malevolent records location ought to address the rate at which social information is developing and advancing every day. Likewise, there is a requirement for better ways to deal with mining substantial interpersonal organization chart for social MUD. Along these lines, this paper predicts scalable *MUD* system as an imperative area for future research. Use of *ensembled-classifiers* can greatly improve the classification accuracy by combining best features of individual classifiers.

Table 2 Classification accuracy results for New York and Tokyo cities

Classifiers-used	New York			Tokyo		
	10F	5F	4F	10F	5F	4F
Decision-Tree classifier	**39.8**	**36.6**	**37.1**	**47.5**	**45.5**	**44.8**
Decision-Tree regressor	**30.8**	**28.2**	28.5	**36.5**	35.1	33.9
KNeighbors classifier	17.8	15.5	15.4	31.8	29.8	29.4
KNeighbors regressor	0.87	0.55	0.66	4.27	3.94	3.46
LDA_MLP classifier	10.6	10.4	**35.6**	35.6	**35.4**	**35.5**
LM_SGD classifier	4.53	3.36	7.41	2.85	3.55	9.77
NB_Bernoulli	8.67	7.04	7.07	35.6	35.4	35.5
NB_Gaussian	14.1	14.1	4.25	3.25	34.3	34.6
NB_Multinomial	1.52	1.46	1.44	3.86	3.67	3.77
Nearest centroid	0.29	0.18	0.22	0.15	0.08	0.11
NN_MLP classifier	11.4	11.7	**35.5**	**35.7**	**35.4**	**35.5**
Perceptron	5.67	1.66	7.41	2.54	3.54	2.70
SVM_SVC	**19.9**	**18.1**	17.1	35.3	35.3	35.4

References

1. Nguyen, H.: Research report: state of social media spam (2013). https://www.proofpoint.com/us/solutions/social-media-protection-and-compliance
2. Javelin Strategy and Research: Identity fraud: Fraud hits an inflection point (2016). https://www.javelinstrategy.com/coverage-area/2016-identity-fraud-fraud-hits-inflection-point
3. Ab Razak, M.F., Anuar, N.B., Salleh, R., Firdaus, A.: The rise of "malware": bibliometric analysis of malware study. JNCA **75**, 58–76 (2016)
4. Van Eecke, P., Truyens, M.: Privacy and social networks. Comput. Law Secur. Rev. **26**(5), 535–546 (2010)
5. Ahn, G.-J., Shehab, M., Squicciarini, A.: Security and privacy in social networks. IEEE Internet Comput. **15**(3), 10–12 (2011)
6. Narudin, F.A., Feizollah, A., Anuar, N.B., Gani, A.: Evaluation of machine learning classifiers for mobile malware detection. Soft. Comput. **1**, 1–15 (2014)
7. Singh, M., Bansal, D., Sofat, S.: Detecting malicious users in twitter using classifiers. In: Proceedings of the 7th International Conference on Security of Information and Networks, pp. 247 (2014)

8. Chu, Z., Gianvecchio, S., Wang, H., Jajodia, S.: Detecting automation of twitter accounts: are you a human, bot, or cyborg? IEEE Trans. Dependable Secur. Comput. **9**(6), 811–824 (2012)
9. Martinez-Romo, J., Araujo, L.: Detecting malicious tweets in trending topics using a statistical analysis of language. Expert Sys. Appl. **40**, 2992–3000 (2013)
10. Almaatouq, A., Shmueli, E., Nouh, M., Alabdulkareem, A., Singh, V.K., Alsaleh, M., Alfaris, A.: If it looks like a spammer and behaves like a spammer, it must be a spammer: analysis and detection of microblogging spam accounts. Int. J. Inf. Secur. **1**, 1–17 (2016)
11. Bishop, E.: 5 threats to your security when using social media (2017). http://www.adweek.com/digital/5-social-media-threats/
12. Mohamed, S., Justin, J.L., Eldawy, A., Mohamed F.M.: LARS*: a scalable and efficient location-aware recommender system. IEEE Trans. KDE **26**(6), 1384–1399 (2014)

An Analysis on Chronic Kidney Disease Prediction System: Cleaning, Preprocessing, and Effective Classification of Data

Ankit, Bhagyashree Besra and Banshidhar Majhi

Abstract *Chronic Kidney Disease* (CKD) is a condition in which the normal functioning of the kidney is highly affected. A prolonged duration in such case may deteriorate the kidney function or *"kidney failure"*. So, it is necessary to detect the disease and its different stages the patient may suffer from. Occurring of *CKD* in some cases may result in complications like anemia, high blood pressure, nerve damage, weak bones, heart, or blood vessel problems, etc. In this manuscript, we have proposed a mechanism that will generate a prediction of *CKD* with higher accuracy value, followed by the estimation of kidney damage percentage. The main objective of this analysis is to automate a prediction system that will diagnose the different stages in *CKD*. It starts with the preprocessing steps, ends with the classification, identifies the correctly classified instances, and then calculate its *Glomerular Filtration Rate* (GFR) value. Hence, it results in estimating the stages in *CKD* so that the patient may be treated well to avoid the kidney failure and help the patient retain a healthy kidney.

Keywords CKD · Prediction mechanism · Machine learning · GFR

1 Introduction

Chronic Kidney Disease (CKD) also termed as a chronic renal disease, is increasing failure of kidney functioning over a period of time. Sometimes, CKD is diagnosed with the risk of diseases like diabetes, high blood pressure, loss of appetite,

Ankit (✉) · B. Besra · B. Majhi
Department of Computer Science and Engineering, National Institute
of Technology Rourkela, Rourkela 769008, Odisha, India
e-mail: ankitguptamiet1500@gmail.com

B. Besra
e-mail: besrabhagyashree@gmail.com

B. Majhi
e-mail: bmajhi@nitrkl.ac.in

© Springer Nature Singapore Pte Ltd. 2019
P. K. Sa et al. (eds.), *Recent Findings in Intelligent Computing Techniques*,
Advances in Intelligent Systems and Computing 707,
https://doi.org/10.1007/978-981-10-8639-7_49

anemia, heart problems, etc. CKD is a prolonged form of kidney disease which may lead to kidney failure. The CKD is a five-staged disease that depends on the percentage of GFR. The higher the percentage of GFR, lesser is the chances of kidney dysfunctioning [1]. Also, the combination of GFR stages and albumin categories may give us a brief knowledge about the mortality of a patient. The higher the GFR range, the higher is the chances of death. The GFR value is estimated either from the two primal equations: (I) the MDRD equation and (II) the CKD-EPI equation. As, the MDRD equation has some limitations among the senior citizens ranging between 70 and order, basic focus lights on the CKD-EPI equation.

In this paper, the proposed preprocessing mechanism will generate more accurate prediction using machine learning and data mining concepts. Data mining has the role to data clean, data preprocessing, data reduction, etc. [2, 3]. These methods will qualify the data resulting in the quality of data mining. Now, the role of machine learning comes after the data mining part, i.e., the classification of class (CKD or NOTCKD), whether the patient suffers from CKD or not [4, 5]. We have implemented few new classification models, that gives better result than the existing ones. Then, we go for the GFR calculation where we can estimate the percentage of decay or the damage of kidney.

This manuscript has been settled into varied section as mentioned: Sect. 2 will provide a brief description about the conditions and constraints required for the materials. Section 3 represents the proposed method in details. Later, the Sect. 4 provides the observations and results justifying the proposed concept. Finally, the paper is concluded in Sect. 5.

2   Materials and Conditions

In this proposed work, the used repository is *UCI machine learning repository* for the dataset called *"chronic_kidney_disease"* with 400 instances × 25 attributes. It has missing attribute that is 1,012 out of 10,000, which is 10.12%, greater than 5%. As per the thumb rule for the number of missing attributes, it should not exceed 5% of the total [5]. For checking the quality as well as the quantity of the dataset that is being chosen needs to have the following factors mentioned in Table 1 below.

We have $400 \times 25 = 10,000$ records, 25 attributes each with more than 10 instances and CKD, NOTCKD have 250, 150 instances, respectively. Therefore,

Table 1 Rule of thumb

Objects	Rules	Status
Records	5,000 or more	✓
Attributes	For each attribute, 10 or more instances	✓
Targets	>100 for each class	✓

all the criteria for thumb rules are satisfied. A brief view of the dataset attributes can be visualized in the subsequent Table 1.

The tool being used is Netbeans IDE 8.2 with JDK 1.8.0, to code the classifiers and the whole phenomenon (data cleaning, data reduction, classification, etc.). After a series of classifications, the highest values are recorded.

3 Proposed Work

The three major steps in the paper are: (I) Preprocessing of the data, (II) Classification of the instances into CKD or NOTCKD, and (III) estimation of GFR percentage, will be discussed in detail.

3.1 Preprocessing

The preprocessing is divided into several techniques; data cleaning, data reduction, etc. In the next lines, we will describe a new approach of data mining that deals with the missing values, qualify the data, etc. The below shown Fig. 1 is the sequence of techniques that possible the accuracy in prediction.

By the rule of thumb; if the dataset has greater than 5% of missing attributes, then the values need to be imputed. So, the first step in this approach is to replace the missing values with either mean or standard deviation, in case of numerical type data and in nominal takes the default value. As mentioned earlier in Sect. 2, the obtained missing value is greater than 5%, therefore, the *replace by value*

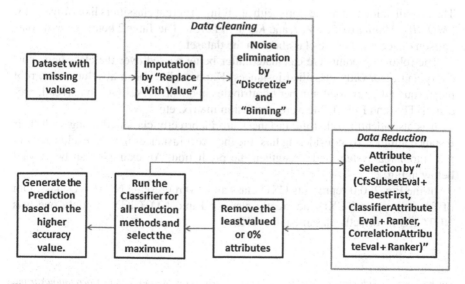

Fig. 1 The proposed preprocessing approach

Table 2 A comparison of classifiers

Classifiers	Percentage of accuracy
Naive-Bayes	99.10
SMO	99.55
IB1	99.90
Multiclassifier	98.30
VFI	99.48
Random forest	99.43

filter is applied on the dataset. The later step is to detect the noise or outliers. For such case, the filtering technique *discretize* is followed by *binning*. This ends the step of data cleaning, later technique is the data reduction. In data reduction, it does not just reduce the dimension of the data but also eliminates the duplicate ones. So, here the *attribute selection* method with a set of *attribute evaluator & search method*, such as *CfsSubsetEval+BestFirst*, *ClassifierAttributeEval+Ranker*, and *Correlation-AttributeEval+Ranker* is used.[1] After applying the *attribute selection* methods, the attributes with 0% content are removed. After the removal, the classifiers are runover the dataset and the results are noted down. Finally, the higher accuracy is counted for the prediction.

3.2 Classification

The classification technique starts with applying different classifiers like *Naive-ayes*, *SMO*, *IB1*, *Multiclassifier*, *VFI*, and *Random Forest*. The Table 2 below gives a comparison of accuracy of the classifiers on the dataset.

The following points are calculated after being passed over the classifiers: correctly and incorrectly classified instances, Kappa value, mean absolute error, root mean squared error, total number of attributes and instances, ROC area, recall, precision, TPR and FPR, F-Measure, confusion matrix, etc.

It is also identified that the instances are incorrectly classified along with their instance number. By considering this, the incorrect instances may be excluded from the automated system and strengthen the prediction. An example can be viewed below in Fig. 2.

In the Fig. 2, 0.0 represents CKD class and 1.0 represents NOTCKD class. Out of four, three were CKD incorrectly classified into NOTCKD and only one was NOTCKD incorrectly classified into CKD.

---

[1]In between *CfsSubsetEval+BestFirst*, *ClassifierAttributeEval+Ranker*, and *CorrelationAttributeEval+Ranker*, the later two gives almost the same result.

```
Correctly classified CKD:        247.0 with accuracy: 61.750000000000001%
Correctly classified NOTCKD:     149.0 with accuracy: 37.25%
Total number of Incorrectly classified elements:      4.0
Total number of Correctly classified elements:  396.0 with accuracy: 99.0%
Inst.    Actual        Predicted       Error                   Probability
60       0.0           1.0             0.13942121943549285     0.8605787805645072
88       1.0           0.0             0.9403163804634049      0.059683619536594984
136      0.0           1.0             0.4309362986693284      0.5690637013306716
184      0.0           1.0             0.003345052110531018    0.9966549478894688
Total number of Incorrectly classified elements: 4
BUILD SUCCESSFUL (total time: 2 seconds)
|
```

Fig. 2 Incorrectly classified instances with the error and probability percentage

3.3 GFR Estimation

The kidney functioning level and stages can be best measured using the test method GFR. The two major different approaches that may help us to determine the GFR number.

1. **The MDRD Equation**:

$$GFR\,(mL/min/1.73\,m^2) = 175\,(S_{cr}) - 1.154\,(Age) - 0.203\,(Gender)(Race) \quad (1)$$

   If in $Gender$, it is male multiply with 1.00 else if female, it is multiplied with 0.742. Race is only applicable for African Americans, if it the case is so, multiply with 1.212 else with 1.00.[2] The results are normalized to 1.73 m^2 of body surface area, as the equation is free from weight or height variables.

2. **The CKD-EPI Equation**:

$$GFR = 141\,min(S_{cr}/\kappa, 1)\alpha \times max(S_{cr}/\kappa, 1) - (1.209)(0.993)(Age)(Gen)(Race) \quad (2)$$

   where,

   - Gen is either male with value 1 or female with value 1.018,
   - $Race$ is 1.159 for black and 1 for others,
   - S_{cr} is serum creatinine in mg/dL,
   - κ is 0.7 for females and 0.9 for males,
   - α is -0.329 for females and -0.411 for males,
   - min indicates the minimum of $(S_{cr}/\kappa, 1)$, and
   - max indicates the maximum of $(S_{cr}/\kappa, 1)$.

   Applying the above two equations, we can be able to categorize them into the several stages of CKD, as such in above-cited Table 3.

---

[2]For patients of 70 and older has not been validated using the MDRD equation.

Table 3 Stages of chronic kidney diseases

Stages of chronic kidney diseases		GFR
Stage 1	Kidney damage with normal kidney function	90 or higher
Stage 2	Kidney damage with mild loss of kidney function	89–60
Stage 3a	Mild to moderate loss of kidney function	59–45
Stage 3b	Moderate to severe loss of kidney function	44–30
Stage 4	Severe loss of kidney function	29–15
Stage 5	Kidney failure	Less than 15

4 Observations and Results

The primal observation that has been obtained here is the estimation of GFR. The following Fig. 3 will give us a brief idea about how the GFR value is been calculated, the stages of CKD been categorized, and the mortality rate is been defined.

The later result can be calculated from the comparison between the GFR value and the albuminuria categories (albumin ranging from 0 to 5) as in Table 4. So, using this mechanism, the instances that has CKD with a particular GFR stage deducing the mortality risk can be found out.

From this above Table 4, it is obtained that the fatality rate in GFR stages along with the albuminuria types A1 is albumin ranging between (0 and 1), A2 is albumin ranging between (2 and 3) and A3 is albumin ranging between (4 and 5). As the stages increases, the fatality chances also increases. Using this method over our dataset, the following results as in the Fig. 4 is found out (Fig. 5).

The final result from the above Fig. 4 is that out of 250 instances, we have 103 instances with low mortality risk, 96 instances with moderate mortality risk, 45 instances with high mortality risk, 3 instances with very high mortality risk, and 3 with the highest mortality risk.

Fig. 3 GFR estimation and CKD stages classification

```
the total number of instances is: 400
kidney diseases are found in: 250 instances
No kidney diseases are found in: 150 instances
the total number of stage1 are: 147
the total number of stage2 are: 52
the total number of stage3a are: 34
the total number of stage3b are: 14
the total number of stage4 are: 1
the total number of stage5 are: 2
BUILD SUCCESSFUL (total time: 0 seconds)
```

Table 4 Mortality rate predictor

		Albuminuria categories		
		A1	A2	A3
GFR Stages	Stage 1	Low risk	Moderate risk	High risk
	Stage 2	Low risk	Moderate risk	High risk
	Stage 3a	Moderate risk	High risk	Very high risk
	Stage 3b	High risk	Very high risk	Very high risk
	Stage 4	Very high risk	Very high risk	Highest risk
	Stage 5	Highest risk	Highest risk	Highest risk

Fig. 4 Mortality rate using the Table 4

```
the total number of instances is: 400
kidney diseases are found in: 250 instances
No kidney diseases are found in: 150 instances
the total number of stage1 are: 147
the total number of stage2 are: 52
the total number of stage3a are: 34
the total number of stage3b are: 14
the total number of stage4 are: 1
the total number of stage5 are: 2
the total number of low mortality rate are: 103
the total number of moderate mortality rate are: 96
the total number of high mortality rate are: 45
the total number of very high mortality rate are: 3
the total number of highest mortality rate are: 3
BUILD SUCCESSFUL (total time: 1 second)
```

Fig. 5 Graph showing the percentage of mortality risk in database

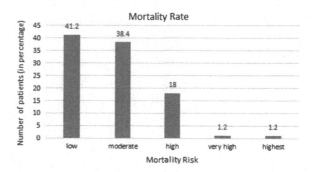

5 Conclusion

This manuscript holds the proposed preprocessing technique, classification with higher accuracy, estimation of GFR and finally, deduce the mortality of patients. Our prediction accuracy has been ranged from 98.30 to 99.55% with reference to the classifier. Out of the total population, only 250 people have been suffering from CKD, and among them, only 1.2% belongs to highest mortality risk. Therefore, in the proposed work, it can plot a patients chance to survive the CKD with respect to the test results, i.e., GFR value, which may help the medical professionals to detect and diagnose the disease more accurately and precisely.

References

1. Jena, L., Kamila, N.: Distributed data mining classification algorithms for prediction of chronic kidney disease. Int. J. Emerg. Res. Manag. Technol. 4(11). ISSN: 2278-9359
2. Durairaj, M., Sivagowry, S.: A pragmatic approach of preprocessing the data set for heart disease prediction. Int. J. Innov. Res. Comput. Commun. Eng. 2(11) (2014)
3. Singhal, S., Jena, M.: A study on weka tool for data preprocessing classification and clustering. Int. J. Innov. Technol. Exploring Eng. (IJITEE) 2(6), 2013. ISSN: 2278-3075
4. Patel, J., Upadhyay, T., Patel, S.: Heart disease prediction using machine learning and data mining technique. IJCSC 7(1), 129–137, Sept 2015–Mar 2016. http://www.csjournalss.com
5. Farhangfar, A., Kurgan, L., Dy, J.: Impact of imputation of missing values on classification error for discrete data. Elsevier 41(12), 3692–3705 (2008)

Part V
Cloud Computing, Distributed Systems, Social Networks, and Applications

Comparison of Fault Tolerance Amid Various Irregular MINs

Ved Prakash Bhardwaj, Piyush Chauhan, Ankit Khare, Alok Jhaldiyal and Nitin

Abstract Multi-Stage Interconnection Networks (MINs) are being used in various parallel processing applications. Main concern of this research work is to check out the fault tolerability of recently proposed MINs and find out the best one among all the considered MINs. Here "faulty" means the switching element (SE) which is in busy state or dead state. One theorem is also proposed in the paper that basically checks the fault tolerability of a MIN. Fault analysis is done for all the MINs, i.e., from three-stage interconnection networks to five-stage interconnection networks. Finally, a highly fault tolerant MIN is obtained among all the discussed MINs.

Keywords Interconnection networks · Multi-stage interconnection network
Switching element · Fault · Fault tolerance

1 Introduction

Parallel computing is one of the vibrant research fields in computer architectures and networking [1]. Interconnection networks (INs) are most widely used in all types of wired network connections [2]. Providing connectivity is the basic functionality of INs [2]. It connects the modules (e.g., memory, processors) either in direct or

V. P. Bhardwaj (✉) · P. Chauhan · A. Khare · A. Jhaldiyal
University of Petroleum and Energy Studies, Dehradun, India
e-mail: ved.juit@gmail.com

P. Chauhan
e-mail: shbichauhan@gmail.com

A. Khare
e-mail: ankit.khare86@gmail.com

A. Jhaldiyal
e-mail: alok.jhaldiyal@gmail.com

Nitin
Jaypee Institute of Information Technology, Noida, India
e-mail: delnitin@gmail.com

© Springer Nature Singapore Pte Ltd. 2019
P. K. Sa et al. (eds.), *Recent Findings in Intelligent Computing Techniques*,
Advances in Intelligent Systems and Computing 707,
https://doi.org/10.1007/978-981-10-8639-7_50

indirect fashions [3, 4]. MIN is a concept that has its significance due to its extraordinary features like low cost, fast data transmission speed, reliability, accuracy, and many more [1, 5]. In recent years, plenty of MINs have been proposed by various researchers [6, 7]. Further, a lot of research is still required on various factors of performance like throughput, cost, and fault tolerance. Here, authors are more focused on fault tolerance. It is one of the essential factors for every MIN [8, 9]. If a MIN is highly fault tolerable however, lacking in efficiency then it can be improved [10, 11]. For this purpose, one can apply new routing strategies and can work on other factors of performance [10, 12]. Hence, in the current research work [13, 14] the authors have put their efforts in order to get a MIN which is highly fault tolerable [6, 7].

In Sect. 2, literature review on fault tolerability is presented. Section 3 is analyzing the fault tolerability of various MINs. Conclusion and future work are given in Sect. 4. References are presented in last section.

2   Literature Review on Fault Tolerability

Fault Tolerance means, the capacity of sustaining the faults. It can be defined as:

> The basic idea for fault tolerance is to provide multiple paths between source and destination so that alternate paths can be used in case of faults [6]

Based on the fault tolerability the authors have categorized the MINs which are as follows.

2.1   Single Switch Fault Tolerant MIN

The MIN which can tolerate only single faulty switch in a specific stage at a particular time comes in this category.

2.2   Double Switch Fault Tolerant MIN

The MIN which can tolerate two faulty switches in a specific stage at a particular time comes in this category.

Table 1 Types of MINs based on number of stages

Name of MIN	Complete name	Type of MIN
IASEN	Irregular Augmented Shuffle Exchange Network	3-stage
IMABN	Irregular Modified Augmented Baseline Network	3-stage
IAON	Irregular Advance Omega Network	3-stage
IASEN-3	Irregular Augmented Shuffle Exchange Network-3	3-stage
AIAMIN-2	Advanced Irregular Alpha Multi-Stage Interconnection Network-2	3-stage
MALN	Modified Alpha Network	4-stage
IASEN-2	Irregular Augmented Shuffle Exchange Network-2	4-stage
AIAMIN	Advanced Irregular Alpha Multi-Stage Interconnection Network	4-stage
MALN-2	Modified Alpha Network-2	5-stage
IMABN-2	Irregular Modified Augmented Baseline Network-2	5-stage
AIASEN	Advance Irregular Augmented Shuffle Exchange Network	5-stage

2.3 N Switch Fault Tolerant MIN

The MIN which can tolerate N faulty switches in a specific stage at a particular time comes in this category, where N is the total number of faulty switches. Further; the authors have taken ample number of MINs and compare them based on their fault tolerability. Table 1 [6–16] has shown all the considered MINs.

All these MINs provide a strong connectivity between every source node and every destination node. Every MIN has a unique functionality for the data transmission.

3 Analysis of Fault Tolerability

In the literature survey [1–16], it is observed that all the discussed MINs are fault sustainable however; it is required to check their fault tolerability in every stage of network. Therefore, the authors have proposed the following theorem:

3.1 Theorem1

If a switching element is connected with n different switching elements then the MIN can tolerate maximum n − 1 faults.

Prove 3.1
MINs consist of a particular number of stages and every stage has a certain number of SEs. Every SE comprises of a set of input lines and a set of output lines. To prove

the theorem, the authors are more focused on output lines of every SE and it is shown by the following equation:

$$FT_{ij} = TOL - 1, \tag{1}$$

where FT_{ij} is the fault tolerability of a switch which exists in i-th stage at j-th position and TOL is the total number of output links of a switch. Therefore, if a switching element s_1 has n output lines it shows that it is connected with n number of switching elements. Hence, if n   1 switching elements are faulty then data can be transmitted through the n-th switching element. It proves the theorem.

3.2   Stage-Wise Analysis of Fault Tolerance

In this section, the authors have checked the fault tolerability of all the MINs which are shown in Table 1. To check this issue, they have applied the theorem from first stage to the second last stage of every MIN of Table 1. In last stage, the data packets can be reached to their given destination with ordinary links or auxiliary links, therefore, it is not considered. After applying the theorem, the authors have got the maximum number of faulty SEs that cannot affect the performance of the MIN.

To analyze the fault tolerability in stage wise manner, the authors have counted the maximum faulty SEs of each stage except the last one. Here, "maximum faulty SEs" means SEs that cannot affect the performance of a MIN, since if all the SEs are faulty in a particular stage then network will get blocked. The Theorem 3.1 is applied on all the MINs which are shown in Table 1 [6–16].

Further, the authors have analyzed all the MINs based on their number of stages. It is shown in Table 2 [6, 8, 11, 15, 16], Table 3 [7, 9, 14] and Table 4 [10, 12, 13]. In every table, CFT (Complete Fault Tolerability of a MIN) has been calculated to find out the highly fault tolerable MIN. Further, the authors have used some symbols (see Tables 2, 3, and 4) which are explained here

MIN   Multi-stage Interconnection Network
FT_1 Fault Tolerability in stage 1
FT_2 Fault Tolerability in stage 2
FT_3 Fault Tolerability in stage 3

Table 2 Fault tolerability of 3-stage interconnection network

MIN	FT_1	FT_2	CFT
IASEN	2	2	4
IMABN	2	4	6
IAON	3	4	7
IASEN-3	2	7	9
AIAMIN-2	2	7	9

Table 3 Fault tolerability of 4-stage interconnection network

MIN	FT$_1$	FT$_2$	FT$_3$	CFT
MALN	2	2	2	6
IASEN-2	2	2	8	12
AIAMIN	2	1	7	10

Table 4 Fault tolerability of 5-stage interconnection network

MIN	FT$_1$	FT$_2$	FT$_3$	FT$_4$	CFT
IMABN-2	2	1	2	7	12
AIASEN	1	2	1	7	11
MALN-2	2	1	7	7	17

3.2.1 Complete Analysis of Fault Tolerability

Here, the authors have presented the fault tolerability of all the discussed MINs. The horizontal axis of Fig. 1 is representing the MINs means the multi-stage interconnection network. The vertical axis shows, the total number of faulty switching elements that a MIN can tolerate.

In category of 3-stage network, IASEN-3 and AIAMIN-2 are better than other MINs of this category. In category of 4-stage network, IASEN-2 is better than other MINs of this category. In category of five-stage network, MALN-2 is better than other MINs of this category. Overall, it can be analyzed from Fig. 1 that MALN-2 is highly reliable as compared to all the considered MIN since it can tolerate 17 faulty SEs and can deliver the data packets from source end to the destination end.

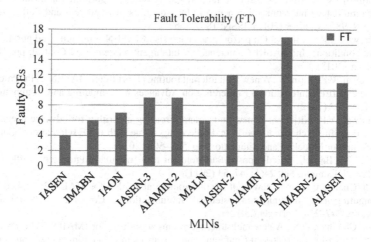

Fig. 1 Fault tolerability of all MINs

4 Conclusion and Future Scope of Work

In this paper, authors have worked on a very sensitive issue, i.e., fault tolerance. Fault creates troubles for the data packets in their routes. Therefore, a strong research has been done in this paper in order to find out a highly fault tolerant network. The complete analysis shows that MALN-2 is better than all the discussed MINs. Now to obtain highly fault tolerant MIN in three-stage and four-stage is a critical challenge. In future, authors will focus on getting highly reliable MINs for three-stage and four-stage networks.

References

1. Kumar, B.V.S., Rao, M.V., Prasad, M.A.R.: Adaptive fault tolerant routing in interconnection networks: a review. Int. J. Adv. Netw. Appl. **02**(06), 933–940 (2011)
2. Sadi, J.A., Day, K., Khaoua, M.O.: A fault tolerant routing algorithm for 3-D torus interconnection networks. Int. Arab J. Inf. Technol. **1**, 69–79 (2003)
3. Chen, C.W.: Design schemes of dynamic rerouting networks with destination tag routing for tolerating faults and preventing collisions. J. Supercomput. **38**(3), 307–326 (2006)
4. Fan, C.C., Bruck, J.: Tolerating multiple faults in multistage interconnection networks with minimal extra stages. IEEE Trans. Comput. **49**(9), 998–1004 (2000)
5. Mahajan, R., Vig, R.: Performance and reliability analysis of new fault-tolerant advance omega network. WSEAS Trans. Comput. **7**(8), 1280–1290 (2008)
6. Bhardwaj, V.P., Nitin, N.: Message broadcasting via a new fault-tolerant irregular advanced omega network in faulty and non-faulty network environments. J. Comput. Electr. Eng. (Hindawi Publishing Corporation) **2013**, 1–17 (2013)
7. Bhardwaj, V.P., Nitin, N.: A new fault tolerant routing algorithm for advance irregular alpha multistage interconnection network. In: Advances in Intelligent and Soft Computing, pp. 979–987. Springer (2012). ISSN: 1867-5662
8. Bhardwaj, V.P., Nitin, N.: On performance analysis of IASEN-3 in faulty and non-faulty network conditions. In: AASRI Conference on Intelligent Systems and Control, pp. 104–109. Elsevier (2013)
9. Bhardwaj, V.P., Nitin, N.: A new fault tolerant routing algorithm for IASEN-2. In: Proceedings of the 2nd IEEE International Conference on Advances in Computing and Communications, pp. 199–202 (2012)
10. Bhardwaj, V.P., Nitin, N.: A new fault tolerant routing algorithm for advance irregular augmented shuffle exchange network. In: Proceedings of the 14th IEEE International Conference on Computer Modeling and Simulation, pp. 505–509 (2012)
11. Gupta, A., Bansal, P.K.: Proposed fault tolerant new irregular augmented shuffle network. Malays. J. Comput. Sci. **24**(1), 47–53 (2011)
12. Nitin, Chauhan, D.S.: A new fault tolerant routing algorithm for MALN-2. In: Eco-friendly Computing and Communication Systems, Communications in Computer and Information Science, pp. 247–254. Springer (2012)
13. Nitin, Chauhan, D.S.: A new fault-tolerant routing algorithm for IMABN-2. In: Proceedings of the 2nd IEEE International Conference on Advances in Computing and Communications, pp. 215–218 (2012)

14. Gupta, A., Bansal, P.K.: Fault tolerant irregular modified alpha network and evaluation of performance parameters. Int. J. Comput. Appl. **4**(1), 9–13 (2010)
15. Ghai, M.: A routing scheme for a new irregular baseline multistage interconnection network. Int. J. Adv. Comput. Sci. Appl. **2**(5), 83–86 (2011)
16. Bhardwaj, A., Shiwani, S.: On performance evaluation of advance irregular alpha multistage interconnection networks-2. Int. J. Comput. Appl. **102**(2), 16–20 (2014)

A Novel Hybrid Algorithm for Overlapping Community Detection in Social Network Using Community Forest Model and Nash Equilibrium

Aparna Sarswat and Guddeti Ram Mohana Reddy

Abstract Overlapping community detection in social networks is known to be a challenging and complex NP-hard problem. A large number of heuristic approaches based on optimization functions like modularity and modularity density are available for community detection. However, these approaches do not always give an optimum solution, and none of these approaches are able to clearly provide a stable overlapping community structure. Hence, in this paper, we propose a novel hybrid algorithm to detect the overlapping communities based on the community forest model and Nash equilibrium. In this work, overlapping community has been detected using backbone degree and expansion of the community forest model, and then a Nash equilibrium is found to get a stable state of overlapping community arrangement. We tested the proposed hybrid algorithm on standard datasets like Zachary's karate club, football, etc. Our experimental results demonstrate that the proposed approach outperforms the current state-of-the-art methods in terms of quality, stability, and less computation time.

Keywords Overlapping community detection · Nash equilibrium · Community forest model · Backbone degree

1 Introduction

Community detection is challenging and an important research area to analyze hidden knowledge and structure of the social network. Community detection can be used in different real-world applications like web graphs, biological networks, and social networks to discover the structure and behavior of the members of the community. Social networks use interactions among the objects so that they can be represented in the form of a network graph where members can be considered as nodes and the

A. Sarswat (✉) · G. Ram Mohana Reddy
National Institute of Technology Karnataka, Surathkal, Mangalore, India
e-mail: sarswataparna@gmail.com

G. Ram Mohana Reddy
e-mail: profgrmreddy@gmail.com

© Springer Nature Singapore Pte Ltd. 2019
P. K. Sa et al. (eds.), *Recent Findings in Intelligent Computing Techniques*,
Advances in Intelligent Systems and Computing 707,
https://doi.org/10.1007/978-981-10-8639-7_51

inter-connections between the objects can be considered as edges of the network graph.

Social networks can have disjoint as well as overlapping community structure. Disjoint community detection defines the boundaries between communities, while overlapping community detection discovers the membership of each node in multiple communities since an individual can be present in several communities in real-world scenario. Disjoint and overlapping community detection can be done using various existing approaches like modularity maximization [1], clique percolation [2], line graph and link partitioning [3], modularity density based optimization [4], etc.

In clique percolation, the resolution for community detection depends on a discrete value k. Hence, the result obtained using this approach varies based on the value of k. Line graph and link partitioning [3] method is based on link partitioning instead of nodes partitioning but the definition of community considered in this method is ambiguous. Hence, this method may not provide higher quality results as compared to the node-based detection. Modularity-based optimization may fail to detect smaller communities, and further it works on the optimization of the objective function but it ignores the detection of the internal structure and boundaries of communities. Also, modularity-based approach requires complicated iterative procedure, and hence this approach cannot be enhanced to work in distributed systems. Thus, a substantial amount of work has already been done to detect the overlapping communities, but none of these works can give a complete and stable insight of community structure in terms of disjoint communities and overlapping nodes within the communities.

These challenges motivated us to propose a novel hybrid algorithm based on community forest model (CFM) and Nash equilibrium. The core idea behind this hybrid approach is to get disjoint and overlapping community structure using the backbone degree and expansion from CFM and then use reassignment function based on stability condition referred to as Nash equilibrium [5]. CFM uses the backbone degree to compute the strength of the edges. Expansion is used to detect boundaries in the form of disjoint communities, and overlapping nodes after a new vertex is added to the community. Then, a reassignment function is used to change the communities while considering the wishes of each vertex. Results obtained finally by a Nash equilibrium state not only provide a way to reach an optimum community structure with considerably less computational time but also provide a stable community structure. This reassignment function based on Nash equilibrium is used to detect and overcome any instabilities in the community membership found so far using CFM.

The key contribution of the proposed work is as follows:

- This is the first hybrid approach to overlapping community detection in social networks which efficiently detects internal structure and boundaries of community with low computation cost and tackles instability in overlapping community detection using Nash equilibrium.

Reminder of this paper is organized as follows. Section 2 deals with summary of existing works on community detection. Section 3 describes the proposed work. Section 4 discusses the performance evaluation in terms of results and analysis and finally, Sect. 5 concludes this paper with future direction.

2 Related Work

There are several existing algorithms related to community detection in complex social networks proposed by different researchers. Some of the important existing approaches related to social network community detection are summarized in Table 1.

There are many more algorithms like clique percolation [2], F-measure metric [10], MMSB model [11], Louvain method [12], backbone degree algorithm [13], and extended modularity Q_m [14]. Clique percolation assumes that a community has sets of fully connected subgraphs, and it explores the adjacent cliques to detect overlapping communities. MMSB model computes the similarity between vertices which can be measured by calculating probability of mixed membership, and thus it helps to determine the probability of membership of a vertex toward a community.

Table 1 Summary of existing works

Author	Methodology	Merits	Demerits
Vidyadhari Jami and G Ram Mohana Reddy [4]	Hybrid bioinspired community detection in social networks.	Good results without any prior information about number of communities	Cannot provide overlapping community structure
Xuyun Wen et al. [6]	A maximal clique-based multi-objective evolutionary algorithm for overlapping community detection	Detects overlapping community with higher partition accuracy and lower computational cost	Genetic algorithm may fall in local maxima
Ismail El-Helw et al. [7]	Parallel approach to detect fast overlapping communities	Fast computation compared to existing approaches	Performance evaluation using standard metrics is required
Yunfeng Xu et al. [8]	Finding overlapping community using community forest model	Simple approach to identify communities without complex objective function maximization	Computation cost is more and parallel processing is required
Michel Crampes and Michel Planti [5]	Overlapping community detection optimization using Nash equilibrium	This approach may lead to a Nash equilibrium in a polynomial time	Adding or deleting policies can give rise to unstable or unsatisfactory situations
Mihai Suciu et al. [9]	Community structure as game equilibrium	Results obtained are better or comparable with other methods in the literature	More appropriate node fitness function is required to capture the contribution of the node in its community

3 Proposed Work

We proposed a hybrid framework based on the community forest model (CFM) and Nash equilibrium to detect the overlapping community structure in social network. The proposed framework can be divided into the following phases.

3.1 Backbone Degree Computation Phase

The goal of this phase is to compute the degree for each backbone in the given network graph using Eq. 1 and add them to a list in the descending order.

Definition 1 A backbone is an edge with two vertices that are connected to the edge.

BD_{uv} is referred to as backbone degree having vertices $(u,v) \in$ edge (uv).

$$BD_{uv} = (W_u + W_v) \times OP_{uv} + \delta \tag{1}$$

where W_u = weight of vertex u in network. W_v = weight of vertex v in network. OP_{uv} = Neighborhood overlap [15]. We let δ as smoothing parameter with value of 0.01 based on experience [13].

3.2 Neighbors Computation and Expansion Phase

In this phase, we select the backbone having maximum degree and add it as the core backbone of the initially empty community. Then, we find the neighbors of nodes of the current community and thus create a neighbor set out of it. The goal of this phase is to compute the sum of backbone degree (SE) for each neighboring vertex (v) in neighbor set using Eq. 2 given by

$$SE_v = \sum_{u \in C and\, v \notin C} BD_{uv} \tag{2}$$

where BD_{uv} is the backbone degree of edge uv and C is the community.

Add vertex with maximum SE to the community. Then, expansion (E_c) is computed using Eq. 3 given by

$$E_c = \frac{|C_{uv}|}{|C|} \tag{3}$$

where $|C_{uv}|$ = number of edges E such that $(u,v) \in$ E,
$u \in C$ and $v \notin C$. $|C|$ = number of nodes in community C.

If the expansion value gets reduced after joining the new neighboring vertex to the community, then we keep on adding the neighboring vertices with the maximum SE. Else we include the neighboring vertex as a part of the boundary set of the community and continue this process of finding the neighboring vertex with maximum SE until the neighbor set is empty. We keep on dividing the rest of the vertices to form other communities, until no backbone with BD_{uv} > threshold (T) is available in the list or the number of the remaining vertices is less than parameter k.

3.3 Nash Equilibrium Computation Phase

In this phase, we collect the resultant communities and create a matrix of size (vertex × communities) and then calculate the reassignment modularity value RE (Eq. 4) given by

$$RE_{i:c1 \to c2} = \frac{(N_{i|2} - N_{i|1})}{m} - \frac{[deg_i^2 + deg_i(s_{c2} - s_{c1})]}{2m^2} \qquad (4)$$

where $RE_{i:c1 \to c2}$ is the reassignment modularity for node i from community c1 to community c2. $N_{i|j}$ is number of edges between vertex i to i' such that i' $\in$ j. deg_i is the degree of node i. s_{c_i} = sum of degree of each node in community c_i. m is equal to |E|.

The maximum RE value corresponding to each vertex is computed. If there is any vertex with positive RE value, then we reassign the vertex to RE column's community. We repeat this approach until no vertex has positive RE value and finally leading to an equilibrium state known as Nash equilibrium [5].

Algorithm 1 shows our proposed hybrid framework based on CFM model and Nash equilibrium.

3.4 Time Complexity of Proposed Hybrid Algorithm

Here, merge sort is used for sorting the backbone list and it takes $O(m.logm)$. Remaining steps of community detection take $O(n + m)$ time. Hence, total running time is $O(m.logm + n + m)$ for n vertices and m edges in network. But we did not consider all the backbones since we have chosen the backbone list based on the threshold T, the size of the filtered backbone list is significantly reduced, and hence the complexity $O(m.logm)$ will be reduced sharply. Hence, CFM has time complexity of $O(n + m)$ approximately. Now, full computation time of RE is in polynomial time and the convergence of algorithm toward the Nash equilibrium is also polynomial. Thus, our proposed hybrid algorithm for overlapping community detection provides a stable optimal solution in polynomial time of $O(n + m)$ approximately.

Algorithm 1 Hybrid Framework using CFM Model and Nash Equilibrium

---

1: **Input** : Network dataset graph G with vertex V and edges E. NVL $\Rightarrow |V|$, BL $\Rightarrow$ sorted list of BD_{uv} greater than T in descending order. k$\Rightarrow \frac{|V|}{10}$. BV $\Rightarrow$ Boundary Vertex Set.

2: **Output** : List with disjoint and overlapping communities(RL).

3: Select e from BL and consider (u,v)$\in$e . set i=0.

4: **for** <l=1 to k > **do**

5:     **if** u $\in$ NVL v $\in$ NVL **then**

6:         $c_i \Leftarrow u, v$

7:         calculate expansion pre_exp using Equation 3 .

8.   calculate neighbor vertex set NB of community i.

9:         BV$\Leftarrow \emptyset$

10:         **if** (NB-BV) is $\emptyset$ **then**

11:             *add c_i to RL, increment i and go to step 3.*

12:         **else**

13:             find neighbor vertices for (NB-BV) and select vertex with maximum SE. add v to community and calculate E_c of the community and save it as new_exp.

14:             **if** new_exp-pre_exp<0 **then**

15:                 remove v from NVL and keep in c_i , go to step 13.

16:             **else**

17:                 remove v from c_i and move v to BV.

18:                 **if** (NB-BV) is $\emptyset$ **then**

19:                     *add c_i to RL, increment i and go to step 3.*

20:                 **else**

21:                     go to step 13.

22:     **else**

23:         go to step 3.

24: collect community list RL and if any remaining vertices which are not members of any community yet.

25: create a matrix Nash_mat vertex$\times$ communities.

26: calculate RE for this matrix.

27: Nash_equlibrium_found=false

28: **while** Nash_equlibrium_found=false **do**

29:     calculate maximum of RE from Nash_mat for each vertex.

30:     **if** maximum RE for any vertex i > 0 **then**

31:         Reassign i to community with positive RE.

32:         compute RE for Nash_mat.

33:     **else**

34:         Nash_equlibrium_found=true

---

4 Performance Evaluation

4.1 Datasets Used

We tested the proposed algorithm on karate [16], football [17], dolphin [18], LFR [19], and Net-Science [20] datasets. Intel(R) core(TM) 2 duo 2.00 GHz processor, 1TB hard disk, and windows 7 64-bit operating system is used for our experiments. Python and NetworkX packages are used for implementation of the algorithm. The threshold value T can be fixed on the basis of experiments and requirements of the

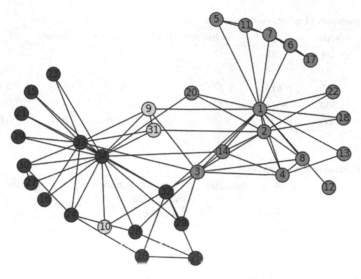

Fig. 1 Zachary's karate club dataset

user. Based on our experiments, $T = 0.3$ is considered for the performance evaluation of our proposed hybrid algorithm.

4.1.1 Zachary's Karate Club

Zachary [16] is a network of 34 friends in karate club. There are 34 vertices and 78 edges in this dataset. We got two real communities and three overlapping nodes as 9, 10, and 31. Figure 1 shows two communities in two different colors in karate dataset. Nodes in yellow color are overlapping nodes.

4.1.2 Football Dataset

Girvan and Newman [17] has 115 vertices and 613 edges. We found 15 communities with 33 overlapping nodes after calculating the backbone degree and reassignment modularity for Nash equilibrium. We found that there is no instability since all RE values are 0 or negative.

4.1.3 Dolphin Dataset

Lusseau et al. [18] contains 62 vertices and 159 edges. We got four communities based on the backbone degree computation and after applying reassignment modularity for Nash equilibrium.

Table 2 Summary of results on different datasets

Dataset	Nodes	Edges	#Overlapping nodes	Precision	Recall	F-score
Karate	34	78	3	1	1	1
Dolphin	62	159	31	0.91	0.90	0.90
Football	115	613	33	0.90	0.89	0.89
LFR500	500	3614	142	0.93	0.74	0.83
LFR1000	1000	7621	316	0.98	0.70	0.81

Table 3 Comparison based on NMI

Dataset	GA-Net algorithm	Two-step genetic algorithm	Our algorithm
Karate	0.7071	0.8910	1.0
Dolphin	0.89	0.8448	0.899
Football	0.8825	0.8810	0.891977

4.1.4   LFR Dataset

LFR [19] program produces the network with overlapping communities as per the input parameters given by the users. We used this algorithm to produce 500 nodes and 1000 nodes of the benchmark network graph.

4.1.5   Net-Science Dataset

Net-Science [20] contains 1589 vertices and 2742 edges. We got 202 communities based on the backbone degree computation and after applying the reassignment modularity for Nash equilibrium.

4.2   Results and Analysis

Table 2 shows the results of different datasets about number of overlapping nodes with precision, recall, and F-score.

Normalized Mutual Information (NMI) Table 3 shows the comparison of our proposed algorithm with two-step genetic algorithm [1] and GA-Net [21] algorithm based on NMI.

Comparison based on Extended Modularity (Q_m) Table 4 shows the comparison of our proposed algorithm with the different algorithms based on extended modularity.

Table 4 Comparison based on Q_m

Dataset	CFM	CPM	Louvain	MMSB	Our algorithm
Karate	0.527	0.501	0.431	Null	0.8294
Football	0.305	0.646	0.594	0.167	0.87753
Net-Science	0.850	0.700	0.521	0.256	0.88451

Table 5 Run-time comparison

Algorithm	CPM	NLA	CFM	Louvain method	GA-Net algorithm	Two-step genetic algorithm	Our algorithm
Time complexity	NP-complete	$O(n^2)$	$O(n+m)$	$O(n.logn)$	$*O(p.s.n')$	$*O(p.s.n')$	$O(n+m)$

*Here p is the population size, s is the size of individuals, and n' is the number of generations

Run-Time Comparison The run-time comparison of our proposed hybrid algorithm (using CFM and Nash equilibrium) with different algorithms discussed in Sect. 2 is shown in Table 5. Here, n is the number of vertices and m is the number of edges.

5 Conclusion and Future Work

In this paper, we mainly focused on overlapping community detection in social networks to understand the network structure. We proposed overlapping community detection based on parameters like expansion and backbone degree which has given quality results. No complicated optimization of the objective function is required in our algorithm, and there is no need of iterative and sampling approach for our algorithm. This makes our approach more efficient in terms of running time without compromising with the quality of results.

This algorithm thoroughly provides the overlapping characteristics of social network and does not require complicated mathematical calculation. To provide a stable solution, we used Nash equilibrium based reassignment function and showed that our approach provides stable and quality overlapping community in polynomial time. Our proposed approach can be applied to any other community detection algorithm in order to get a stable equilibrium in a polynomial time. Finally, we observed that computation time of our approach is relatively less compared to existing approaches. Our future research will enhance the reassignment methods used for Nash equilibrium and identify how frequent adding or deleting new nodes in very large-scale network can be tackled so that instability can be avoided.

References

1. Meena, J., Devi, V.S.: Overlapping community detection in social network using disjoint community detection. In: 2015 IEEE Symposium Series on Computational Intelligence. IEEE (2015)
2. Palla, G., et al.: Uncovering the overlapping community structure of complex networks in nature and society. Nature **435.7043**, 814–818 (2005)
3. Evans, T.S., Lambiotte, R.: Line graphs, link partitions, and overlapping communities. Phys. Rev. E **80**(1), 016105 (2009)
4. Jami, V., Ram Mohana Reddy, G.: A hybrid community detection based on evolutionary algorithms in social networks. In: 2016 IEEE Students Conference on Electrical, Electronics and Computer Science (SCEECS). IEEE (2016)
5. Crampes, M., Planti, M.: Overlapping community detection optimization and nash equilibrium. In: Proceedings of the 5th International Conference on Web Intelligence, Mining and Semantics. ACM (2015)
6. Gong, M.-G., et al.: Community detection in dynamic social networks based on multiobjective immune algorithm. J. Comput. Sci. Technol. **27.3**, 455–467 (2012)
7. El-Helw, I., Hofman, R., Bal, H.E.: Towards fast overlapping community detection. In: 16th IEEE/ACM International Symposium on Cluster, p. 2016. Cloud and Grid Computing (CCGrid), IEEE (2016)
8. Xu, Y., et al.: Finding overlapping community from social networks based on community forest model. Knowl.-Based Syst. (2016)
9. Kohlberg, E., Mertens, J.-F.: On the strategic stability of equilibria. Econometrica **54**(5), 1003–1037 (1986)
10. Banerjee, A., et al.: Model-based overlapping clustering. In: Proceedings of the Eleventh ACM SIGKDD International Conference on Knowledge Discovery in Data Mining. ACM (2005)
11. Airoldi, E., M., et al.: Mixed membership stochastic blockmodels. J. Mach. Learn. Res. 1981–2014, 9 Sep 2008
12. Blondel, V.D., et al.: Fast unfolding of communities in large networks. J. Stat. Mech. Theory Exp. **10**, 10008 (2008)
13. Xu, Y., Hua, X., Zhang, D.: A novel disjoint community detection algorithm for social networks based on backbone degree and expansion. Expert Syst. Appl. **42**(21), 8349–8360 (2015)
14. Li, J., Wang, X., Cui, Y.: Uncovering the overlapping community structure of complex networks by maximal cliques. Phys. Stat. Mech. Appl. **415**, 398–406 (2014)
15. Easley, D., Kleinberg, J.: Networks, Crowds, and Markets: Reasoning About a Highly Connected World. Cambridge University Press (2010)
16. Zachary, W.W.: An information flow model for conflict and fission in small groups. J. Anthropol. Res. **33.4**, 452-473 (1977). http://www-personal.umich.edu/~mejn/netdata/
17. Girvan, M., Newman, M.E.J.: Community structure in social and biological networks. In: Proceedings of the National Academy of Sciences, vol. 99.12, pp. 7821–7826 (2002). http://www-personal.umich.edu/~mejn/netdata/
18. Lusseau, D., et al.: The bottlenose dolphin community of Doubtful Sound features a large proportion of long-lasting associations. Behav. Ecol. Sociobiol. **54.4**, 396–405 (2003). http://www-personal.umich.edu/~mejn/netdata/
19. Lancichinetti, A., Fortunato, S., Radicchi, F.: Benchmark graphs for testing community detection algorithms. Phys. Rev. E **78**(4), 046110 (2008)
20. Newman, M.E.J.: Finding community structure in networks using the eigenvectors of matrices. Phys. Rev. E **74.3**, 036104 (2006). http://www-personal.umich.edu/~mejn/netdata/
21. Pizzuti, C.: Ga-net: a genetic algorithm for community detection in social networks. In: International Conference on Parallel Problem Solving from Nature. Springer, Berlin, Heidelberg (2008)

Key Researcher Analysis in Scientific Collaboration Network Using Eigenvector Centrality

Anand Bihari and Sudhakar Tripathi

Abstract The scientific impact of an individual is measured by the citation count of their articles. Several citation-based indices and centrality measures are present for the evaluation of scientific impact of individual. In the research community, generally every author gets full credit of citation count of an article, but rarely their contribution is equal. To resolve this issue, we used the Poisson distribution to distribute the share credit of authors in multi-authored article and for evaluation of scientific impact of an individual, the eigenvector centrality has been used. In centrality measures, the eigenvector centrality is a good measure to evaluate the scientific impact of individual, because it uses the scientific impact of collaborators as well as the collaborators of collaborated researchers. For calculation of eigenvector centrality, first we set that the initial amount of influence of every node (author) is the total number of normalized citation count and the collaboration weight is the correlation coefficient based on individual normalized citation count. To validate the proposed method, an experimental analysis has been done on the collaboration network of 186007 scholars.

Keywords Social network · Scientific collaboration · Eigenvector centrality
Correlation coefficient

A. Bihari (✉) · S. Tripathi
Department of Computer Science and Engineering, National Institute
of Technology Patna, Patna, Bihar, India
e-mail: anand.cse15@nitp.ac.in; csanandk@gmail.com

S. Tripathi
e-mail: stripathi.cse@nitp.ac.in; p.stripathi@gmail.com

© Springer Nature Singapore Pte Ltd. 2019
P. K. Sa et al. (eds.), *Recent Findings in Intelligent Computing Techniques*,
Advances in Intelligent Systems and Computing 707,
https://doi.org/10.1007/978-981-10-8639-7_52

1 Introduction

The scientific product of an individual or a group of researchers is the research articles. The research articles are generally based on previous study. It is either reviews, analysis or proposed article. The evaluation of the scientific impact of an individual or a group of authors is based on the scientific impact of all those types of article published. In the recent era, generally the research work is completed by the group of researchers who may from the same subject areas or from different subject areas and they form a network of scientific collaboration called scientific collaboration network. In this network, the nodes represent individual researcher and an edge between nodes represents the scientific collaboration. For construction of collaboration network, numerous research works had been conducted by the eminent scientist [1] as well as finding prominent scientist [2, 3] using social network analysis metrics and citation-based indices. In this article, we discuss the Poisson distribution to share credit between all authors in multi-authored articles and for finding a key or prominent researcher, we used eigenvector centrality. In eigenvector centrality, the initial amount of influence is the total normalized citation count based on Poisson distribution and the collaboration weight is the correlation coefficient based on the individual normalized citation count.

2 Related Work

Farkas et al. [4] discussed the weighted collaboration network and used geometric mean of citation count for finding collaboration weight between nodes. Abbasia et al. [1, 5] used the total number of publication instead of total number of citation count as a collaboration weight and social network analysis metrics for scientific evaluation of an individual author. Liu et al. [6] discuss the weighted co-authorship network of the Digital library (DL) research community and analyse the community using social network analysis metrics. In this article, author proposed the author rank based on Pagerank algorithm for evaluating the scientific impact of individual and mentioned that the author rank is a good measure than the social network analysis metrics. Liu et al. [7] proposed a new method for constructing a collaboration network. Here, author used geometric series to calculate the share credit to all authors in a particular article and the collaboration weight is calculated based on the law of gravity. Then, convey the Pagerank for scientific evaluation and mentioned the advantages of Pagerank over the other centrality measures of social network. Bihari et al. [2, 3] discussed the social network analysis metrics and citation-based indices for evaluating the scientific impact of an individual scientist. For conducting the experiment, publication data have been downloaded from IEEE Xplore and form the collaboration network of available datasets. In this network, author used a total number of citation count of all those articles, which has published together as a collaboration weight. Bihari et al. [8] discussed the importance of the eigenvector centrality in the scientific evaluation of the individual author in the community and modified the eigenvector centrality algorithm of network for refinement of the calculation process of eigenvector centrality. In this article, authors set the initial amount of influence which is the degree centrality of a particular author instead of 1/n, where n is the total number of node present in the network and the collaboration weight is the total number of citation count earned by all those articles which are published together. Bihari et al. [9] discussed the importance of weak and strong edge in collaboration network and mentioned that the some of the edge has very less or marginal influence in scientific evaluation. For removing the less influence edge, author used the maximum spanning tree and covey the social network analysis metrics for scientific evaluation.

3 Methodology

In this section, we discussed the methodologies which are used in this analysis.

3.1 Correlation Coefficient

To find the cross-correlation between any two entities, the correlation coefficient can be used. The basic objective of this is to compute the correlation impact between two entities based on individual input [10]. Mathematically, it is defined as follows:

$$CR(m, n) = \frac{\sum_{i=1}^{k}(m_i - \overline{m})(n_i - \overline{n})}{(t - 1)Std_m Std_n} \tag{1}$$

where m_i, n_i represents the single unit, $\overline{m}$ and $\overline{n}$ represent the mean value of m and n, respectively, t is the total unit and Std_m, and Std_n represents the standard deviation of m and n.

3.2 Poisson Distribution

To compute probabilities of the success of individual based on the mean number of success, Poisson distribution can be used [11, 12]. The number of phenomena that occurs on a per unit basis is often are well approximated by the Poisson distribution. The function of Poisson distribution is given as follows:

$$P(x) = \frac{e^{-\lambda} \times \lambda^x}{x!} \tag{2}$$

where e is the base of the natural logarithm, λ is the mean number of success and x is the number of success.

3.3 Eigenvector Centrality

In social network analysis, eigenvector centrality (proposed by [13] in 1987) is one of the most important measures to compute the rank of a node with the help of its direct neighbour nodes. Basically, it is the extension of degree centrality, but one of the basic differences between eigenvector centrality and the degree centrality is that it uses the importance of neighbour nodes rather than counting the total number of neighbours [13, 14]. The base of eigenvector centrality is the eigenvalue (λ) and eigenvector of the graph. Let us we consider a graph Gph = (N, E), where |N| represents the total number of nodes. Mathematically, it is defined as follows:

$$ECT_p = \frac{1}{\lambda} \sum_{q \in M(p)} ECT_q = \frac{1}{\lambda} \sum_{q \in Gph} Adj_{(p,q)} ECT_q \tag{3}$$

where ECT_p is the eigenvector centrality of node p, M(p) is the total number of direct neighbours of node p, $Adj_{(p,q)}$ is the adjacency value of node pair p and q and the λ is a constant.

The gist of the eigenvector vector centrality is to evaluate the impact of a node in a graph based on the impact of neighbour nodes. If a node have connection with less number of high influential node, then may get high centrality value than the node have connection with relatively more number of nodes with less influential [15].

4 Data Collection and Cleansing

For construction and analysis of collaboration network and evaluation of scientific impact of individual, the collaboration data are required. To do so, we extract article details with their authors' name, citation count, publication year and publication type from IEEE Xplore [16] for the period of January 2010 to July 2016 including journal articles, conference proceedings and the transaction article. The article details are extracted with the computer science and engineering keywords. The extracted data are in CSV format with numerous field. For this analysis, we required only the publication ID, publication authors and the citation count. So we filtered those field data from raw data. In raw data, some publication details are incomplete or unreadable, so simply we remove all those publication details. After successfully cleaning of publication details, only 96,503 articles and 186,007 authors are available for analysis. In this dataset, 47.15% articles were published by the exact two authors, 4% article was published by one author and rest of the article published by the more than two authors. So we conclude that the mostly researcher work is done more than two authors.

5 Our Proposed Method

In the research community, the evaluation of the scientific impact of an individual author is based on the citation count of articles and all authors get full credit. But rarely they contribute equally, so it will require a mathematical model to share credit between all authors in multi-authored articles. de Solla Price et al. [17] used the fractional credit (i.e. 1/k where k is the total number of authors) to share the credit to all authors. Egghe et al. and Liu et al. [7, 18] used the geometric series, Trueba et al. [19] used the arithmetic counting and Hagen et al. [20] used harmonic credit to share credit to all authors in multi-authored articles. Instead of all those credit allocation mechanism, in this article, we proposed a new method to share credit to all authors based on the Poisson distribution. The basic formula for the Poisson distribution is

$$P(x) = \frac{e^{-\lambda} \times \lambda^x}{x!} \tag{4}$$

where e is the base of the natural logarithm, λ is the mean number of success and x is the number of successes. In our proposed method, we slightly modify the formula of the Poisson distribution. In our calculation, x represents the rank of an author, and λ is the mean number of authors (i.e. 1/k, where k is the total number of author). For example, suppose an article published by the four authors; then the share credit for all authors is calculated as follows: $P(0) = \frac{e^{-0.25} \times 0.25^0}{0!} = 0.77$ for the first author, $P(1) = \frac{e^{-0.25} \times 0.25^1}{1!} = 0.19$ for the second author, $P(2) = \frac{e^{-0.25} \times 0.25^2}{2!} = 0.024$ for the third author and for the last or fourth author $P(3) = \frac{e^{-0.25} \times 0.25^3}{3!} = 0.002$. The sum of the total share credit is 1. The total citation score of kth author in p^{th} article is defined as

$$P(k) = \frac{e^{-\lambda} \times \lambda^k}{k!} \times cit_p \tag{5}$$

The total citation score of an author is the sum of the score of all articles.

$$SC = \sum_{k=1}^{m} P(k) \tag{6}$$

After that, calculate the collaboration weight between every author pair using correlation coefficient based on individual citation share credit. Finally, we used eigenvector centrality to evaluate the scientific impact of individual. In eigenvector centrality, we used the total normalized citation count (Eq. 6) as an initial amount of influence of every node, and the adjacency weight of an author pair is the correlation coefficient value.

6 Analysis and Result

To validate the proposed method, we made an experimental analysis of available publication dataset. First, calculate the share credit of all author and then calculate the eigenvector centrality for all authors using Python and NetworkX [21], and selected top 10 authors are shown in Table 1.

After that, we made a comparative analysis of the proposed method with traditional fashion which is implemented in [22]. In traditional method, the initial amount of influence is 1/n, where n is the total number of node in the network and the adjacency weight is the total number of citation count of all those articles which is published together. The top 10 authors from traditional fashion are shown in Table 2.

The results of comparative analysis are shown in Table 3 and found that none of the authors is present in both types of experimental analysis. After the analysis of

Table 1 Eigenvector centrality report of top 10 researchers in the network

No	Name of author	Eigenvector centrality
1	Bowyer, K. W.	0.408175229
2	Grother, P.	0.407998366
3	Sarkar, S.	0.407813063
4	Phillips, P. J.	0.407714304
5	Liu, Z.	0.407714304
6	Vega, I. R.	0.407714304
7	Cucchiara, R.	0.013761
8	Piccardi, M.	0.013057854
9	Prati, A.	0.012744659
10	Farag, A. A.	0.012215692

Table 2 Eigenvector centrality report of top 10 researchers based on traditional fashion in the network

No	Name of author	Eigenvector centrality
1	Mizuno, T.	0.1334739264
2	Kamae, T.	0.1333693085
3	Fukazawa, Y.	0.1295166699
4	Grove, J. E.	0.1288715982
5	Kuss, M.	0.1284210454
6	Schaefer, R.	0.1284210454
7	Ozaki, M.	0.1284210454
8	Dubois, R.	0.1284210454
9	Thompson, D. J.	0.1284210454
10	Lauben, D.	0.1284210454

Table 3 Comparative analysis of top 10 authors

No	Name of author	Traditional centrality value	Proposed centrality value
1	Bowyer, K. W.	0.0024140822	0.4081752295
2	Grother, P.	0.0024140822	0.4079983663
3	Sarkar, S.	0.0000000001	0.4078130630
4	Phillips, P. J.	0.0000000001	0.4077143037
5	Liu, Z.	0.0000000070	0.4077143037
6	Vega, I. R.	0.0001908312	0.4077143037
7	Cucchiara, R.	0.0001908308	0.0137601581
8	Piccardi, M.	0.0001906779	0.0130578541
9	Prati, A.	0.0001906824	0.0127446591
10	Farag, A. A.	0.0001906779	0.0122156922
11	Schaefer, R.	0.1284210454	0.0000000611
12	Ozaki, M.	0.1284210454	0.0000000603
13	Kuss, M.	0.1284210454	0.0000000232
14	Lauben, D.	0.1284210454	0.0000000232
15	Grove, J. E.	0.1288715982	0.0000000232
16	Kamae, T.	0.1333693085	0.0000000037
17	Dubois, R.	0.1284210454	0.0000000036
18	Mizuno, T.	0.1334739264	0.0000000036
19	Fukazawa, Y.	0.1295166699	0.0000000035
20	Thompson, D. J.	0.1284210454	0.0000000035

proposed and traditional method results, it seems that the proposed method gives better results than traditional one.

7 Conclusion and Future Work

In this paper, we have investigated the scientific collaboration network of scientist and discovered the prominent actor in research community using eigenvector centrality. For this, first we used the Poisson distribution to share credit to all authors in multi-authored articles. After that, we calculated the correlation coefficient between author pairs based on the individual normalized citation count for correlation weight. Finally, the eigenvector centrality has been used for scientific evaluation of an individual. For calculation of eigenvector centrality, first we set that the initial amount of influence of every node is the total normalized citation count earned from all publications and the collaboration weight is a correlation coefficient value based on individual normalized citation score. After that, we made a comparative study of the traditional one and the proposed one and found that the proposed method gives better result than the traditional one. But our proposed Poisson distribution method penalized all authors who published a number of co-authored articles.

References

1. Abbasi, A., Altmann, J.: On the correlation between research performance and social network analysis measures applied to research collaboration networks. In: 44th Hawaii International Conference on System Sciences (HICSS), 2011, pp. 1–10. IEEE (2011)
2. Bihari, A., Pandia, M.K.: Key author analysis in research professionals relationship network using citation indices and centrality. Procedia Comput. Sci. **57**, 606–613 (2015)
3. Pandia, M.K., Bihari, A.: Important author analysis in research professionals relationship network based on social network analysis metrics. In: Computational Intelligence in Data Mining, vol. 3, pp. 185–194. Springer (2015)
4. Farkas, I., Ábel, D., Palla, G., Vicsek, T.: Weighted network modules. New J. Phys. **9**(6), 180 (2007). http://stacks.iop.org/1367-2630/9/i=6/a=180
5. Abbasi, A., Hossain, L., Uddin, S., Rasmussen, K.J.: Evolutionary dynamics of scientific collaboration networks: multi-levels and cross-time analysis. Scientometrics **89**(2), 687–710 (2011)
6. Liu, X., Bollen, J., Nelson, M.L., Van de Sompel, H.: Co-authorship networks in the digital library research community. Inf. Process. Manag. **41**(6), 1462–1480 (2005)
7. Liu, J., Li, Y., Ruan, Z., Fu, G., Chen, X., Sadiq, R., Deng, Y.: A new method to construct co-author networks. Physica A: Stat. Mech. Appl. **419**, 29–39 (2015)
8. Bihari, A., Pandia, M.K.: Eigenvector centrality and its application in research professionals' relationship network. In: 2015 International Conference on Futuristic Trends on Computational Analysis and Knowledge Management (ABLAZE), pp. 510–514. IEEE (2015)
9. Bihari, A., Tripathi, S., Pandia, M.K.: Key author analysis in research professionals' collaboration network based on MST using centrality measures. In: Proceedings of the Second International Conference on Information and Communication Technology for Competitive Strategies, p. 118. ACM (2016)
10. https://www.mathsisfun.com/data/correlation.html
11. Grimmett, G., Stirzaker, D.: Probability and Random Processes. Oxford University Press (2001)
12. Krishnan, V., Chandra, K.: Probability and Random Processes. Wiley (2015)
13. Bonacich, P., Lloyd, P.: Eigenvector-like measures of centrality for asymmetric relations. Soc. Netw. **23**(3), 191–201 (2001)
14. Newman, M.E.: The New Palgrave encyclopedia of economics. Math. Netw. **2**, 1–12 (2008)
15. Ding, D., He, X.: Application of eigenvector centrality in metabolic networks. In: 2nd International Conference on Computer Engineering and Technology (ICCET), 2010, vol. 1, pp. V1–89. IEEE (2010)
16. http://ieeexplore.ieee.org/xpl/opac.jsp
17. de Solla Price, D.J.: Multiple authorship. Science **212**, 986 (1981)
18. Egghe, L., Rousseau, R., Van Hooydonk, G.: Methods for accrediting publications to authors or countries: consequences for evaluation studies. J. Am. Soc. Inf. Sci. **51**(2), 145–157 (2000)
19. Trueba, F.J., Guerrero, H.: A robust formula to credit authors for their publications. Scientometrics **60**(2), 181–204 (2004)
20. Hagen, N.T.: Harmonic allocation of authorship credit: source-level correction of bibliometric bias assures accurate publication and citation analysis. PLoS ONE **3**(12), e4021 (2008)
21. Swart, P.J., Schult, D.A., Hagberg, A.A.: Exploring network structure, dynamics, and function using NetworkX. In: Proceedings of the 7th Python in Science Conference (SciPy 2008)
22. Bihari, A., Tripathi, S.: A new method for key author analysis in research professionals' collaboration network. In: Proceedings of the 3rd International Doctoral Symposium on Applied Communication and Security Systems (ACSS). Springer (2017) (In Press)

A Protected Cloud Computation Algorithm Using Homomorphic Encryption for Preserving Data Integrity

Prakhar Awasthi, Sanya Mittal, Sibeli Mukherjee
and Trupil Limbasiya

Abstract Cloud computing is growing very rapidly because it can drastically cut down the costs of hosting data, is highly scalable, and increases the number crunching capacity manifold, and it is also a revolutionary concept. Cloud computing delivers us from this handicap and de-couples the need for computing resources to be present on site to access them. Thus, it democratizes access to computing and storage. The data on the cloud may be safe while computing, but the security of data while a transfer is not ensured en route. Our aim is to build a fully homomorphic encryption method that supports mathematical operations in addition to being efficient, quick, and utilizing minimum resources. Now, once we have a fully homomorphic encryption, the problem becomes to ensure how the user can know for sure that whether the operation on encrypted data achieved is correct or not.

Keywords Cloud computing · Encryption · Integrity · Modification · Security

1 Introduction

Cloud computing is a process of using a network of remote servers hosted on the Internet for the purpose of computing and storing data, rather than on a local server.

P. Awasthi (✉) · S. Mittal · S. Mukherjee · T. Limbasiya
NIIT University, Neemrana, Rajasthan, India
e-mail: prakhar.awasthi@st.niituniversity.in

S. Mittal
e-mail: sanya.mittal@st.niituniversity.in

S. Mukherjee
e-mail: sibeli.mukherjee@st.niituniversity.in

T. Limbasiya
e-mail: limbasiyatrupil@gmail.com

© Springer Nature Singapore Pte Ltd. 2019
P. K. Sa et al. (eds.), *Recent Findings in Intelligent Computing Techniques*,
Advances in Intelligent Systems and Computing 707,
https://doi.org/10.1007/978-981-10-8639-7_53

509

The paradigm of cloud computing allows services like data storage and processing. An usage of cloud computing became necessity into the various online activities in last few years. Because we have reached at the end of Moore's law, it is no longer economically viable or technologically possible to shrink the size of transistors. People wish most of services within a limited period so that facilities can be helpful to them in achieving something in a different manner. Authors [1] described issues regarding the data privacy in cloud computing environments. The concerns arise in both technical and strategic involvements during the transactions. They categorized these problems according to the degree of data involvement.

According to authors in [2–4], there are various security measures taken in each type of cloud services. Each type has their own advantages, and hence most industries now look for hybrid setups and multi-cloud strategies so that risks are reduced. In [1], authors worked on a hybrid approach allowing an end-to-end solution, which is policy based, to preserve data in the cloud system. This approach learns from the drawbacks of the existing schemes.

Homomorphic encryption is a form of encryption that allows computations to be carried out on ciphertext, thus generating an encrypted result which, when decrypted, matches the result of operations performed on the plaintext. In other words, homomorphic encryptions allow complex mathematical operations to be performed on encrypted data without compromising the encryption. One of the main advantages of homomorphic encryption is that it provides computing on a private data which we do not want to share with anyone else. For example, you are having the transactions of your bank account and you want the computing to happen on this data which is indeed private for you. Maybe you trust your cloud provider, but what if trust is not enough? Is there a way to get the compelling economic benefits of cloud computing while keeping your data secure? Hence, the homomorphic encryption comes in the picture.

Figure 1 represents the simple overview of secure cloud computing using an encryption method for providing an advanced level of data protection. With the help

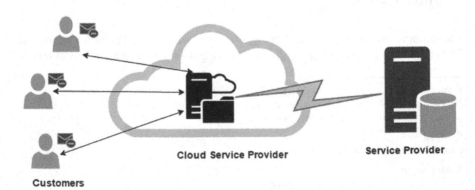

Fig. 1 A standard structure of secure cloud computing

of general structure, the readers can understand the needs of a secure model for data enumeration when customers use the services from the cloud service providers.

As we know, most of the systems have different advantages as well as disadvantages. At the same point, we have various drawbacks of cloud computing and homomorphic encryption systems. As we all know, there are mainly two types of homomorphic encryption: Additive homomorphic and multiplicative homomorphic. For example, authors in [5] stated that RSA is multiplicative homomorphic since it permits multiplication operation on encrypted data without the need of private key. Similarly, Pallier cryptosystem is additively homomorphic. One of the biggest drawbacks of homomorphic encryption is that the multiplicative homomorphic variant of RSA is not used semantically in a secure manner. Homomorphic encryption is prone to alteration of data in case your trusted party doing the computation untruthfully. For example [6], we practice the encryption system so that the other party should not know the actual value, which we passed to perform the computation(s). But in case the second party doing computations is corrupt or even if not corrupt but doing computations in a wrong way, there is no way we will come to know of this concern and we will get some value after decrypting the final ciphertext. And this will not let you know if it has been derived in the way you wanted it to be.

2 Literature Review

The main work of homomorphic encryption is to compute the operation on the encrypted data for more secure process. In [7], authors talked about the fully homomorphic public key encryption. They mainly framed the idea of using lattices and also mentioned about the security of the lattices. They explained how fully homomorphic encryption can be derived from the theory of algebraic number field. Here, they described the different schemes and has given analysis on the key generation algorithm, encryption algorithm, and decryption algorithm for each of the schemes.

Many improvements have been made to the protocols, which are practiced for the calculations on the data using the encryption system [8]. Authors mainly described that the low overhead is enough to make the protocol potentially in a practical manner. At the cost of additional communication, the cloud provides high and increased security among the distribution of operations on various clouds at different locations. Here, authors also discussed the distributed and pre-computation encryption and decryption with the help of homomorphic strategy. Authors also explained that the security of the decryption of data completely relies on the outsourcing parts of the key which is a secret and not allowing the decryption of ciphertext or any other part of it.

The main reason for the use of cloud and homomorphic encryption is to keep data more secure. As discussed in [9], the practical application of homomorphic encryption is to provide cloud services in the departments such as medical, advertising organizations, financial sectors, and many more. They also talked about the facts that all statistical calculations use more additive and very less multiplicative operations, and it can be computed in less time consumption.

As author said in [10] that the main improvement of this methodology with respect to Gentry's method was that it was simpler. Thus, it was resulted in the size of the public key, which was too large to implement meaningful operations. Therefore, there was a demand to decrease the key size while at the same time maintaining the correctness, and it was conceptual simplicity. Authors were successful in their endeavor by encrypting with a quadratic form in the public key elements, instead of the typical linear form. After that, authors were able to shrink the public key size and they had to also ensure that it was semantically secure. It was based on a stronger variant of the approximate-GCD problem, which was already considered by [11]. Consequently, they were able to describe the rest implementation of the resulting fully homomorphic scheme while simultaneously ensuring that the level of efficiency is not compromised.

We know RSA is multiplicative homomorphic and Paillier is additively homomorphic. Since we know that an encryption can be called homomorphic if it supports operations on encrypted data, we need a cryptographic system that can support both. One way to succeed in this objective is to keep the noise down. This is because every ciphertext has a noise component in it and any operations on it yields more noise, and if the noise crosses a threshold then it does not decrypt correctly rendering the whole exercise futile. Hence, noise acts as a limiting factor on the degree of polynomial upon which we can operate. This is achieved through an operation called "ciphertext refresh". We try to minimize the degree of the decryption polynomial because if it is small enough, then the resulting noise in this new ciphertext is considerably smaller than the original. Once we have two of these, then we can apply a homomorphic function, which was not feasible previously due to the threshold noise. The condition being that the degree of polynomial that can be evaluated exceeds the degree of the decryption polynomial, since we must take into account the future computations. This method was employed by [12]. Here, authors divided an ingenious strategy to reduce the size of public key by storing only a small subset of the public key. However, the final key was generated by combining elements in small subset multiplicative. Scientists [12] also did not recommend using probabilistic decryption circuit, since the error probability would be too high for our set of parameters. The difficulty also lies in determining a secure set of concrete parameters. Their approach was to implement the known attacks, measure their running time, and extrapolate for large parameters. The concrete parameters are fixed according to the desired level of security. Authors practiced four security levels, namely, toy, small, medium, and large, where each one of them corresponds to 42, 52, 62, and 72 bits of security, respectively. While encryption takes few minutes, decryption is almost instantaneous.

3   Problem Statement

The operations in homomorphic encryption, as evident from the theories, have been performed in encrypted form. The security in this manner is perpetuated during the transfer of data. The main user, after receiving the message, can decrypt it using the private key it possesses. For example, if a person wants to perform the addition operation of two variables (say, α and β), she/he does not know how to perform addition s/he will send data through the cloud, encrypted with some key(s). After encrypting with a key κ, say, the number became $\alpha + \kappa$ and $\beta + \kappa$. The server performs the operation on the latter, followed by sending the outcome to the user. After that, the output is decrypted using their private key, and the result will be obtained by the user. But the problem here lies that if a person does not know that the answer whatever she is getting correct or not. There is a possibility that incorrect computation may have been made or some malicious persons might have interrupted the process while computation. In this paper, we suggest a system to overcome this weakness, which will help the user for verifying the operation. This will enable every user to receive the correct output and not to encounter the problem.

4   Proposed System

4.1   Methodology

As we have seen in the problem statement that if a user wishes to keep all this original data hidden from the outer environment. Then, it is possible to do so by the application of homomorphic encryption. If one user does not know a particular operation, then it can easily be outsourced from different systems but only concerned is that the data exchanges between the user and the other end(s) should be kept confidential. This confidentiality factor is resolved by the application of homomorphic encryption because it allows customer to transfer the encrypted data, and the computation is done on the encrypted data. Homomorphic encryption terminologies need for the procedure are as follows:

1. **Key generation**: Key generation helps in encryption of the data when the user is transmitting the data for a particular operation. It is also need for decrypting the final outcome received from the outsider.
2. **Security**: Cloud computing helps organizations to provide them with a huge space for saving the data, which can be accessible from different locations easily. Even though it is needed to be secure from all dimensions, various security issues can be encountered, which can be related to privacy, data, integrity, confidentiality, and authentication. But with the application of homomorphic encryption, security problems can be handled because it works on the encrypted data.

3. **Correctness**: The main and the most important part of homomorphic encryption is the correctness of the final output. If a user data is from C_1 to C_t, then we can say ...

$$C_1 \leftarrow Enc_{pk}(m_1)$$
$$C_2 \leftarrow Enc_{pk}(m_2)$$

.

.

.

$$C_t \leftarrow Enc_{pk}(m_t)$$
$$C^* = Eval_{pk}(f, C_1, \ldots, C_t)$$
$$Dec_{sk} = f(m_1, \ldots, m_t)$$

4. **Compactness**: The complexity of the computing a function on an encrypted data is more complex than decrypting the final output using the secret key. So the complexity of decrypting C^* does not depend on complexity of the function, which the user wants to compute.

4.2 Correctness of the Proposed Algorithm

We have stated a step-by-step procedure to verify justification of the proposed algorithm, which is beneficial in recognition of the suggested method:

- The input message is named as p (plaintext). Plaintexts consist of multiple data: $p_1, p_2, \ldots, p_t$.
- Plaintext can be encrypted using a key k_1 such that plaintext gets converted to ciphertext consisting of $c_1, c_2, \ldots, c_t$. Note that k_1 must encrypt the plaintext using the same operation to be performed.
- This ciphertext goes to the cloud for an operation to be performed on them. Now, the ciphertext undergoes an operation say addition.

$$\therefore \quad c_1, c_2, \ldots, c_t = \sum_{i=1}^{t} C_i$$

- The C_i is sent to the user and the customer decrypts it with the help of key k_2 and gets the answer for the plaintext, that is, $\sum_{i=1}^{t} P_i$, which is the final solution of the same.
- The user can verify whether P_i is the correct summation of the plaintext using the original key k_1 as $C_i - 2 \cdot k_1 = P_i$.

5 Proposed Protocol Analysis

In Layman terms, addition is considered the opposite operation of subtraction and multiplication is the opposite of division. Here, we performed addition on the data and subtraction to verify the decrypted outcome similarly. We have practiced division to verify multiplication. We got two equations as the time taken for the computation of $C2 \cdot k_1 = P_i$ and $\dfrac{C_i}{K_2} = P_i$. A generalization of these equations ($Y = 1 - 2X$ and $Y = \frac{1}{X^2}$) has been shown in Fig. 2 and Fig. 3, respectively. We can identify that a value after the decryption is going down in negative value when increasing the key value in Fig. 2. However, a value after the decryption is also touching down zero in case of increasing the key value in Fig. 3.

As we know that computation of multiplicative takes more time than additive so for most of the operations, which use multiplication such as standard deviation, logistical regression is done with additive method and the time complexity is less than time taken by performing multiplicative. The time complexity of addition of input of two n-digit numbers N and N giving the output as one $n + 1$ digit number is $\Theta(n)$, and time complexity of multiplicative of two n-digit numbers giving the output as one $2n$-digit number is $O(n^2)$ [2]. Θ gives the worst-case complexity and O gives the average-case complexity. Hence, an additive is more preferred over multiplicative. Finally, we can say that the encryption of addition will take lesser time than the encryption of multiplicative operations.

However, we have also one challenge again. We have seen that multiplicative and additive operations use division and subtraction for decrypting the ciphertexts to check the correctness. Performing the same in a reverse manner, that is, encryption by subtraction and decryption by addition as well as encryption with division and division with multiplication, does not work correctly.

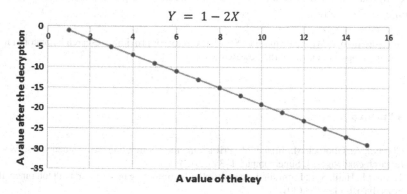

Fig. 2 The graph for subtraction operation ($Y = 1 - 2X$)

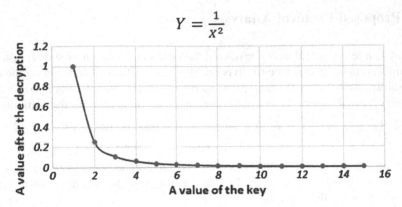

Fig. 3 The graph for multiplicative operation ($Y = \frac{1}{X^2}$)

6 Conclusion

Cloud computing is the future of the technological industry. It is by far one of the most helpful creations in the information-driven industry. It helps us to access and store data in huge amounts without worrying about having large physical devices. The variety of data available on cloud makes it hard to find a generic solution to all threats. However, if our data is being stored on the cloud, then one is certainly worried about the safety and security of the data that is sent. We have discussed issues that may occur while transferring information and accessing data for different computations. Additionally, we have recommended a scheme focused on addition, subtraction, multiplication, and division functions, and it helps the customer to verify that the data has not been disclosed and has not been modified on the way or during the computations. We have also discussed experimental outcomes of the suggested scheme to clarify it.

Acknowledgements We are thankful to NIIT University for providing research laboratory facility and funding support to carry out this research work.

References

1. Ghorbel, A., Ghorbel, M., Jmaiel, M.: Privacy in cloud computing environments: a survey and research challenges. J. Supercomput. 1–38 (2017)
2. Tebaa, M., Hajji, S.E.: From single to multi-clouds computing privacy and fault tolerance. IERI Procedia **10**, 112–118 (2014)
3. Tebaa, M., El Hajji, S., El Ghazi, A.: Homomorphic encryption applied to the cloud computing security. Proc. World Congr. Eng. **1**, 4–6 (2012)
4. Tebaa, M., Hajji, S.E.: Secure cloud computing through homomorphic encryption (2014). arXiv:1409.0829

5. El Makkaoui, K., Ezzati, A., Hssane, A.B.: Challenges of using homomorphic encryption to secure cloud computing. In: 2015 IEEE International Conference on Cloud Technologies and Applications (CloudTech), pp. 1–7 (2015)
6. Atayero, A.A., Feyisetan, O.: Security issues in cloud computing: the potentials of homomorphic encryption. J. Emerg. Trends Comput. Inf. Sci. 2(10), 546–552 (2011)
7. Smart, N.P., Vercauteren, F.: Fully homomorphic encryption with relatively small key and ciphertext sizes. In: International Workshop on Public Key Cryptography, pp. 420–443. Springer, Berlin, Heidelberg (2010)
8. Bouti, A., Keller, J.: Towards practical homomorphic encryption in cloud computing. In: IEEE Fourth Symposium on Network Cloud Computing and Applications (NCCA), pp. 67–74 (2015)
9. Naehrig, M., Lauter, K., and Vaikuntanathan, V.: Can homomorphic encryption be practical? In: Proceedings of the 3rd ACM Workshop on Cloud Computing Security Workshop, pp. 113–124 (2011)
10. Coron, J. S., Mandal, A., Naccache, D., Tibouchi, M.: Fully homomorphic encryption over the integers with shorter public keys. In: Annual Cryptology Conference, pp. 487–504. Springer, Berlin, Heidelberg (2011)
11. Van Dijk, M., Gentry, C., Halevi, S., Vaikuntanathan, V,: Fully homomorphic encryption over the integers. In: Annual International Conference on the Theory and Applications of Cryptographic Techniques, pp 24–43. Springer, Berlin, Heidelberg (2010)
12. Gentry, C., Sahai, A., Waters, B.: Homomorphic encryption from learning with errors: conceptually-simpler, asymptotically-faster, attribute-based. In: Advances in Cryptology CRYPTO, pp. 75–92. Springer, Berlin, Heidelberg (2013)

Circle-Time Packing: Visualization and Decision Support for Digital Cloud Computing

Mithileysh Sathiyanarayanan and Odunayo Fadahunsi

Abstract The Digital World (D-World) has changed the human life completely since the invention of the Internet, especially the cloud computing in the recent time which has brought an incremental development in, one of the key technological areas, data analytics, towards the realization of the D-World. As the D-World will create new and big business opportunities in many areas, data analytics will effectively help in examining data and inform business decisions. In this paper, we propose a small multiple circle-time packing identifier for visualizing and decision support in digital cloud computing (preliminary results are discussed). We formulate our proposed structure around potential business opportunities and will help in conducting technical feasibility studies. Circle-time packing is a visualization revolution that will offer intuitive yet valuable insight for users of all levels. This research will examine the challenges facing digital cloud service provision (storage) and will identify problems that require human supervision and guidance of automated agents with visualization support. Once identified, visualization methodologies will be applied to design interactive visualizations that can communicate the current system state and strategies for improving system performance. Addressing this problem will yield improved understanding of how humans can work together with advanced automation.

Keywords Circle packing · Visualization · Decision support · Cloud computing

M. Sathiyanarayanan (✉) · O. Fadahunsi
City, University of London, London, UK
e-mail: Mithileysh.Sathiyanarayanan@city.ac.uk

O. Fadahunsi
e-mail: Odunayo.Fadahunsi.1@city.ac.uk

© Springer Nature Singapore Pte Ltd. 2019
P. K. Sa et al. (eds.), *Recent Findings in Intelligent Computing Techniques*,
Advances in Intelligent Systems and Computing 707,
https://doi.org/10.1007/978-981-10-8639-7_54

519

1 Introduction

The term cloud computing was first introduced by professor Ramnath Chellappa in 1997 during a lecture. Cloud computing enables efficient management of data centres, time-sharing and virtualization of resources with a special emphasis on the business model through Infrastructure-as-a-Service, Platform-as-a-Service and Software-as-a-Service models on demand from anywhere in the world.

With the continuing growth and adoption of cloud computing, management of computing infrastructure is becoming an increasingly complex challenge. Data centres themselves are changing with both mega data centres, and distributed micro-data centres emerging to satisfy the needs of different markets. At the same time, customers are requiring faster turnaround times for service deployment and fast reactions to changing workloads. This creates new challenges for cloud service provisioning and orchestration including support for highly scalable systems, modelling across layers of the software stack, dynamic changes to services and supporting infrastructure, heterogeneity of service level agreements, deployments and workloads, and dealing with unexpected events. Management of these systems will require increased levels of automation. These will be driven by heuristics and high-level policies, but understanding the consequences of changes to these policies will be difficult for humans to assess and comprehend. Interactive visualizations can provide a means for human operators to observe and direct the actions of intelligent automated agents in a control system. In this manner, automation can become a "team player" in a joint cognitive system that combines the processing power of sophisticated automation with human flexibility and resilience. In order to design such a system, a visual decision support strategy is required. This strategy involves

1. Identifying and presenting important system data to an operator, and
2. Representing this data using graphical encodings that support identification, investigation, interpretation and resolution of system goals.

This research will examine the challenges facing cloud service provision and will identify problems that require human supervision and guidance of automated agents. Once identified, visualization methodologies will be applied to design interactive visualizations that can communicate the current system state and strategies for improving system performance. Addressing this problem will yield improved understanding of how humans can work together with advanced automation.

The research brings together expertise from a number of different domains, with both the academic and enterprise partner bringing specialist skills to the research. The skills of the researchers in information visualization and software development will be enhanced through the expertise of the academic and enterprise partners in producing complex information visualizations to support specific tasks. The collaboration with the cloud services lab in particular will provide us with training on the operation and orchestration of computing infrastructure in order to better understand the work of the human operator. The enterprise partner also brings specific knowledge and software components to the research for machine learning for software infrastructure management and the processing of large-scale graph databases.

2 Description

The research lies on the intersection of Human–Computer Interaction (HCI), Information Visualization (IV) and Machine Learning (ML), bringing these disciplines together in order to solve real-world problems in the management of new classes of complex, large-scale computing systems. The central aim of the research is to improve human collaboration with machine learning agents through the use of interactive visualizations. However, the design of these visualizations must mesh seamlessly with the capabilities and challenges faced by the human operator and the complementary capabilities of the automation. This can only be achieved by first constructing a thorough understanding of the work performed by the human operator and models representing it. Building on this understanding, the project will develop design patterns for the management of virtualized infrastructure and make methodological advances by extending the ecological interface design framework [1] to accommodate virtual resources and flexible constraints.

1. Identification and modelling of suitable orchestration problems that involve machine learning and human input.
2. Developing test case simulations for cloud orchestration problems that can be run on a test bed.
3. Designing visualizations that highlight goals, constraints and conflicts associated with orchestration problems.
4. Performing empirical studies on these visualizations in order to understand their role and effectiveness in supporting collaboration with automated agents.

These visualizations will be evaluated using real-world data by subject matter experts. The central research question to be addressed is 'can novel interactive visualizations improve human interaction with machine learning algorithms and in turn improve the quality of decision-making outputs when controlling dynamic complex systems?'.

The research started building on previous work in information visualization [2–4]. Some existing work looks at the use of data analysis within visualization, e.g. for dimensionality reduction [5]. While many commercial visualization and business intelligence tools exist, these are generally focussed towards exploratory data analysis rather than system monitoring and are better suited to certain uses [6]. Nevertheless, there is evidence to support the use of visualization within industrial contexts—for example, authors of the paper [7] present a study in which visualizations outperform tabular displays for ERP tasks. The authors of another paper [8] present a detailed breakdown of opportunities within manufacturing, and look at the existing literature in terms of visualization functionalities and domain requirements. For many years, the 'Information seeking mantra' of overview, zoom and filter, details on demand [9] have been influential within the visualization community, with this interaction pattern used in many applications. Visual displays used within complex industrial settings do not generally fit this pattern, however, for a number of reasons. For example, many applications are used for control as well as information gathering and monitoring, making task-specific displays more common, likewise decision-making may require data to be obtained from multiple related views on the system and this can be difficult within large-scale 'full system' visualizations. In order to develop visualizations more suited to control, the project will begin with a state-of-the-art analysis, before progressing on the development of user stories in collaboration with domain experts. The different modes of supervision and control represented within these user stories will be investigated through the rapid proto-typing of interactive visualizations and subsequent empirical investigation of these. This will involve obtaining feedback both on low-fidelity prototypes produced early in the process, as well as high fidelity and final prototype visualizations, following the standard human-centred design process. Operator performance will be explored both through quantitative measures such as accuracy and time on task, but also qual-itatively in terms of their thought process, trust in the automation (and understanding of it), as well as satisfaction.

3 Preliminary Results

As a starting point, we developed circle packing changing over time in the form of small multiples. A circle packing is an arrangement (packing) of circles inside a big circle such that there are only containment and no overlaps to represent the hierarchy (hierarchical data). Although circle packing is not as space efficient as a tree map, it better reveals the hierarchy. The hierarchical layers can be implemented in such a way to facilitate the rapidly developing data and meet the following ten requirements (discussed in detail in the extended version of the paper) (Fig. 1):

- **Availability,**
- **Reliability,**
- **Feasibility,**
- **Scalability,**
- **Communicability,**
- **Interpretability,**
- **Interactivity,**

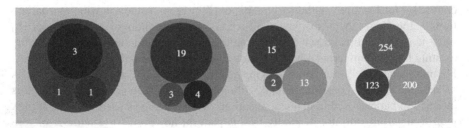

Fig. 1 Screenshot of the D3 circle-time packing representing data from 2013–2016 (left to right)

- **Manageability,**
- **Performance, and**
- **Cost.**

In cloud computing, we focus on storage and its changes over time. We considered Data-Driven Documents (D3) to implement the web version of the circle-time packing as a small multiple. This helps us to understand the hierarchical changes in the storage over time. As an extended version, we will be considering advanced applications and management in the cloud computing.

4 Conclusion and Future Work

Software-Defined Infrastructure (SDI) is an emerging technology that will radically change the way in which ICT infrastructure is provided, managed and consumed. The research will have a strong HCI component, starting with end-user interviews, observations and analysis, feeding into the design of visualizations to support the work.

Infrastructure management is a highly specialized activity that is currently carried out using a wide array of software tools. This domain has its roots in systems administration with the result that many of the tools are command line interfaces and have been slow to adopt visualization methods. There is no substantial body of HCI research concerning this work domain or the form of support envisaged for it. Where visualization has been applied to infrastructure, it has tended to be illustrative in nature, focusing on making current relationships explicit but having minimal support for scenario planning or interaction with machine learning agents. This research will be built on current work on machine learning for orchestration in any of the cloud services lab. The research will aim to visualize aspects of machine learning algorithms and allow for manipulation of algorithm inputs and inspection of results [10–13]. The work in human–computer interaction, and information visualization will build on the research conducted at various areas, both in terms of domain-specific issues within HCI, the design of information visualizations to support work [14], and the use of methods which help to leverage models of the work performed within this design process. The project will make a contribution to knowledge by bringing together the fields of human–computer interaction, information visualization and machine learning in order to address the challenges posed by an emerging form of large-scale computing system. As well as this domain-specific contribution, it will also advance our understanding of human interaction with (and supervision of) advanced automation based on machine learning which is an increasingly important class of technology. Machine learning's greatest potential across organizations includes improving forecasting, predictive analytics and potential to deliver real-time optimization for designing effective data architecture and technology infrastructure [15].

Acknowledgements We would like to thank Gavin Doherty of Trinity College Dublin for his insights on Digital Cloud Computing. We would also like to thank our university for supporting us in this work.

References

1. Burns, C.M., Hajdukiewicz, J.: Ecological Interface Design. CRC Press (2004)
2. Spence, R.: Information Visualization, vol. 1. Springer (2001)
3. Chen, C.: Top 10 unsolved information visualization problems. IEEE Comput. Graph. Appl. **25**(4), 12–16 (2005)
4. Chintalapani, G., Plaisant, C., Shneiderman, B.: Extending the utility of treemaps with flexible hierarchy. In: Proceedings of the Eighth International Conference on Information Visualisation, 2004: IV 2004, pp. 335–344. IEEE (2004)
5. Börner, K., Chen, C., Boyack, K.W.: Visualizing knowledge domains. Ann. Rev. Inf. Sci. Technol. **37**(1), 179–255 (2003)
6. Kobsa, A.: An empirical comparison of three commercial information visualization systems. In: IEEE Symposium on Information Visualization, 2001 (INFOVIS 2001), pp. 123–130. IEEE (2001)
7. Parush, A., Hod, A., Shtub, A.: Impact of visualization type and contextual factors on performance with enterprise resource planning systems. Comput. Ind. Eng. **52**(1), 133–142 (2007)
8. Sackett, P.J., Al-Gaylani, M., Tiwari, A., Williams, D.: A review of data visualization: opportunities in manufacturing sequence management. Int. J. Comput. Integr. Manuf. **19**(7), 689–704 (2006)
9. Shneiderman, B.: The eyes have it: a task by data type taxonomy for information visualizations. In: Proceedings of the IEEE Symposium on Visual Languages, 1996, pp. 336–343. IEEE (1996)
10. Sathiyanarayanan, M.: Multi-channel deficit round-robin scheduling for hybrid tdm/wdm optical networks (2012)
11. Sathiyanarayanan, M., Kim, K.S.: Multi-channel deficit round-robin scheduling for hybrid tdm/wdm optical networks. In: Proceedings of the 4th International Congress on Ultra Modern Telecommunications and Control Systems (ICUMT 2012), pp. 552–557, St. Petersburg, Russia, Oct 2012
12. Sathiyanarayanan, M., Abubhakar, B.: Dual mcdrr scheduler for hybrid tdm/wdm optical networks. In: Proceedings of the 1st International Conference on Networks and Soft Computing (ICNSC 2014), pp. 466–470, Andra Pradesh, India, Aug 2014
13. Sathiyanarayanan, M., Abubakar, B.: Mcdrr packet scheduling algorithm for multi-channel wireless networks. In: Proceedings of 3rd International Conference on Advanced Computing, Networking and Informatics, pp. 125–131. Springer (2016)
14. Sathiyanarayanan, M., Mohammad, A.: Euler-time diagrams: a set visualisation technique analysed over time. In: Proceedings of 4th International Conference on Advanced Computing, Networking, and Informatics (2016)
15. Achunala, D., Sathiyanarayanan, M., Abubakar, B.: Traffic classification analysis using omnet++. In: Proceedings of 4th International Conference on Advanced Computing, Networking, and Informatics (2016)

Resource Allocation for Video Transcoding in the Multimedia Cloud

Sampa Sahoo, Ipsita Parida, Sambit Kumar Mishra,
Bibhdatta Sahoo and Ashok Kumar Turuk

Abstract Video content providers like YouTube and Netflix cater their content, i.e., news and shows, on the web which is accessible anytime anywhere. The multi-screens like TVs, smartphones, and laptops created a demand to transcode the video into the appropriate video specification ensuring different quality of services (QoS) such as delay. Transcoding a large, high-definition video requires a lot of time, computation. The cloud transcoding solution allows video service providers to overcome the above difficulties through the pay-as-you-use scheme, with the assurance of providing online support to handle unpredictable demands. This paper presents a cost-efficient cloud-based transcoding framework and algorithm (CVS) for streaming service providers. The dynamic resource provisioning policy used in framework finds the number of virtual machines required for a particular set of video streams. Simulation results based on YouTube dataset show that the CVS algorithm performs better compared to FCFS scheme.

Keywords Cloud · Video transcoding · Merger · Video stream · QoS

S. Sahoo (✉) · I. Parida · S. K. Mishra · B. Sahoo · A. K. Turuk
National Institute of Technology, Rourkela, Rourkela, India
e-mail: sampaa2004@gmail.com

I. Parida
e-mail: ipsitaparida07@gmail.com

S. K. Mishra
e-mail: skmishra.nitrkl@gmail.com

B. Sahoo
e-mail: bibhudatta.sahoo@gmail.com

A. K. Turuk
e-mail: akturuk@gmail.com

© Springer Nature Singapore Pte Ltd. 2019
P. K. Sa et al. (eds.), *Recent Findings in Intelligent Computing Techniques*,
Advances in Intelligent Systems and Computing 707,
https://doi.org/10.1007/978-981-10-8639-7_55

1 Introduction

Now the content viewing has shifted from traditional TV system to video streaming in laptops, smartphones, etc. through the Internet. Content providers like YouTube, Netflix, and Hulu as well as TV channels cater their content, i.e., news, shows, live events, and user-generated content on the web which is accessible anytime anywhere. According to Cisco Visual Networking Index (VNI) report published in the year 2016, video streaming will escalate up to 82% of the total network traffic by 2020, up from 70% in 2015 [1]. The variation in the user's demand concerning resolution, bit rates, frame rate, or a combination of these makes the job of media professionals critical to managing it. It is not possible to store videos with all possible formats, resolutions, and frame rates as it requires massive storage and computational resources. This process will also increase the financial cost of the video content providers. One of the solutions is to store some (e.g., popular) video in popular formats and transcode unpopular videos on-demand [2]. The rapid growth of mobile devices, user Preferences, and networks have created a requirement for video transcoding into the appropriate specification such as resolution, quality, bit rate, video format, etc. and simultaneously ensuring different quality of services (QoS) such as delay. Converting a compressed video into another compressed video is termed as video transcoding [3]. But transcoding of videos in real time is a time-consuming and challenging task as it holds a strict delay requirement.

Video transcoding not only reduces the video file size but also gives an opportunity to select from an extensive set of options. It makes the video viewable across platforms, devices, and networks. Usually, transcoding requires a high-quality mezzanine file to start with and convert it into a form supported by the targeted device. Transcoding a large, high-definition video to a diverse set of screen sizes, bit rates, and quality requires a lot of time, computation, and storage capacity. To overcome the difficulty associated with the transcoding process content providers is using the cloud services. A user only needs to specify its requirements and subscribes the services provided by the cloud, and the rest of the task, i.e., time-consuming transcoding process, will be performed using cloud resources at the back end. The advantage of transcoding in the cloud is lower cost, virtually unlimited scalability, and elasticity to counter peak demand in real time. The cloud transcoding solution allows video service providers to pay as they use, with the assurance of providing online support to handle unpredictable demands [4]. Cloud-based video transcoding reserves resources based on current workload to satisfy predefined QoS. However, online transcoding in the cloud has its challenges. The first key problem is the hard delay (streaming and transcoding) requirement. The second challenge is balancing resources available and the demand while ensuring cost and QoS constraints. The insufficient resource reservation for transcoding may cause delay of the video playback. Video transcoding in the cloud can be done in the following ways: through a dedicated VM or using different VMs for different video segments simultaneously. The first approach requires a significant number of VMs for a large set of video stream, whereas the second method can transcode several video streams simultaneously reducing the number of VMs [2]. In this paper, we assume the second approach and discuss the implementation of a cloud-based platform for transcoding of videos.

The rest of the paper is organized as follows: Sect. 2 summarizes the related works. Cloud-based video transcoding architecture and resource allocation policy are presented in Sects. 3 and 4, respectively. Section 5 analyzes the simulation results. The paper is concluded in Sect. 6.

2 Literature Study

Researchers proposed and implemented various transcoding frameworks where video transcoding can be performed partially or on-demand. Few works on cloud-based transcoding are presented here. Researchers have worked on energy-efficient and real-time task allocation [5, 6]. Zhao et al. address the cost issue of multi-version video-on-demand system in the cloud. The decision whether to store or transcode is made based on the popularity of video, storage, and computation cost [7]. Zhang et al. proposed an energy-efficient algorithm to route transcoding jobs in the multimedia cloud [8]. Lei et al. presented an analytical model of a cloud-based online video transcoding system to predict the minimum resource reservation for specific QoS constraints (i.e., minimum system delay, targeted chunk size) [9]. Gao et al. used a partial transcoding scheme to minimize the operational (i.e., storage + computation) cost of content management in media cloud. Based on the user, viewing pattern decision is made whether to cache or transcode online a video segment [10]. Li et al. introduced a cloud-based video streaming service (CVSS) architecture for on-demand transcoding of video streams using cloud resources. CVSS architecture gives a cost-efficient platform to streaming service providers for using cloud resources to meet QoS demands of video streams [11]. Li et al. designed a cloud transcoder to reduce the download time and improve the data transfer rate. Transcoding is executed based on video popularity and transcoder status, i.e., below the certain threshold [12]. Fareed et al. presented a prediction-based dynamic resource allocation algorithm to allocate and deallocate VMs for video transcoding service with the aim of achieving cost efficiency in infrastructure as a service cloud [2]. Chen and Chang implemented a cloud-based, scalable, and cost-effective video streaming service platform to serve all the transcoding requests [13].

3 System Model

We propose the cloud-based on-demand video transcoding (CVT) framework as shown in Fig. 1. The framework shows the sequence of actions taken place when a user requests videos from a streaming service provider. The functionalities of various components are as follows: The *streaming service providers* like YouTube and Netflix

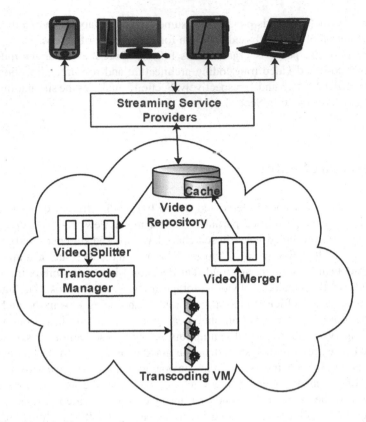

Fig. 1 Cloud-based video transcoder

accept user's request and check if required video is present in video repository or not. If the video is present in its desired format, then starts streaming the video. If the video is not in a format that is requested, online transcoding is done using cloud resources. To perform transcoding, video is divided into small chunks by the *video splitter* and then *transcode manager* maps the video chunks to appropriate transcoding VMs based on certain QoS. The queue formed by video streams near transcode manager has two parts: start-up and batch. Start-up part of the queue consists of first few group of pictures (GOPs) of each video stream, and rest streams are in batch queue part. For a video request, video streams in start-up queue are assigned to VM first and then the streams present in the batch queue. The *transcoding VM* transcodes the source videos into targeted videos with desired video specification concerning format, resolution, quality, etc. with certain QoS constraints. Each transcoding VM is capable of processing one or more simultaneous transcoding task. *Video merger* is used to place all the video streams in the right order to create the resulting transcoded stream. A copy of the transcoded video is stored in video repository to save time and computation cost. All the possible forms of popular and frequently accessed videos are stored in *cache storage*, a part of video repository. The unpopular video requested by the user is transcoded online and served to the user.

Algorithm 1: CVS Resource allocation algorithm

Input : $V_{DS} \leftarrow V_1, V_2, ..., V_n$: List containing n videos

$nI \leftarrow$ number of I frames of each video

$Q \leftarrow$ Queue contains 1st video chunk, 2nd video chunk, ... of $V_1, .., V_n$.

$size \leftarrow$ size of the video

$len_{sq} \leftarrow$ length of queue associated with start-up queue(SQ)

$len_Q \leftarrow$ length of Q

$len_{bq} \leftarrow$ length of batch queue(BQ)

$maxTH \leftarrow$ Maximum threshold

Output : $nVMs \leftarrow$ *the number of VMs required*

Initialise: $nVMs \leftarrow 0$

---

1 $nIFrames \leftarrow nI$ *of* V_i

2 *Enqueue* len_{sq} *no of video chunks from Q to SQ*

3 *Allocate SQ to VM*

4 $nVMs \leftarrow nVMs + 1$

5 $len_{bq} \leftarrow len_Q - len_{sq}$

6 **for** $i = 1$ *to* len_{bq} **do**

7   *calculate avgQlen of VMs* ▷ /*avgQlen() calculates the average of filled portion of queue of each VM*/

8   **if** $avgQlen() \geq maxTH$ **then**

9   *activate new VM*

10        $nVMs = nVMs + 1$

11   **end**

12     $VM_{min} \leftarrow$ *Select VM with shortest filled queue*

13   *enqueue* $BQ[i]$ *to* VM_{min}

14 **end**

---

4 Resource Allocation Policy

The proposed algorithm, Algorithm 1, predicts the number of VMs for transcoding the videos demanded by users. The videos are divided into smaller chunks at GOP level. Therefore, the number of smaller chunks possible is equal to number of I frames present in a particular video. The queue Q is formed by taking ith video chunk from each video and en-queuing them sequentially where $i = 1, 2, ..., n$ and n is the number of videos. In steps 2 and 3, a start-up queue (SQ) is formed and allocated to VM. The start-up queue will transcode some initial chunks of videos to provide faster response to users. So we always have 1 or more VMs for transcoding. Once the start-up queue is completed, the corresponding VM can be used with other

VMs for transcoding the batch queue (BQ). Before allocating a video chunk to any VM, the average load (queue length of each VM) of active VMs is calculated. If the average length is more than the threshold value, a new VM is activated. The VM with least load is selected (V_{min}) and the new video chunk is en-queued to V_{min}.

5   Simulation Results

We have uniformly selected videos from YouTube dataset collected from UCI repository—Online Video Characteristics and Transcoding Time Dataset Data Set for simulation. Out of 10 fundamental video characteristics, the attributes which we have used here are video ID, duration, codec, number of I frames, and total transcoding time. We have made certain assumptions to correlate with the result better.

1. Number of I frames in a particular video is same as the number of GOPs.
2. The transcoding time of each GOP is by dividing total transcoding time with number of I frames.
3. The VMs allocated are considered as homogeneous.
4. The queue associated with each VM has same length (say 100). The maximum threshold value is assumed to be 70% of the VM's queue length.

Figure 2 demonstrates the number of VMs required for transcoding the corresponding number of requests. We see that the resources allocated for the users' requests are quite cost-effective as the number of VMs active is less because we perform the initial check whether to activate another VM as per need. In Fig. 3, there is a comparison of our proposed resource allocation algorithm with existing scheduling algorithm. It is observed that the number of VMs required in FCFS is quite high in comparison to the proposed algorithm.

Fig. 2 Resource prediction by the proposed algorithm

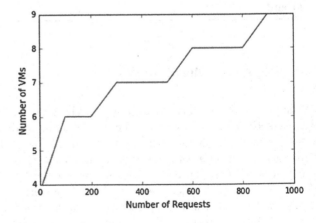

Fig. 3 Comparison of the proposed algorithm with FCFS

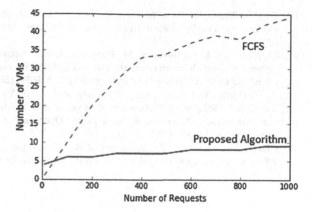

6 Conclusion

In this paper, we propose an on-demand CVT architecture that provides a cost-efficient platform to transcode video streams. The dynamic resource allocation scheme present in CVT predicts the appropriate number of VMs so that resource provisioning cost is reduced. For cost efficiency, a video is broken into several video streams so that multiple streams can be transcoded on a single VM, thus reducing the VM requirement. This architecture can be useful for video streaming providers to utilize cloud resources and improve user's satisfaction with low cost.

References

1. Forecast and methodology, 2015–2020 white paper. Cisco visual networking index (2016)
2. Jokhio, F., Ashraf, A., Lafond, S., Porres, I., Lilius, J.: Prediction-based dynamic resource allocation for video transcoding in cloud computing. In: 21st Euromicro International Conference on Parallel, Distributed, and Network-Based Processing, pp. 254–261. IEEE (2013)
3. Sahoo, S., Sahoo, B., Turuk, A.K.: An analysis of video transcoding in multi-core cloud environment (2017)
4. http://download.sorensonmedia.com/pdfdownloads/lowres/whitepaper.pdf (2011)
5. Mishra, S.K., Deswal, R., Sahoo, S., Sahoo, B.: Improving energy consumption in cloud. In: 2015 Annual IEEE India Conference (INDICON), pp. 1–6. IEEE (2015)
6. Mishra, S.K., Deswal, R., Sahoo, S., Sahoo, B.: Improving energy consumption in cloud. In: 2015 Annual IEEE India Conference (INDICON), pp. 1–6. IEEE (2015)
7. Zhao, H., Zheng, Q., Zhang, W., Du, B., Li, H.: A segment-based storage and transcoding trade-off strategy for multi-version VOD systems in the cloud. IEEE Trans. Multimed. 149–159 (2017)
8. Zhang, W., Wen, Y., Cai, J., Wu, D.O.: Toward transcoding as a service in a multimedia cloud: energy-efficient job-dispatching algorithm. IEEE Trans. Veh. Technol. 2002–2012 (2014)
9. Wei, L., Cai, J., Foh, C.H., He, B.: Qos-aware resource allocation for video transcoding in clouds. IEEE Trans. Circuits Syst. Video Technol. 49–61 (2017)

10. Gao, G., Zhang, W., Wen, Y., Wang, Z., Zhu, W.: Towards cost-efficient video transcoding in media cloud: insights learned from user viewing patterns. IEEE Trans. Multimed. 1286–1296 (2015)
11. Li, X., Salehi, M.A., Bayoumi, M., Buyya, R.: CVSS: a cost-efficient and QoS-aware video streaming using cloud services. In: 16th IEEE/ACM International Symposium on Cluster, Cloud and Grid Computing (CCGrid), pp. 106–115. IEEE (2016)
12. Li, Z, Huang, Y., Liu, G., Wang, F., Zhang, Z.-L., Dai, Y.: Cloud transcoder: bridging the format and resolution gap between internet videos and mobile devices. In: Proceedings of the 22nd International Workshop on Network and Operating System Support for Digital Audio and Video, pp 33–38, ACM (2012)
13. Chen, K.-B., Chang, H.-Y.: Complexity of cloud-based transcoding platform for scalable and effective video streaming services. Multimedia Tools and Applications, pp. 1–18 (2016)

A Trust Network Driven User Authorization Scheme for Social Cloud

Stephy P. Susan and Greeshma Sarath

Abstract Social cloud leverages the real-world relationship in online social networking by sharing the resources with pre-established trust among the online users. It creates a heterogeneous resource sharing environment in the context of social networks. Since social cloud offers a real-world environment which enhances multi-user collaboration, sharing resources among obscure users is a challenging task, but for exploiting the advantage of social cloud it is necessary to associate with unknown users and this anonymity leads to threats. Building a trust network among the users could be a solution to this problem. This paper proposes a novel method to authorize unknown users with an initial trust based on their existing relation in the social network for secure sharing of resources. Trust network is created from existing relationships in the social network for evaluating trust of users for granting access limits over a resource. The proposed method is validated using simulations.

Keywords Cloud computing · Social cloud · Social networking · Trust evaluation · Trust network · User authorization · Cloud security

1 Introduction

Cloud computing and social networking have recently gained a lot of research intrigue on account of the wide usage of well-known cloud services providers like AWS (Amazon Web Service), Google App Engine, Microsoft Azure, and social media like Twitter, Facebook, and so forth. On consolidating the upside of both, social cloud is getting more prominent. A social cloud [1] is a resource and service sharing framework utilizing relationships established between members of a social network.

S. P. Susan (✉) · G. Sarath
Department of Computer Science and Engineering, Amrita School of Engineering,
Amritapuri, Amrita University, Amrita Vishwa Vidyapeetham, Kollam, India
e-mail: stephypsusan@gmail.com

G. Sarath
e-mail: greeshmasarath@am.amrita.edu

© Springer Nature Singapore Pte Ltd. 2019
P. K. Sa et al. (eds.), *Recent Findings in Intelligent Computing Techniques*,
Advances in Intelligent Systems and Computing 707,
https://doi.org/10.1007/978-981-10-8639-7_56

533

Social cloud is not only a social network, but also a cloud that builds upon a social network's advanced digital representation of real-world relationships as a means to capture, assess, and interpret relationships between individuals.

Resources in a social cloud are physical or virtual entities or capabilities such as computing capacity, services, and software that are useful to other members in the group. Users in the social cloud can simultaneously own a resource and access the resources owned by other users. The sharing or access of resources inside the social cloud is in the basis of trust between the users [2]. In a social network like Facebook, the relationships are in the form of friends, friends of friends, and not a friend. The trust among these relations varies with respect to the context of their interactions. Resources shared by users in a social cloud are heterogeneous and it depends on context of the user. For example, some users may share storage, while the others may share access to compute resources and services or they could offer combinations of resources, capabilities, and services. Since the users are interacting and sharing resources directly, there exist different possibilities of attacks from malicious users inside the cloud. Therefore, all connections are not equally trusted, so it is necessary to monitor and control the users for proper working of trusted social cloud.

Trust is not a computational value, rather it is a conjecture that a trustee will act as how he/she is relied upon to do. It depends on the confirmation of the trustee's behavior [3]. In software engineering, trust is classified into two: trust of the "user" and trust of the "system". The concept of "user" trust is developed from psychology and sociology, with a definition as "A subjective expectation an entity has about another entity's future conduct." This suggests that trust is inherently customized in light of the input of past interaction between users. In this sense, trust is relational. On the off chance that two users interact with each other frequently like users in Facebook, Twitter, etc., their relationship fortifies and trust develops in light of their cooperation. In the event that the experience between the clients is positive and then trust increases and vice versa. The rate at which trust increases or decreases relies upon user behavior. Thus, relational trust between users built upon interactions over time. Therefore, an anonymous user did not acquire any trust when he signs up into the social cloud.

Revoking access privileges to the new users based on the anonymity cause loss of user's interest over the technology. But a resource owner cannot grant higher access rights to an unknown user, because at any time user can induce attacks. Trust-based access control methods are used to authorize users in cloud but the trust has to develop over time. In social cloud scenario, even users with lower access rights (due to lack of direct interaction with resource owner) are able to contribute to the technology, and therefore trust developing time for new users will be a hindrance in the usage of social clouds. In this paper, we represent social cloud as a trust network where the edges represent users and links connecting individuals which implies a the degree of trust. So that users who are unknown to the resource owner will also get a chance to acquire a higher trust value and thus higher access rights based on the recommendation trust. We propose a trust model for evaluating trust based on direct and indirect (friend of friend, acquaintance) connections where indirect connections are treated as recommendations which propagate [4] through the network.

2 Related Work

Social clouds are categorized as social storage cloud [5], Social CDN (Content Delivery Network), and social compute cloud. In social CDNs, for example, Facebook or Google plus, users build scalable cloud-based applications hosted by cloud services like Amazon Web Services. There is no literature related to trust-based social cloud authorization. For the smooth working of these facilities, the trust and security will be a great concern.

One of the major security issues experienced in social cloud environments is anonymity [6]. With the help of anonymity, the users involve in malicious activities. This creates security risk to the resources and services provided in the cloud. In [7], they proposed an access control mechanism for cloud data by providing tokens to the cloud users. This mechanism involves issuing tokens to the users based on their needs and authorization level. But this method has a limitation in scalability of social cloud scenario. A trust-based authorization mechanism is proposed by [8], a set of policies are defined for user authorization and they assumed that there exists a predefined trust and they do not specify how it is evaluated. There exist systems that recommend friends to users. Carullo et al. [9] proposed one such system in social network where friends are recommended based on similarity factors and an approach based on hubs and authorities. An automated grouping system [10] proposed a method to group friends into best friends, normal friends, and visitors by computing the closeness degree (trust) between the user and his/her friends. For recommendation of a node to a node, [11] proposes a trust-based recommendation system for peer-to-peer (p2p) systems from past interactions. This system decreases the trust with the malicious activity in a p2p system by giving recommendations based on the services provided by a peer.

In all the existing works, anonymous or new users did not get any kind of trust in the initial stage, and they have to gain trust among the peers over time. Granting access rights to anonymous users introduces vulnerabilities, and also revoking the access rights because of anonymity will not give benefits to social clouds. For getting the advantage of the existing relation in the social network to a newcomer, we proposed a trust-based authorization through the trust network.

3 Proposed Solution

An anonymous user cannot be given a higher access right in the social cloud as it is possible that he will induce some kind of attacks which crashes the entire system in worst case. At the same time, it is not good to assign the lowest access right to him as it is not possible to achieve the complete utilization of his capabilities, for example, a resource owner can limit the access only to his trusted circle [12]. So there should be a means to assign a fair level of authorization to an anonymous user on his initial request for a resource. In this method, an initial access right is assigned to an anonymous user based on the recommendation given by direct relations of the resource owner. In social cloud scenario, there exist different user groups and each group has a common interest like Facebook groups. Suppose a Facebook group initiates a volunteer computing project on a social cloud, and users in the group can be a part of the project and use the resources. The peer group contains thousands of users and each one may not be even known to each other. When a new user from the group requested for a resource, he should get an authorization from the resource owner. A resource owner specifically offers authorization to a user, who is a companion of his/her from a trusted gathering.

Social cloud environment creates links connecting users within a social network that implies a certain degree of trust in view of social relationships and actual collaborative activities being performed by them. All connections in the network are not equally trusted, and users categorize connections either individually or into groups, customizing them to have separated perspectives of their social cloud. Since the trust is transitive in nature, the current trusted associations are utilized to assess the trustworthiness of an obscure client. We are using the concept of trust network for evaluating trustworthiness of users in social cloud environment.

3.1 Trust Network

Trust is not an independent value in a social scenario; it is subjective, as well as transitive. Trust of one person toward other is depended on another user's trust toward him. So the best way to control the flow of trust is by building a trust network. Trust network is created for the social cloud with users as the nodes and relation between them as the edges. Relations are denoted as a trust value. Figure 1 shows a representation of real-time social cloud network. The trust is defined for every relation and it will be directed. Trust of each person is determined by analyzing the activities of users or by the recommendation of a trusted user, and the mutual trust will be built on the basis of this trust.

To formulate the problem mathematically, let us consider U_i where i = 1, 2, ..., n as n number of cloud users and T_u be the trust of the resource owner or other users toward each cloud user. Let us consider that the cloud users and resource owners as a network with users as nodes and edges represent the connection between the users. Each user U_i will have a close circle of trusted users represented as tu. Users in group tu will have a direct connection with the all resources owned by U_i. Trust toward each user (Weighted edges) in tu_i will be different, as shown in Fig. 2. Trust values are in the range $[-1, 1]$. The trust network is created using Algorithm 1.

When a user gives a request to the cloud service, he will be given an access right based on the direct trust value of resource owner's trust toward him. If the user is not directly connected with the resource owner, he will be given a trust value based on the recommendations provided by users in the trusted circle of resource owner. It is not mandatory that the newcomer should have an immediate connection with any user in tu_i. He will get a recommendation using the propagative property of trust

Algorithm 1 Trust-Network Creation

INPUT: Users (U_i), Social Circle (SC)
OUTPUT: Trust-Network
1: $Nodes = \{ \}, Edges =\{ \}$
2: **for** each U_i **do**
3:    **if** $U_i \notin Nodes$ **then**
4:      $Nodes.add(create\_vertex(U_i))$
5:    **end if**
6:    **for** each SC of U_i **do**
7:      **for** each relation $R_i \in SC_i$ **do**
8:        **if** $R_i \notin N$ **then**
9:          $Nodes.add(create\_vertex(R_i))$
10:          $Edges.add(U_i, R_i, T_i(SC_i))$
11:        **end if**
12:      **end for**
13:    **end for**
14: **end for**

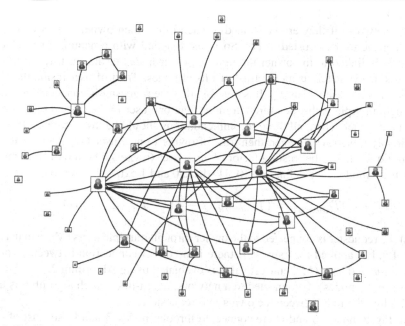

Fig. 1 Social cloud network. Nodes represent cloud users and edges show relation between users

even if he is indirectly connected to the owner through several users. If the requester is entirely new to the owner, he will be authorized with a neutral trust value. A role will be assigned to the user based on the initial trust value. The authorizations given to the users with initial trust value (neutral) should be such that they wont be able to make any harm by inducing an attack.

Since trust is asymmetric, it is possible that user A trusts B with a trust value of 0.5 but user B might not trust user A with same amount of trust as shown in Fig. 2. Let

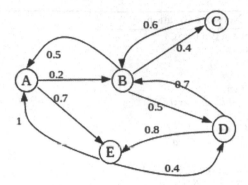

Fig. 2 Trust network

us consider a user U_i (D in Fig. 2) request for a resource owned by resource owner (ow) A. Then, we have to check whether the owner and user is somehow linked in the network; if they are not linked to each other, then owner cannot determine whether the user is trusted or not. So U_i is assigned with a neutral trust value. If the user is linked to the owner through any other nodes in the trust network, then his trust is evaluated using the trust of linked nodes. Trust of user inside the close circle tu is represented as T_{tu}. It is possible that the requester is connected to owner through multiple paths. i.e., through more than one user in the trusted circle of the resource owner. Then, all these recommendations should be taken care of. If a user is directly linked with owner, then trust of owner to new user (T_{ow-u}) is evaluated using Eq. 1. "α" is the trusting factor of each user (owner), i.e., the rate at which a user trust another. We define $\alpha \in [0, 1]$ where 0 and 1 extreme cases.

$$T_{(ow-u)} = (T_{(ow-tu_i)}) - (1 - T_{(tu_i-u)}) * \alpha)$$ (1)

If the requester is connected to the owner through multiple users (More than two), then Eq. 1 is applied recursively. If there is no direct connection between resource owner ow and user U_i, then trust value is evaluated using Algorithm 2.

Since the nodes are connected to a network, depth-first search algorithm is used for finding the link between the owner and requester.

In Fig. 2, nodes A and D are connected through nodes B and E, so trust of A to D will be dependent on trust of A to B, B to D, and A to E, E to D by propagative property of trust. If α value increases, trust value decreases. α is high for all users who trust another person in a slow rate and low for all users who trust another person in high rate. Users are categorized based on trust value as shown in Table 1. Access control strategies are defined for a resource based on the trust level of the user and security aspects needed by a resource. If a resource needs a higher level of security, then access rights on that resource will be given only to the user level "Trusted".

Anonymous users are benefited from our trust model by getting an extra privilege at the time of initial authorization itself instead of building the trust value gradually.

Algorithm 2 Trust Evaluation

INPUT: Trust-Network (T), Resource owner (U_{ow}), Requester (U_i), Neutral Trust
OUTPUT: Trust Value (T_u)
1: $T_u = 1$, $P = \{\ \}$
2: Find all paths from U_{ow} to U_i using DFS.
3: Add all paths to P
4: **if** $|P| = 0$ **then**
5:    $T_u = Neutral\ Trust$
6: **else**
7:   **for** each $P_i \in P$ **do**
8:      Find $T_u(P_i)$ using Equation 1.
9:      $T_u = Min\{T_u, T_u(P_i)\ \}$
10:   **end for**
11: **end if**

Table 1 Trust level of users

Trust value (T)	Level of trust	Description
$T < 0.5$	Malicious	Highly untrusted
$T > -0.5$ and < 0	Distrust	Lowest possible trust
$T = 0$	Neutral	No trust or distrust
$T > 0 < 0.5$	Mean	Average trustworthiness
$T > 0.5$	Trusted	Completely trusted

Later, their behavior should be monitored and their trust should vary based on their behavior. When an anonymous user requests for a resource, then the resource owner checks for their initial trust and he gives permission based on the result obtained.

4 Experimental Results

We have developed a Java-based program to validate our proposed system. Based on the trust values, the user is categorized into trusted or malicious. Equation 1 is validated using sample values as shown in Table 2.

Table 2 shows the trust gained by the different users based on the neighbors' trust recommendation. Cases 1 and 2 show that owner is completely trusting the trusted user and he in turn trusts the requester completely with a trust value 1, then by Eq. 1, owner trusts the requester with a trust value "1". As both the trust values are maximum, α does not have any significance. In all other cases, trust value will decrease as α increases, and vice versa. From this, we infer that α is inversely proportional to the trust value.

We simulated a trust network using Google Plus dataset. This dataset includes users and their friend list based on various circles, i.e., friends, family, acquaintance, etc. Based on these circles, the trust toward a person is initialized. For all users in

Table 2 Initial trust

Cases	T_{ow-tu}	T_{tu-us}	α	T_{ow-us}
1	1	1	1	1
2	1	1	0.5	1
3	1	0	0.5	0.5
4	1	0	1	0
5	0	1	1	0
6	0	1	0.5	0
7	0.5	0.5	0.5	0.25
8	0.5	0.5	1	0
9	0.9	0.9	0.5	0.85

family circle, we gave 0.8 as the trust value, for friends 0.5, and for acquaintance 0.1. It includes 107614 nodes and 13673453 edges, nodes represent the users and edges represent the relation between the user. The edges are weighted edges where weight value represents trust between the users. Family, friends, and acquaintance of all users in the dataset is represented as a graph forming the initial trust network. In social networks like Facebook and Google Plus, friend of friend is also considered as a friend. Here, we are evaluating the trustworthiness of unknown friends from the trust network using Algorithm 1 based on the recommendation.

The results are validated using trust network formed with 10000 selected nodes from the Google Plus dataset. Then, 50 users are added as newcomers who are not part of the trust network. 100 nodes are designated as resource owners and others used to give requests to access the resources. The resource owner and requester may or may not be connected with each other in the network. Here exist four cases.

Case 1: A requester who is directly linked with the trusted user of owner.
Case 2: A requester who is linked through malicious users to the owner.
Case 3: A requester who is linked through both trusted and malicious users.
Case 4: A requester who is not at all linked to the owner through any paths.

In case 1, a requester is requesting a resource to owner and he is a friend of friend of owners trusted friend circle. According to Algorithm 1, he gains an initial trust value on the behalf of his connections. If he is connected with more than one path to the user, then the minimum trust among all path should be taken. In case 2, our model gives a negative trust to the requester as he is connected with malicious users. In case 3, trust model evaluated the trust toward the requester based on all the paths (trusted and untrusted), since our model is taking the minimum trust among different paths, he gets a minimum trust according to his malicious links. In case 4, the newcomer is not connected to any of the preexisting trusted or untrusted users, then his trust cannot be determined from the network. Therefore, he will be assigned a neutral trust value.

Table 3 Trust based on neighbors

# Links	# links in each trust level				Trust
	Malicious	Distrust	Mean	Trusted	
120	0	40	40	40	−0.6
20	0	0	0	20	0.3
70	40	20	10	0	−0.9
100	50	50	0	0	−1
80	0	60	20	0	−0.4
70	0	0	50	20	0.1
120	0	0	80	40	0.3

Table 3 shows the result obtained from the testing of test cases described above, based on different types of neighbors connected with the requester in the trust network. The trust of preexisting users in the network is evaluated by either monitoring the behavior [13] of the user or user is to directly categorize the user to friends, family, etc.

From the experiments, it is observed that the requester will get an initial trust from the trust network based on the recommendation of preexisting users. This initial trust depends on the trust of owner's neighbor, their category, and how much he trusts the requester. From the results in Table 3, it is observed that a newcomer will gain a minimum initial trust of the level which whom he/she is mostly connected with. Suppose a requester is connected with any one of the users in "Distrust level", then he only gets initial trust close to −0.5. Therefore, he cannot induce any kind of attacks to the social cloud resources. But if the user is only linked with users in "Trusted" category, then he will achieve a minimum initial trust close to 0.5. No new requester will get an initial trust greater than his neighbors, and therefore new requester will not attain a trust level and cannot induce any kind of malicious activities. It is possible that the recommendation provided by a user in the trusted circle is not accurate, as trusting factor (α) is different for different users. A continuous user behavior monitoring is necessary in this case.

5 Conclusion

Social cloud provides a platform for users where they can participate and contribute in various ways. Since it is based on social network, access privileges are given to a user based on his trust value. In this paper, we have proposed a novel method for authorizing an anonymous user based on the recommendation obtained from the trust network. An anonymous requester gains an initial trust value greater than neutral trust based on the recommendations provided for him. Therefore, he will get a fair chance to access the resource with a high privilege. Since we assigned minimum trust value by considering all the recommendations, probability of a malicious user gaining high privilege is less. The proposed method is validated by constructing a trust network of cloud users, and we observed that the constructed trust network authorizes users according to their trust value. Since our proposed method does not filter inaccurate recommendations provided by a user, a thorough monitoring of user behavior is required. This would be a direction for the future research.

References

1. Chard, K., Bubendorfer, K., Caton, S., Rana, O.F.: Social cloud computing: a vision for socially motivated resource sharing. IEEE Trans. Serv. Comput. **5**(4), 551–563 (2012)
2. Gayathri, K.S., Thomas, T., Jayasudha, J.: Security issues of media sharing in social cloud. Procedia Eng. **38**, 3806–3815 (2012)
3. Adali, S.: Trust as a computational concept. Modeling Trust Context in Networks. Springer, pp. 5–24 (2013)
4. Yadav, A., Chakraverty, S., Sibal, R.: A survey of implicit trust on social networks. In: 2015 International Conference on Green Computing and Internet of Things (ICGCIoT), pp. 1511–1515. IEEE (2015)
5. Chard, K., Simon, C., Rana, O.F., Bubendorfer, K.: Social clouds: a retrospective. IEEE Cloud. Comput. **2**(6), 30–40 (2016)
6. Shyamala, C.K., Hemaashri, S.: An enhanced design for anonymization in social networks. Indian J. Sci. Technol. **9**(30) (2016)
7. Khaled, A., Husain, M.F., Khan, L., Hamlen, K.W., Thuraisingham, B.: A token-based access control system for rdf data in the clouds. In: 2010 IEEE Second International Conference on Cloud Computing Technology and Science (CloudCom), pp. 104–111. IEEE (2010)
8. Zahoor, E., Perrin, O., Bouchami, A.: CATT: a cloud based authorization framework with trust and temporal aspects. In: 2014 International Conference on Collaborative Computing: Networking, Applications and Worksharing (CollaborateCom), pp. 285–294. IEEE (2014)
9. Carullo, G., Castiglione, A., De Santis, A.: Friendship recommendations in online social networks. In: 2014 International Conference on Intelligent Networking and Collaborative Systems (INCoS), pp. 42–48. IEEE (2014)
10. Shyamala, C.K., Hemaashri, S., Swetha, R.: An improved recommendation system for social networks. Int. J. Control Theory Appl. (ICSCS) **8**(5), 1903–1910 (2015). IJCT A International Science Press
11. Can, A.B., Bhargava, B.: Sort: a self-organizing trust model for peer-to-peer systems. IEEE Trans. Dependable Secure Comput. **10**(1), 14–27 (2013)
12. Chard, K., Caton, S., Rana, O.F., Bubendorfer, K.: Cloud computing in social networks. Social cloud. IEEE Cloud **10**, 99–106 (2010)
13. Tian, L.-Q., Lin, C., Ni, Y.: Evaluation of user behavior trust in cloud computing. In: 2010 International Conference on Computer Application and System Modeling (ICCASM), vol. 7, pp. V7–567. IEEE (2010)

QTM: A QoS Task Monitoring System for Mobile Ad hoc Networks

Mamata Rath, Bibudhendu Pati, Chhabi Rani Panigrahi and Joy Lal Sarkar

Abstract This paper presents monitoring system for Quality of Service (QoS) based task module called QoS Task Monitoring (QTM) in Mobile Adhoc Networks (MANET) using mobile agent as basic element. Currently MANET is one of the most promising and advanced solution for wireless networks due its significant performance in resuming connectivity in drastic situations. In such environment, there is maximum chance of network disconnection and possibility of immediate set up of network is almost impossible. The fundamental routing process in a MANET involves facilitating uninterrupted communication in the network system between two mobile stations at any point of time and the basic key concern being selection of the most suitable forwarding node to advance the real-time packets from source towards destination so that the optimization of the network can be achieved by maximum utilization of available resources. Transmission of real-time applications is one of the most challenging issue in MANET due to transportation of high volume of data including audio, video, images, animation, and graphics. This paper presents a monitoring approach for checking the Quality of Service (QoS) task modules during competent routing with the use of mobile agents. An intelligent mobile agent is proposed in QTM System which has been designed in the QoS-based platform for checking and controlling the processing tasks using longest critical path method at the forwarding node to select it as the best option out of all neighbor nodes. Simulation result shows higher packet delivery ratio and uniform jitter variation which suits favorably to multimedia and real time applications.

M. Rath (✉)
C. V. Raman College of Engineering, Bhubaneswar, India
e-mail: mamata.rath200@gmail.com

B. Pati · C. R. Panigrahi
Department of Computer Science, Rama Devi Women's University, Bhubaneswar, India
e-mail: patibibudhendu@gmail.com

C. R. Panigrahi
e-mail: panigrahichhabi@gmail.com

J. L. Sarkar
Department of CSE, C. V. Raman College of Engineering, Bhubaneswar, India
e-mail: joylalsarkar@gmail.com

© Springer Nature Singapore Pte Ltd. 2019
P. K. Sa et al. (eds.), *Recent Findings in Intelligent Computing Techniques*,
Advances in Intelligent Systems and Computing 707,
https://doi.org/10.1007/978-981-10-8639-7_57

543

Keywords MANET · QoS · Real-time applications · Mobile agent
AODV protocol

1 Introduction

The basic objective of an efficient network is to provide best required service to the user during communication [1]. In QoS provision, the network behavior becomes more deterministic in order to achieve better throughput with maximum utilization of resource [2]. In a wireless network, a QoS service request normally includes better communication with minimum bandwidth consumption, maximum permissible delay for application specific tasks, a particular maximum delay variance (jitter), and packet loss rate up to an extent [3]. The network can provide QoS flow only if it satisfies the required configuration of network parameters in the QoS service request, then only there is continuous flow of communication during transmission of data with specified criteria under the agreement. So, a network offers a set of service guarantees during the period of flow. As per the agreement with the service provider, the QoS service request is given to the network by the user as a QoS requirement specification [4]. The network finds an appropriate path from source to destination with resource availability to satisfy the QoS requirements for the specified service [5]. This practice is called QoS routing. After a proper path is found, a specific resource reservation protocol is applied to reserve required resources' dedicated to that path. Improved resource reservation methods are used by protocols to schedule the processing in a way that resources are allocated correctly to required services without fail. To ensure QoS flow in the network many improved techniques are used in different layers such as intelligently accessing and capturing the channel for real-time applications in MAC layer, priority scheduling in multimedia applications. In this research work, a QoS task scheduling system with controlling and continuously monitoring of activities in dedicated QoS routing has been designed in the network layer with an intension of providing best suitable QoS service within minimum permissible delay [6].

2 Literature Review

Viewpoint of the prominent AODV protocol in MANET follows that the topographic configuration is communicated to other adjacent station on demand. A Comparative analysis of improved AODV proposals has been carried out in [7]. For QoS-based routing it is essential to minimize the end-to-end delay and optimization of power consumption so that the throughput can be improved. Therefore survey and analysis of power optimized protocols [8] is essential before new research proposal. An improved protocol design with inter communication between data link layer and network layer [9] increases network life time. A Quality of service platform including

Fig. 1 Block diagram of the proposed approach

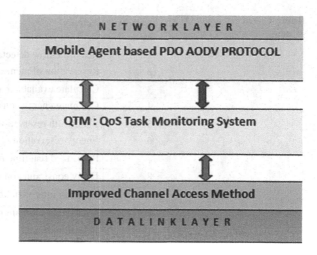

optimized network layer protocol with superior channel access method at MAC layer has been implemented [10] to achieve better packet delivery ratio. A mobile agent based platform has been designed in [11] for supporting agent service among mobile devices by selecting a suitable service provider dynamically, the job of which is to forward the packets between two agents which are not able to communicate with one another. Based on the blue tooth protocol, a prototype model has been implemented to ensure proper agent management and the simulation result shows better network performance. Based on link stability between mobile nodes an on demand routing protocol has been designed using few sets of agents such as route discovery agent, node manager agent, link manager agent, and route discovery agent. This approach is based on estimation of some valuable time such as link expiry time, neighbor identification time, time of response to hello messages by the adjacent station and then proper communication among agents to select next forwarding node.

3 The Proposed System

To solve the challenge of scheduling the QoS functional activities, an improved algorithm is used in this approach based on our previous research work done [12] using Critical Path Method (CPM) and the new system is called QTM system. According to this method, when any application prefers QoS routing, it includes some QoS-related tasks, among which there is a definite begin task and finish task involved to be checked at every intermediate node. The tasks are defined as per the QoS requirement of the source node. The functionality of the designed mobile agents in this system are quite challenging and their job is distributed for computation of allowable routing delay at every stage of activities (Fig. 1).

Table 1 Activity list in QTM system

Activity no	Activity	Completion time (ms)
1	Real time flow detection	6
2	Reservation of memory	2
3	Calculate available bandwidth	3
4	Calculate energy consumption	2
5	Bandwidth reservation	4
6	Energy reservation	1
7	Prioritized real time policy	1
8	Delay constraint control	6
9	Uniform jitter variation	3
10	Total delay estimation	1
11	Selection of next node	1

Basic necessity of QoS activity monitoring in mobile stations in wireless network emerges from the fact that real time data transmission generally get heavily influenced by issues such as packet arrival delay, video streaming problem, bandwidth constraints, link failures and cyber threats. In the proposed research module, a mobile agent has been utilized for Quality of Service checking at every intermediate node during packet transmission in order to check if the QoS criteria are satisfied in the channel during routing. To achieve this, the Mobile Agent visits from node to node and at every node it validates the correctness of activities listed out in Table 1.

All the activities are part of proposed QTM system and they are prioritized with their probable time of processing depending on many network parameters which are stored in the mini database [3] of the concerned node's routing engine that gets updated by the static agent of the concerned station. Using the heuristic method [13] probable processing time for every activity is computed for each intermediate node by the mobile agent. Then the total possible delay calculation is carried out using the longest CPM [14], if the total delay estimated is below the constraint range of QoS criteria, then that node will be selected for the next forwarding node. This processing logic has been implemented with incorporation of Mobile agent in our proposed system. Figure 2 depicts the functional mechanism of selection of next hop node. The Source Node (SN) attempts to send a Route Request (RREQ) [15] packet towards the destination. So, it needs to forward the packet to an intermediate node. Being a real-time packet transmission, the next hop node should be chosen carefully with rich resource availability and it must satisfy all the QoS criteria specified by the Source Node at the Mobile agent during Registration with the MA. So the MA gets the neighbouring nodes list from the routing table information of the source node through the static agent and visits every next hop node (here referred to as Target Node) to check the suitability of QoS activities (Table 1).The Mobile Agent calculates the total threshold value In terms of period (P1, P2, P3) which indicates a measure of chance of that node being selected. Factors considered under QoS criteria

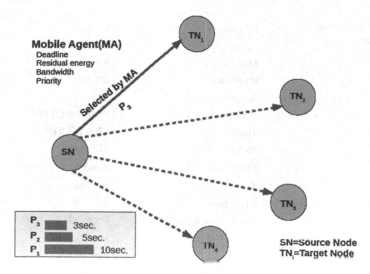

Fig. 2 Selection of next node in QTM mechanism

are Deadline, Residual Energy, Bandwidth and Priority. Deadline depicts the time duration in which the packet should reach the destination.

4 Simulation and Results

The proposed system has been simulated using the simulation tool NetSim Ver 8.3. It is an efficient discrete event simulation software that supports finite state machine approach of modeling projects and research utilities. It can be used for both wired and wireless networks and it has versatility of applications. It supports re-configuration of most popular protocols of wireless sensor networks and Mobile ad hoc networks. Table 2 shows the simulation parameters used in the proposed system.

Bandwidth can be distinctly denoted as the transmission of number of bits per second in a channel. For example when the capacity of a network is said to be 100 Mbps, then it can send 100 Mbps. Following section illustrates few results when simulated for various network performance parameters. Figure 3 shows the bandwidth comparison with node speed and Fig. 4 depicts the PDR comparison with node mobility. Bandwidth consumption proportionately increases with increase in node mobility and node speed. As the number of mobile nodes increases, there is more processing task at the intermediate nodes to forward, send and receive packets along with handling the challenges of route failure, reduced battery power and decreasing transmission capacity. When compared with other similar MA-based improved strategy

Table 2 Simulation
parameters

Parameters	Values
Area	1000×1000
Mac	802.11 e
Radio range	250 mt
Simulation time	50 s
Routing protocol	Proposed QTM
Traffic source	CBR and VBR
Packet size	512 byte
Mobility model	Random way point
Speed	5–10 m/s
Pause time	5 s
Interface queue type	Drop tail
Network interface type	Wireless phy
Simulator	Netsim 8.3

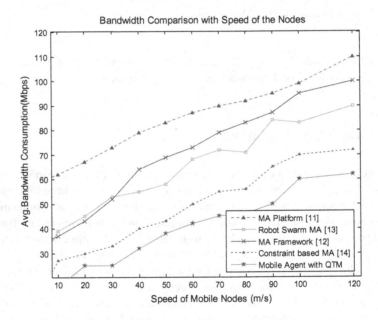

Fig. 3 Bandwidth comparison with node speed

to handle QoS, our proposed mobile agent based strategy MA with QTM consumes
comparatively less bandwidth and efficiently transmits the real-time data packets.

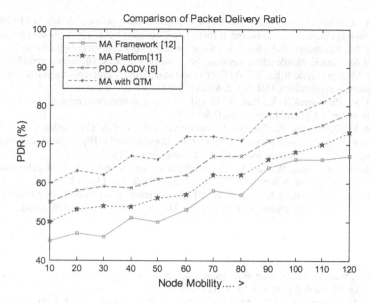

Fig. 4 PDR comparison with node mobility

5 Conclusion

The genuine issue of real-time traffic during data transmission in wireless mode has been partially solved in this network research work. A MA has been configured in the proposed QTM system that controls the QoS activities in a real-time platform based on the previous research work by the author. Basic function of the mobile agent is to calculate optimized delay estimation of QoS activities at the neighboring nodes before forwarding the real-time packets specifically multimedia transmission towards the load balanced path so that the challenging issues of real-time transmission such as video streaming problem, variation of jitter and deadline mission, etc., can be handled effectively. The Minimum but longest possible critical path method has been employed at the next hop station of a specific node by the mobile agent with assistance of the static agent. Simulation analysis, evaluation and results shows better performance of the proposed system in terms of reduced bandwidth consumption, an improvement in network life time and reduced end-to-end transmission delay.

References

1. http://www.cs.nccu.edu.tw/~ttsai/netprotocol/Chapter10.pdf. Accessed 9Apr 2017
2. Pattanayak, B., Rath, M.: A mobile agent based intrusion detection system architecture for mobile adhoc networks. J. Comput. Sci. **10**, 970–975 (2014)

3. Rath, M., Kumar Pattanayak, B.: Energy competent routing protocol design in MANET with real time application provision. Int. J. Bus. Data Commun. Netw. **11**(1), 50–60 (2015)
4. Rath, M., Pattanayak, B.K. Rout, U.: Study of challenges and survey on protocols based on multiple issues in mobile adhoc network. Int. J. Appl. Eng. Res. **10**, 36042–36045 (2015)
5. Rath, M., Pattanayak, B.K., Pati, B.: Energy efficient MANET protocol using cross layer design for military applications. Def. Sci. J. **66**(2) (2016)
6. Rath, M., Pattanayak, B.K., Pati, B.: MANET routing protocols on network layer in realtime scenario. Int. J. Cybern. Inform. (IJCI) **5**(1) (2016)
7. Rath, M., Pattanayak, B., Pati, B.: Comparative analysis of AODV routing protocols based on network performance parameters in mobile adhoc networks. Foundations and Frontiers in Computer, Communication and Electrical Engineering (2016)
8. Rath, M., Pattanayak, B., Pati, B.: A contemporary survey and analysis of delay and power based routing protocols in MANET. ARPN J. Eng. Appl. Sci. **11**(1) (2016)
9. Rath, M., Pattanayak, B.K, Pati, B: Inter-layer communication based QoS platform for real time multimedia applications in MANET. In: The Proceedings of IEEE WiSPNET, Chennai, India, pp. 613–617 (2016)
10. Rath, M., Pati, B., Pattanayak, B.K.: Cross layer based QoS platform for multimedia transmission in MANET. In: 3rd International Conference on Electronics and Communication Systems, Coimbatore, India, pp. 3089–3093 (2016)
11. Park, J., Youn, H., Lee, E.: A mobile agent platform for supporting ad-hoc network environment. Int. J. Grid Distrib. Comput. **1**(1), 9–16(8) (2008)
12. Rath, M., Pattanayak, B.K.: MAQ: a mobile agent based quality of service platform for MANETs. Int. J. Bus. Data Commun. Netw. **13**(1) (2017)
13. http://www2.cs.uni-paderborn.de/cs/ag-monien/PERSONAL/SENSEN/Scheduling/icpp/node4.html. Accessed 20 Apr 2016
14. Hendrickson, C., Tung, A.: Advanced Scheduling Techniques. Project Management for Construction. cmu.edu, 2.2 edn. Prentice Hall (2008). ISBN 0-13-731266-0
15. Kakkasageri, M.S., Manvi, S.S., Goudar, B.M.: An agent based framework to find stable routes in mobile adhoc networks (MANETs). In: TENCON IEEE Region 10 Conference, Hyderabad, pp. 1–6 (2008)

A Study on Smart Device Application Platform

Manikandan Shanmugam and Monisha Singh

Abstract Cloud Computing is considered to be one of the hottest research areas as it provides an approach through which the data is stored and accessed over the Internet in a virtual environment. The main idea to adapt this technology is that it shares the available resources rather than having separate local servers. This technology plays a crucial role in the healthcare sector as the healthcare industries believe that by incorporating cloud services within the healthcare sector it could provide quality services to the patients. Many industrial specialists suggest ways of converting the huge amount of data collected from the healthcare into meaning information and later sharing this valuable information to the user at the right time. The smart device is an electronic rig that is efficient to answer, sympathize and interact mutually with its users and other smart devices, one of the upcoming smart devices are smart shirts. Smart shirts allow the user to share information like Facebook or LinkedIn profile details. This paper focuses on providing wearable devices to the user in order to have monitored over his/her health.

Keywords Smart device · Cloud computing · Healthcare industry · Sensors

1 Introduction

With the traditional healthcare systems and with limited medical facilities the burden of handling ever growing population is increasing. The migration of healthcare industry to cloud is inevitable. We need to adapt technologies like cloud computing and smart devices in order to bring a change in healthcare sector over the traditional method. Cloud computing and health care has gone through a rapid development in

M. Shanmugam (✉)
Master of Computer Applications, Christ University, Bengaluru, India
e-mail: manikandan.s@mca.christuniversity.in

M. Singh
Department of Computer Science, Christ University, Bengaluru, India
e-mail: monisha.singh@christuniversity.in

© Springer Nature Singapore Pte Ltd. 2019
P. K. Sa et al. (eds.), *Recent Findings in Intelligent Computing Techniques*,
Advances in Intelligent Systems and Computing 707,
https://doi.org/10.1007/978-981-10-8639-7_58

551

recent years, so much powerful and far-reaching applications are available. Cloud computing and mobile devices when integrated together can change the culture of health care by transforming direct care services to mobile cloud computing services [1]. Due to insufficient exercise and unhealthy food habits, the cause for diseases keeps increasing within the society. In developing countries such as India, the smart device which collects health-related details should be cost effective so that the market accepts. The data regarding one's health will be collected with the help of various sensors like accelerometers, electrocardiography, body temperature and made available to the user at the right time. The proposed system will be useful to elderly people who needs constant monitoring over their health state as well as youngsters who are into fitness [2–4].

2 Various Issues in Existing System

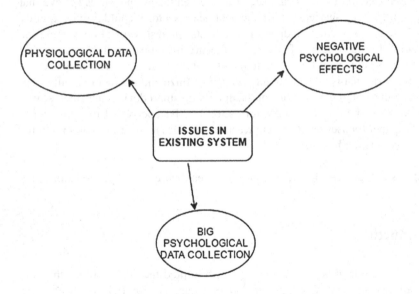

2.1 Physiological Data Collection

User mobility should be taken into consideration while collecting data or else the user must wear multiple sensors or devices for physiological data collection. On the other hand, some of the traditional smart devices provide inaccurate data which cannot be used for data mining process [3, 5].

2.2 Negative Psychological Effects

Users might end up with a negative impression of getting health issues by wearing sensors on them which might further cause stress. In order to avoid this adverse effect in psychological data collection, we need to rethink the design of the traditional data collection method and innovate an efficient model [3, 4].

2.3 Big Psychological Data Collection

In today's market wearable devices such as smart glasses, smart watches, and fitness bands are widely accepted, however, the data collected from these devices cannot be considered as big data. So this data holds only a minimum reference value in prediction and diagnosis of chronic diseases. When a user is diagnosed with a help of ECG monitoring accurate results can be drawn but the user cannot stay in comfort for a longer period of time when diagnosed in this method [3].

2.4 Anti-Wireless for Body Area Networking

In traditional system, rather than using wired cables, wireless technology was encouraged. But by adopting this method there might be challenges in collecting physiological data due to user mobility and surrounding environment, another major concern energy consumption of wireless networking. In the upcoming generations of body area network in order to promote green communication, wireless technology should be avoided [3, 4].

3 Working of Smart Device Application Platform

The architecture of the Smart device application platform consists of two parts, namely front end and back end. The front end provides a visual interface in the smart device which contains information regarding one's health whereas the back end is a place where the actual data collection, data transmission, data analysis and preprocessing takes place.

3.1 Explanation

The smart device will collect health-related data using sensors and make it available
to the user or his friends at the time of emergencies like accidents, heart attacks [6].
The sensors which will be incorporated into the smart device are

Accelerometers, which are used to detect the activities and different postures related
body.
Electrocardiography (ECG), which are used to detect the heart rates.
Respiration sensors are used to detect the breathing process of the user.
Finally, the body temperature sensors [4].

These are the sensors which are used to capture various health-related activities.
When a smart device is integrated with multiple sensors it will be easier to predict
complex health condition.

The basic functions of the smart device are collecting health-related data, transmitting data to a cloud server, perform various preprocessing operations and making
it available to the user at the right time [6]. The smartphones act as a communication
link between smart device and smartphone and shares data. Many signal processing
tasks will be performed in the smartphone and the processed digital signals can be
sent to a specialist for further diagnosis and appropriate prediction of diseases can
be made [7].

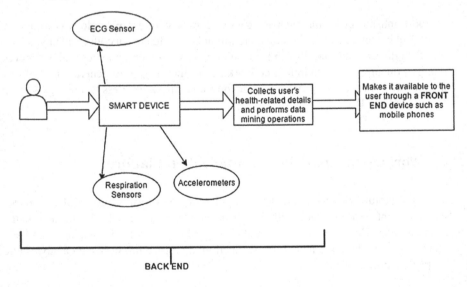

4 Proposed System

In the present market, most of the people are moving towards digitalization and want
most of their tasks done with a simpler and smaller device. The present-day's smart
devices which are used for health monitoring causes a lot of discomforts to the user,

so a company called "Arrow" has come up with smart shirts. At present smart shirts provide information regarding one's address, LinkedIn, and Facebook profile details. In future, information regarding one's health can also be embedded in the chip so that with the help an NFC-enabled smartphone one can access others health details easily and efficiently, this feature plays a key role when a person had met with an accident an needs immediate treatment schemes.

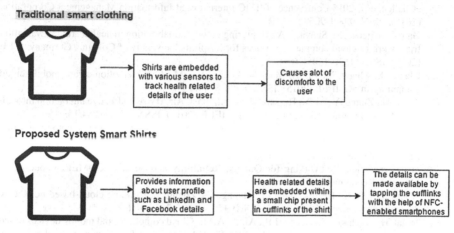

Traditional smart clothing

Shirts are embedded with various sensors to track health related details of the user → Causes alot of discomforts to the user

Proposed System Smart Shirts

Provides information about user profile such as LinkedIn and Facebook details → Health related details are embedded within a small chip present in cufflinks of the shirt → The details can be made available by tapping the cufflinks with the help of NFC-enabled smartphones

5 Conclusion

In this work, the author provided an overviewed the various functionalities and design issues faced by present day smart devices and concludes by proposing an efficient method of collecting user's related data, processing the collected data and making it available to the user in form of a smaller and simpler smart device called smart shirts. This system reliable mode and causes less harm when compared to traditional method of monitoring user's health-related details.

References

1. Lo'ai, A.T., Mehmood, R., Benkhlifa, E., Song, H.: Mobile cloud computing model and big data analysis for healthcare applications. IEEE (2016). ISSN: 2169-3536
2. Ma, J., Peng, C., Chen, Q.: Health information exchange for home-based chronic disease self-management—a hybrid cloud approach, Zhejiang University of Technology, Hangzhou, China. IEEE (2014). ISSN: 978-1-4799-4284-8/14
3. Chen, M., Ma, Y., Li, Y., Wu, D., Zhang, Y., Youn, C.-H.: Wearable 2.0: enabling human-cloud integration in next generation healthcare systems. IEEE Commun. Mag. (2017)
4. Ma, Y.-C., Chao, Y.-P., Tsai, T.-Y.: Smart-clothes—prototyping of a health monitoring platform. In: 2013 IEEE Third International Conference on Consumer Electronics, Berlin (ICCE-Berlin). ISSN: 978-1-4799-1412-8/13

5. Doukas, C., Pliakas, T., Maglogiannis, I.: Mobile healthcare information management utilizing cloud computing and android OS. In: 32nd Annual International Conference of the IEEE EMBS Buenos Aires, Argentina, 31 Aug–4 Sept 2010. IEEE. ISSN: 978-1-4244-4124-2/10

6. Lin, K., Xia, F., Wang, W., Tian, D., Song, J.: System design for big data application in emotion-aware healthcare. IEEE (2016). ISSN: 2169-3536

7. Mvelase, P., Dlamini, Z., Dludla, A., Sithole, H.: Integration of smart wearable mobile devices and cloud computing in South African healthcare. In: Cunningham, P., Cunningham, M. (Eds) eChallenges e-2015 Conference of IIMC International Information Management Corporation (2015). ISSN: 978-1-905824-53-3

8. Gao, F., Thiebes, S., Sunyaev, A.: Exploring cloudy collaboration in healthcare: an evaluation framework of cloud computing services for hospitals University of Cologne, Germany. IEEE (2016). ISSN: 1530-1605/16

9. Chen, M., Zhang, Y., Li, Y., Mao, S., Leung, V.C.M.: EMC: emotion-aware mobile cloud computing in 5G. IEEE (2015). ISSN: 0890-8044/15

10. Chen, M., Zhang, Y., Li, Y., Hassan, M.M., Alamari, A.: AIWAC: affective interaction through wearable computing and cloud technology. IEEE (2015). ISSN: 1536-1284/15

11. Yang, C.-N., Wu, F.-H.: E-health services for elderly-care based on Google cloud messaging. IEEE (2015). ISSN: 978-1-5090-1893-2/15

12. Zhang, Y., Wang, L., Hu, L., Wang, X., Chen, M.: In: 2014 10th International Conference on Heterogenous Networking for Quality, Reliability and Securtiy COMER: Cloud-based Medicine Recommendation

13. Zhang, P., Hu, S., He, J., Zhang, Y., Huang, G., Zhang, J.: Building cloud-based healthcare data mining services. IEEE (2016). ISSN: 978-1-5090-2628-9/16

14. Jemal, H., Kechaou, Z., Ayed, M.B., Alimi, A.M.: Cloud computing and mobile devices based system for healthcare application. In: 2015 IEEE International Symposium on Technology in Society (ISTAS) Proceedings. ISSN: 978-1-4799-8283-7

15. Ahn, Y.W., Cheng, A.M.K., Baek, J., Jo, M., Chen, H.-H.: An auto-scaling mechanism for virtual resources to support mobile, pervasive, real-time healthcare applications in cloud computing. IEEE (2013). ISSN: 0890-8044/13

A Combinational Approach for Optimal Packing of Parallel Jobs in HPC Clusters

Dhirendra Kumar Verma, Alpana Rajan, Amit Paraye and Anil Rawat

Abstract Job scheduling plays important role for efficient utilization of compute resources viz. High-Performance Computing Clusters (HPCC) and Supercomputers. In such environments different users execute computing jobs requiring varying execution time ranging from few minutes to couple of weeks. If jobs are not packed carefully, computing resources will lose computational time due to scheduling gaps. Backfilling is a common way to overcome this problem. In this paper we propose a combinational approach for backfilling of non-preemptive parallel jobs in HPCCs. In this approach the job scheduler picks a combination of jobs in waiting that can optimally pack the current backfill window. Proposed approach is an effort towards optimizing utilization of compute resources in HPCCs. In Proposed approach EASY backfilling has been used along with window policy of Maui Scheduler. In this paper different scenarios have been included and proposed approach has been simulated. It is observed that choosing a combination of jobs, instead of a single job through Firstfit or Bestfit gives more optimal results for backfilling.

Keywords Job scheduling · Parallel jobs · Backfill · HPCC

D. K. Verma (✉) · A. Rajan · A. Paraye · A. Rawat
Computer Division, Raja Ramanna Centre for Advanced Technology, Indore, India
e-mail: dhirendrav@rrcat.gov.in

A. Rajan
e-mail: alpana@rrcat.gov.in

A. Paraye
e-mail: aparaye@rrcat.gov.in

A. Rawat
e-mail: rawat@rrcat.gov.in

© Springer Nature Singapore Pte Ltd. 2019
P. K. Sa et al. (eds.), *Recent Findings in Intelligent Computing Techniques*,
Advances in Intelligent Systems and Computing 707,
https://doi.org/10.1007/978-981-10-8639-7_59

1 Introduction

In cluster computing parallel jobs run simultaneously on several processors. Parallel jobs can be classified as rigid, moldable, evolving, and malleable. In rigid and moldable jobs allocated processors remains fixed during the execution of the job whereas evolving and malleable jobs can change their processor requirements during execution [1]. Most of the workload on HPC clusters or supercomputers consists of rigid and moldable jobs. In this paper we considered only rigid parallel jobs for backfilling. We also assumed that jobs are submitted along with their estimated run time.

Backfilling offers significant scheduler performance improvement. In a typical large system, enabling backfill can increase system utilization by around 20% and improve turnaround time by an even greater amount [2]. Two main variants of backfilling include Conservative and EASY backfilling. EASY (Extensible Argonne Scheduling system) backfilling is a popular variant of backfilling [3].

EASY backfilling provides reservation only to the top priority job which is at the head of the job queue. Thus it provides more opportunity for backfilling. The Conservative backfilling allows each job to reserve the resources it needs, when it is inserted into the job queue [4]. Providing reservations to all the jobs reduces the opportunity of backfilling. In this paper we used EASY backfilling with combinational approach along with window policy of Maui scheduler.

2 Related Work

Backfilling increases system utilization by running lower priority jobs in scheduling gaps before higher priority jobs. It takes place when higher priority jobs are not delayed. Fattened Backfilling proposed in [5] provides more backfilling opportunities by making first job of the queue delay, although the delay is not more than average waiting time of the already finished jobs. Backfilling has been used by various free as well as commercial software schedulers including Maui, LSF, PBS, SLURM and others [6]. Lookahead optimizing scheduler in [7] introduces amount of lookahead into the queue and shows that using a lookahead window of 50 jobs better packing of jobs can be achieved. It chooses a set of jobs using dynamic programming to maximize resource utilization. In [8] pair-wise combinations with the largest region of resource requirement is selected for backfilling. In [7] and [8] all the idle regions are combined and considered as a single window for optimization.

In proposed approach the scheduler picks a job combination along with window concept of Maui. A job combination may contain a single job or multiple jobs which can optimally pack the current backfill window. Maui determines backfill windows by analyzing idle nodes in the scheduling gaps to find largest node-time rectangle. Once the backfill windows are determined, Maui begins to traverse them by widest

Table 1 Jobs for backfilling

Job_Id	J1	J2	J3	J4	J5
No. of nodes	4	3	3	5	3
Run time	9	10	12	10	9

window first policy by default. As each backfill window is evaluated, backfilling is applied using Firstfit, Bestfit, or any other policy [9].

3 Proposed Approach

The basic purpose of proposed approach is to maximize utilization of compute resources in HPCC systems and to increase the throughput of the system. Proposed approach uses exhaustive search to find a job combination for optimal packing of current backfill window. The scheduler first examines scheduling gaps and creates the first widest backfill window. The scheduler then examines all possible combinations of waiting jobs that can be considered for backfilling and chooses the one which can optimally pack the current backfill window. After packing current backfill window the next backfill window is created in time sequence along with any unused compute resources of the previous backfill window.

4 Performance Analysis

In this section we took three different scenarios and analyzed performance of proposed combinational approach, Bestfit, and Firstfit policies for backfilling. In each scenario scheduling gaps are represented through dotted region and all three policies are applied to pick jobs for backfilling. Jobs that are chosen for backfilling will never delay the start time of top priority job (job at the head of the waiting queue) and jobs having advance reservation.

Resource requirement of a job is calculated by multiplying compute resource requirement with pre estimated run time that is COMPUTE RESOURCE X RUN TIME. Also the degree of fit inside a scheduling gap has been evaluated by COMPUTE RESOURCE X TIME parameter. A COMPUTE RESOURCE may be a single processing core, a processor, or a compute node in the cluster. We preferred wider jobs over long tailed jobs when both results in same resource requirement.

First Scenario: Table 1 shows five jobs considered for backfilling. Initially at t = 0 (time) six COMPUTE RESOURCES are free as depicted in Fig. 1.

The scheduler first examines scheduling gaps and creates the first widest backfill window having six COMPUTE RESOURCES for 10 time units (t = 0 to t = 10). The Firstfit policy examines waiting jobs and picks the first jobs that can be fitted in first

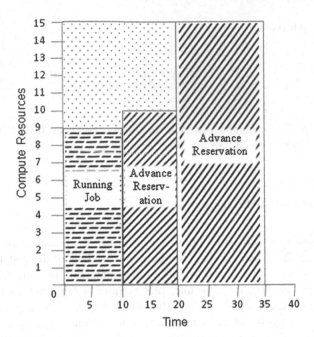

Fig. 1 Scheduling gap at t = 0

backfill window. In this scenario job J1 is picked. After scheduling J1 the scheduler creates next widest backfill window and looks for remaining jobs for backfilling using Firstfit policy. If no job can fit in the present backfill window, then scheduler looks for next widest window in time sequence. Only one job is picked at a time and is scheduled for execution.

Backfilling of jobs using Firstfit is shown in Fig. 2.

Bestfit policy picks the largest job from the waiting queue that can be fitted in the present backfill window. After scheduling first largest job which is J4 the scheduler then creates the next backfill window and picks next largest job using Bestfit. Backfilling of jobs using Bestfit is shown in Fig. 3.

Proposed combinational approach considers a job combination for backfilling. First it created all possible combinations of waiting jobs and then observes resource requirements of each combination. The combination which can optimally pack the present backfill window is picked and scheduled for execution. The major difference between earlier policies and proposed approach is the selection of job combination instead of picking a single job and creating the next backfill window. Figure 4 shows backfilling of jobs using proposed combinational approach.

It is clear from Figs. 2, 3 and 4 that combinational approach optimally packs scheduling gaps and results in minimum wastage of resources. If we calculate wastage of resources in terms of COMPUTE RESOURCE X TIME then wastage of resources by Firstfit will be 44 ($2 \times 20 + 1 \times 1 + 3 \times 1$), wastage of resources by Bestfit will be

Fig. 2 Backfilling of jobs using Firstfit

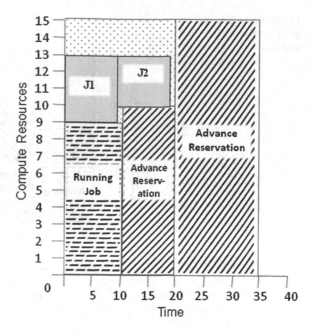

Fig. 3 Backfilling of jobs using Bestfit

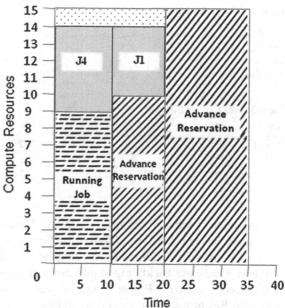

24 $(1 \times 20 + 4 \times 1)$, whereas proposed combinational approach results in wastage of 3 (1×3) resources only (Table 2).

Second scenario:

Fig. 4 Backfilling of jobs using combinational approach

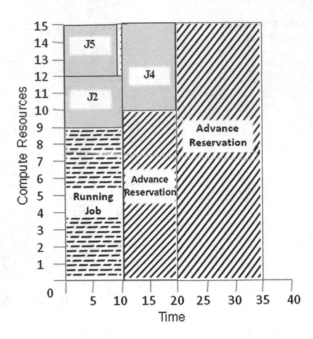

Table 2 Jobs for backfilling

Job_Id	J1	J2	J3	J4	J5	J6	J7
No. of nodes	3	3	4	2	5	8	4
Run time	8	10	15	12	20	8	10

Table 3 Jobs for backfilling

Job_Id	J1	J2	J3	J4	J5	J6
No. of nodes	5	4	3	2	2	6
Run time	15	20	20	25	18	10

Initial state of scheduling gaps is represented in Fig. 5.

Figure 6 shows backfilling of jobs using Firstfit policy.

Figure 7 shows backfilling of jobs using Bestfit policy.

Figure 8 indicates backfilling of jobs using proposed combinational approach.

For second scenario Firstfit results in wastage of 32 ($1 \times 2 + 2 \times 10 + 2 \times 5$) resources, Bestfit results in wastage of 40 ($1 \times 15 + 1 \times 25$) resources, whereas proposed approach results in wastage of 16 ($1 \times 13 + 1 \times 3$) resources only as evident from Figs. 6, 7 and 8 (Table 3).

Third scenario

Initial state of scheduling gaps is represented in Fig. 9.

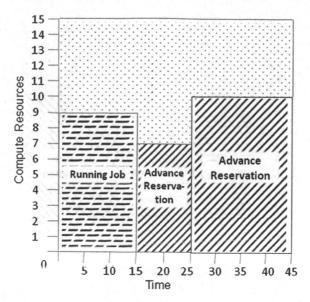

Fig. 5 Scheduling gap at t = 0

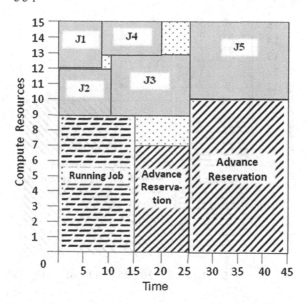

Fig. 6 Backfilling of jobs using Firstfit

Backfilling of jobs using Firstfit policy is depicted in Fig. 10.
Backfilling of jobs using Bestfit policy is depicted in Fig. 11.
Backfilling of jobs through combinational approach is depicted in Fig. 12.

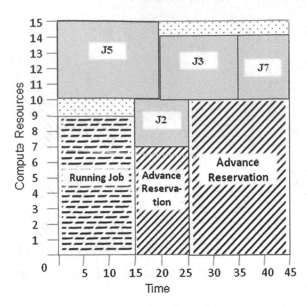

Fig. 7 Backfilling of jobs using Bestfit

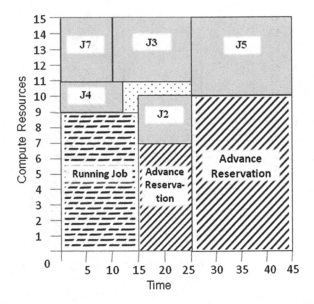

Fig. 8 Backfilling of jobs using combinational approach

For third scenario Firstfit results in wastage of 25 $(3 \times 5 + 2 \times 5)$ resources, Bestfit results in wastage of 20 (1×20) resources, whereas proposed approach results in wastage of 4 (2×2) resources only as evident from Figs. 10, 11 and 12.

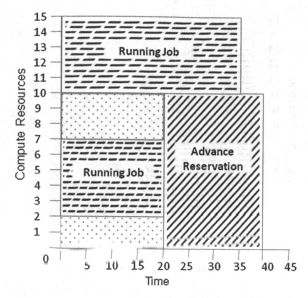

Fig. 9 Scheduling gap at t = 0

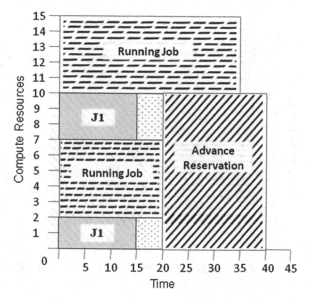

Fig. 10 Backfilling of jobs using Firsfit

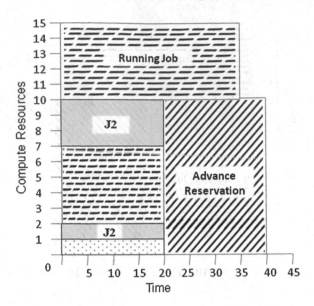

Fig. 11 Backfilling of jobs using Bestfit

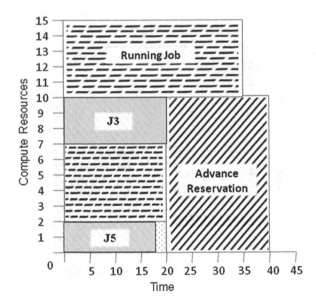

Fig. 12 Backfilling of jobs using combinational approach

After observing the performance of all three policies it is clear that proposed combinational approach gives more optimal results for backfilling than using Firstfit or Bestfit policies.

5 Conclusion and Future Work

In HPC clusters Job scheduling is a critical task, which directly influences the satisfaction of users and system administrator. Administrators are always interested in maximizing resources utilization whereas users are interested in getting higher throughput and minimum response time for their jobs. Proposed approach gives very encouraging results to meet these goals. Proposed approach is found to consistently perform better or equivalent to the Bestfit and Firstfit policies for a large number of different scenarios.

We considered only the jobs having prior estimation of run time in all the scenarios. Fairness and starvation of big jobs is also beyond the scope of the present paper and is left for future work. We are further working towards the implementation of working model of proposed approach to maximize utilization of compute resources in HPC Clusters.

References

1. Feitelson, D.G., Rudolph, L.: Toward convergence in job schedulers for parallel supercomputers. Job Scheduling Strategies for Parallel Processing, vol. 1162. Lecture Notes in Computer Science, pp. 1–26 (1996)
2. Maui Administrator's Guide. Maui 3.2 (2016). http://docs.adaptivecomputing.com/maui/pdf/mauiadmin.pdf
3. Lifka, D.: The ANL/IBM SP scheduling system. In: 1st Workshop on Job Scheduling Strategies for Parallel Processing (JSSPP), pp. 295–303 (1995)
4. Feitelson, D, Rudolph, L., Schwiegelshohn, U.: Parallel job scheduling—a status report (2004)
5. Martin, C.G., Vega-Rodriguez, M.A., Gonzalez-Sanchez, J.L.: Fattened backfilling: an improved strategy for job scheduling in parallel systems. J. Parallel Distrib. Comput. **97**, 69–77 (2016)
6. Etsion, Y., Tsafrir, D.: A short survey of commercial cluster batch schedulers. Technical Report, School of Computer Science and Engineering, The Hebrew University of Jerusalem (2005)
7. Shmueli, E., Feitelson, D.G.: Backfilling with lookahead to optimize the performance of parallel job scheduling. J. Parallel Distrib. Comput. **65**(9), 1090–1107 (2005)
8. Yi, S., Wang, Z., Ma, S., Che, Z., Liang, F., Huang, Y.: Combinational backfilling for parallel job scheduling. **2**, v2 112–v2 116 (2010)
9. Jackson, D., Snell, Q., Clement, M.: Core algorithms of the maui scheduler, Brigham Young University, Provo, Utah 84602. http://scitas.epfl.ch/pdf/tech/docs/corealgs.pdf

Vulnerability Assessment for Virtual Machines in Virtual Environment of Cloud Computing

Rajendra Patil and Chirag Modi

Abstract In this paper, we design an automated framework for vulnerability assessment of virtual machines in cloud computing. It uses multithread model to scan newly joined VMs for potential vulnerabilities. We define ranking score to find the critical vulnerabilities based on impact score, exploitability score, and risk score. All the scanned vulnerabilities are analyzed for ranking score and the critical vulnerabilities are sent to the patching system.

Keywords Cloud computing · Virtualization · Virtual machines · Vulnerability assessment · Security

1 Introduction

Despite of various benefits of virtualization, it has many vulnerabilities which bring additional security challenges to virtualization-based computing technologies. These vulnerabilities allow an attacker to disturb availability, integrity and confidentiality of the underlying resources and services [1, 2]. The most relevant threats for the cloud computing have been investigated by CSA [3] and ENISA [4]. Help Net Security [5] stated cloud security challenges such as DDoS attacks, data breaches, data loss, insecure access points and notifications and alerts. Due to the flexibility provided by virtualization, new VMs can be added dynamically on the physical server. When new VM joins, it may have potential vulnerabilities. For example, newly joined VM typically starts with open ports, default services and default active protocols [6]. These vulnerabilities can be exploited in future resulting different system level and network level attacks. VM is the potential target for an attacker to gain access to the

R. Patil (✉) · C. Modi
Department of Computer Science and Engineering, National Institute of Technology Goa,
Farmagudi, Ponda 403401, India
e-mail: rajendrapatil@nitgoa.ac.in

C. Modi
e-mail: cnmodi@nitgoa.ac.in

© Springer Nature Singapore Pte Ltd. 2019
P. K. Sa et al. (eds.), *Recent Findings in Intelligent Computing Techniques*,
Advances in Intelligent Systems and Computing 707,
https://doi.org/10.1007/978-981-10-8639-7_60

hypervisor or other installed VMs [6, 7]. Once new VM joins the virtual environment, attackers try to exploit its vulnerabilities to gain full control. Thus, to offer secure computing services over the Internet, all VMs should be secured. However, scanning all the VMs at the same time is very difficult and time consuming. It is efficient to detect and prevent attacks from newly joined VM rather than scanning whole virtual environment.

In this paper, we present an automated vulnerability assessment framework to prevent both system level and network level attacks which can be performed through co-hosted VMs, internal networks and external networks in cloud. It looks for the potential vulnerabilities from a newly joined VM. Newly joined VM goes through the secure tunnel where it is scanned and patched for known vulnerabilities.

Rest of this paper is organized as follows: Sect. 2 presents the existing proposals for vulnerability assessment. A detailed discussion of the proposed framework is given in Sect. 3, while Sect. 4 analyzes the results. Finally, Sect. 5 concludes our research work with references at the end.

2 Related Work

Kotikela et al. [8] have used ontological vulnerability database (OVDB) and performs many-to-many mapping between vulnerabilities of the OVDB and scripts of the attack script database. It is capable of assessing the vulnerabilities in well-known software. Kamongi et al. [9] have modeled the security vulnerabilities and classifies them based on vulnerability ontology. An ontology knowledge base (OKB) is developed using the data provided by NVD [10]. OKB is used to study and assess security vulnerability of individual or component parts of the cloud environment system. Kamongi et al. [11] have proposed a methodology for ranking cloud vulnerabilities. It starts with vulnerability discovery process and ends up with unified and ranked vulnerabilities. It generates the attack path and defines a Weighted Average Mean metric based on CVSS to return the rank of the vulnerability. Kamongi et al. [12] have proposed an automated architecture for threat modeling and risk assessment for cloud. It uses OKB to model the threats and assess the risks. The OKB captures the relationship between vulnerabilities, defenses mechanisms and attacks. Zineddine [13] has proposed a vulnerability assessment and mitigation technique. It uses Cuckoo search algorithm with Levy flights as random walk identify an optimal set of available solutions to an identified vulnerabilities.

It has been observed that the vulnerability assessment on running cloud environment has performance overhead. The flexibility of VM is not taken into consideration. The proposed framework performs the vulnerability assessment on VM when it becomes active in the virtual environment.

Fig. 1 Deployment of the proposed framework in virtual environment

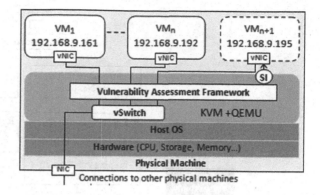

3 Proposed Framework

The objective is to create a more proactive virtual environment. It should identify the vulnerabilities from the VMs, prioritize them and patch them before an attacker exploit them. As shown in Fig. 1, a virtual environment includes running VMs, Hypervisor and Host OS. Here, we consider Type 2 hypervisor. The proposed framework (Vulnerability Assessment Framework) is installed on hypervisor. A VM with IP "192.168.9.195" is considered as a newly joined VM and other VMs (192.168.9.161-192.168.9.192) are running VMs. The framework maintains a database storing information for all VMs. It has scanning interface (SI) to the newly joined VM. The proposed framework continuously looks for the new joining of the VM and scans the vulnerabilities.

3.1 Multithreaded Model for VM Management

The multithreaded model is used to manage the flexibility of VMs. It uses a universal unique identifier (UUID) to uniquely identify the VM. When VM is created first time, it is automatically assigned a UUID. The UUID remains same throughout the lifecycle of the VM. Thus, in case of DHCP-based IP address allocation, even the IP address is changed, the VM can be uniquely identified using UUID. The multithreaded model assigns a new thread to each of the newly joining VM. The working at each thread is presented in Fig. 2.

3.2 Early Vulnerability Scanning

It performs vulnerability scan for potential vulnerabilities using Nexpose [14]. The entries for all VMs are available in VMD database. Each VM is identified using

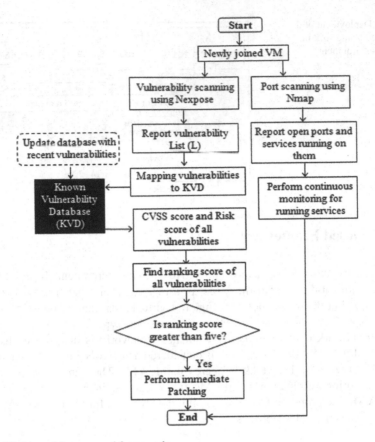

Fig. 2 Working of the proposed framework

UUID. The entry in database corresponds to the VM and stores information such as UUID (primary key), IP address, MAC address, VM name, OS running, last scanned date and time, last updated date and time, number of vulnerabilities found, number of vulnerabilities patched, risk score. When new VM joins the virtual network, its IP address and UUID will be used to scan the VM for known vulnerabilities. The scanning process identifies guest OS type and software of VM, and returns the list of known vulnerabilities in VMD database for corresponding VM.

In addition, it performs the port scan using Nmap [15] to find open ports in target VM and the services running on those ports. It sends probes to various ports and classifies the responses to determine the current state of the port. The port scan tests is performed on Standard ports, Alternative ports and Application specific ports.

3.3 Mapping Vulnerability to Nexpose Database

The discovered vulnerabilities are mapped to known vulnerability database (KVD). Nexpose database stores information such as vulnerability ID, title, description, date_published, CVSS_vector, CVSS_score, severity level, risk_score. The CVSS_vector defines the six base metrics grouped into the access metrics and the impact metrics. The access metrics include Attack Vector, Access Complexity, and Authentication. The impact metrics include Confidentiality Impact, Integrity Impact and Availability Impact. All these metric values are finalized in NVD [10]. In addition, another crucial attribute is risk_score.

3.4 Vulnerability Evaluation and Ranking

After successful mapping of vulnerabilities to Nexpose database, the CVSS metric and risk score are returned. For each vulnerability, it calculates the Impact score and Exploitability score using predefined formula of NVD as given below.

$$Impact\_score = 10.41 * (1 - (1 - C\_Impact) * (1 - I\_Impact) * (1 - A\_Impact))$$
$$Exploitability\_score = 20 * AccessComplexity * Authentication * AccessVector$$
$$Risk\_score = risk\,score/100$$

In addition, we define vulnerability_exploit_rate and vulnerability_impact_rate, which reflect the number of times the vulnerability has been exploited and the cost of mitigation of the vulnerability respectively. These two parameters are used to calculate the modified Impact (M_I_Score) and Exploitability score (M_E_Score).

$$M\_I\_Score = Impact\,score * vulnerability\_impact\_rate$$
$$M\_E\_Score = Exploitability\,score * vulnerability\_exploit\_rate$$

We define the ranking score for the vulnerabilities, which is calculated based on the M_I_Score, M_E_Score and Risk_score as shown below

$$Ranking\_Score = (M\_I\_Score + M\_E\_Score + Risk\_score)/3$$

4 Experimental Results

We have tested the proposed framework on KVM and performed vulnerability scanning on newly joined VM. The total 11 vulnerabilities are found (refer Table 1). The ranking score with greater than five is considered as critical vulnerabilities. In port scanning, total eight open ports and respective services are reported (refer Table 2).

Table 1 Vulnerabilities, CVSS vector and corresponding ranking score

SN	Vulnerability name	CVSS_vector	Risk score	E_S	I_S	Ranking score
1	SMB signing disabled (cifs-smb-signing-disabled)	AV:A/AC:M/Au:N/C:C/I:C/A:N	8.11	5.5	9.2	7.6
2	TLS/SSL birthday attacks on 64-bit block ciphers (SWEET32) (ssl-cve-2016-2183-sweet32)	AV:N/AC:L/Au:N/C:P/I:N/A:N	4.32	10	6.9	7.07
3	SMB signing not required (cifs-smb-signing-not-required)	AV:A/AC:H/Au:N/C:C/I:C/A:N	8.03	3.2	9.2	6.8
4	TLS/SSL server is enabling the BEAST attack (ssl-cve-2011-3389-beast)	AV:N/AC:M/Au:N/C:P/I:N/A:N	4.54	8.6	2.9	5.3
5	TLS/SSL server supports RC4 cipher algorithms (CVE-2013-2566) (rc4-cve-2013-2566)	AV:N/AC:M/Au:N/C:P/I:N/A:N	4.05	8.6	2.9	5.1
6	TLS server supports TLS version 1.0 (tlsv1_0-enabled)	AV:N/AC:M/Au:N/C:P/I:N/A:N	3.35	8.6	2.9	4.95
7	TLS/SSL server supports the use of static key ciphers (ssl-static-key-ciphers)	AV:N/AC:H/Au:N/C:P/I:N/A:N	2.54	4.9	2.9	3.44
8	TCP timestamp response (generic-tcp-timestamp)	AV:N/AC:L/Au:N/C:N/I:N/A:N	0	10	0	3.3
9	NetBIOS NBSTAT traffic amplification (netbios-nbstat-amplification)	AV:N/AC:L/Au:N/C:N/I:N/A:N	0	10	0	3.3
10	TLS/SSL server supports 3DES cipher suite (ssl-3des-ciphers)	AV:N/AC:H/Au:N/C:N/I:N/A:N	0	4.9	0	1.6
11	ICMP timestamp response (generic-icmp-timestamp)	AV:L/AC:L/Au:N/C:N/I:N/A:N	0	3.9	0	1.3

(E_S = Exploitability score, I_S = Impact score)

Table 2 Open ports and services running on newly joined VM

Port	Protocol	State	Service	Version
135	TCP	Open	Msrpc	Microsoft windows RPC
139	TCP	Open	Netbios-ssn	Microsoft windows netbios-ssn
445	TCP	Open	Microsoft-ds	Windows 7 SP 1 microsoft-ds
1026	TCP	Open	Msrpc	Microsoft windows RPC
1027	TCP	Open	Msrpc	Microsoft windows RPC
1028	TCP	Open	Msrpc	Microsoft windows RPC
1029	TCP	Open	Msrpc	Microsoft windows RPC
22350	TCP	Open	CodeMeter	CodeMeter runtime

5 Conclusions

We have proposed a vulnerability assessment framework to minimize the risk by performing an entry-level assessment of VM vulnerabilities. We define the ranking score to rank the vulnerabilities. In addition, the proposed framework performs the port scanning. We have implemented and evaluated the proposed framework on KVM based cloud environment. The critical vulnerabilities are given as input to patching system and services running on open ports are given as input to continuous monitoring system. Thus, it minimizes the risk at early stage of VM. In future, this framework can be extended to prevent possible intrusions in cloud environment.

References

1. Metzler, J.: Virtualization: Benefits, Challenges, and Solutions. Riverbed Technology, San Francisco (2011)
2. Li, S.H., Yen, D.C., Chen, S.C., Chen, P.S., Lu, W.H., Cho, C.C.: Effects of virtualization on information security. Comput. Stand. Interfaces 1–8 (2015)
3. The notorious nine: cloud computing top threats in 2013 v.1.0. Cloud Security Alliance (2013). http://cloudsecurityalliance.org/research/top-threats
4. Marinos, L., Sfakianakis, A.: ENISA threat landscape responding to the evolving threat environment. Report by European network and Information Security Agency (2012). http://www. enisa.europa.eu
5. Five principal cloud security challenges. Help Net Security (2015). http://www.net-security.o rg/secworld.php?id=18763

6. Candid, W.: Threats to virtual environments. Symentac White Paper (2014). www.symantec.c om/.../whitepapers/threats_to_virtual_environments.pdf

7. Hashizume, K., Rosado, D.G., Fernández-Medina, E., Fernandez, E.B.: An analysis of security issues for cloud computing. J. Internet Serv. Appl. **4**(1) (2013)

8. Kotikela, S., Kavi, K., Gomathisankaran, M.: Vulnerability assessment in cloud computing. In: International Conference on Security and Management (SAM) (2012)

9. Kamongi, P., Kotikela, S., Kavi, K., Gomathisankaran, M., Singhal, A.: Vulcan: vulnerability assessment framework for cloud computing. In: 7th International Conference on Software Security and Reliability (SERE), pp. 218–226 (2013)

10. National Vulnerability Database NIST (2017). https://nvd.nist.gov/

11. Kamongi, P., Kotikela, S., Gomathisankaran, M., Kavi, K.: A methodology for ranking cloud system vulnerabilities. In: 4th International Conference on Computing, Communications and Networking Technologies (ICCCNT), pp. 1–6 (2013)

12. Kamongi, P., Gomathisankaran, M., Kavi, K.: Nemesis: automated architecture for threat modeling and risk assessment for cloud computing. In: 6th International Conference on Privacy, Security, Risk and Trust (PASSAT), pp. 1–10 (2014)

13. Zineddine, M.: Vulnerabilities and mitigation techniques toning in the cloud: a cost and vulnerabilities coverage optimization approach using Cuckoo search algorithm with Lévy flights. Comput. Secur. 1–18(2015)

14. Nexpose Vulnerability Scanner. Nexpose—Rapid7. https://www.rapid7.com/products/nexpos e/

15. Nmap: the network mapper. https://nmap.org

A Learning Technique for VM Allocation to Resolve Geospatial Queries

Jaydeep Das, Arindam Dasgupta, Soumya K. Ghosh
and Rajkumar Buyya

Abstract Provisioning of virtual machines for efficient geospatial query management on cloud is an interesting and challenging work. The aim of this paper is to distribute workloads of different types of spatial queries into suitable virtual machine efficiently. To increase the effectiveness of the system serving geospatial queries, we use real-time geospatial query pattern learning methodology. This methodology is used to train the application specific properties, and the system will learn which type of the geospatial query should be allocated to what type of virtual machine automatically. The learning methodology gives knowledge about the resource required by each type of geospatial query. Using this understanding, various geospatial query templates are stored in the query template repository for further assistance. By this way, fast and robust assignment of virtual machine for the geospatial queries is possible which reduces their waiting time.

Keywords Geospatial query · VM allocation · Cloud computing

J. Das (✉) · A. Dasgupta
Advanced Technology Development Centre,
Indian Institute of Technology Kharagpur, Kharagpur, India
e-mail: jaydeep@iitkgp.ac.in

A. Dasgupta
e-mail: adgkgp@gmail.com

S. K. Ghosh
Department of Computer Science and Engineering,
Indian Institute of Technology Kharagpur, Kharagpur, India
e-mail: skg@iitkgp.ac.in

R. Buyya
Cloud Computing and Distributed Systems (CLOUDS) Lab School of Computing
and Information Systems, University of Melbourne, Melbourne, Australia
e-mail: rbuyya@unimelb.edu.au

© Springer Nature Singapore Pte Ltd. 2019   577
P. K. Sa et al. (eds.), *Recent Findings in Intelligent Computing Techniques*,
Advances in Intelligent Systems and Computing 707,
https://doi.org/10.1007/978-981-10-8639-7_61

1 Introduction

With the major advances of various sensor technologies and wireless techniques, a substantial amount of heterogeneous data with geospatial context has been accumulated by different organizations. Using these geospatial data, the organizations generate meaningful information by processing them. Besides, the use of geospatial information with the advent of smartphone technology has been drastically increased among common people. Thus, in order to facilitate such requirement, a scalable infrastructure is always in demand. In geospatial query operation, a huge amount of geospatial data is required to be accessed which are processed through multiple geospatial operations. For any geospatial query, the amount of processing varies from one location to another. Therefore, for the same query template, the computational resource requirement is always variable.

Cloud computing paradigm provides such computing and storage resources as and when required to meet Quality of Service (QoS) requirements of applications and users varying with time. The cloud computing platforms provide cost-effective, scalable, and minimal management strategy [1, 2]. It is the technology which is used in various geospatial domains. However, the provisioning of cloud resources may not be utilized in efficient way. As a result, the cost of geospatial query becomes high. Therefore, the cloud resources should be assigned for resolving a geospatial query in such a way that the cost of the query can be minimized. In any geospatial query, a lot of geospatial services are needed in particular order at different instances of time as the services have been provided by different organizations. Whenever a geospatial query is requested by a user, the query resolver generates a parse tree to know the required data services and processing applications. Again, if same query is requested by other users, from different geographical areas of interest, then different types of workload have been raised.

In most of the previous work, the virtual machine (VM) utilization cost is optimized by considering any one of the strategies such as query scheduling [3, 4], resource provisioning [5] and deployment of the query [6, 7] in particular type of VM. However, to optimize the query performance with budget constraints, all the strategy must be considered in judicious manner. In this work, a framework has been developed which distributes the cloud resources based on incoming geospatial queries. A learning technique has been applied for selection of VM and query provisioning. A VM allocation has been developed to learn the geospatial service composition pattern. It extracts the most effective heuristic strategy for executing the geospatial query with optimized cost.

2 Related Work

Virtual machine scheduling for workload management is a mature problem. Many works had been done for VM scheduling to serve query resolution. In the literature, workloads are dealt with resource provisioning, query placement and query scheduling [8] to minimize the cost. Some authors propose query response time minimization to achieve optimal VM scheduling.

As a weighted profitable within deadline scheduling problem is NP-complete [9], Chi et al. [3] proposed a greedy scheduling algorithm based on a service-level agreement tree (SLA-Tree) which helps in scheduling, dispatching and capacity planning. According to SLA, profits of each query are varied based on the query response time. The algorithm is built-in assuming that scheduled queries will happen in near future. The amount of profit lost or gained after postponement or expedition of actual scheduled query is measured through SLA-Tree.

Cost-based scheduling (CBS) [9] and incremental cost-based scheduling (iCBS) [4] algorithms are based on continuous changing of priority score of queries and execute queries accordingly. iCBS governs the cost depending upon query response time and follows piecewise linear SLA. For many piecewise linear SLAs, the time complexity in iCBS reduces to $O(\log N)$ in contrast to $O(N)$ time complexity of CBS.

In [6], two approximation algorithms of SLA violation penalty for uniform and non-uniform query processing are proposed, where approximation algorithm with dynamic programming is used to improve the quality and reduce the cost of query processing.

3   System Architecture

In order to resolve the spatial query in cloud environment, it is needed to identify the relevant services and consequently selects the suitable VMs. Most of the geospatial queries coming from general users are very specific in nature. Only the data services can be varied in different contexts. Same types of geospatial queries are grouped together and make a specific template and stored in a repository. Geospatial queries are parsed by query parse tree generator. From the parse tree, system can understand which types of geospatial services are required to resolve those geospatial queries. Geospatial service registry provides the appropriate VM details information where the geospatial query can get execute. A VM allocator has been developed to learn the service patterns of the queries. In this framework, a VM pool contains multiple VMs with different configurations. According to geospatial query from users, the appropriate VMs will be dispatched to resolve the query. In Fig. 1, the framework for geospatial query resolution on cloud has been depicted. The major component of the framework has been described below.

- *Spatial query service resolver*: The purpose of this component is to receive geospatial query from users in SQL like format through geospatial query interface. It feeds the geospatial query-to-query parse tree generator and get back parsed query tree. It accesses the geospatial web service registry to find the location of producer of those services. It also generates a service chain which describes the accessing logic of the geospatial web services to resolve the query.

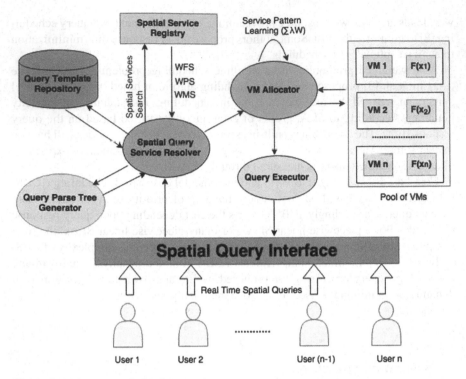

Fig. 1 Spatial query resolution framework

- *Query parse tree generator*: It generates a geospatial service query tree to identify the spatial feature services and processing services.
- *Query template repository*: Several types of geospatial query templates are deposited here. Templates are basically different combinations of geospatial services.
- *Spatial service registry*: A spatial service registry is a software component. It supports the run-time discovery and evaluation of resources such as geospatial services, geospatial datasets and geospatial application schemes. It is often associated with managed repositories. Spatial query service resolver gets information about available spatial services from spatial service registry.
- *VM allocator*: It schedules and allocates VMs according to availability of geospatial services which helps to resolve the geospatial queries. VM allocator analyses and captures the prevalent service patterns from the historical log. After using optimization algorithm, VMs are allocated for geospatial query resolution. It takes the weighted service requests from the query solver, i.e. $< \lambda, w >$, where the (λ) term denotes the timestamp value when the request is logged and w denotes the weight vector, which represents the computing resource requirement based on the applications. After receiving geospatial queries from the spatial query resolver, it assigns VMs by analysing service patterns. After dispatching VMs to query executor, it

checks the VM pools for the availability of VMs and dispatches if feasible. Otherwise, it stores the VM schedules in a queue and again after any completion of process check for availability. It is observable that sometimes VM scheduling is not possible according to the VM scheduler. It happens because in many real-life scenarios, it may take more time to complete a task than the expected time. A feedback function $F(x) = F(x_1), F(x_2), \ldots F(x_n)$ is proposed to take the real-time stamp values and provide feedback to the scheduler to improve the scheduling decisions.

- *VM executor*: After getting the details of allocated VMs and processing schedule geospatial queries, it executes geospatial queries with the help of spatial services in one or several pre-allocated VMs. Results of the geospatial queries return back to the users through spatial query interface.

4 VM Assignment Methodology

In Algorithm 1, initially assign a XLarge VM for a new query. If XLarge VMs are not available, geospatial query sends into a waiting queue. After execution of a new geospatial query, a feedback contains storage, memory and processing unit details, and will create a new geospatial query template and post it into query template repository. If incoming geospatial query is matched with existing geospatial query template, then it assigns to same type of VM for execution. Otherwise, geospatial queries are sent into a waiting queue for further execution.

Algorithm 1: Algorithm for VM assignment of spatial query

Input : SpQry, An incoming spatial query
Output: SpVM, The selected VM for spatial query
$m \leftarrow checkQueryType(SpQry)$;
if *(m = null)* **then**
  $VM \leftarrow searchVM(XLarge)$;
  if *(VM = null)* **then**
  waitingQueue.add(SpQry);
  while $((VM \leftarrow searchVM(XLarge)) = null)$;
else
  $VMType \leftarrow searchVMType(m)$;
  $SpVM \leftarrow searchVM(VMType)$;
end
if $(SpVM = null)$
$waitingQueue.add(SpQry)$;
while$((SpVM \leftarrow searchVM(VMType)) = null)$;
return (SpVM);

- *Resource utilization data capturing*: A geospatial query template is generated with storage amount, processing cores and memory utilization after execution of each geospatial query. This template is getting stored into query template repository.

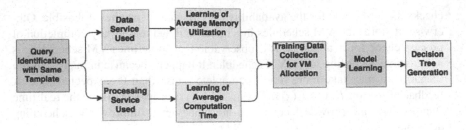

Fig. 2 Decision tree generation technique

When similar kind of geospatial query will come for execution, then it will be easy to allocate a suitable VM for that geospatial query execution.

- *Decision tree generation*: After getting the Information, i.e. storage amount, processing cores and memory utilization, a decision tree will generate to take the decision about the VM-type (XLarge, Large, Regular, Small and XSmall) allocation. This is done by the VM-type function in Algorithm 1 and it is shown in Fig. 2.
- *VM selection*: After generation of a decision tree, an appropriate VM is allocated for new type of geospatial query. If existing geospatial query template is available, then VM allocator allocates VM accordingly for geospatial query execution.

5 Performance Evaluation

We have done our experiment in Meghamala,[1] private cloud of IIT Kharagpur, with different types of VMs. In order to evaluate the performance of the framework, a prototype of geospatial services along with query resolution component has been implemented in the cloud environment. The data services provide the geospatial data such as road network, land use/land cover, drainage and settlements. The processing services provide various data processing features such as shortest path, buffer, nearest neighbour and geospatial set operations. These services have been accessed after parsing the geospatial queries. A set of sample geospatial queries based on different geospatial query templates has been created. A programme has been deployed to trigger these geospatial queries continuously to the query resolver. For each geospatial query, the trigger time and the response time have been considered to calculate the waiting time through that programme and collected in a file. The number of geospatial query trigger has been increased gradually from 20 to 500, whereas waiting time is increased to 1150 ms and later it decreased to 640 ms and it varies in between 650 and 700 ms. From Fig. 3, we observe that initially geospatial query waiting time is increased, and later it is decreased by increasing the number of geospatial queries.

---

[1]http://www.sit.iitkgp.ernet.in/Meghamala/.

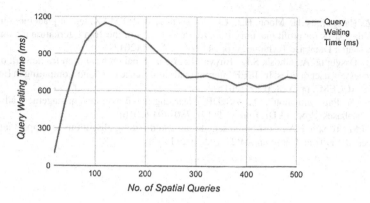

Fig. 3 No. of spatial query versus query waiting time

6 Conclusions and Future Work

In this paper, we proposed a learning-based algorithm for VM allocation to resolve geospatial queries with the help of geospatial query templates in cloud platform. From our experiment result, we observe that initially waiting time increases because the geospatial queries are new to the system and VM assignment takes more time. After learning the geospatial queries and generating the query templates, it becomes easy to assign VMs for similar kind of geospatial queries. In future work, we would like to include the priority value which will depend on the frequency of each geospatial query, and monetary cost which will depend on types of VM. It may improve the efficiency of VM allocation technique for geospatial query. We would like to perform these experiments in various public cloud platforms such as Amazon EC2, Microsoft Azure and IBM Bluemix.

References

1. Dastjerdi, A.V., Buyya, R.: An autonomous time-dependent SLA negotiation strategy for cloud computing. Comput. J. **58**(11), 3202–3216 (2015)
2. Mansouri, Y., Toosi, A.N., Buyya, R.: Cost optimization for dynamic replication and migration of data in cloud data centers. IEEE Trans. Cloud Comput. (2017)
3. Chi, Y., Moon, H.J., Hacigümüş, H., Tatemura, J.: SLA-tree: a framework for efficiently supporting SLA-based decisions in cloud computing. In: Proceedings of the 14th International Conference on Extending Database Technology, pp. 129–140. ACM (2011)
4. Chi, Y., Moon, H.J., Hacigümüş, H.: iCBS: incremental cost-based scheduling under piecewise linear sLAS. Proc. VLDB Endow. **4**(9), 563–574 (2011)
5. Jalaparti, V., Ballani, H., Costa, P., Karagiannis, T., Rowstron, A.: Bridging the tenant-provider gap in cloud services. In: Proceedings of the Third ACM Symposium on Cloud Computing, p. 10. ACM (2012)

6. Liu, Z., Hacıgümüş, H., Moon, H.J., Chi, Y., Hsiung, W.P.: PMAX: tenant placement in multi-tenant databases for profit maximization. In: Proceedings of the 16th International Conference on Extending Database Technology, pp. 442–453. ACM (2013)
7. Das, J., Dasgupta, A., Ghosh, S.K., Buyya, R.: A geospatial orchestration framework on cloud for processing user queries. In: IEEE International Conference on Cloud Computing in Emerging Markets (CCEM), pp. 1–8. IEEE (2016)
8. Marcus, R., Papaemmanouil, O.: WiSeDB: a learning-based workload management advisor for cloud databases. Proc. VLDB Endow. 9(10), 780–791 (2016)
9. Peha, J.M., Tobagi, F.A.: Cost-based scheduling and dropping algorithms to support integrated services. IEEE Tran. Commun. 44(2), 192–202 (1996)

A Trust-Based Secure Hybrid Framework for Routing in WSN

Komal Saini and Priyanka Ahlawat

Abstract Most of the routing protocols designed for wireless sensor network (WSN) still suffer from some disadvantages such as high path breakage ratio, low throughput, limited energy, end to end delays, etc. In this paper, we aim to enhance the security of the energy based routing protocol for WSN. We present a trust-based secure hybrid framework for energy-efficient routing (TSER) which is a hybrid framework that establishes a secure, trustful, and energy efficient route with proper authentication of the nodes. The route is formed on the basis of minimum hop count and residual energy to eliminate delays and during data transmission the nodes authenticate each other by exchanging key messages and then verifying them from the base station. The TSER not only ensures that data transmission is secure but also balances out the end to end delays between the sources to destination path.

Keywords Wireless sensor networks · End-to-end delays · Residual energy

1 Introduction

A Wireless Sensor Network (WSN) is an ad hoc network which is composed of large number of cheap, low power and densely deployed sensors [1]. In WSN, secure routing is very important due to its fully distributed nature and constrained computational resources. The network is vulnerable to the node capturing attacks, in which the attacker seize the node and obtain all the information and wreck the confidentiality and security of the network [2, 3]. Depending upon the attacker behavior, attacks can be classified as stated below [4, 5]:

K. Saini (✉) · P. Ahlawat
Department of Computer Engineering, National Institute of Technology, Kurukshetra,
Kurukshetra 136119, India
e-mail: komalsaini492@gmail.com

P. Ahlawat
e-mail: priyankaahlawat@nitkkr.ac.in

© Springer Nature Singapore Pte Ltd. 2019
P. K. Sa et al. (eds.), *Recent Findings in Intelligent Computing Techniques*,
Advances in Intelligent Systems and Computing 707,
https://doi.org/10.1007/978-981-10-8639-7_62

585

- Blackhole attack: In this attack, an attacker node discards all the packets that it has to forward.
- Greyhole attack: A malicious node will drop certain type of packets and forwards only a part of them in this attack.
- Sinkhole attack: In Sinkhole attack a mischievous node tries to misguide nearly all the traffic and disguise itself as a sink node.
- Sybil attack: In this attack, the malicious node can create and impersonate different nodes to manipulate the recommendations and promote itself as trustful node.

Sensor nodes have a limited amount of energy thus energy consumption is an important factor during the network running course. We focus on two issues security and energy efficiency and present a trust-based secure hybrid framework for energy efficient routing (TSER) which is trustful, more secure and energy efficient. The paper is arranged as follows: Sect. 2 discusses the related routing algorithms. Section 3 consists of the proposed hybrid scheme. Section 4 contains results and discussion. Finally, we conclude our paper in Sect. 5.

2 Related Work

Several routing algorithms have been proposed for secure routing based on trust evaluation, energy consumption and security. These three elements are very essential for a reliable communication between nodes in the network. Duan et al. [6] proposed a lightweight trust derivation and trust computation algorithm and has potential to resist various attacks. Gong et al. [7] designed energy efficient trust-aware routing protocol for energy efficiency and security of WSN. The limitation of this approach is that as the number of compromised nodes increases its performance regarding security decreases. Kaur et al. [8] proposed a scheme which develops a secure and trustworthy route relying upon the past and present node to node interactions by. Random Password Generation (RPC) [9] algorithm focuses on the various traffic levels and security during data transmission in WSN. Compare and Match Position Verification Method [10] use tables to store information about the nodes and whenever the nodes want to communicate they authenticate each other with that table. Message Authentication and Passing method [10] is used to authenticate a node while data transmission. We find that a single routing algorithm cannot meet all the desired requirements. Thus, there is need for some hybrid framework. So, we have proposed a hybrid framework for efficient and secure transmission of data.

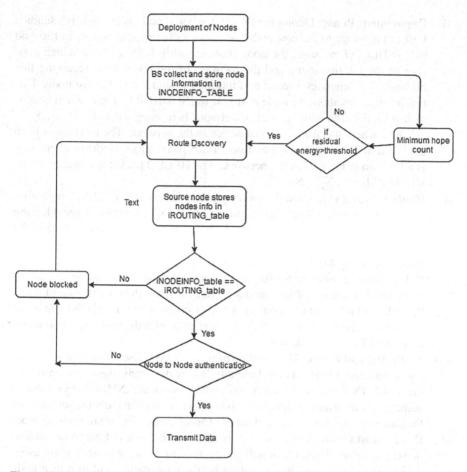

Fig. 1 Flow chart for proposed hybrid framework for energy efficient secure routing

3 The Proposed Scheme

Network consists of **N** nodes which are randomly deployed under the control of a base station (BS). These nodes are promising, well configured and energy efficient. It is assumed that during deployment of network, no attack has been performed. The BS and source node are assumed to be secure. Figure 1 shows the flow chart for proposed hybrid framework for energy efficient secure routing in WSN. The proposed algorithm works in four different phases as follows:

(i) **Deployment Phase**: During deployment when a node is created, BS sends a HELLO message to the new node along with the timestamp in it. In respond to this HELLO message, the node replies with a RES message which consists of the ID, timestamp and the location of the node. After receiving this message, BS generates a secret and unique key and sends it to the node. This full information about the node is stored in iNODEINFO_*table* which is controlled by BS. The entire model of network is represented as $G = \{(n_1, n_2, ..., n_n), BS\}$ where n is the number of nodes in the network. The location of each deployed node is represented as $Location(n_i) = (rand(x), rand(y))$ where x, y is a location in the area of the network. The HELLO packet is represented as: $HELLO(Msg, \tau) \sum_{i=0}^{n} N_I$.

(ii) **Route Discovery**: In route discovery phase from S to D in TSER, a path with minimum hop count is selected from set of all possible paths. After selecting the path with minimum number of hop counts, the S will ask every node in the path about their residual energy and their current information (ID, timestamp, location). A threshold is set by the BS for the residual energy and is stored in S. The nodes respond the S with their current residual energy and information. If the residual energy of intermediate nodes is more than the threshold value then, the full information about the node is stored in the iROUTING_*table* of S otherwise the path is discarded and next path with the minimum hop count is selected for route formation.

(iii) **Route Authentication**: The duration of time between discovering the route and data transmission is assumed to be small. Before data transmission the entries of the iROUTING_*table* are verified with the entries of the iNODEINFO_*table* to authenticate the transmission. This algorithm ensures that the route selected for the data transmission consists of all the trusted nodes for secure transmission.

(iv) **Data Transmission**: During data transmission, the nodes of the route authenticate each other. The node n_i will send a request message to node n_j by using its key as $msg(ki)$, which was issued by the base station and in return node n_j also sends an acceptance message using its key message as $msg(kj)$. Later these keys are confirmed by BS and an ok signal is sent for data transmission.

The pseudocode for this algorithm is given below in detail.

Algorithm 1: Algorithm for the proposed scheme

1. **Let** G = {n_1, n_2, n_3,..., n_n}
2. **Let** **BS** be the well configured nodes
3. **for** i = 1 to n   //nodes are placed randomly
4. n_i ← Location(rand(x), rand(y))
5. ID(n_i ← i);
6. **End** loop
7. **Let** **L**i be the set of links between the pair of nodes
8. n_i → **HELLO(Msg,)**   //for every node n_i
9. n_i **RES** BS; RES = ID, , x,y;
10. BS generates key **k**i //for every node n_i
11. BS send ki to every node (**k**i → n_i)
12. RES + ki → **iNODEINFO_table**
13. **End** loop
14. **for** i = S to D   //route discovery
15. Find all possible paths from S → D
16. Choose path **P**i with minimum hope count
17. Check for energy.
          If energy N$_{curr}$ < E$_{thresh}$
          Then choose next path with minimum hope count.
18. **Else**
          S ← RES (ID, , x,y)   //every node send its current information to source
19. RES → **iROUTING_table**
20. **End** loop
21. **for** i = S to D   //data transmission
22. if(currentNi.info==iROUTING_table==iNODEINFO_table) then
23. n_i → **send(request)** → n_i + 1
24. n_i + 1 → **send(acceptance)** → n_i
25. **if** (msg(n_i), msg(n_i + 1) exists (iNODEINFO_table)) then
26. n_i → **Send** (dataP → n_i + 1)
27. **Else**
28. n_i + 1 is blocked as malicious node
29. **End** if
30. Choose next optimal path

This algorithm is so secure that it is almost impossible for a node to perform any attack. If a path breaks due to energy deficiency or a malicious node, the S node after selecting the next optimal path with minimum hop count, verify the nodes of the route from its trusted nodes. These trusted nodes are the nodes which are once authenticated by the S node. The nodes including the S node do caching when they authenticate one another. So instead of asking the base station, verification can be done in the neighborhood itself hence it takes less time and energy. A threshold for

Fig. 2 Comparing end to end delay between existing method CAM-PVM and MAP with TSER

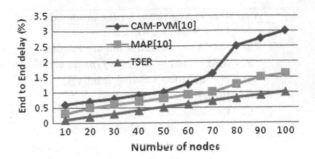

the maximum number of nodes to be asked for verification of a node is set by the S node. If the BS hop count is greater than the threshold then it will verify from its neighbors whose hop count is less than the threshold value otherwise it will ask the BS for verification.

4   Results and Discussion

One simulator setup is used for simulation of the system model. All the nodes are controlled by a single BS. The network efficiency is evaluated by comparing the before and after the implementation throughput of the proposed hybrid approach. The scheme is analyzed by calculating the end to end delay.

Figure 2 shows that TSER has least delay. It is because we are using the shortest path for sending the data packets. The route is selected on the basis of minimum hop counts so that it takes least time for a packet to reach the destination and hence, reduce the end-to-end delays. Also, due to the caching of the nodes, it saves much time as the S node does not have to visit the BS again and again for the verification, which is time-consuming and takes a lot of energy.

5   Conclusion and Future Plan

The TSER is used for a secure communication in the network by detecting and preventing Sybil attacks. In TSER, the nodes are verified by the BS in the beginning and after that the neighboring nodes can also verify them. The route established using TSER is more trustworthy, more secure and more reliable. Also it saves time and energy as it does not have to visit the BS every time to verify a node. End-to-end delay is least as the path selection is done on the basis of the hop counts. The complexity of the algorithm is reduced by caching the nodes. In future, we plan to optimize the values of threshold.

References

1. Lewis, N., Foukia, N., Govan, D.G.: Using trust for key distribution and route selection in wireless sensor networks. In: Network Operations and Management Symposium, pp. 787–790. IEEE (2008)
2. Ahlawat, P., Dave, M.: An improved hybrid key management scheme for wireless sensor networks. In: 4th International Conference on Parallel Distribution and Grid Computing (PDGC 2016) (2016)
3. Ahlawat, P., Dave, M.: A hybrid approach for path vulnerability matrix on random key pre-distribution for wireless sensor networks, wireless personal communications (2016). Springer. https://doi.org/10.1007/s11277-016-3779-6
4. Yu, Y., Li, K., Zhou, W., Li, P.: Trust mechanisms in wireless sensor networks: attacks analysis and countermeasures. J. Netw. Comput. Appl. 867–880 (2012). Elsevier
5. Lopez, J., Roman, R., Agudo, I., Fernandez-Gago, C.: Trust management systems for wireless sensor networks: best practices, computer communications 1086–1093 (2010). Elsevier
6. Duan, J., Yang, D., Zhu, H., Zhang, S., Zhao, J.: A trust—aware secure routing framework in wireless sensor networks. Int. J. Distrib. Sens. Netw. **2014**, 14 pages. Article ID 209436 (2014). Hindawi Publishing Corporation
7. Gong, P., Chen, T.M., Xu, Q.: ETARP: an energy efficient trust-aware routing protocol for wireless sensor networks. J. Sens. **2015**, 10 pages. Article ID 469793 (2015). Hindawi Publishing Corporation
8. Kaur, J., Gill, S.S., Dhaliwal, B.S.: Secure trust based key management routing framework for wireless sensor network. J. Eng. **2016**, 9 pages. Article ID 208714 (2016). Hindawi Publishing Corporation
9. Amuthavalli, R., Bhuvaneswaran, R.S.: Detection and prevention of sybil attack in wsn employing random password comparison method. J. Theor. Appl. Inf. Technol. **67**, 236–246 (2013)
10. Dhamodharan, U.S.R.K., Vayanaperumal, R.: Detecting and preventing sybil attacks in wireless sensor networks using message authentication and passing method. Sci. World J. **2015**, 7 pages. Article ID 841267 (2015). Hindawi Publishing Corporation

PDLB: An Effective Prediction-Based Dynamic Load Balancing Algorithm for Clustered Heterogeneous Computational Environment

Devendra Thakor and Bankim Patel

Abstract Load balancing is one of the major requirements of distributed systems for the effective utilization of resources. Existing dynamic load balancing algorithms take the corrective actions once the system becomes unbalanced. It could be more appropriate to design algorithm which can schedule incoming jobs in such way that system remains in balance state without increasing unnecessary system overhead. In this paper, prediction-based dynamic load balancing algorithm is designed which can balance load between clusters using predicted status of clusters. The designed algorithm is applied on priority scheduling algorithm. Our experimental result shows that the prediction-based load balancing algorithm improves the cluster utilization as compared to existing scheduling algorithm.

Keywords Load balancing · Distributed system · Dynamic load balancing
Cluster computing

1 Introduction

The distributed computing environment is one of the preferable areas due to technological advancement in the way of computation and communication. Distributed computing platform allows to access geographically scattered resources. It is open to add or delete resources and is able to tolerate any kind of fault in the system [1]. The geographically scattered resources are grouped into virtual groups known as cluster. The incoming jobs which are generated by different users are assigned to cluster for the execution. This process of creating clusters and allocating incoming jobs to

D. Thakor (✉)
Department of Computer Engineering and IT, Chhotubhai Gopalbhai Patel
Institute of Technology, Uka Tarsadia University, Bardoli, India
e-mail: devendra.thakor@utu.ac.in

B. Patel
Shrimad Rajchandra Institute of Management and Computer Application,
Uka Tarsadia University, Bardoli, India
e-mail: bankim_patel@srimca.edu.in

© Springer Nature Singapore Pte Ltd. 2019
P. K. Sa et al. (eds.), *Recent Findings in Intelligent Computing Techniques*,
Advances in Intelligent Systems and Computing 707,
https://doi.org/10.1007/978-981-10-8639-7_63

the specific cluster is known as cluster computing. The biggest challenge in cluster computing is uneven utilization of clusters. It has been noted in the literature that gradually some of the clusters become overloaded, while the other clusters remain mediumloaded or underloaded in cluster computing system [2, 3]. Such situation leads toward wastage of the available resources and reduces the overall system's performance.

Load balancing is the process of harmonizing load between clusters of distributed systems [2–5]. Load balancing algorithms improve performance of system by distributing or redistributing equal load to each processing unit. This improves the performance by minimizing response time and maximizing throughput and resource utilization. The load balancing algorithms can be categorized into two types—static and dynamic [3–5]. The static algorithm balances the load on the basis of information which is available at compile time, while dynamic algorithm balances load by considering run-time information [3–5]. The proper understanding and analysis of all components of load balancing algorithm like location, transfer, and selection policies is required for designing effective dynamic load balancing algorithm. The analysis of components of existing dynamic load balancing algorithms plays important role in pinpointing the limitations of current research and justifies the need of designing novel efficient dynamic load balancing algorithm [6, 7].

Existing dynamic load balancing algorithms are divided into two categories like sender initiative and receiver initiative [6–8]. In sender initiative approach, overloaded node initiates load balancing algorithm and passes the jobs to lightly loaded nodes. In receiver initiative approach, lightly loaded node starts the load balancing algorithm by informing current status to overloaded nodes [8]. In both the approaches, the balancing activity starts after system becomes unbalanced. The better approach is to run balancing algorithm at the time of process scheduling. The effectiveness of balancing algorithm when it is running with process scheduling depends on the predicted future status of cluster. Thus, a predictive algorithm can be designed which learns from the observation and predicts the future status of clusters in distributed system. The designing of predictive load balancing algorithm is not addressed properly in the literature [3–7]. Once the predictive algorithm for dynamic load balancing is designed successfully, system can take corrective actions on the basis of prediction and can avoid the problem of unbalance load among the clusters of distributed systems.

One of the main objectives of study is to design prediction algorithm which can predict future status of cluster in heterogeneous computational environment and to design load balancing algorithm which can balance load between clusters using predicted values of the prediction algorithm. Second objective is to apply designed prediction-based load balancing algorithm on scheduling algorithm. The priority scheduling algorithm is selected for testing and checking the performance of prediction-based load balancing algorithm.

The rest of the paper is organized as follows. In Sect. 2, related work is discussed. The system model is defined in Sect. 3. The prediction and dynamic load balancing algorithms are proposed in Sect. 4. Section 5 explains the implementation scenario and simulation setup. The simulation results and analysis of result are shown in Sect. 5. Finally, conclusions and future work are stated in Sect. 6.

2 Related Work

The issue of load balancing in distributed system is addressed well in literature. There are many approaches for load balancing in distributed system proposed by different researchers which can be found in literature [6–12]. In [6], authors explored and explained the essential components for designing an effective dynamic load balancing algorithm for distributed computing. The comparative analysis of information policy and location policy is discussed in the paper. Authors proposed modified demand-driven information policy (MDDIP) and limited broadcast location policy (LDLP) which further reduces the numbers of messages passed between the nodes. Author is focused on three points related to dynamic load balancing algorithm for distributed systems in [7]. First, the author discussed about the key issues for designing effective dynamic load balancing algorithm, second, comparative analysis of various existing dynamic load balancing algorithms is done and lastly, author proposed new dynamic load balancing algorithm which improves response time by reducing communication overhead. An adaptive decentralized sender-initiated load balancing algorithm that utilizes the load estimation approach is presented in [8]. The algorithm is adaptive as it can estimate different types of strongly influencing system parameters.

A dynamic threshold-based scheduling and load balancing algorithm are proposed in [9]. The algorithm dynamically updates the threshold values with respect to run-time changes in resource workload. It has considered workload of resources as load index instead of number of tasks in queue because weights of tasks are different. The proposed threshold-based load balancing algorithm (TBLBA) gives better results in terms of task response time, level of load balancing, and resource utilization compared to min-min and ant colony scheduling algorithms. An algorithm for cluster formation and super node selection is explained with simulation result in [10]. Authors proposed two super node selection algorithms which focused on multi-criteria optimization techniques, and the performance of hybrid algorithm is evaluated under different cluster configurations, different load scenarios, and different network topologies. In [11], authors proposed prediction-based dynamic load balancing technique for heterogeneous clusters. The proposed technique tries to predict three different types of resource requirement of incoming job. They considered processor, memory, and IO requirements of coming job. The dynamic load balancing algorithm which can predict neighbor's load information is proposed in [12]. Authors proposed and simulated load prediction and heuristic neighbor selection algorithms. The simulation results show that the proposed algorithm reduces the average response time of each job.

3 System Model

The distributed system consists of a large number of geographically dispersed heterogeneous computing resources, namely $N_1, N_2, N_3, \ldots, N_n$. The system also has a different set of users that are connected to system via network. It is assumed that there are maximum N nodes in the system. The notations used in the paper are tabulated in Table 1.

The following properties of the distributed system are considered.

- Users generate job which may be of even different sizes.
- Study is limited to non-preemptive scheduling which reduces system overhead as there is no process migration.
- High capacity of queue size is selected, which eliminates the possibility of rejecting the job due to unavailability of queue.
- Clusters are heterogeneous with respect to number of machines, cost of processing, processing speed, operating system, memory capacity, and I/O capacity.
- The load L_i of each node N_i is computed using the question $L_i = C_i + M_i + IO_i$. The status of each node is defined as follows [13]:
  $L_i = 0 \rightarrow node\ is\ idle$
  $L_i < T_j \rightarrow lightly\ loaded\ node$
  $T_j < L_i < T_j \rightarrow medium\ loaded\ node$
  $L_i > T_j \rightarrow highly\ loaded\ node$

Table 1 List of notations

Notation	Description
N	Total number of nodes in the system
N_i where $1 \le i \le N$	ith node of the system
C_i	Processor load of N_i
M_i	Memory load of N_i
IO_i	I/O load of N_i
L_i	Total load of the node N_i
M	Total number of clusters
g_j	jth cluster, where $1 \le j \le m$
k_j	Size of the cluster g_j
CT_j	Threshold value jth cluster
ST	Threshold value of system
$Overloaded_j$	Overloaded status counter of jth cluster
$Underloaded_j$	Underloaded status counter of jth cluster
OL	List of overloaded clusters
UL	List of underloaded cluster
ML	List of mediumloaded cluster

- Cluster threshold CT is defined as the average load of the cluster which can be computed using the following equation [13]. For cluster g_j,

$$CT_j = \frac{\sum_{l=1}^{k_j} Li}{k_j} \quad Where 1 \leqslant j \leqslant m \tag{1}$$

- System threshold ST is defined as the average threshold of clusters which can be computed using the following equation [14]:

$$ST = \frac{\sum_{i=1}^{m} CT_i}{m} \tag{2}$$

4 Prediction-Based Dynamic Load Balancing Algorithm

As part of research work, two algorithms are proposed. First one is the prediction algorithm and second one is load balancing algorithm. The proposed prediction-based algorithm takes load information of cluster for the defined period of time as an input. It maintains the load status counter for each cluster per day on hourly basis. It learns from the observations and predicts the future load status of cluster in distributed system. The proposed load balancing algorithm selects the best suitable cluster for job execution which balances load in distributed computing system. The selection of best suitable cluster depends on two parameters, first is cluster having underloaded or mediumloaded status as predicted value and second is cluster having capabilities to execute specific incoming jobs. The algorithm allocates jobs for execution to the cluster having underloaded status for particular duration in past which makes the future status of current underloaded clusters as mediumloaded clusters and it may also change future status of current overloaded cluster as mediumloaded. Thus, dynamic load balancing is done by the combination of both the algorithms, which enables system to take actions on the basis of prediction and avoids the problem of unbalance load among the clusters of distributed systems.

In the prediction algorithm, step1 defines observation cycle. The duration of 1 month is selected as observation cycle. Steps 2, 3, and 4 execute algorithm for each hour, day, and week of a month. Steps 5 and 6 calculate system threshold (ST) value and cluster threshold value (CT) by equations discussed in [14]. The steps 6–8 find load status of each cluster for 24 h in a day. The rules have been defined for checking the status of cluster for particular hour. The cluster is considered as an overloaded if the cluster threshold value is high compared to system threshold value and vice versa. The step 9 sets the status of cluster for particular day on the basis of threshold policy [9]. The algorithm considers the predefined double threshold values as threshold policy. The defined threshold values are T1 = 8 h and T2 =16 h in a day.

The following rules are defined based on threshold policy for setting the status of cluster in a day.

Overloaded   if cluster remains overloaded for more than T2 value in a day.
Underloaded   if cluster remains underloaded for more than T2 value in a day.
Mediumloaded   if cluster remains overloaded for more than T1 value and under-
                   loaded for less than T2 value in a day.

The load balancing algorithm takes the output of proposed prediction algorithm as an input and finds suitable cluster for executing incoming job. The prediction algorithm prepares the list of underloaded, mediumloaded, and overloaded clusters. The load balancing algorithm first selects the cluster from the list of underloaded clusters for job execution. It selects mediumloaded node only if list of underloaded clusters is empty. At last, algorithm submits incoming job to one of the overloaded clusters as it is extreme situation where all the clusters in system are overloaded.

Pseudocode for Cluster Load Status Prediction [14]

```
Algorithm: Prediction Algorithm
    Input: load information of clusters
    Output: predicted status of clusters
1. define observation cycle
2. for each week  in month do
3.       for each day in week do
4.             for each hour in day do
5.                  calculate system threshold value ST
//Following steps finds the status of each cluster hour wise
6.                  for j = 1 to m
7.                        calculate cluster threshold value CTj
8.                        If(CTj > ST)
                              Statusj = overloaded
                              Overloadedj++
                          else(CTj < ST)
                              Statusj= underloaded
                              Underloadedj++
//Following steps finds the status of each cluster day wise
9.       if(Overloadedj > 16)
              Statusj = overloaded
          elseif(Underloadedj > 16)
              Statusj = underloaded
          else
              Statusj = mediumloaded
10. update load status of cluster for next execution cycle
```

Pseudocode for Load Balancing [14]

```
Algorithm: Load Balancing
    Input: predicted load status of clusters
    Output: best suitable cluster for job execution
1. for all incoming job in a day do
2. read predicted load status value of clusters
3. prepare UL, ML, and OL list
4. if(size of UL > 0)
        randomly select one cluster from UL list
      submit job to the selected cluster
6. elseif(size of ML > 0)
        randomly select one cluster from ML list
      submit job to the selected cluster
7. else
     randomly select one cluster
     submit job to the selected cluster
8. update load status of clusters
```

5 Implementation and Simulation Setup

In order to demonstrate improvement in load balancing, simulation is done using ALEA [15]. ALEA is GridSim [16]-based job scheduling simulator. The GridSim is Java-based simulation tool which provides facilities for modeling and simulating the entities like users, heterogeneous resources, resource load balancers, and applications. The GridSim is used to evaluate the load balancing algorithms. The ALEA can deal with job scheduling problems like heterogeneity of job and resources, dynamic changes in incoming job or machine failure in cluster, and grid environment [15]. The prediction-based load balancing algorithm is implemented with the help of ALEA. The real-time dataset from the Gaia cluster log is selected for testing the prediction algorithm. The Gaia cluster is one of the four clusters operated by the University of Luxembourg HPC Center (ULHPC) initially released in 2011 [17]. The Gaia is a heterogeneous cluster that has been upgraded several times. The selected dataset contains 3 months of data from May to August 2014. It is used mainly by biologists working with large data problems and engineering people working with physical simulations. They created total 51,987 jobs during the 3 months period and submitted to Gaia cluster having 151 nodes, manufactured by Bull and Dell, with a total of 2004 cores for job execution [17].

Table 2 describes the specifications of the Gaia cluster. The Gaia cluster is the collection of totally eight heterogeneous clusters. Each cluster has different numbers of machines and processers as mentioned in Table 2. It contains totally 2004 processors to execute incoming jobs. The performance of prediction-based dynamic load

Table 2 Cluster configuration

Cluster id	Cluster name	Nodes	CPUs	Total CPUs
1	gaia-[1–60]	60	12	720
2	gaia-[61–62]	2	12	24
3	gaia-[63–72]	10	12	120
4	gaia-73	1	160	160
5	gaia-74	1	36	36
6	gaia-[75–79]	5	16	80
7	gaia-[80–119]	40	12	480
8	gaia-[120–151]	32	12	384
Total		151	272	2004

balancing algorithm (PDLB) is evaluated using the simulation setup described and is compared with the traditional priority scheduling algorithm (PS). The percentage of cluster usage is selected as performance parameter. Table 3 shows the average utilization of each cluster and average utilization of all clusters together. The results show that average utilization of available resources is increased by almost 6% in PDLB algorithm as compared to PS algorithm. The utilization of individual cluster is improved in six clusters out of eight clusters in the proposed approach. The utilization of first cluster is 70.67%, and on the other hand utilization of fifth cluster is only 8.58% in priority scheduling which indicates unbalanced load in between clusters. The utilization of second, third, fourth, fifth, and sixth clusters is very less in existing approach. It shows that balancing is required among the clusters. The problem of unbalanced load between clusters is resolved in prediction approach. The PDLB resolved problem by improving utilization of totally five clusters that is second to sixth cluster by 10% which increased the average utilization of system by 6%. In the

Table 3 Cluster utilization

Cluster id	Cluster name	PS (%)	PDLB (%)
1	gaia-[1–60]	70.67	63.86
2	gaia-[61–62]	10.01	19.71
3	gaia-[63–72]	13.91	24.63
4	gaia-73	16.28	27.38
5	gaia-74	8.58	19.05
6	gaia-[75–79]	10.05	21.42
7	gaia-[80–119]	42.68	41.30
8	gaia-[120–151]	33.19	35.76
Total		205.37	253.11
Average utilization		25.67	31.64

existing approach, the utilization of the first cluster having highest amount of processing power is 70.67%, which indicates that most of the incoming jobs are executed by the first cluster. In PDLB algorithm, the utilization of first cluster reduces by 6% means utilization of other clusters increased in the system. The experimental result shows that the level of load balancing is increased in PDLB algorithm compared to PS algorithm.

Figures 1 and 2 show the cluster usage on day-to-day basis in PS and PDLB, respectively. The each row in graph represents individual cluster. It shows that there are totally eight rows as the dataset is having eight clusters. The green, yellow, and red colors represent less, medium, and high utilization of clusters. Figure 1 shows that the first cluster having highest amount of computing power remains busy most of the days. The first, seventh, and eighth clusters are utilized more as compared to other clusters. Figure 2 shows that PDLB can predict the loaded situation of the first, seventh, and eighth clusters and scheduled the incoming jobs in such a way

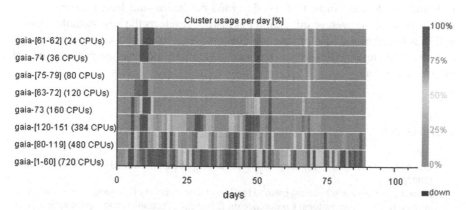

Fig. 1 Cluster utilization per day in priority scheduling

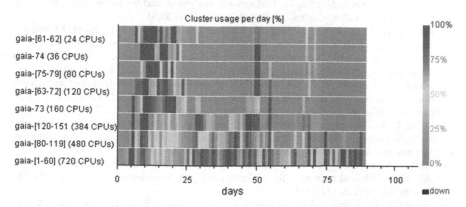

Fig. 2 Cluster utilization per day in PDLB

that utilization of other clusters increases and thus increases the overall utilization of available resources. Figure 2 shows that, for first 35 days, there is an improvement in terms of load balancing in the prediction approach.

6 Conclusion

Authors have designed prediction and load balancing algorithms for distributed system. The prediction algorithm predicts the status of clusters in advance. The load balancing algorithm uses the predicted status of clusters for scheduling incoming job. The predicted information plays an important role in selection of the best destination cluster for incoming jobs. The traditional priority-based scheduling algorithm and prediction-based dynamic load balancing algorithm are simulated on real-time dataset of 3 months having 51987 jobs in real-time cluster configuration. The cluster utilization is calculated day-wise, and the average utilization of available clusters is found out. Results show that the designed prediction and load balancing algorithms improve the average utilization of clusters. It selects the best suitable clusters and thus balances the load among the clusters. As a future work, an improvement in individual cluster utilization is planned which will increase the performance of distributed system.

References

1. Liu, M.L.: Distributed Computing: Concepts and Applications, 4th edn. Pearson Education (2009)
2. Datta, L.: A new task scheduling method for 2 level load balancing in homogeneous distributed system. In: 2016 International Conference on Electrical, Electronics, and Optimization Techniques (ICEEOT), pp. 4320–4325, Mar 2016
3. Sharma, S., Singh, S., Sharma, M.: Performance analysis of load balancing algorithms. Int. J. Comput. Electr. Autom. Control Inf. Eng. 2(2), 367–370 (2008)
4. Kameda, H., Li, J., Kim, C., Zhang, Y.: Optimal Load Balancing in Distributed Computer Systems. Telecommunication Networks and Computer Systems. Springer, London (2012)
5. Rajguru, A.A., Apte, S.: A comparative performance analysis of load balancing algorithms in distributed system using qualitative parameters. Int. J. Recent Technol. Eng. 1(3) (2012)
6. Mehta, M.A., Jinwala, D.C.: Analysis of significant components for designing an effective dynamic load balancing algorithm in distributed systems. In: 2012 Third International Conference on Intelligent Systems Modelling and Simulation, pp. 531–536, Feb 2012
7. Mehta, M.A.: Designing an effective dynamic load balancing algorithm considering imperative design issues in distributed systems. In: 2012 International Conference on Communication Systems and Network Technologies, pp. 397–401, May 2012
8. Shah, R., Veeravalli, B., Misra, M.: On the design of adaptive and decentralized load balancing algorithms with load estimation for computational grid environments. IEEE Trans. Parallel Distrib. Syst. 18(12), 1675–1686 (2007)
9. El-Zoghdy, S.F., Elnashar, A.I.: A threshold-based load balancing algorithm for grid computing systems. J. High Speed Netw. 21(4), 237–257 (2015)

10. Mehta, M.A., Jinwala, D.C.: A hybrid dynamic load balancing algorithm for distributed system. JCP **9**(8), 1825–1833 (2014)
11. Chandra, P.K., Sahoo, B.: Prediction based dynamic load balancing techniques in heterogeneous clusters. In: Proceedings of International Conference on Computer Science and Technology, pp. 189–192 (2010)
12. Lim, J.W., Hoong, P.K., Yeoh, E.T.: Neighbor's load prediction for dynamic load balancing in a distributed computational environment. In: Proceedings of TENCON, pp. 1–6. IEEE (2012)
13. Mehta, M., Jinwala, D.: A hybrid dynamic load balancing algorithm for heterogeneous environments. In: Proceedings of International Conference on Grid Computing and Applications, pp. 61–65 (2011)
14. Thakor, D., Patel, B.: Prediction based dynamic load balancing algorithm for distributed system. Natl. J. Syst. Inf. Technol. **9**(2), 67–76 (2016)
15. Klusáček, D., Rudová, H.: Alea 2: job scheduling simulator. In: Proceedings of the 3rd International ICST Conference on Simulation Tools and Techniques (2010)
16. Buyya, R., Murshed, M.: Gridsim: A toolkit for the modeling and simulation of distributed resource management and scheduling for grid computing. Concurr. Comput. Pract. Exp. **14**(13–15), 1175–1220 (2002)
17. Emeras, J.. Parallel Workloads archive: University of Luxembure GAIA cluster. http.//www. cs.huji.ac,il/labs/parallel/workload/l_unilu_gaia/index.html. Accessed 01 Jan 2017

An Agent-Based Approach for Dynamic Load Balancing Using Hybrid NSGA II

Vishnuvardhan Mannava, Sai Swanitha Kodeboyina, Swathi Bindu Bodempudi and Chandrika Sai Priya Addada

Abstract To date, there have been many observations about load balancing on different machines. Many researchers identified load balancing as a key component for scheduling problems, as strongly NP-hard. Techniques to solve scheduling problems are frequently unfeasible. Metaheuristic techniques are more generic and applicable to solve wider range problems. We choose NSGA II in genetic algorithms because they are known for parallelization, use probabilistic selection techniques, and multi-objective evolutionary algorithm enriches the dynamic performance. This analysis sheds light on using agents to perform decision-making and scheduling operations. In this paper, we propose and implement ANSGA II which we believe is the first of its kind, an agent-based hybrid model which provides Pareto front solutions whose individual solutions will satisfy multi-objectives. This multi-agent-based approach fastens the algorithm performance. Computational experiments and results for our proposed agent-based model thrive in results toward optimality, in scheduling.

Keywords Load balancing · Job shop scheduling · Workforce planning
Metaheuristic · Genetic algorithms · NSGA II · Agents

V. Mannava (✉) · S. S. Kodeboyina · S. B. Bodempudi · C. S. P. Addada
Department of Computer Science and Engineering, K L University, Guntur 522502,
Andhra Pradesh, India
e-mail: vishnu@kluniversity.in

S. S. Kodeboyina
e-mail: swanitha.kodeboyina@gmail.com

S. B. Bodempudi
e-mail: swathichowdary@gmail.com

C. S. P. Addada
e-mail: chandrikasaipriya@gmail.com

© Springer Nature Singapore Pte Ltd. 2019
P. K. Sa et al. (eds.), *Recent Findings in Intelligent Computing Techniques*,
Advances in Intelligent Systems and Computing 707,
https://doi.org/10.1007/978-981-10-8639-7_64

1 Introduction

Load balancing means distribution of work in an efficient means so that the total processing time of the work will be minimized [15]. Load balancing problem became a vital factor in job scheduling. Distribution of work (jobs) can be done by load balancer. Load balancing problem can be stated as follows: consider a set of M machines $(M_1, M_2, M_3, \ldots, M_n)$ which can process at most one job at a time and set of n jobs. Assume the processing time for each job (j) is t_j and j(k) be the subset of jobs assigned to machine k. Here, in load balancing problem, we need to minimize makespan [15] (minimum time to complete all the jobs).

There are three rules that define the process of load balancing. They are location rule, distribution rule, and selection rule [15]. Based on these three rules, we consider job shop scheduling and workforce planning. Load balancing is challenging problem in task scheduling, task allocation in parallel environments which need an effective load balancing techniques [5].

The genetic algorithm is one of the metaheuristic techniques for scheduling. Genetic algorithms contain different terms like chromosome, crossover, and mutation. Chromosome is the array of bits or digits, characters. These chromosomes are the encoded solution to the problems [2]. Different fitness functions can be used for the problems, and depending on the solution there are operators and genetic algorithms for the process to be carried [3]. The operators are selection, crossover, and mutation. The crossover operation is followed by the mutation. Mutation takes place if the randomness of the obtained offspring is not much differentiable than the parents [4]. For multi-objective optimization, the genetic algorithms serve the best solutions [15], whereas in this paper we propose a modified and advanced genetic algorithm that supports for parallel processing which inbuilt have some adaptive metrics for dynamic load balancing type of problems and we compare with other algorithms [5, 14]. Genetic algorithms can be implemented in parallel using agents. We introduced an agent-based parallel approach to reduce makespan [15], completion time, waiting time, and idle time for job shop scheduling. Agents can be better used for proper load balancing [5]. We used JADE (Java Agent Development Environment) middleware for the communication between agents.

This paper is organized as follows: Sect. 2 focuses on the literature survey and Sect. 3 states the problem description. Proposed methodology is described in Sect. 4. Section 5 presents the computational results. Section 6 deals with discussion. Lastly, Sect. 7 presents the conclusion and future scope.

2 Literature Survey

Few researches have addressed the problem of dynamic load balancing by scheduling, heuristic techniques, and one of the oldest and best metaheuristic techniques that are evolutionary algorithms. In the past few decades, these EA algorithms are wide spreading on many applications with their ability to handle large search spaces, providing better approximate solutions and optimizing the problem in minimum time. There are many metaheuristic techniques like Tabu search, simulated annealing, ant colony, genetic algorithms [4], and neural networks which are used. Among these techniques, genetic algorithms are used to find better solutions for optimization and decision-making for dynamic load balancing and it can be easily paralleled to work on different machines. The adaptive behavior of developed programs is essential. Load distribution design patterns using genetic algorithms were proposed by [11]. Nowadays, various types of hybrid genetic algorithms are used for solving hard problems [13]. NSGA II (non-dominated sorting genetic algorithms) is most popularly used technique for hard problem solving with multi-objective [3]. In this paper, NSGA II is used instead of NSGA. These both algorithms do not have much similarities, NSGA uses tournament selection with rank-based method, whereas NSGA II uses crowding distance technique which is much efficient than NSGA.

Dynamic load balancing is of many types centralized, decentralized, dynamic, and static, and these different dynamic load balancings are classified into scheduling problems based on three rules: location rule, distribution rule, and initiation rule. Key issues in load balancing techniques adopted are load sharing, task migration, load distribution, information exchange, and load balancing operation. The algorithm considers load balancing issues such as threshold policies, inter-processor communication and process, and information exchanged criteria. The other issues of load balancing using genetic algorithms are given by Albert Y Zomaya [15]. In autonomic computing systems, load balancing problem can be solved by dynamic composition of web services using SOA, and genetic algorithms are proposed by [9, 10]. There are many genetic algorithms for optimization among which most recently popularized genetic algorithm is the NSGA II and when compared to the algorithms among genetic algorithms, NSGA II is opted for its non-dominated searching techniques and optimal solutions in multi-objective optimization problems [3]. Dynamic load balancing discussed in this paper is multi-objective problem and NSGA II provides a Pareto optimal set which contains the optimal solutions that satisfy all the objectives rather than satisfying one or two solutions. The decision-based approach can be obtained using the concept of multi-agents, where several applications are applicable for agent-based approach along with genetic algorithms, for test generations by making seamless integration agents with genetic algorithms [2]. Communication between agents can be done in many ways. Aspect-oriented programming is one in which agents can communicate, which is explained in [8]. Automatic tuning for agent-based approach is adopted along with genetic algorithms in which agent-based simulation (ABS) is made through genetic algorithms. Interestingly, the authors implemented dynamic adaptability and invocation of JADE agent services from P2P JXTA [12]. This effective composite working model of genetic algorithms along with multi-agent system is also applicable for our problem agent-based approach very effectively and jobs are scheduled using a hybrid architecture of genetic algorithms [1].

3 Problem Description

The job shop scheduling with several parallel machines is a difficult task. Processors in each stage are unrelated. Till now, nobody adopted adaptive metrics in solving the scheduling problems. But in our paper, we proposed solution to scheduling problems using adaptive metrics and agents.

So far, there is no better approach or a pattern for solving scheduling problem with agents along with decision-making strategies. Till now, there is no optimal and efficient approach to solve scheduling problems using NSGA II and agents. By using NSGA II along with agents is the best approach for solving scheduling problems. We define job shop scheduling problem formally with the below definitions:

1. $J = \{J_1, J_2, J_3, \ldots, J_n\}$ is set of n jobs.
2. $M = \{M_1, M_2, M_3, \ldots, M_n\}$ is set of m machines.
3. Each job J has k operations, let Ji be a job then operations in $J_i = \{J_{i_1}, J_{i_2}, J_{i_3}, \ldots, J_{i_k}\}$.
4. Let T be the total processing time of all the jobs.
5. Jmi be the set of jobs that are scheduled on machine Mi. As job shop scheduling problem follows precedence, scheduled operation is such that all the preceding operations have finished. Then, set of all scheduled operations on machine $J_m = \{J_{m_1}, J_{m_2}, J_{m_3}, \ldots, J_{m_n}\}$.
6. N_j be set of jobs which are unfinished.
7. F_j be set of finished jobs.

MK_j is the makespan of jobs which should be less than maximum time. If i followed by j, then machine J should be allocated with i followed by j. If I is a job with a set of operation i_1, i_2, i_3 followed by precedence, then they should be arranged in the order of precedence in waiting queue. M_{i_j} be a job I with operation j on machine m, then it should not be processed more than one time in the same machine. If same job scheduled on machine twice, then the value of computation is taken from previous computation of job.

As referred by [3] that Kamran Zamanifar solved job shop scheduling problem with genetic algorithms and agents, that is not much efficient in terms of average completion time, waiting time, and machine idle time. There is no proper decision-making strategy. Here, the agents communicate via communication channel. The appropriate decision-making strategy is proposed in this paper. We use NSGA II to schedule the problem with agents. By using this, we will get very efficient schedule and also by using this schedule agents can make a strategy plan how to assign jobs to the machine by maintaining load balancing.

The three main constraints are as follows: No two operations should be executed on the same machine at a time; No two machines should process the same operation twice; and Operations in the job should maintain precedence, i.e., if op_i and op_j are two operations where i followed by j then j should not be scheduled earlier than i. Up to our knowledge, till now there is no such combinatory efficient usage of scheduling problem with NSGA II agents. This pattern can solve almost all scheduling problems.

4 Proposed Methodology and Algorithm

The initial population is created by using the method proposed in [7].

Fitness function gives the chances of survival of the chromosome in the next generation. In this paper, the fitness function is adopted from the paper [6].

$Fitness = minimum\,make\,span \times average\,utilization \times \#\,waiting\,queues$
$\#\,processing\,queues$

The selection process in the NSGA II algorithms is carried out through calculating crowding distance. In order to calculate crowding distance, non-dominated sorting is carried out.

NON-DOMINATED SORT:

After initial population is generated, the population is sorted.

1. Initialize $Dp = \varnothing$. This set contains the individuals dominated by the individual.
2. Initialize $Np = \varnothing$. This set represents the individuals which dominate the individual.
3. Let i be an individual which is in population that dominates j then add j to the set of Dp $Dp = Dp \cup \{j\}$.
4. If k be an individual that dominate i, then increment $Np = Np + 1$.
5. If $Np = 0$, then there is no element that dominates the individual i then it belongs to first front. Then, rank the individual to one, i.e., rank = 1. Then, update the first front set by adding this individual i to first front. Front1 = Front1 + 1.
6. This is done for all the individuals in the population.

Then tournament run is applied on the individuals on the crowding distance and individuals with less distance to the objective, i.e., the best individuals more nearer to the fittest are chosen and sent for the crossover operator.

4.1 Hybrid Parallel NSGA

NSGA (Non-dominated sorting genetic algorithms) is the popular genetic algorithm which is non-domination based on dynamic load balancing. The initial population is generated randomly, and then the population is sorted into sets which are non-dominated; these sets are the Pareto sets in which each set contains elements which are non-dominated elements but they dominate the elements of other sets. These Pareto sets together form the set of feasible solutions. The selection is made from this Pareto set; rank is assigned to these based on the Pareto set and then depending on these ranks the fitness value is assigned. NSGA II.

```
Input: problemsize,populationsize,pcrossover,pmutation
output: jobschedule
Population  InitializePopulation(Populationsize, ProblemSize)
EvaluateAgainstObjectiveFunctions(Population)
FastNondominatedSort(Population)
```

```
Selected <- SelectParentsByRank(Population,Populationsize)
Offspring <- CrossoverAndMutation(Selected,Pcrossover,Pmutation)
While (StopCondition())
     EvaluateAgainstObjectiveFunctions(Offspring)
     Merge <- Merge(Population, Offspring)
     Frontres <- FastNondominatedSort(Merge)
     Parents
     For (Frontsresifrontsres)
         CrowdingDistanceAssignment(Frontsresi)
         If (Size(Parents)+Size(Frontsresi>Populationsize)
           FrontsresL<- i
             Break()
         Else
             Parents   Merge(Parents,Frontsresi)
         End
     End
     If (Size(Parents)<Populationsize)
         SortByRankAndDistance()
         For (P1ToPPopulationsize-sizeFrontresL)
             Parents   <- Pi
         End
     End
     Selected   SelectParentsByRankAndDistance(Parents,Populationsize)
     Population  Offspring
     Offspring   CrossoverAndMutation(Selected, Pcrossover,Pmutation)
End
Return (Offspring)
```

The algorithms proposed above give NSGA II functionality, and to make it parallel we used an agent-based architecture with hybrid model of genetic algorithms. Different models of genetic algorithms in parallel approach are given [2], in which a hybrid model is developed by adopting master–slave and island model (coarse-grained). As we use island model, first the population is divided among different host machines, where each host is assigned an agent. The agents named are as follows as referred [14]. Here, we have proposed NSGA II algorithm with agents known as ANSGA II algorithm.

MA (Master Agent)
WA_i (Worker Agent)
PA (Processor Agent)
SA (Synchronizing Agent)

The parallel NSGA II algorithm developed in this model consists of two phases, the execution phase and migration phase. The execution phase is carried out in which once the population created by the master and worker agents, the master agents divide the population into subpopulations and it is distributed to different processor agents through island model. Then, migration phase takes place where migrants move from one agent to another agent. This migration is coordinated by Synchronizing Agent (SA) which receives message from each processor agent upon completion of the execution. Once SA agents receive the message from all the agents, then it broadcasts message to all the agents to start migrations, and then all the agents exchange the best migrants and replace them with the weak migrants; here, migrants are nothing but the best chromosomes, and then the population contains best chromosomes.

We proposed agents in assigning the jobs to the machines. Here, we presented the computational results for normal job scheduling problem with GA and another one is the JSS with agents. The result is that job shop scheduling with agents is very efficient. The another comparison is the job shop scheduling problem with agents and NSGA II with agents. In this comparison, the result is job shop scheduling with NSGA II with agents much more efficient than the job shop scheduling problem with GA. The algorithm is Agent Decision-Making algorithm (ADM). In this algorithm, first the agent will initialize the initial beliefs and desires. Then, the beliefs are updated via brf (), belief revision function and the desires are selected via options function, and next finally the intentions are updated via filter (). By observing the environment again the beliefs, desires and intentions are updated. By reconsider (), the agent will decide whether the new beliefs are to be considered, i.e., updated or not. Finally, agent decision-making will take place through this algorithm.

The proposed methodology organizes as follows:

Algorithm 1: AOJ	Algorithm 2: ADM

```
{scriptsize}
Jobs_assigning(n,m)
{
/* i is the job number*/
  for(int i=0;i<n;i++)
  {
    /* j is the operation number*/
    for(int j=0;j<m;j++)
    {
      Calculate completion time Oij
      Queue[]=Oij;
      /*completion time is given to an array*/
    }
  }
  for(int p=0;p<i;p++)
  {
  for(int q=i+1;q<j;q++ )
    {
    If(queue[p]<queue[q])
      {
        a=queue[p];
        Queue[p]=queue[q];
        Queue[q]=a;
      }
    }
/*operation a is assigned to machine*/
  assign(a) ;
  repeat until queue is empty;
```

```
{scriptsize}
Adm(X, Y)
{
/* X0 are the initial beliefs*/
  X    X0;
  Y    Y0;
While true do
  Get the next percept through the sensors;
  X    brf(X,A);
  Y    options(X, Y);
  Z    filter(X, Y, Z);
  /*Ac is the set of many actions*/
  Pi   plan(X, Y, Ac);
  While not (empty (Pi) or succeeded(X, Y) or impossible
    b    first element of Pi;
    Execute (b);
    Pi    tail of Pi;
    Observe environment to get next percept
    X    brf(X, K)
    If reconsider(X, Y) then
      Y    options(X, Y);
      Z    filter(X, Y, Z);
    End-if
    If not sound (Pi, Y, Z) then
      Pi   plan(X, Y, Ac)
    End-if
  End-while
End-while
}
```

5 Adaptive Metrics

Initialization: Consider H is the initial population generated randomly. Every chromosome includes some gene. It is used to produce random number from the below equation. Through this equation, the feasibility is tested. If it satisfies the feasibility or if it is said to be feasible, then it would be the part of population. This is repeated for random number of time until we get random H population with the feasible chromosomes. Rather than starting genetic algorithms with random population when GA is started with the estimated feasible solution, then the optimal fitness can be obtained in known time.

$$\Omega = \{X_1 \wedge X_x / \mu \le X_1 \le W_1, \wedge, \mu_x \le X_x \le W_x\} \tag{1}$$

Cross operator:
Instead of using fixed pc, we adjust it adaptively based on the following formula:

$$p_c = \begin{cases} P_{c_1} = P_{c_1} - P_{c_2} f^1 - f_{avg} / f_{max} - f_{avg}, & f^1 \ge f_{avg} \\ P_{c_1}, & f^1 \le f_{avg} \end{cases} \tag{2}$$

f_{max} is the highest fitness value in the population;
f_{avg} is the average fitness value in every population;
f^1 is higher fitness value between two individuals;

In addition, we set Pc1 = 0.9, Pc2 = 0.6. Mutation operator:
Instead of using fixed pm, we adjust it adaptively based on the following formula:

$$p_m = \begin{cases} P_{m_1} = P_{m_1} - P_{m_2} f - f_{avg} / f_{max} - f_{avg}, & f \ge f_{avg} \\ P_{m_1}, & f \le f_{avg} \end{cases} \tag{3}$$

where f1 is higher mutationtness value.

6 Computational Results

The following emergent results were identified from the analysis of applying of dynamic load balancing to scheduling problems using NSGA II and agents. The tables presented below show the experimental results of the proposed algorithms and also empirical study is done. The results are generated for dynamic multi-objectives like make span, i.e., average completion time of jobs, waiting time of jobs, starvation, and idle time of the machines. The schedules are generated by running the algorithm in [4]. After schedule has obtained, then by implementing our proposed ANSGA II algorithm and ADM algorithm, we are going to assign jobs to the machines in such a way that we get very efficient results. The formulas for obtaining the objectives are

$$W_{avg} = \frac{\sum W_{ji} \sum O_{ji_k}}{\sum N_{ji} \sum O_{ji_k}} M_{avg} = \frac{\sum M_{ji} \sum O_{ji_k}}{\sum N_{ji} \sum O_{ji_k}} I_{avg} = \frac{\sum I_{ji} \sum O_{ji_k}}{\sum N_{ji} \sum O_{ji_k}} \tag{4}$$

In addition, we set the following:
Crossover probability values: $P_c 1 = 0.9$, $P_c 2 = 0.6$. Mutation probability values: $P_m 1 = 0.1$, $P_m 2 = 0.001$. Maximum utilization ∞ 1/idle time.

Starvation time = waiting time avg > waiting time max.

	Make span	Waiting time	Idle time
Job shop scheduling	4.3	1.2	0
GA	4.8	0.8	0.6
GA with agents	4.2	0.6	0.6
ANSGA II	2.8	0.5	0

The table indicates the comparison between the computational results of JSS, GA, and GA with agents and the proposed ANSGA II. The parameters considered are make span, waiting time, and idle time. From this table, it is very clear that our proposed NSGA II with agents, i.e., ANSGA II is very efficient from our computational results.

	Make span	Waiting time
ANSGA II	2.8	0.5
MOPSO [13]	5.6	1.3
NSGA II [13]	3.4	1.12
GA[1]	3.5	2

7 Discussion

One of the main goals of this experiment was an attempt to find a way to predict the best efficient solution for solving scheduling problems using non-dominated Pareto set solutions by agents and decision-making strategy. From our findings, we conclude that by using adaptive metrics and best decision-making strategy algorithm using agents, we found a very efficient solution to scheduling problems, for a small set of jobs. The findings of Asadzadeh, Leila, and Kamran Zamanifar [14] are very useful in collecting the data required for our paper. In that paper, they discussed about GA with agents and stated that it is the best approach, but in our paper we proposed NSGA II with agents (ANSGA) and compared ANSGA II with GA with agents and found that our proposed algorithm is the best one. From our findings, we conclude that the ADM is the best decision-making algorithm for agents. Our results are compared with the results of [14, 15]. The values of [15] are taken for a small set of values from graph and compared with our results. Our research focused on adapting agents in scheduling problems, but we could not develop a pattern. In fact, the inclusion of pattern development in our paper might be very efficient, as it provides a pattern and it can be applicable to any scheduling problem.

8 Conclusion and Future Scope

In this paper, it has been shown that how load balancing can be applied to a wide range of scheduling problems. We have underlined the importance of using agents in scheduling problems. The findings of this study indicate that NSGA II with agents provides better solution than genetic algorithms with agents. The results are tabulated. The final result of our paper is compared with the results of [14]. The findings of our study imply that NSGA II with agents have better results than GA with agents. We minimized the average waiting time in the queue, average completion time of jobs, and also the average machine idle time. We also focused on the concept of starvation. There is no starvation time while assigning jobs to the machines using agents. For decision-making strategy, we have used the standard decision-making algorithm, by which the agents communicate and coordinate with each other. There is much scope in the future in the direction of this field, i.e., solving scheduling problems with agents. The future scope is that we can generate a design pattern using this ANSGA II algorithm by which we can solve any scheduling problems. Any other metaheuristic search techniques can be used in our pattern.

References

1. Asadzadeh, L., Zamanifar, K.: An agent-based parallel approach for the job shop scheduling problem with genetic algorithms. Math. Comput. Modell. **52**(11), 1957–1965 (2010)
2. Coello Coello, C.A., Lamont, G.B., Van Veldhuizen, D.A., et al.: Evolutionary Algorithms for Solving Multi-objective Problems, vol. 5. Springer (2007)
3. Deb, K., Pratap, A., Agarwal, S., Meyarivan, T.A.M.T.: A fast and elitist multiobjective genetic algorithm: NSGA-II. IEEE Trans. Evol. Comput. **6**(2), 182–197 (2002)
4. Della Croce, F., Tadei, R., Volta, G.: A genetic algorithm for the job shop problem. Comput. Oper. Res. **22**(1), 15–24 (1995)
5. Jiang, Y.: A survey of task allocation and load balancing in distributed systems. IEEE Trans. Parallel Distrib. Syst. **27**(2), 585–599 (2016)
6. Joyce, K.E., Hayasaka, S., Laurienti, P.J.: A genetic algorithm for controlling an agent-based model of the functional human brain. Biomed. Sci. Instrum. **48**, 210 (2012)
7. Luque, G., Alba, E.: Parallel Genetic Algorithms: Theory and Real World Applications, vol. 367. Springer (2011)
8. Mannava, V., Ramesh, T.: A novel way of invoking agent services using aspect oriented programming via web service integration gateway. In: Trends in Network and Communications, pp. 675–684. Springer (2011)
9. Mannava, V., Ramesh, T.: An adaptive design pattern for genetic algorithm-based composition of web services in autonomic computing systems using SOA. In: International Conference on Grid and Pervasive Computing, pp. 98–108. Springer (2012)
10. Mannava, V., Ramesh, T.: Load distribution composite design pattern for genetic algorithm-based autonomic computing systems. Int. J. Soft Comput. **3**(3), 85 (2012)
11. Mannava, V., Ramesh, T.: Load distribution design pattern for genetic algorithm based autonomic systems. Proc. Eng. **38**, 1905–1915 (2012)
12. Mannava, V., Ramesh, T., Vasireddy, P.: A novel way of providing dynamic adaptability and invocation of JADE agent services from P2P JXTA using aspect oriented programming. In: International Conference on Network Security and Applications, pp. 552–563. Springer (2011)

13. Rashidi, E., Jahandar, M., Zandieh, M.: An improved hybrid multi-objective parallel genetic algorithm for hybrid flow shop scheduling with unrelated parallel machines. Int. J. Adv. Manuf. Technol. **49**(9), 1129–1139 (2010)
14. Saeidi, S.: A multi-objective mathematical model for job scheduling on parallel machines using NSGA-II (2016)
15. Zomaya, A.Y., Teh, Y.-H.: Observations on using genetic algorithms for dynamic load-balancing. IEEE Trans. Parallel Distrib. Syst. **12**(9), 899–911 (2001)

A Secure VM Consolidation in Cloud Using Learning Automata

Sambit Kumar Mishra, Bibhudatta Sahoo and Sanjay Kumar Jena

Abstract Cloud computing system is a progression of distributed system that has been adopted by worldwide scientifically and commercially. For optimal utilization of cloud's potential power, effective and efficient algorithms are expected, which will select best resources from available cloud resources for different applications. This allocation of user requests to the cloud resources can optimize several parameters like energy consumption, makespan, throughput, etc. In this paper, we have proposed a learning automata based algorithm to minimize the makespan of the cloud system and also to increase the resource utilization that holds secured resource allocation. We have simulated our algorithm, $ALOLA$ with the help of CloudSim simulator in a heterogeneous environment. During the comparison of the algorithm, we provide a finite set of tasks to the $ALOLA$ algorithm once and estimate the makespan of the system. We have compared our proposed technique ($ALOLA$), i.e., with learning automata and without learning automata (random allocation algorithm), and show the system performance.

Keywords Cloud computing · DVFS · Learning automata · Makespan
Resource · Task allocation · VM

S. K. Mishra (✉) · B. Sahoo · S. K. Jena
National Institute of Technology Rourkela, Rourkela, India
e-mail: skmishra.nitrk@gmail.com
URL: http://www.springer.com/lncs

B. Sahoo
e-mail: bibhudatta.sahoo@gmail.com

S. K. Jena
e-mail: skjena@nitrkl.ac.in

© Springer Nature Singapore Pte Ltd. 2019
P. K. Sa et al. (eds.), *Recent Findings in Intelligent Computing Techniques*,
Advances in Intelligent Systems and Computing 707,
https://doi.org/10.1007/978-981-10-8639-7_65

1 Introduction

Cloud computing is an evolution of grid computing. In the current scenario, cloud computing is a buzzword and acquires more and more concentration from users. Cloud computing affords a vast pool of shareable resources (CPU, memory, storage, workstations, operating system, energy, network throughput, network loads and delays, etc.) which delivers scalable on-demand resources as a service over the Internet with the cooperation of virtualization technique. In other words, a Cloud is a collection of parallel and distributed system which is interconnected and virtualized. These virtualized resources allotted to the customer according to their respective Service-Level Agreement (SLA) between the customer and supplier. The virtualization technique virtualizes the physical resources of the physical hosts in the form of virtual machines (VMs).

In this modern time, every user wants to get their services in less time. Therefore, the task allocation performs a significant role [1]. The number of users managing over the Internet is much more and increasing with day-by-day. To afford services to these huge number of users is a challenging task and one of the best solutions is cloud computing [2]. Task consolidation problem is a combinatorial optimization problem in the field computer science. Allocation of the task in cloud computing is an NP-Hard problem [1]. Different researchers have been worked on this allocation problem and presented various heuristic algorithms [3–6]. The aim of task consolidation problem is to assign a finite set of tasks to cloud resources in such a way that it will approach to minimize makespan, minimize the execution cost, and maximize the utilization of resources [5, 7]. In the current time, cloud computing is earning lots of interest in various domains by processing big data [8, 9]. In that, data are gathered from several sources such as sensor networks, social networks, and vehicles [10]. There is still scope to address security matter of data collected from above sources to cloud data center.

To have a suitable model for a realistic environment, agents need to be able to illustrate the flexible behaviors in various situations, because, in agreements, agents struggle with complex environments, unusual deadlines, and inadequate information regarding the opponents [11]. Therefore, an agreement system must be designed with efficient learning tools to be ready to adopt a proper strategy. Learning through setting out the selection of actions of the cloud providers based on the gathered information over time could satisfy such design. The complication of the learning effect in a multi-agent system, more because the agents have to learn the influences of their activities and also, coordinates with other agents. Learning automaton is an adaptive agent applied for delivering decisions [12]. Here, we propose an adaptive resource allocation mechanism where each cloud resource (or VM) has an assigned priority for each task.

The main aim of writing this paper is to introduce a heuristic algorithm for minimizing the makespan and maximizing the resource utilization. We are viewing resources as heterogeneous in nature. The remaining of the paper is constructed as follows. Section 2 outlines a summary of learning automata; Sect. 3 describes a

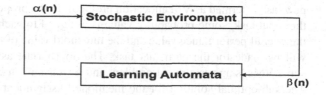

Fig. 1 Interaction of learning automata and environment

brief idea about the problem statement. Section 4 describes our proposed algorithm to minimize the makespan of the heterogeneous computing environment along with the secured migration. Section 5 explains about simulation and results, utilization of VMs and effectiveness of our algorithm followed by the conclusion of the paper in Sect. 6.

2 Learning Automata

The idea of learning automaton is identified as an importance of the effort on illustrating experimental execution on the arrangement of activities from the past studies. The word learning is the divergence in the system performance dependents upon the past studies with respect to time. The environment generates different solutions to make the automated system. The learning automaton selects one action randomly from a set of actions. The selection will be done randomly because all actions have same probability value. The action probabilities are modified according to the response supplied by the environment. This process is repeated continuously to settle the optimal action by choosing the action with the largest probability value. Learning automata has attained applications in designing automated system, modeling biological system, computer vision, transportation, particle swarm optimization, etc. Several learning automata and their utilization have studied in [13, 14]. The most significant feature of a learning system is its adaptable capability to magnify the efficiency over time. The system is represented as $E = \{\alpha, \beta, \gamma\}$ in which α is set of inputs received by the environment, β is set of outputs, and γ is the penalty value. Figure 1 presents the relationship between learning automata and the environment.

3 Problem Statement

The task allocation and VM allocation problems play a significant role in the efficient utilization of cloud resources. There are n number of heterogeneous input tasks to the cloud system and m number of heterogeneous VMs in terms of resource capacity (like processing speed, main memory, secondary storage, and bandwidth) in the cloud system. The mapping of tasks to the available VMs is termed as task allocation problem, and the allocation of new VM to the available host for the execution of the new task is termed as VM allocation problem. In the proposed method, according to the input task, some probability value is assigned to each VM. Then, according to the overall performance value and the threshold value of the VM, the VM selection will be made for the particular task. The appropriate assignment of new tasks to some VMs is an assignment problem and a well-known NP-complete problem [6]. The sub-optimal solution for the mentioned assignment problem with the aim of minimizing the makespan of the system is the objective of this work.

4 Algorithm

The following presented algorithm is used to achieve an optimal allocation of the set of user requests (tasks) and the cloud resources (VMs). Initially, all the VMs have an equal probability which implies that any of the VMs is fairly likely to be preferred. So, the first VM is picked randomly. Then, the overall performance (op) of the VM is estimated and compared with the threshold value. The probability value of that VM is improvised if the op is more than the threshold otherwise degraded the probability value. Here, α and β are reward and penalty constants, respectively. The learning system examines the threshold of VM with the overall performance of that VM. For each task, if the op value is crossed the threshold for a VM, then that VM is discarded by the learning system.

4.1 Security

At the time of designing a moving target (VM migration) protection scheme, we must assure that the enemies are difficult to overcome the actual internal infrastructures; however, the approach is not confidential [15]. In this scheme, the tasks are initially allocated to some VMs randomly and then according to the objective, the tasks are moving from one VM to another. This migration of VMs and tasks should be secured; this information should secretly available to the cloud service provider only. As long as the enemies cannot reach the information of all cloud resources (VMs), our approach is always safe.

We remark that during VM migration, there might be some information leakage. But, in this paper, the design of VM migration systems is not within the range, we prospect more secured VM migration schemes. Virtualization technology provides security services with high assurance by confining them into separate protected VMs. It supports security applications to have perfect visibility to raw memory state of other VMs.

Algorithm 1 : $AOALA$: Algorithm for Optimal Allocation using Learning Automate

---

Input: T: set of n tasks sorted in descending order of their length; T_j: Threshold set for each tasks; $P_i j$: Action probability of ith task on jth VM;

Output: Mapping of Tasks and VMs

1: **for** each task T_i, $1 \le i \le n$ **do**
2:     Initialize same action probability for all VMs;
3:     Randomly, select one of the VM;
4:     Calculate op_j;
5:     **if** $op_j < T_j$ **then**
6:         $P_{ij}(k+1) = P_{ij}(k) + \alpha(1 - P_{ij}(k))$;
7:         $P_{il}(k+1) = (1 - \alpha) \times P_{il}(k), i \ne j, 1 \le l \le m$;
8:     **else**
9:         $P_{ij}(k+1) = (1 - \beta) \times P_{ij}(k)$;
10:        $P_{il}(k+1) = \frac{\beta}{n-1}(1 - \beta) \times P_{il}(k), i \ne j, 1 \le l \le m$;
11:     **end if**
12:     SV = Select the VM with highest P_{ij} value;
13:     **if** $|SV| = 1$ **then**
14:         Allocate T_i to V_j;
15:     **else**
16:         Randomly select a VM from SV;
17:         Goto Step-4;
18:     **end if**
19: **end for**

---

5 Experimental Result

The simulation is conducted on the cloud computing environment tools: CloudSim [16]. CloudSim is one of the simulation tools of cloud environment which provides evaluation and testing of cloud services and infrastructure before the development of the real world. One data center is created with default properties as it mentioned by the CloudSim designer. All hosts are running on a single data center. To simulate our approach, we have some assumptions as follows:

- Total number of VMs will be proportional to the available cores in a host.
- All hosts, as well as VMs, have heterogeneous resource capacity.
- Each host has a finite number of VMs.
- Tasks are non-preemptive in nature.
- Each task is independent in nature, and also the resource requirement of each task is independent of each other.

In scenario-1, the number of virtual machines is set to 40, and the number of tasks increases from 20 to 200 in the gap of 20. The comparison graph for the calculation of makespan of the system using the learning automata technique ($AOALA$) and without using the learning automata technique is shown in Fig. 2a. From the fig, the consumption of makespan value in the $AOALA$ algorithm is less. In scenario-2, the number of tasks is set to 100, and the number of virtual machines increases from 10 to 100 in the gap of 10. The comparison graph for the calculation of makespan of

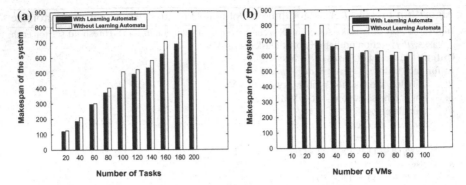

Fig. 2 Comparison of makespan of the system using learning automata $AOALA$ and without using learning automata, where the number of VM is fixed and task varies (as shown in **a**), and the number of task is fixed and VM varies (as shown in **b**)

the system using the learning automata technique ($AOALA$) and without using the learning automata technique is shown in Fig. 2b. From the figure, the consumption of makespan value in the $AOALA$ algorithm is less.

6 Conclusion

In this paper, we have studied the learning-based task scheduling approaches in homogeneous and heterogeneous cloud environments proposed by different researchers. We have proposed a task-based heuristic algorithm with the help of learning automata to make the system automated for dynamic task allocation in heterogeneous cloud computing environment. We have discussed the learning-based system model. We have proposed the $ALOLA$ allocation algorithm to implement the mapping of cloud tasks to VMs. The proposed algorithm is simulated in CloudSim simulator, and then we compared the algorithm with learning automata and without learning automata assigning priorities to the VMs. The simulation result shows that $ALOLA$ algorithm has improved makespan.

Acknowledgements This research work is partially supported by Information Security Education and Awareness Project, Phase-II (ISEA-II) funded by Ministry of Electronics and Information Technology (MeitY), Government of India.

References

1. Krishna, P. V.: Honey bee behavior inspired load balancing of tasks in cloud computing environments. In: Applied Soft Computing, pp. 2292–2303 (2013)
2. Devi, C., Uthariaraj, V.: Load balancing in cloud computing environment using improved weighted round robin algorithm for non preemptive dependent tasks. Sci. World J. (2016)
3. Misra, S., Krishna, P.V., Kalaiselvan, K., Saritha, V., Obaidat, M.S.: Learning automata-based QoS framework for cloud IaaS. IEEE Trans. Netw. Serv. Manag. 1(11), 15–24 (2014)

4. Granmo, O.C., Oommen, B.J.: Solving stochastic nonlinear resource allocation problems using a hierarchy of twofold resource allocation automata. IEEE Trans. Comput. **59**(4), 545–560 (2010)
5. Mishra, S.K., Deswal, R., Sahoo, S., Sahoo, B.: Improving energy consumption in cloud. In: Annual IEEE India Conference (INDICON), pp. 1–6, Dec 2015
6. Li, K., Xu, G., Zhao, G., Dong, Y., Wang, D.: Cloud task scheduling based on load balancing ant colony optimization. In: Sixth Annual China Grid Conference, Liaoning, pp. 3–9 (2011)
7. Sahoo, S., Nawaz, S., Mishra, S.K., Sahoo, B.: Execution of real time task on cloud environment. In: Annual IEEE India Conference (INDICON), pp. 1–5 (2015)
8. Puthal, D., Nepal, S., Ranjan, R., Chen, J.: Threats to networking cloud and edge datacenters in the Internet of Things. IEEE Cloud Comput. **3**(3), 64–71 (2016)
9. Puthal, D., Nepal, S., Ranjan, R., Chen, J.: DLSeF: a dynamic key-length-based efficient real-time security verification model for big data stream. ACM Trans. Embed. Comput. Syst. (TECS) **16**(2), 51 (2016)
10. Puthal, D., Mir, Z.H., Filali, F., Menouar, H.: Cross-layer architecture for congestion control in Vehicular Ad-hoc Networks. In: IEEE International Conference on Connected Vehicles and Expo (ICCVE), pp. 887–892, Dec 2013
11. Matos, N., Sierra, C., Jennings, N.R.: Determining successful negotiation strategies: an evolutionary approach. In: International conference in Paris, France, pp. 182–189 (1998)
12. Thathachar, M.A., Sastry, P.S.: Networks of Learning Automata: Techniques for Online Stochastic Optimization. Springer, New York, NY, USA (2004)
13. Torkestani, J.A., Meybodi, M.R.: Finding minimum weight connected dominating set in stochastic graph based on learning automata. Inf. Sci. **200**, 57–77 (2012)
14. Esnaashari, M., Meybodi, M.R.: Deployment of a mobile wireless sensor network with k-coverage constraint: a cellular learning automata approach. Wirel. Netw. **19**(5), 945–968 (2013)
15. Zhang, Y., Li, M., Bai, K., Yu, M., Zang, W.: Incentive compatible moving target defense against VM-colocation attacks in clouds. In: IFIP International Information Security Conference, pp. 388–399. Springer, Berlin, Heidelberg (2012)
16. Calheiros, R.N., Ranjan, R., Beloglazov, A., De Rose, C.A., Buyya, R.: CloudSim: a toolkit for modeling and simulation of cloud computing environments and evaluation of resource provisioning algorithms. Softw.: Pract. Exp. **41**(1), 23–50 (2011)

Author Index

Printed in the United States
By Bookmasters